C0-ASX-401

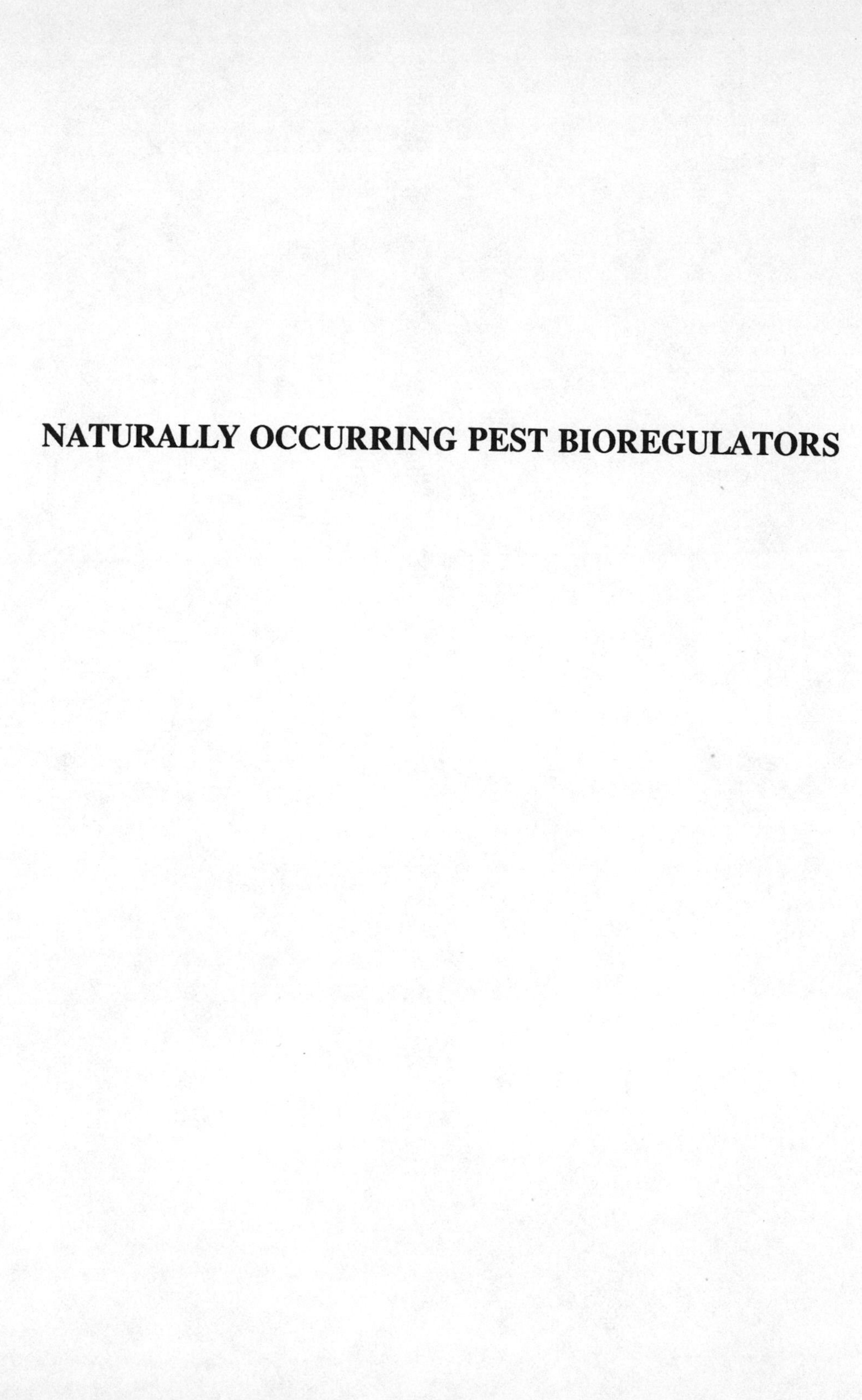

NATURALLY OCCURRING PEST BIOREGULATORS

ACS SYMPOSIUM SERIES **449**

WIDENER UNIVERSITY LIBRARY CHESTER, PA.

Naturally Occurring Pest Bioregulators

Paul A. Hedin, EDITOR
U. S. Department of Agriculture

Developed from symposia sponsored by the Division of Agrochemicals at the 197th and 198th National Meetings of the American Chemical Society and the International Chemical Congress of Pacific Basin Societies at the 1989 International Chemical Congress of Pacific Basin Societies

American Chemical Society, Washington, DC 1991

Library of Congress Cataloging-in-Publication Data

Naturally occurring pest bioregulators
Paul A. Hedin, editor

p. cm.—(ACS Symposium Series, 0097–6156; 449)

"Developed from a symposium sponsored by the Division of Agrochemicals and the 1989 International Chemical Congress of Pacific Basin Societies at the 197th and 198th national meetings of the American Chemical Society, and the 1989 International Chemical Congress of Pacific Basin Societies, Honolulu, Hawaii, December 17–22, 1989."

Includes bibliographical references and index.

ISBN 0–8412–1897–8

1. Natural pesticides—Congresses. 2. Bioactive compounds—Congresses. I. Hedin, Paul A. (Paul Arthur), 1926– . II. American Chemical Society. Division of Agrochemicals. III. International Chemical Congress of Pacific Basin Societies (1989: Honolulu, Hawaii) IV. Series

SB951.145.N37N38 1991
632'.95—dc20 90–22914
CIP

The paper used in this publication meets the minimum requirements of American National Standard for Information Sciences—Permanence of Paper for Printed Library Materials, ANSI Z39.48–1984. ∞

Copyright © 1991

American Chemical Society

All Rights Reserved. The appearance of the code at the bottom of the first page of each chapter in this volume indicates the copyright owner's consent that reprographic copies of the chapter may be made for personal or internal use or for the personal or internal use of specific clients. This consent is given on the condition, however, that the copier pay the stated per-copy fee through the Copyright Clearance Center, Inc., 27 Congress Street, Salem, MA 01970, for copying beyond that permitted by Sections 107 or 108 of the U.S. Copyright Law. This consent does not extend to copying or transmission by any means—graphic or electronic—for any other purpose, such as for general distribution, for advertising or promotional purposes, for creating a new collective work, for resale, or for information storage and retrieval systems. The copying fee for each chapter is indicated in the code at the bottom of the first page of the chapter.

The citation of trade names and/or names of manufacturers in this publication is not to be construed as an endorsement or as approval by ACS of the commercial products or services referenced herein; nor should the mere reference herein to any drawing, specification, chemical process, or other data be regarded as a license or as a conveyance of any right or permission to the holder, reader, or any other person or corporation, to manufacture, reproduce, use, or sell any patented invention or copyrighted work that may in any way be related thereto. Registered names, trademarks, etc., used in this publication, even without specific indication thereof, are not to be considered unprotected by law.

PRINTED IN THE UNITED STATES OF AMERICA

ACS Symposium Series

M. Joan Comstock, *Series Editor*

1991 ACS Books Advisory Board

V. Dean Adams
Tennessee Technological University

Paul S. Anderson
Merck Sharp & Dohme Research Laboratories

Alexis T. Bell
University of California—Berkeley

Malcolm H. Chisholm
Indiana University

Natalie Foster
Lehigh University

Dennis W. Hess
University of California—Berkeley

Mary A. Kaiser
E. I. du Pont de Nemours and Company

Gretchen S. Kohl
Dow-Corning Corporation

Michael R. Ladisch
Purdue University

Bonnie Lawlor
Institute for Scientific Information

John L. Massingill
Dow Chemical Company

Robert McGorrin
Kraft General Foods

Julius J. Menn
Plant Sciences Institute, U.S. Department of Agriculture

Marshall Phillips
Office of Agricultural Biotechnology, U.S. Department of Agriculture

Daniel M. Quinn
University of Iowa

A. Truman Schwartz
Macalaster College

Stephen A. Szabo
Conoco Inc.

Robert A. Weiss
University of Connecticut

Foreword

THE ACS SYMPOSIUM SERIES was founded in 1974 to provide a medium for publishing symposia quickly in book form. The format of the Series parallels that of the continuing ADVANCES IN CHEMISTRY SERIES except that, in order to save time, the papers are not typeset, but are reproduced as they are submitted by the authors in camera-ready form. Papers are reviewed under the supervision of the editors with the assistance of the Advisory Board and are selected to maintain the integrity of the symposia. Both reviews and reports of research are acceptable, because symposia may embrace both types of presentation. However, verbatim reproductions of previously published papers are not accepted.

Contents

Preface

WHILE KNOWLEDGE ABOUT NATURAL PRODUCT STRUCTURES with biological activities toward pests has increased rapidly in recent years, agricultural scientists have generally depended upon their traditional methods for controlling plant and animal pests. Even though highly active compounds have been identified, only in a few situations have they been used for control or, as in the case of crop plants, for selecting resistant varieties. However, agents such as pheromones, antifeedants, and insect- and plant-growth regulators have found some commercial application. Also, the use of traditional pesticides is being circumscribed by environmental concerns. Growing resistance of pests to present pesticides has given greater urgency to the search for better, safer compounds and better, safer delivery systems. The need to treat more precisely has also provided additional opportunities for the use of natural products.

These biologically active natural products can be considered to be bioregulators, the name modeled in part on the plant-growth hormones that were shown to regulate various plant growth and development processes. They are found to be active at ever decreasing concentrations. This has, in turn, contributed to the development of increasingly more sensitive analytical procedures to measure and identify the compounds. In fact, the elucidation of a vast array of chemical structures with diverse biological activities has been achieved, and is continuing to occur, with ever increasing frequency.

This book, based on presentations at three American Chemical Society symposia, has been organized into five divisions dealing with the control of insects and other animals, diseases, and weeds. The compounds, their activities, their biosynthesis, and their mechanisms of action are explored.

The first section, *Bioregulation of Insect Behavior and Development*, includes chapters on arthropod and insect repellents, the identification of a beetle pheromone, nonparalyzing factors from hymenoptera, endogenous regulation of pheromone biosynthesis and mating, and systems for controlled release of pheromones.

Mechanisms of Plant Resistance to Insects, the second section, focuses on the role of trichomes in resistance to potatoes and tomatoes, enzymatic antinutritive defenses of tomatoes, and the roles of phytoalexins in insect control.

The next section, *Allelochemicals for Control of Insects and Other Animals*, discusses insect resistance factors in petunias, geraniums, corn, centipede grass, sunflowers, and neem, as well as insecticidal activity of monoterpenoids and fish toxins from mangrove plants.

Phytoalexins and Phototoxins in Plant Pest Control presents a model approach to the binding of a phytoalexin elicator to DNA, studies of phytoalexins in cotton and peanuts, chapters on phototoxic metabolites of tropical plants and on photosensitizing porphyrins as herbicides.

The final section, *Allelochemicals as Plant Disease Control Agents*, looks at black shank fungus in tobacco, suppression of Fusarium Wilt and other fungi by microorganisms, antifungal and antibacterial compounds in Peruvian plants, and the description of a countercurrent chromatographic separation of complex alkaloids in tall fescue.

I hope that this book will contribute to the understanding and the subsequent adoption of additional criteria and research strategies for the control of pests. The compounds should be effective at low concentrations, selective in activity against specific pests, of limited toxicity to nontarget organisms, and environmentally nonpersistent. We also need a better understanding about their mechanisms of action.

I am grateful to all of the participants for their contributions to the book. Finally, I thank the Agricultural Research Service of the U.S. Department of Agriculture for providing me with the opportunity and the support to organize the symposia and compile the book.

PAUL A. HEDIN
U.S. Department of Agriculture
Mississippi State, MS 39762

September 5, 1990

Chapter 1

Use of Natural Products in Pest Control

Developing Research Trends

Paul A. Hedin

Crop Science Research Laboratory, Agricultural Research Service, U.S. Department of Agriculture, Mississippi State, MS 39762

The rapid growth of knowledge of natural products with biological activities toward pests now provides an option for treatment, a clearer understanding of biochemical mechanisms, and a basis for biorational approaches to the design of pest control agents. Compounds that modify insect behavior are also valuable for pest control because they are normally not toxic to the target insect or to the environment.

Natural products are often a key to understanding ecological systems. When the compounds are identified, progress can be made in understanding the metabolic cycles, the enzymes that lead to their biosynthesis, and the underlying genetic controls. An understanding of the interactions should make it possible to devise minimum changes that yield maximum benefits. Some developing research trends in the utilization of natural products for pest control are reviewed.

Particularly over the past 25 years, there has been much activity directed to chemical work on the isolation and identification of a wide array of biologically active natural products that in some way affect the behavior, development and/or reproduction of pests such as insects, diseases and the growth of weeds. However with regard to crop plants, agronomists, entomologists, and other agricultural scientists have generally depended on traditional methods for selecting resistant varieties with adequate yield properties. Even though highly active allelochemicals have been identified, only in a limited percentage of situations has chemical guidance been the leading factor in screening for the properties. On the other hand, agents such as pheromones, antifeedants, insect and plant growth regulators, to name a few, have found some commercial application. Also, synthetic analogues based on the activity of some natural products have found an even wider market.

This chapter not subject to U.S. copyright

Published 1991 American Chemical Society

However, perhaps the day of the biologically active natural product has come. Environmental concerns are circumscribing the use of traditional pesticides. Growing resistance of pests to present pesticides has given added urgency to the search for better, safer compounds, and better, safer delivery systems. The need to treat more precisely has also provided additional opportunities for the use of natural products. It should be stated, however, that resistance can be expected to develop to natural products that have a single mode of action.

Biologically active natural products can be considered to be bioregulators, the name modeled in part on the plant growth hormones that were shown to regulate various plant growth and development processes. Today, the levels required for activity are ever smaller and smaller. For example, the pheromones have been found to be active at picogram concentrations and below. Now, the neurohomones have been found to be just as active, and these are just two examples.

The elucidation of compounds of exquisitely unique structures becomes more and more common as access to powerful instrumental techniques rapidly increases. As one surveys the avalanche of new, brilliant elucidations, it is thrilling to realize that so much power and capability has been developed, particularly in the past decade.

The advent of new capabilities in biotechnology forecasts another quantum leap as the capability grows to insert genes into crop plants. The present prominence of _Bacillus thuringiensis_ (BT) might suggest that the natural products chemist can be dispensed with, but it can be anticipated that pests will develop resistance to BT sooner or later. Therefore, the biotechnologist apparently will have to rely on the chemist to identify other systems to express. It thus appears that there is a future for the natural products chemist.

Pests can most broadly be classified as those attacking animals or plants. Those pests that attack animals are often arthropods, usually insects. Other pests of animals are microbes and nematodes. Their treatment is generally considered a veterinary activity, although natural products including antimicrobials are often used. Thus, animal health concerns, unless they are insect related, are generally considered separately, and will not be emphasized in this survey. The protection of plants from pests has been divided into three categories; control of (1) insects, (2) diseases, and (3) weeds. While pest control has traditionally been accomplished over the past 40-50 years with pesticides of synthetic origin, this was not the case prior to the development of synthetic pesticides such as DDT. In recent years, a number of natural products including those from fermentation have been finding a niche in the market place, because of the increasing concerns about classical pesticides, and more stringent constraints about their use. Accordingly, an effort has been made to identify those natural products that have been found useful for the control of a diversity of pests.

It was recognized that the value of this review would be increased by the inclusion of as many literature citations as possible. Many of the citations are of recent meeting abstracts, particularly ACS meetings. These should provide an entrance into the literature. An effort has been made to achieve a relatively equal emphasis for all areas that have been identified, but this is intended as a survey to show impact and not to be an exhaustive review.

Control of Insects

Sex, Assembling, and Other Pheromones. The recent literature (1-4) contains hundreds of citations attesting to the specificity, uniqueness, and efficacy of these compounds, a few of which are listed in Chapters 2 and 3 of this volume. Two of the major drawbacks to utilization of pheromones are that marketing opportunities are limited and that the control achieved is usually less than complete. Pheromones may occasionally be suitable for control, but may be most applicable in survey applications or as part of integrated pest management programs. Pheromones have repeatedly been used to demonstrate the presence of insects where scouting and other physical procedures are ineffective. It has been demonstrated that pheromones can be used both for attraction and for confusion. For confusion, either large quantities of the pheromone or isomers can be used. It has been further suggested that where several insects attack a crop, all of the pheromones may be formulated together. Where several components are required for an insect response, confusion may be achieved by altering the ratio or withholding one or more components.

In addition to continuing work to identify pheromones, considerable work on pheromone biosynthesis is in progress. Two recent symposia each include multiple papers describing various aspects (4, 5). Chapters 4-6 in this volume describe regulatory factors of pheromone biosynthesis that when more completely understood may find field application similar to that of insect growth regulators.

Allelochemicals. Food attractants, repellents, feeding stimulants, feeding deterrents (antifeedants), oviposition stimulants and deterrents, toxicants, and nutritional factors are examples of allelochemicals. They are usually secondary plant constituents (non-nutritional chemicals affecting insect behavior and development). Allelochemicals giving the host plant an adaptive advantage are called allomones, and factors giving the insect an adaptive advantage are called kairomones. A number of books and symposia have described work on their isolation, identification, and application (6-13).

For many years, the U.S. Department of Agriculture conducted a research program to identify attractants and repellents by bioassaying plant extracts and synthetic compounds, many related to known plant compounds. Attractants for the Mediterranean fruit fly, oriental fruit fly, and Japanese beetle and many others were selected as the result of these programs and widely used. However, in recent years, food attractants (kairomones) have found only infrequent applications. On the other hand, antifeedant compounds with very high activity, notably azadirachtin from neem

(13), have recently become the subject of renewed interest, and several have now been registered for commercial use. The interests of several industrial companies have been instrumental in this new trend. The antifeedants tested to date are sometimes specific for certain insects. These variable activities may serve as the basis both for structure-activity studies and mechanistic studies at the receptor level.

The use of antifeedants has an added advantage over that of toxicants in that the insect consumes little if any of the plant, thus limiting damage. Pest control materials which are specific for insects and which affect metabolic pathways or anatomical structures (including cellular) which are unique to insects may possess less toxicity for mammals, and thus may be superior as pesticides. It has also been speculated that highly specific lipolytic and proteolytic enzymes will be found to control specific processes in insects, as they do in mammals; thus they will be important targets for research.

Chapters 2 and 9-20 in this volume discuss the control of insects with allelochemicals, and the mechanisms by which plants biosynthesize these compounds and use them to become resistant to their pests.

Toxicants. The discussion of this subject has been separated from that of allelochemicals to emphasize that novel compounds possessing strong toxic activity, and often accompanying antifeedant action, have been isolated from plants that are often tropical or desert in origin (12-15). One of the apparent explanations for the large number of tropical and desert plants possessing highly toxic compounds is that they must coexist with higher insect and disease infestations than plants grown in temperate climates. In evolutionary time, plants that are most able to coexist (in part because of their biosynthesis of toxicants) have been selected. An increase in the accessibility of these plant materials will depend largely on the activity of botanical and pharmaceutical collection programs. Chapters 21 and 28 in this volume are pertinent examples of directed efforts to identify natural toxins from tropical plant sources.

A prominent example of a synthetic pesticide whose structure was based on a natural product is the family of pyrethroids (16). After the elucidation of the natural product structure, a search for compounds with optimum properties has continued for many years and has resulted in wide commercialization.

Naturally Occurring Hormonal Agents from Insects and Plants. The juvenile, molting, brain, and diapause hormones, antihormones such as the precocenes, and the prostaglandins are examples of these agents. Several books and symposia (5, 10, 12, 17) provide a summary of the status of these classes of compounds. It is probable that additional hormonal and antihormonal agents with high activity and specificity will be isolated from plants or insects. Some will act in a direct manner while others may promote biosynthesis of a secondary compound(s) with toxicity or other activity toward the insect. The high activity and great specificity of these compounds make it probable that field applications can be carried out under optimal conditions, often as a part of pest management systems.

During the past few years, a number of peptides have been isolated from insects and plants with very powerful hormonal or toxic effects (18, 19). Chapter 7 in this volume describes work on the effects of enzymes on cuticle tanning that are hormonal like in actions and have activity at low concentrations.

Microbial and Viral Agents for the Control of Insect Pests. Microbial agents traditionally have been visualized as contributing to biological control of pests. Many of these microbial agents can now be defined chemically. Work is also progressing on fly control in animal production and antigens for nematodes and insects (19).

The development of BT preparations for control of insect pests has accelerated greatly in the past 3-5 years, particularly since formulations have been developed to extend field life. Other microbial and viral agents are also being developed and show varying degrees of promise for field applications (18-21).

Natural Products as Inducers of Insect Resistance. Plant growth regulators have been shown to increase the biosynthesis of certain secondary plant constituents that in turn decrease plant attack by insects. α-Naphthaleneacetic acid, for example, elicits increased terpene biosynthesis in citrus, thus decreasing attack by fruit flies. The approach of using both natural and synthetic plant growth regulators may continue to find applications in insect control.

A general term for compounds whose biosynthesis is elicited or induced in plants as the result of attack by a pest is "phytoalexin". These higher plant metabolites are antibiotic to certain potential plant pathogens and also on occasion to insects (Chapter 13 of this volume). The phytoalexins typically are biosynthesized in greater concentrations when the plant is subjected to stress. Therefore, the attacking agent (fungi, bacteria, or viruses in most work to date) elicits the initiation or increased synthesis of phytoalexins (antibiotic compounds). It has recently been shown that attack by insects and nematodes also can elicit the formation of phytoalexins. A number of chemical elicitors of phytoalexins have also been identified (11, 12, 22, 23).

Induced Autointoxication. The goal of this approach is to apply precursors that develop activity only in the target species. Although photosensitization is not always required to develop activity, this type of sensitization was reported a number of years ago as occurring when houseflies were fed various dyes (24). Acetylenes and furanocoumarins have been shown to affect the insect melanins and they are phototoxic (25). The advantage of this approach for practical application is that it may permit foliar application of compounds with low toxicity and perhaps high specificity for the target insect. Continued studies of the mechanisms of activation can be expected to generate more efficient precursors. Much of the recent work has been summarized in two symposia and a subsequent book (26, 27), and in Chapters 25 and 26 of this volume.

Control of Weeds

Weed control is defined as the selective application of stress agents that can be chemical (synthetic or natural), tillage, fertilizer, cultural practice, or other. It has been suggested and is generally recognized that control of weeds must follow the approach of integrated pest management. Some of the desired results are that the system be energy efficient, conservative to soil, environmentally safe, inexpensive, effective, broad spectrum where desirable, and selective.

The control of weeds has generally been considered to be the domain of synthetic chemicals, but natural products are beginning to have an impact (28, 29).

Natural Products that Regulate Plant Growth. Natural products that possess growth regulating activity can be broadly categorized into two groups: (1) growth substances such as auxins, gibberellins, cytokinins, abscisic acid, ethylene, and their synthetic analogues or mimics, and (2) the so-called secondary plant growth substances such as the phenols, aliphatic and aromatic carboxylic acids and their derivatives, steroids, alkaloids, terpenoids, amino acids, and lipids. Of the secondary plant growth substances, some of the unsaturated lactones, terpenoids, steroids, and alkaloids tend to have limited species distribution, are produced in small quantities, but may possess some specific activity. Nearly all of the so-called secondary plant growth substances can be viewed as originating from the acetate and shikimic acid pathways. Mandava (30) has listed a number of these compounds with their specific activities. Duke (Chapter 26 in this volume) reported on several tetrapyrrole intermediates of heme biosynthesis that generate high levels of singlet oxygen. Many compounds that effect heme and/or chlorophyll pathways are strongly herbicidal due to accumulation of phytotoxic levels of these tetrapyrroles.

Investigations of the role that these natural products play in the metabolism of the plants where they are produced and the extent to which they control growth and developmental processes appear of interest. Also, the nature of the control mechanisms and their efficacy for herbicidal activity may warrant investigation. The recent interest in brassinosteroids as plant growth regulators may also lead to an application for weed control (31).

Allelopathic Agents. Weeds (and other plants) secrete chemicals from the roots and tops that inhibit seed germination and growth of proximate plants. These chemicals can be shown to be very effective in the laboratory and their field effects in-situ are evident, but adaptation for commercial use has been slow to follow. Nevertheless, there is a continuing effort to identify allelopathic agents. Two comprehensive books by Thompson (28) and Waller (29) summarize much of the recent literature.

Disease Control

Diseases in plants may be categorized as resulting chiefly from viruses, bacteria and fungi. It has been stated that modern organic fungicides have in effect biorationally evolved from the past use of sulfur, copper sulfate, and copper sulfate hydrated lime. Over the past 30-40 years, a succession of synthetic fungicides and other disease control agents have been developed. They have captured most of the market and continue to hold it. Nevertheless, large numbers of natural products have been screened for various disease control activities (12, 30, 32).

Fungicides from Tropical and Subtropical Plants. As discussed earlier, plants from tropical regions of the world are subjected to severe disease pressures, partically because of the heat and humidity (other disease organisms require cool, humid conditions). These plants may adapt to pressure of diseases in evolutionary time by developing defense systems. One mechanism of defense is the biosynthesis of highly active antifeedants for insects. A number have been identified (14, 32) which suggests that other types of biological activity such as antifungal activity also may be found in these plants. The availability of plant material from the tropics has increased because of industrial and public sector groups securing them for examination of their pharmaceutical and anticancer properties. The work of Miles et. al. (Chapter 28 in this volume), who screened Peruvian plants for several antimicrobial activities, is an example of this approach. Snook and Chortyk (Chapter 27 of this volume) found several tobacco root phenolics to be active against the black shank fungus. Hasegawa et al (Chapter 29 of this volume) isolated antagonistic microorganisms from the rhizosphere soil of Adzuke-bean root that controlled Fusarium wilt and isolated several antibiotic constituents. Also, a number of antifungal constituents were reported in the symposium book of Cutler (12).

Antibiotics. The successful use of antibiotics against bacterial diseases of humans has led to large-scale screening of antibiotics for plant disease control. They are normally produced commercially for use by microbiological processes. Some of the antifungal antibiotics are cycloheximide, griseofulvin, blasticidin S, kasugamycin, polyoxin, ezomycin, and validamycin A. The antibacterial antibiotics include streptomycin, cellocidin, chloramphenicol, and novobiocin. In another area, Strobel and Myers (33) isolated a bacterium normally found on leaves of wheat, barley, and oats that can defeat the fungus responsible for Dutch elm disease. The bacterium is a pseudomonad that produces fungus-killing antibiotics. Alternatively, the bacteria could be directly used or the bacterial antibiotic could be microbiologically produced and used for control. The success with Dutch elm disease fungus suggests that suitable screening programs may identify still other disease-killing agents.

Elicitation of Phytoalexins. Phytoalexins (previously discussed briefly with regard to activities against insects) are generally defined as low molecular weight products of plant biosynthesis that have antibiotic properties to one or several groups of microorganisms. The preformed levels of phytoalexins are generally low or non-detectable in healthy plant tissue, but they accumulate to high levels at the site of attack of the plant by an invading microorganism. Bell et al.(34) defines those components of resistance that are present in healthy plants as part of a "constitutive defense". Chemicals that effect active defense responses are referred to as elicitors, and may be either biotic or abiotic in origin. Bell et al. (34) have identified a number of abiotic elicitors of terpenoid and tannin synthesis in cotton such as chilling injury, UV irradiation, cupric ions, polymers from microbial cell walls such as polysaccharides, and pesticides. In a sense, many of those bioregulators, natural and synthetic, that increase plant resistance to pests can be viewed as elicitors of phytoalexins.

Of special interest are the polysaccharides that act as elicitors. Various heterogeneous polymers obtained from fungal and bacterial cell walls also act as elicitors. These include extracellular polysaccharides, lipoprotein polysaccharides, and glycoproteins.

Studies with other plant-pathogen systems have shown that oligomers from chitosan and β-1,3-glucans can be potent elicitors (35). Both chitin and the glucans are common constituents of fungal cell walls and are cleaved by chitinases and β-1,3-glucanases that occur in plants; concentrations of these enzymes increase in many plants soon after infection. A specific oligomer size (usually 5-8 subunits) of chitosan or β-1,3-glucans is required for appreciable activity. Is is possible that such sizes are cleaved from lipoprotein polysaccharides, extracellular polysaccharides, and dead cells of fungal and bacterial pathogens to activate defense responses in cotton. A β-N-acetylglucosaminidase has been demonstrated in cotton tissue (34). Pectate oligomers also may act as elicitors and can be formed by the action of fungal or host pectinase enzymes on pectin (33).

Summary. Naturally occurring pest bioregulators have been isolated from a wide diversity of sources, often from geographically stressed regions which favor rapid growth of pest populations. In order for plants (or animals) to survive, they must develop mechanisms to avoid or cope with these pests. These mechanisms often involve chemicals that deter feeding, that are toxic, or that slow growth and maturation of the pest.

Natural products are also a key to understanding ecological systems. With the identification of the constituents, progress has been made in understanding the metabolic cycles, the enzymes that lead to their biosynthesis, and the underlying genetic controls. For application of these natural products to pest control, an understanding of the physiology and behavior are essential to elucidate the underlying chemically mediated interactions between host and pest.

Natural products have provided leads to new pesticides. Knowledge of the biosynthetic processes of natural products and related information has often led to an understanding of the mode of action. A knowledge of the required functional groups has led to the design of synthetic pesticides with high activity and specificity.

Natural products may also find their place in the market place. Economically feasible bioregulators may be produced by microbiological processes such as fermentation. In other instances, processing of plant material may provide commercially adequate yields. For other applications such as the production of pheromones, hormones, or peptides, synthesis of the natural products may be the best strategy. Another application of natural products to pest control may be through the insertion of genes into plants expressing the compound. This approach is just now receiving increased emphasis.

When a biologically active natural product is identified, often more is gained than knowledge about the specific structure. An understanding of how the national product is biosynthesized and by which mechanism(s) it expresses its activity should make it possible to devise additional minimal changes that yield maximum benefits. Thus, those changes that benefit people may be selected which do minimum or no damage to the ecosystem.

Literature Cited

1. Beroza, M. Pest Management with Insect Sex Attractants; ACS Symposium Series No. 23; American Chemical Society: Washington, DC, 1976; 192 pp.
2. Inscoe, M. N. In Insect Suppression with Controlled Release Pheromone Systems; Kydonieus, A. E., Beroza, M., Eds.; CRC Press: New York, Vol. II, p. 201-295.
3. Leonhardt, B. A.; Beroza, M. Insect Pheromone Technology: Chemistry and Application; ACS Symposium Series No. 190, American Chemical Society: Washington, DC, 1982; 260 pp.
4. Hedin, P. A.; Menn, J. J.; Editors, Insect Chemical Communication: Unifying Concepts; Special Issue, J. Chemical Ecology 1988, 14, 1979-2145.
5. Carlson, D. A. Symposium on Biosynthesis and Catabolism of Insect Pheromones and Hormones; Abstracts of Papers, 199th National Meeting of the American Chemical Society: Boston, MA, April 1990, AGRO 36-40, 55-60, 74-79.
6. Wallace, J. W.; Mansell, R. L. Biochemical Interactions Between Plants and Insects; Recent Advances in Phytochemistry, Vol. 10, Plenum Press, N. Y., 1976; 425 pp.
7. Hedin, P. A. Host Plant Resistance to Pests; ACS Symposium Series No. 62; American Chemical Society: Washington, DC, 1977; 286 pp.
8. Rosenthal, G. A., Janzen, D. H. Herbivores: Their Interaction with Secondary Plant Metabolites; Academic Press: New York, 1979; 718 pp.
9. Hedin, P. A. Plant Resistance to Insects; ACS Symposium Series No. 208; American Chemical Society: Washington, DC, 1983; 375 pp.
10. Hedin, P. A. Bioregulators for Pest Control; ACS Symposium Series No. 276; American Chemical Society: Washington, DC, 1985; 540 pp.

11. Green, M. B.; Hedin, P. A. Natural Resistance to Pest: Roles of Allelochemics; ACS Symposium Series No. 296; American Chemical Society: Washington, DC, 1986; 243 pp.
12. Cutler, H. G. Biologically Active Natural Products: Potential Uses in Agriculture; ACS Symposium Series No. 380; American Chemical Society: Washington, DC, 1988; 483 pp.
13. Morgan, E. D.; Mandava, N. B. Handbook of Natural Pesticides: Vol. VI, Insect Attractants and Repellents; CRC Press: Boca Raton, FL.; 249 pp.
14. Kubo, I.; Nakanishi, K. In Host Plant Resistance to Pests; Hedin, P. A., Ed.; ACS Symposium Series No. 62; American Chemical Society: Washington, DC, 1977; pp 165-178.
15. Mabry, T. J.; Burnett, W. C.; Jones, S. B.; Gill, J. E. In Host Plant Resistance to Pests; Hedin, P. A., Ed., ACS Symposium Series No. 62; American Chemical Society: Washington, DC, 1977; pp. 179-184.
16. Elliott, M. Synthetic Pyrethroids; ACS Symposium Series No. 42; American Chemical Society: Washington, DC, 1977; 229 pp.
17. Gilbert, L. I. The Juvenile Hormones; Plenum Press: New York, 1976; 572 pp.
18. Osburn, B. I. Symposium on Natural Antimicrobial Peptides and their Application in Agriculture; Abstracts of Papers, 197th National Meeting, American Chemical Society: Dallas, TX, April 1989, AGRO 13-17, 30-33.
19. Menn, J.J.; Kelly, T. J.; Masler, E. P.; Edwards, J. V.; Cherry, J. P; Eto, M., Hammock, B. D.; Okai, H. Symposium on Biologically Active Peptides in Insects, Plants, and Food; Abstracts of Papers, 197th International Chemical Congress of Pacific Basin Societies: Honolulu, HI, December 1989, AGRO 43-47, 82-88, 105-110, 133-138, 140-145, 382-390, 398-404.
20. Cross, B.; Cody, S. Symposium on Formulation of Proteins for Agricultural Applications; Abstracts of Papers, 200th National Meeting, American Chemical Society: Washington, DC, August 1990, AGRO 5-8, 14-18, 40-44.
21. Hedin, P. A.; Menn, J. J.; Hollingworth, R. M. Biotechnology for Crop Protection; ACS Symposium Series No. 379; American Chemical Society: Washington, DC, 1988, 471 pp.
22. Keen, N. T.; Bruegger, B. In "Host Plant Resistance to Pests"; Hedin, P. A., Ed.; ACS Symposium Series No. 62; American Chemical Society: Washington, DC, 1977; pp 1-26.
23. Hedin, P. A.; Jenkins, J. N.,; Thompson, A. C.; McCarty, J. C.; Smith, D. H.; Parrott, W. L.; Shepherd, R. L. J. Agric. Food Chem. 1988, 36, 1055-1061.
24. Carpenter, T. L; Heitz, J. R. "Abstracts of Papers", 178th National Meeting, American Chemical Society: Washington, DC, Sept. 1979; PEST 34.
25. Berenbaum, M. Science (Washington, DC): 1978, 201, 532-533.
26. Heitz, J. R.; Downum, K. R. Light Activated Pesticides; ACS Symposium Series No. 339; American Chemical Society; Washington, DC, 1987; 355 pp.
27. Towers, G. H. N.; Kubo, I.; Tang, C. S.; Mizutani, J. Symposium on Chemical Aspects of Plant-Microogaranism, Plant-Insect, and Plant-Plant Interactions; Abstracts of Papers,

1989 International Chemical Congress of Pacific Basin Societies: Honolulu, HI, December 1989, AGRO 164-171, 195-202.

28. Thompson, A. C. The Chemistry of Allelopathy: Biochemical Interactions Among Plants; ACS Symposium Series No. 268; American Chemical Society: Washington, DC, 1985; 471 pp.
29. Waller, G. R. Allelochemicals: Role in Agriculture and Forestry; ACS Symposium Series No. 330; American Chemical Society: Washington, DC, 606 pp.
30. Mandava, N. B. In "Plant Growth Substances"; Mandava, N. B., Ed.; ACS Symposium Series No. 111; American Chemical Society: Washington, DC, 1979; pp 135-213.
31. Cutler, H. G. Symposium on Brassinosteroids; Abstracts of Papers, 200th National Meeeting, American Chemical Society: Washington, DC, August 1990, AGRO 76-79, 100-102, 119-122, 131-133.
32. Jacobson, M. In Host Plant Resistance to Pests; Hedin, P. A., Ed.; ACS Symposium Series No. 62; American Chemical Society: Washington, DC, 1977; pp 153-164.
33. Strobel, G.; Myers, D. F. Science News (Washington, DC): 1980, 117, 362.
34. Bell, A. A.; Mace, M. E.; Stipanovic, R. D. In Natural Resistance of Plants to Pests: Roles of Allelochemicals; Green, M. B, Hedin, P. A., Eds.; ACS Symposium Series No. 296; American Chemical Society; Washington, DC, pp 36-54.
35. Keen, N. T.; Yoshikawa, M.; Want, M. C. Plant Physiol. 1983, 71, 466-471.

RECEIVED September 17, 1990

BIOREGULATION OF INSECT BEHAVIOR AND DEVELOPMENT

Chapter 2

Arthropod Natural Products as Insect Repellents

Murray S. Blum[1], Daniel M. Everett[2], Tappey H. Jones[3], and Henry M. Fales[3]

[1]Department of Entomology and [2]Department of Computer Science, University of Georgia, Athens, GA 30602
[3]Laboratory of Chemistry, National Heart, Lung, and Blood Institute, National Institutes of Health, Bethesda, MD 20892

The defensive allomones of arthropods have been evolved to blunt the attacks of a variety of predatory species. These natural products function primarily by repelling adversaries and thus usually constitute the first line of defense. An examination of the deterrent efficacies of compounds produced by honey bees, thrips, and ants demonstrates that these exocrine products are highly effective repellents against a diversity of ant species. A potpourri of natural products including aliphatic and aromatic ketones, esters, fatty acids, and alkaloids has been determined to possess well-developed repellent properties at physiological concentrations. These results emphasize the great potential of insect-derived compounds as an outstanding source of repellents in the never-ending battle with species of pest arthropods.

Among animals, arthropods are distinguished by their utter dominance in terms of both numbers of species and individuals. The virtual ubiquity of these populous organisms guarantees that they will be subject to great predatory pressure from the invertebrates and vertebrates with which they share their fragile world. Not surprisingly, arthropods themselves constitute a major group of predators, and among these, it is not unlikely that ants are dominant. Indeed, ants are probably the major predatory animals in the world, with most of their 10,000-15,000 species (1) exhibiting a carnivorous propensity combined with an efficient system of prey acquisition (2). For most arthropods, ants probably represent the most frequently encountered predators with which they must contend. Defense against ants and the legions of other predatory animals thus becomes a sine qua non for the survival of arthropodous species in a variety of ecological contexts.

0097–6156/91/0449–0014$06.00/0
© 1991 American Chemical Society

Arthropods synthesize an incredible diversity of natural products in their exocrine glands (3) which are utilized with great effectiveness to blunt the attacks of their adversaries. The defensive secretions of these invertebrates can often be delivered with great accuracy, thus ensuring that aggressive predators are subjected to the full impact of these secretory effronteries (4). Significantly, these exudates usually manifest their defensive efficacy as effective repellents and thus constitute the first line of defense of the prey species. The deterrent value of the natural products in these secretions is inestimable, since it can permit their producers to escape predators without injurious physical confrontations. This is especially important in encounters with ants, since these social insects can quickly launch devastating en masse attacks.

A variety of studies has demonstrated that ants may be rapidly deterred by the defensive secretions of diverse arthropods on which they attempt to prey (5, 6). For example, many of the small and delicate species of thrips (Thysanoptera) produce anal secretions that are directed against the ants with which they frequently have encounters (7). Although the defensive secretions of very few thrips species have been analyzed, evaluation of the compounds present in a few of these exudates indicates that they are effective deterrents for ants (8, 9). These results further suggest that repellents for ants may be commonly encountered in the defensive secretions of a variety of arthropodous species. Furthermore, since ants have frequent antagonistic interactions with other species of ants, it could be anticipated a priori that these formicids would have evolved powerful ant deterrents themselves.

Both field and laboratory studies have documented the deterrent efficacies to ants of the venomous secretions of these insects (10, 11, 12). For example, it has been demonstrated that the raiding modus vivendi of one ant species was made possible by the utilization of a venom-derived alkaloid that is a powerful repellent for workers of raided ant species (13). Although ants synthesize an incredible diversity of venomous alkaloids (14), their activities as repellents for these formicids have only been described preliminarily (15). Significantly, ants appear to be typical of social insects in generating deterrent compounds in their exocrine glands, and examples of other hymenopterans producing repellents, of great social significance, have recently been reported (16).

Honey bee queens produce a rectal secretion which repels aggressive workers in the colony as an example of an intraspecific repellent (16). If such pheromonal repellents are commonly produced by social insects, a cornucopia of deterrent natural products awaits identification in the exocrine secretions of these arthropods.

In the present report, we present results of very recent investigations on the chemistry and deterrent activities to ants of the diverse compounds identified as exocrine compounds of thrips. In addition, we evaluate the repellency of some honey bee natural products to honey bee workers as an example of how pheromones themselves can be candidates as deterrents for the species that produce them. Finally, detailed analyses of the repellencies of a host of nitrogen heterocycles identified in the venoms of ants are described.

Chemistry and Repellency of Thrips Natural Products

The first compound identified as a natural product of thrips was γ-decalactone (I), a product of *Bagnalliella yuccae* (8). The anal exudate, which is discharged when the thrips are disturbed, does not contain any other detectable volatiles. Both laboratory and field studies demonstrated that the secretion effectively deterred ant workers either as a topical irritant or by repellency per se. Workers of *Monomorium minimum* quickly withdrew after contact with the anal fluid and avoided sites at which the exudate had been released (8). The repellency of the anal exudate was clearly identified with γ-decalactone, two thrips equivalent of this compound evoking 50% repellency for *M. minimum* workers. Similar results were obtained with *M. pharaonis* and *I. humilis*, two other species of predatory ants. The results of field studies corroborated the laboratory findings (8).

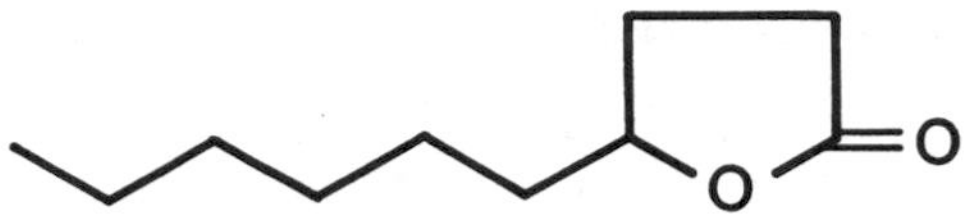

γ-Decalactone

The chemistry and repellent efficacy of the anal exudate of the Cuban laurel thrips, *Gynaikothrips ficorum*, a gall-inhabiting species, have also been examined (9). The exudate is dominated by a 1:1 ratio of hexadecyl acetate and pentadecane; tetradecyl acetate, tridecane, tetradecane, and heptadecane constitute minor concomitants. The anal secretion of *G. ficorum* is both an effective contact deterrent and repellent for aggressive ants.

Workers of *Wasmannia auropunctata* are rapidly repelled by an anal droplet from a thrips that is within a millimeter of the ants (9). Topical treatment of ants with anal exudate corresponding to 0.25 thrips equivalents resulted in 100% of the workers dragging themselves away from the scene of the encounter in much the same manner as occurs when these ants contact the thrips under field conditions. Both hexadecyl acetate and pentadecane caused dragging behavior after being applied to ant workers, which responded in a dose-dependent manner. Significantly, although hexadecyl acetate is more active than pentadecane, a combination of the two compounds is considerably more effective than the equivalent amounts of either of the two compounds alone. Therefore, it is evident that the ester and hydrocarbon interact synergistically to augment the deterrent efficacy of the exudate (9).

Recently, the chemistry of total extracts of another species of *Gynaikothrips* was examined (17). *G. uzeli* is reported to produce the same hydrocarbons and acetates as *G. ficorum*, but in addition synthesizes a novel monoterpene, β-acaridial, a compound recently identified as a natural product of a mite (18). β-Acaridial has also been identified as an important compound produced by two other species of thrips, *Varshneyia pasanii* and *Liothrips kuwanai* (17). The function of β-acaridial, which is very unstable, is not known.

Another monoterpene, perillene, has been identified as a product of thrips in several genera. This furanomonoterpene accompanies β-acaridial in extracts of *V. pasanii*, *L. kuwanai*, and *L. piperinus* (19, 20). Perillene has also been detected as a major product in extracts of *Teuchothrips longus*, *Arrhenothrips ramakrishnae*, and *Schedothrips* sp. (21). Vapors of this compound produced a dose-dependent repellency in workers of two ant species, *Monomorium carbonarium* and *Iridomyrmex humilis*, under laboratory conditions. Thus, perillene functions as a repellent and its presence in the anal exudates of diverse thrips species clearly augments the deterrent efficacies of these discharges. On the other hand, it has been suggested that high dosages of perillene (20-200 μg) can function as an alarm pheromone (17).

A third monoterpene, rose furan, has been identified as the major constituent produced by the thrips *Arrhenothrips ramakrishnae* (21). Rose furan is accompanied by perillene and in addition, hexadecyl acetate is quantitatively significant. Two aromatic compounds, phenol and phenylacetaldehyde, are minor concomitants in the anal exudate.

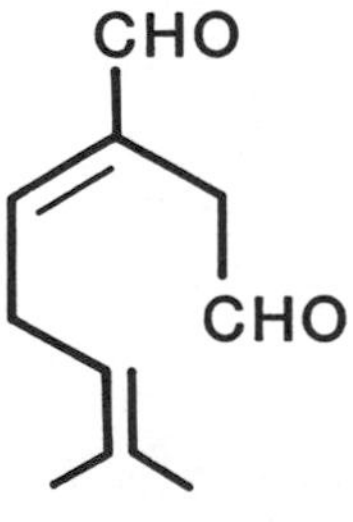

CHO
CHO

β-Acaridial

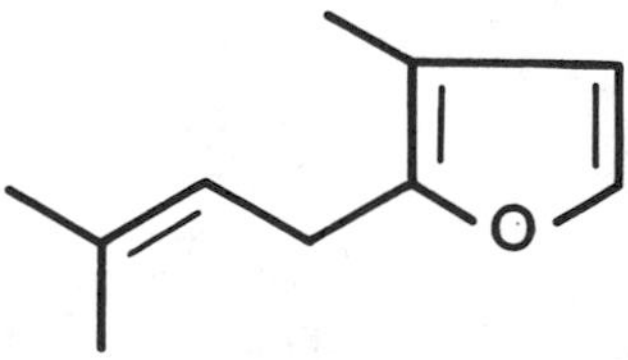

O

Rose Furan

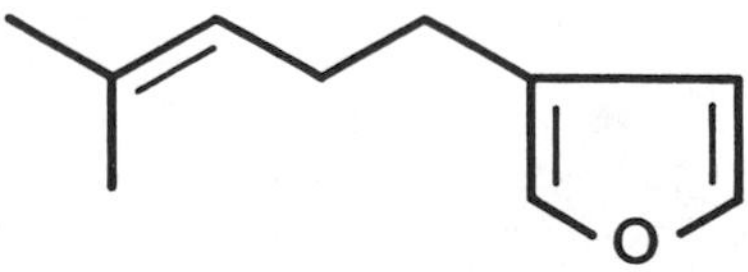

O

Perillene

Fatty acids are also characteristic natural products of species of thrips in a diversity of genera. A total extract of *Hoplothrips japonicus* was dominated by (*E*)-3-dodecenoic acid which is considered to be an alarm-aggregation pheromone; (*Z*)-5-dodecenoic acid is a minor concomitant (*22*). 2-Methylbutyric acid, a defensive allomone of swallowtail larvae (*23*), is the only free fatty acid detected in the previously described anal exudate of *Varshneyia pasanii* (*20*). It is believed that the anal discharge of this species, thoroughly dominated by alkanes, acetates, and oxygenated monoterpenes, exhibits an increased repellent "punch" because of the presence of the C_5 acid.

The anal fluid of a *Dinothrips* sp. contains only isovaleric and decanoic acids in equal quantities (*24*). A combination of the acids is highly repellent to workers of the fire ant *Solenopsis invicta*. Decanoic acid is the major allomone present in the anal fluid of *Euryaplothrips crassus* and is accompanied by dodecanoic acid as a quantitatively important product (*21*). The exudate of *E. crassus* also contains 2-phenylacetaldehyde, phenol, and 4-octadec-9-enolide as minor constituents.

Dodecanoic acid has also been identified as a major compound in the anal exudate of *Elaphrothrips tuberculatus*, and it is accompanied by two other acids, (*Z*)-5-tetradecenoic acid and 5,8-tetradecadienoic acid (*25*). On the other hand, a novel animal natural product, juglone, provides this exudate with a particularly distinctive exocrine chemistry. Juglone, which is synthesized *de novo* by *E. tuberculatus*, is an outstanding repellent for ant species in several genera. The allelopathic effects of juglone, a product of black walnut (*Juglans nigra*), are well established (*26*), and it is really not surprising that this highly phytotoxic and reactive quinone should exhibit considerable repellent activity for ants as a consequence of assaulting their chemoreceptors.

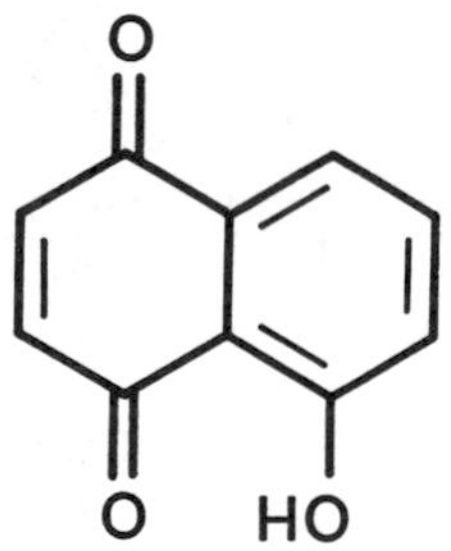

O
O HO

Juglone

Honey Bee Pheromones as Repellents

Some pheromones of social insects which function as excitants and attractants at low (=physiological) concentrations are powerful repellents when present at super threshold concentrations (27). The ability of these insects to be deterred by high levels of their own pheromones can provide a means of detecting new repellents for eusocial arthropods. Beyond this consideration, it is possible that these "pheromonal repellents" might serve as repellents for a wide range of insects.

Honey bee workers utilize a mandibular gland pheromone, 2-heptanone, as a low level alarm pheromone at the hive entrance (28). This compound, which is a repellent for foraging worker bees (29), produces abnormal behavioral reactions when workers are exposed to high concentrations of it (30), in much the same way as is observed with 2-heptanone-producing ants (31). Significantly, when this methyl ketone is applied to the hands as a 0.5, 1.0, or 2% aerosol solution, it is possible to manipulate the bees in a hive without them reacting aggressively. Workers are repelled by the ketonic super threshold concentration and retreat from its source in a nonagitated state. The high concentration of this pheromone thus effectively disarms the workers and provides a repellent "shield" for the pheromone emission source. Since some cockroaches utilize 2-heptanone as a defensive allomone (32), it is possible that this compound may be repellent to a broad spectrum of insect species.

Young honey bee queens produce a repellent pheromone that effectively tranquilizes workers that may interact antagonistically with these virgin females (16). The active compound, o-aminoacetophenone, is a minor component in the anal exudate that is discharged by the molested queens (33). This compound is also a defensive allomone of an ant species (34), raising the possibility that it may possess general deterrent activity for arthropods.

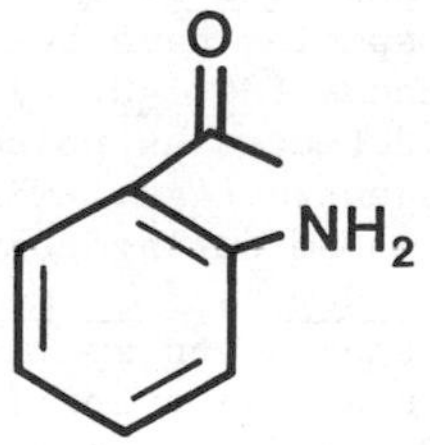

O

NH_2

o-Aminoacetophenone

The Repellency of Venomous Alkaloids of Ants

Ants in the myrmicine genera *Solenopsis* and *Monomorium* are distinctive in producing venoms that are dominated by alkaloids rather than proteins (35). Fire ants (*Solenopsis* spp.) in the subgenus *Solenopsis* characteristically synthesize poison gland secretions that are dominated by the *cis*- and *trans*-isomers of 6-alkyl- or 6-alkylidene-2- methylpiperidines (36). On the other hand, species in the subgenus *Diplorhoptrum* frequently produce 2,5-dialkylpyrrolidines as venomous constituents (37), as well as 3,5-dialkylpyrrolizidines (38), and 3,5-dialkylindolizidines (39). The venoms of *Monomorium* species typically contain 2,5-dialkylpyrrolidines that are frequently accompanied by 3,5-dialkylpyrrolizidines and 3,5-dialkylindolizidines (40, 41, 42). With the exception of the dialkylpiperidines (reviewed in 14), virtually nothing is known about the biological activities of the other classes of venomous nitrogen heterocycles.

In view of the reported deterrent activities of *Solenopsis* and *Monomorium* venoms under field conditions (10, 11, 12), it seemed worthwhile to evaluate the comparative repellencies to ants of some of these venom-derived alkaloids. In this investigation, the deterrency of these nitrogen heterocycles to a variety of ant species was determined by using a feeding bioassay in which the reactions of hungry ant workers to alkaloid-treated food were quantified (43). Selection of a variety of aggressive ant species in combination with a diversity of candidate compounds, belonging to the main classes of venomous alkaloids, presented an opportunity to examine these formicid natural products in terms of their activities as repellents.

Queenright (queen-containing) colonies of 10 species of ants, belonging to two major subfamilies, were utilized for repellency studies. Members of the subfamily Myrmicinae included *Solenopsis invicta*, *Crematogaster ashmeadi*, *Pheidole dentata*, *Monomorium minimum*, *M. viridum* and *M. pharaonis*. The subfamily Dolichoderinae was represented by *Iridomyrmex pruinosus*, *I. humilis*, *Tapinoma sessile*, and *T. melanocephalum*. Whereas *S. invicta* and the *Monomorium* species have been demonstrated to synthesize alkaloid-rich venoms (36, 40, 41), these nitrogen heterocycles have not been detected as poison gland products of any of the other myrmicine genera (44). Since we have frequently observed workers secreting venom during competitive interactions between *M. pharaonis*, *M. viridum*, and *T. melanocephalum* in southern Florida, and the other seven species in northern Georgia, these species seemed particularly appropriate for evaluating the potencies of these alkaloids as ant repellents.

Both foraging and feeding were stimulated by not providing the ant colonies with food for 48 hours. Food was then offered to the ants as droplets of honey to which were added 1 or 2 μg of alkaloids in 2 μl of absolute ethanol. In one study, both treated droplets and colonies were randomized for each replicate which compared the repellencies of four alkaloids that included a *Solenopsis* alkaloid, *trans*-6-undecyl-2-methylpiperidine (36), two *Monomorium* dialkylpyrrolidines (41), and a *Solenopsis* dialkylpyrrolizidine (38). In this investigation, all species of ants were utilized except M. pharaonis, T. melanocephalum, and I. humilis. A second replicate of the measurements was performed 24 hours later, with the same colonies and baits randomized in a different pattern.

In a second study, the comparative deterrencies of eight Solenopsis dialkylpiperidines (36) and two Solenopsis dialkylindolizidines (39) were compared. Test species included two alkaloid producers, M. pharaonis and S. invicta, and two species that do not synthesize alkaloidal poison gland secretions, I. humilis and T. melanocephalum (45, 46). This study, which was implemented the same way as described for the previous test, was also designed to compare the deterrent activities of the cis- and trans-2,6-dialkylpiperidines (relation of substituents at C-2 and C-6). Included were compounds in which the 6-alkyl group was normal C_9, C_{13}, and C_{15}; the 6-alkylidene group was either Z-4-tridecenyl or Z-6-pentadecenyl. Compounds are referred to as cis or trans in combination with an abbreviation for the chain length (plus unsaturation) of the group attached to C-6. Thus, cis-6-nonyl-2-methylpiperidine (I) is designated as cis-C_9 and 2-trans-6-(4-tridecenyl)-2-methylpiperidine (II) is designated trans-$C_{13:1}$ (Figure 1). Two dialkylindolizidines, (5Z,9Z)-3-hexyl-5-methylindolizidine (Hex Ind) and (5Z,9Z)-3-ethyl-5-methylindolizidine (Et Ind), were included for comparison. Three replicates of the measurements were performed in the study, using different random combinations of colonies and treatments, at 7-10 day intervals.

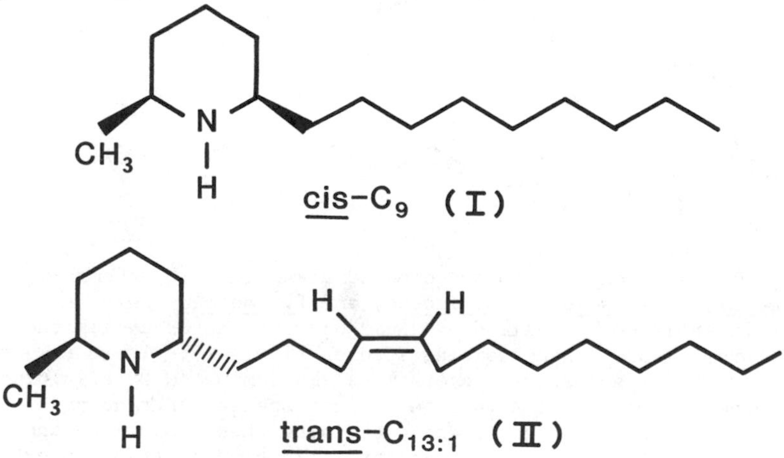

Figure 1.

For the most part, each of the seven ant species exhibited characteristic responses to the four alkaloids evaluated in the first study. For all species, a 2 µg dosage of an alkaloid was significantly more repellent than 1 µg. However, workers of four of the species, C. ashmeadi, I. pruinosus, M. minimum, and M. viridum, were repelled by each of the alkaloids of both concentrations. Two of the compounds, 2-(1-hex-5-enyl)-5-nonylpyrrolidine (III) and 2-(1-hex-5-enyl)-5-(1-non-8-enyl)pyrrolidine (IV), were significantly more repellent than

(5*Z*,8*E*)-3-heptyl-5-methylpyrrolizidine (V) for all species (Figure 2). The *trans*-C_{11} piperidine (VI) was intermediate in repellency between the pyrrolidines III and IV, and the pyrrolizidine (V), but was not significantly different from either class of alkaloids (43).

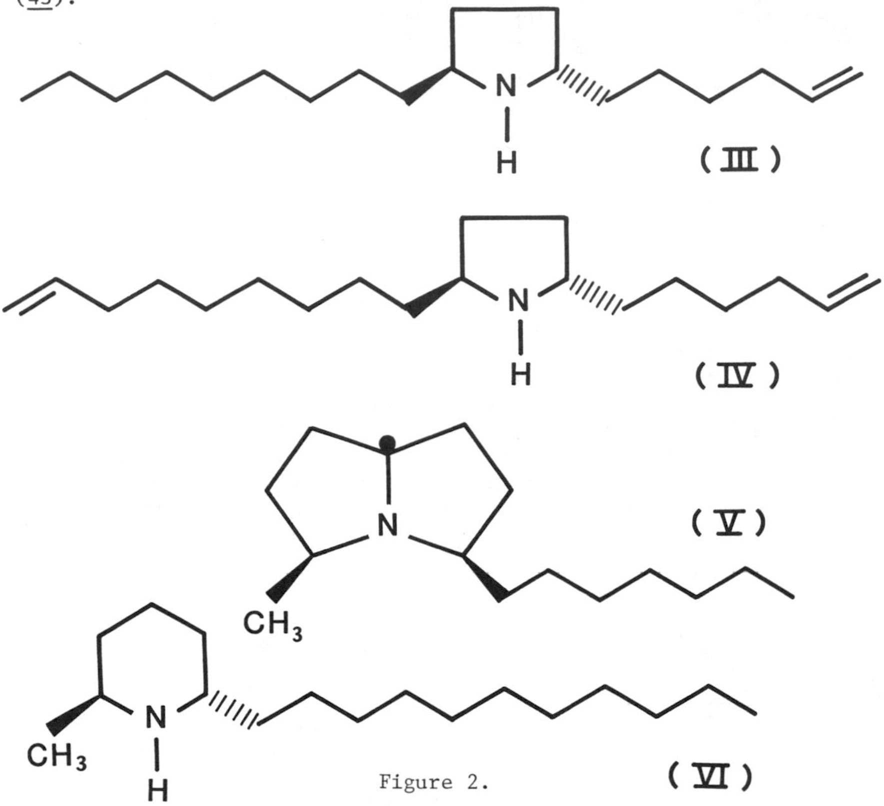

Figure 2.

The species that do not produce alkaloid-rich venoms--*C. ashmeadi*, *T. sessile*, *I. pruinosus*, and *P. dentata*--were significantly more repelled by these nitrogen heterocycles than the three alkaloid-synthesizing species (Table 1). These results suggest that species that generate venoms dominated by alkaloids may have a competitive advantage against species that do not produce poison gland secretions containing these volatile--and repellent--natural products. Recent field studies (47) support this conclusion.

In the second study, which utilized two species not known to produce alkaloidal venoms--*I. humilis* and *T. melanocephalum*--and two alkaloid-producing species, *M. pharaonis* and *S. invicta*, dose-dependent responses were only observed for two of the species with selected dialkylpiperidines (43). The responses of workers of *M. pharaonis* to the *cis*-C_9 piperidine and the *cis*-C_{13} piperidine were clearly dose dependent, as was the response of workers of *T. melanocephalum* to the *cis*-C_{13} piperidine.

Table 1. Mean number of ant workers (as percent of controls) of alkaloid- and nonalkaloid-producing species feeding on baits treated with four ant-derived alkaloids.

	Alkaloid			
	Pyrrolidine[a]	Pyrrolidine[b]	Pyrrolizidine[c]	Piperidine[d]
Alkaloid producers	46.8 ± 6.6	46.1 ± 5.3	60.5 ± 8.0	53.2 ± 8.0
Nonalkaloid producers	15.5 ± 3.4	15.2 ± 3.3	36.4 ± 1.6	28.9 ± 5.3
All species	28.0 ± 3.8	27.6 ± 3.4	46.0 ± 5.1	37.4 ± 4.7

[a]2-(1-hex-5-enyl)-5-nonylpyrrolidine.
[b]2-(1-hex-5-enyl)-5-(1-non-8-enyl)pyrrolidine.
[c](5*Z*, 8*E*)-3-heptyl-5-methylpyrrolizidine.
[d]*trans*-6-undecyl-2-methylpiperidine.

Workers of both T. melanocephalum and I. humilis, species that do not produce alkaloidal venoms, were especially sensitive to the deterrent effects of different alkaloids; S. invicta workers were not repelled by any of the alkaloids (Table 2). Workers of M. pharaonis were not repelled as effectively as those of the two nonalkaloid-producing species (44). The repellencies of the alkaloids for the sensitive species were: cis-$C_{13:1}$ > cis-C_{13} > cis-C_9 > trans-C_{13} > cis-$C_{15:1}$ > trans-$C_{15:1}$ > Hex Ind > trans-C_{15} > cis-C_{15} > Et Ind > EtOH (significant at 5% level). These rankings were obtained using Duncan's Multiple Range Test on the numbers of ants feeding at each bait, expressed as percentages of the average number of ants from their colony feeding at control baits (see Table 1).

Pairwise comparison of the repellencies of the cis and trans-isomers of the C_{13}, C_{15} and $C_{15:1}$ dialkylpiperidines demonstrated that the cis isomers of the C_{13} and $C_{15:1}$ alkaloids are more repellent than their trans counterparts. On the other hand, no significant differences in the deterrencies of the cis- and trans-C_{15} compounds were observed.

Table 2. Mean number of ant workers (as percent of controls [EtOH]) of alkaloid- and nonalkaloid-producing species feeding on baits treated with 10 ant-derived alkaloids.

Alkaloid[a]	Alkaloid Producer	Nonalkaloid producer
cis-C_{13}	61.1 ± 8.2	7.5 ± 2.1
cis-C_9	69.9 ± 12.3	3.8 ± 2.2
cis-$C_{15:1}$	68.8 ± 5.8	25.1 ± 3.8
trans-C_{13}	73.1 ± 10.4	14.8 ± 3.6
cis-C_{15}	86.7 ± 7.1	49.9 ± 6.0
EtOH	99.5 ± 7.1	103.8 ± 6.0
trans-$C_{15:1}$	92.0 ± 8.0	30.4 ± 5.2
trans-C_{15}	86.0 ± 9.0	41.1 ± 5.0
EtOH	100.5 ± 9.0	96.2 ± 7.7
Hex Ind	84.3 ± 6.8	40.8 ± 6.3
cis-$C_{13:1}$	57.2 ± 8.9	0.5 ± 0.2
Et Ind	86.7 ± 5.8	94.4 ± 7.6

[a]See text for full names of alkaloids.

Conclusions

Repellent compounds appear to commonly fortify the exocrine secretions of a diversity of arthropods. As illustrated by the deterrents synthesized by thrips, bees, and ants, considerable structural eclecticism characterizes these natural products. The great variety of defensive allomones which has been evolved to blunt the assaults of multifarious predators attests to both the biosynthetic virtuosity of insects and the wealth of potential candidate repellents.

It would be no exaggeration to state that the incredible success of insects is in no small way due to their ability to effectively counter, with chemical arsenals, the predatory effronteries of their enemies. These allomonal deterrents have been "tried and tested" in the fullness of evolutionary time, and offer humankind a treasure trove of proven repellents with which to challenge pest species. It is time to give these arthropods credit for having perfected chemical warfare, and to exploit the defensive products of their success for society's benefit.

Literature Cited

1. Brown, W. L.; Taylor, R. W. In "The Insects of Australia." Univ. Melbourne Press, Melbourne, 1970; pp. 951-959.
2. Wilson, E. O. The Insect Societies; Harvard Univ. Press: Cambridge, 1971.
3. Blum, M. S. Chemical Defenses of Arthropods; Academic Press: New York, 1981.
4. Eisner, T. In "Chemical Ecology"; Sondheimer, E.; Simeone, J. B., Eds. Academic Press: New York, 1970.
5. Nutting, W. L.; Blum, M. S.; Fales, H. M. Psyche 1974, 81, 167.
6. Wallace, J. B.; Blum, M. S. Ann. Ent. Soc. Am. 1969, 62, 509.
7. Lewis, T. Thrips; Academic Press: London, 1973.
8. Howard, D. F.; Blum, M. S.; Fales, H. M. Science 1983, 220, 335.
9. Howard, D. F.; Blum, M. S.; Jones, T. H.; Fales, H. M.; Tomalski, M. D. Phytophaga 1987, 1, 163.
10. Hölldobler, B. Oecologica. 1973, 11, 1971.
11. Bhatkar, A.; Whitcomb, W. H.; Buren, W. R.; Callahan, P. S.; Carlyse, T. Environ. Ent. 1972, 1, 274.
12. Baroni-Urbani, C.; Kannowski, P. B. Environ. Ent. 1974, 3, 755.
13. Blum, M. S.; Jones, T. H.; Hölldobler, B.; Fales, H. M.; Jaouni, T. Naturwissenschaften 1980, 67, 144.
14. Jones, T. H.; Blum, M. S.; Fales, H. M. Tetrahedron 1982, 38, 1949.
15. Blum, M. S. In "Biologically Active Natural Products"; Cutler, H.G., Ed.; ACS Symp. Ser. No. 380, 1988, pp. 438-449.
16. Post, D. C.; Page, R. E.; Erickson, E. H. J. Chem. Ecol. 1987, 13, 583.
17. Suzuki, T.; Haga, K.; Leal, W. S.; Kodama, S.; Kuwahara, Y. Appl. Ent. Zool. 1989, 24, 222.
18. Leal, W. S.; Kuwahara, Y.; Suzuki, T. Agric. Biol. Chem. 1989, 53, 875.
19. Suzuki, T.; Haga, K.; Kuwahara, Y. Appl. Ent. Zool. 1986, 21, 461.
20. Suzuki, T.; Haga, K.; Kodama, S.; Watanabe, K.; Kuwahara, Y. Appl. Ent. Zool. 1988, 23, 291.
21. Blum, M. S.; Fales, H. M. Ananthakrishnan, T. Unpublished results, 1988.
22. Haga, K.; Suzuki, T.; Kodama, S.; Kuwahara, Y. Appl. Ent. Zool. 1989, 24, 242.
23. Eisner, T.; Meinwald, Y. C. Science 1965, 150, 1733.
24. Blum, M. S.; Whitman, D. W.; Fales, H. M. Unpublished results, 1989.

25. Blum, M. S.; Crespi, B.; Jones, T. H.; Howard, D. F.; Howard, R. W. Unpublished results, 1988.
26. Rietveld, W. J.; Schlessinger, R. C.; Kessler, K. J. J. Chem. Ecol. 1983, 9, 1119.
27. Blum, M. S.; Warter, S. L. Ann. Ent. Soc. Am. 1966, 59, 774.
28. Shearer, D. A.; Boch, R. Nature 1965, 206, 530.
29. Simpson, J. Nature 1966, 209, 531.
30. Boch, R.; Shearer, D. A.; Petrasovits, A. J. Insect Physiol. 1970, 16, 17.
31. Blum, M. S.; Warter, S. L.; Monroe, R. S.; Chidester, J. C. J. Insect Physiol. 1963, 9, 881.
32. Moore, B. P.; Brown, W. V. J. Aust. Ent. Soc. 1979, 18, 123.
33. Page, R. E.; Blum, M. S.; Fales, H. M. Experientia 1988, 44, 270.
34. Blum, M. S.; Brand, J. M.; Amante, E. Experientia 1981, 37, 816.
35. Jones, T. H.; Blum, M. S. In "Alkaloids: Chemical and Biological Perspectives"; Pelletier, S. W. Ed.; John Wiley and Sons, Inc., 1983; Vol. 1, pp. 33-84.
36. MacConnell, J. G.; Blum, M. S.; Fales, H. M. Tetrahedron 1971, 26, 1129.
37. Pedder, D. J.; Fales, H. M.; Jaouni, T.; Blum, M. S.; MacConnell, J. G.; Crewe, R. M. Tetrahedron 1976, 32, 2275.
38. Jones, T. H.; Blum, M. S.; Fales, H. M.; Thompson, C. R. J. Org. Chem. 1980, 45, 4778.
39. Jones, T. M.; Highet, R. J.; Blum, M. S.; Fales, H. M. J. Chem. Ecol. 1984, 10, 1233.
40. Ritter, F. J.; Rotgans, I. E. M.; Talman, E.; Verwiel, E.; Stein, F. Experientia 1973, 29, 530.
41. Jones, T. H.; Stahly, S. M.; Don, A. W.; Blum, M. S. J. Chem. Ecol. 1988, 14, 2197.
42. Jones, T. H.; Highet, R. J.; Don, A. W.; Blum, M. S. J. Org. Chem. 1986, 51, 2712.
43. Blum, M. S.; Everett, D. M.; Tomalski, M. D.; Jones, T. H. Unpublished results, 1989.
44. Blum, M. S. In "Insect Poisons, Allergins, and Other Invertebrate Venoms"; Tu, A. T. Ed.; Maurice Dekker, Inc., 1984; pp. 225-242.
45. Cavill, G. K. W.; Houghton, E. Aust. J. Chem. 1974, 27, 879.
46. Tomalski, M. D.; Blum, M. S.; Jones, T. H.; Fales, H. M.; Howard, D. F.; Passera, L. J. Chem. Ecol. 1987, 13, 253.
47. Andersen, A.; Blum, M. S.; Jones, T. H. Unpublished results, 1989.

RECEIVED May 16, 1990

Chapter 3

Aggregation Pheromone of *Carpophilus lugubris*

New Pest Management Tools for the Nitidulid Beetles

Robert J. Bartelt, Patrick F. Dowd, and Ronald D. Plattner

Northern Regional Research Center, Agricultural Research Service, U.S. Department of Agriculture, 1815 North University Street, Peoria, IL 61604

The existence of a male-produced aggregation pheromone in the nitidulid beetle, *Carpophilus lugubris* Murray, was demonstrated using a wind tunnel (flight) bioassay. The pheromone attracts both sexes. It was obtained by trapping volatiles from feeding beetles. The key purification step was HPLC on a silver-nitrate coated silica column. Based on the mass and ultraviolet spectra of the pheromone, the mass spectra of the hydrogenated derivatives, and knowledge of the pheromone of a related nitidulid, four synthetic compounds were prepared as pheromone candidates. One of these, (2*E*,4*E*,6*E*,8*E*)-7-ethyl-3,5-dimethyl-2,4,6,8-undecatetraene matched the pheromone by all spectral and chromatographic criteria. The synthetic compound was active in the wind tunnel and also in the field. A possible biosynthetic route is discussed.

The dusky sap beetle, *Carpophilus lugubris* Murray (Coleoptera: Nitidulidae), is a small (ca. 4 mm), dark colored species that occurs throughout the temperate and tropical portions of the Western hemisphere (1). Like most nitidulid species, *C. lugubris* frequents sites where plant materials are fermenting or decomposing, but these beetles also have the unfortunate tendency of infesting ears of corn in the field (1). In the United States, *C. lugubris* is responsible for the rejection of large amounts of sweet corn at canneries (2). Furthermore, beetles of this family are able to transmit fungi into corn which can cause the subsequent buildup of mycotoxins (3). *C. lugubris* is also commonly found in oak woods and is capable of vectoring the fungus which causes the oak wilt disease (4).

Pheromones are now routinely used as tools in the monitoring and control of a variety of insect pests, but until lately, nothing was known about the pheromones of nitidulid beetles. An aggregation pheromone was recently discovered in *C. hemipterus* (5), a close relative of *C. lugubris*, and it includes two tetraene hydrocarbons which were previously unknown to science: (2*E*,4*E*,6*E*,8*E*)-3,5,7-trimethyl-2,4,6,8-decatetraene (1) and (2*E*,4*E*,6*E*,8*E*)-3,5,7-trimethyl-2,4,6,8-undecatetraene (2). The pheromone is produced by male

This chapter not subject to U.S. copyright
Published 1991 American Chemical Society

1 2

beetles, and it attracts both sexes. It appears to work primarily as a synergist of food source volatiles. In the field, we believe it facilitates the assembly of groups of beetles at suitable feeding sites so that mating and oviposition can take place.

The research reported here is a continuation of the study of nitidulid pheromones. The previous knowledge of the pheromone of *C. hemipterus* allowed identification of the *C. lugubris* pheromone from a very small natural sample.

Beetles, Bioassays, and Pheromone Collection

The *C. lugubris* beetles for these studies were field collected during the summer of 1988 in oak woods and corn fields near Bath, Illinois. They were attracted to traps baited with whole-wheat bread dough inoculated with baker's yeast. The collected beetles were then maintained in the laboratory on an artificial diet as described previously for *C. hemipterus* (6). Adult beetles lived as long as 6 months under these conditions. The availability of field-collected adults obviated the development of an egg-to-adult rearing procedure.

The laboratory pheromone bioassays were conducted in a wind tunnel as described previously for *C. hemipterus* (5). Briefly, ca. 200-400 adult beetles were placed, without food, into the wind tunnel about 16 hr prior to beginning the bioassays. As with *C. hemipterus*, the beetles flew readily only after being starved for a number of hours. A bioassay was conducted by applying a test solution to a 7-cm disk of filter paper (Whatman 541), which was folded into quarters and hung, along with a control filter paper, in the upwind end of the wind tunnel, the two baits being separated by 30 cm. The number of beetles alighting on each bait during a three-min test period was recorded. Usually, 20-30 three-min bioassay tests could be conducted during a day with one group of beetles. The tests were separated in time by 2-5 min. An active preparation was always run every third or fourth test to ensure that the beetles remained in a responsive condition. The age of bioassay beetles was not controlled, but satisfactory results were obtained from the time the beetles were collected in the field until they were ca. 6 months old. The bioassay counts were subjected to analysis of variance after transformation of the data to the $\log(X+1)$ scale. In the tables, significant differences from the controls at the $P = 0.05$, 0.01, and 0.001 levels are denoted by *, **, and ***, respectively.

Pheromone collections were also made as described for *C. hemipterus* (5). Tenax porous polymer was used to trap the volatiles emitted from groups of beetles feeding on pinto-bean diet. Parallel samples were derived from males, from females, and also from the diet medium alone, for later chromatographic and bioassay comparisons. Counts of beetles were kept so that the pheromone production could be quantified in terms of beetle-days. (A beetle-day is the average amount of pheromone collected from one beetle in one day). Beetles were separated by sex and set up for pheromone collection within a week of trapping them in the field. The following results were based on a collection of ca. 5000 beetle-days.

Initial Bioassays and Chromatography

Beetle-derived samples were always bioassayed at a dose of 5 beetle-days. In preliminary bioassays, the Tenax extract derived from males was clearly more active than that from females, which paralleled the previous result with *C. hemipterus*. In seven 3-min paired comparisons, the collection from males attracted an average of 14.1 beetles while that from females attracted only 1.4. Both sexes responded.

The extract from males was fractionated on silica gel to separate components by polarity, as described previously (5). The solvents and bioassay activities of the fractions are presented in Table I. The hexane fraction was the only one that showed clear activity, suggesting that the pheromone of *C. lugubris* was a hydrocarbon. The slight activity of the most polar fraction appears to have been due to the artificial diet on which the insects were fed during pheromone collection.

Table I. Wind Tunnel Bioassays with Chromatographic Fractions Derived from *C. lugubris* males

A. Silica gel fractions from Tenax collection

Fraction	Mean bioassay count (n=4) Fraction	Control
Hexane	28.4 ***	0.2
5% Ether-hexane	0.0	0.6
10% Ether-hexane	0.2	0.2
50% Ether-hexane	1.2	0.2
10% MeOH-methylene chloride	2.7 *	0.0

B. $AgNO_3$-HPLC fractions from hexane fraction (above)

Fraction (ml after injection)	Mean bioassay count (n=4) Fraction	Control
3.0-4.5	0.0	0.2
4.5-5.0	0.0	0.0
5.0-5.5	0.2	0.4
5.5-6.0	35.8 ***	0.0
6.0-6.5	8.5 ***	0.2
6.5-7.0	2.3 **	0.0
7.0-7.5	0.0	0.4

The active hexane fraction was subjected to HPLC on a silver-nitrate coated silica column (7) as described for *C. hemipterus* (5); 25% toluene-hexane was used as the solvent. Fractions 0.5 ml in volume were collected and bioassayed (Table I). The activity eluted primarily in a fraction 5.5-6.0 ml after injection. This elution volume was well beyond the solvent front (3.0 ml); thus, there was evidence for unsaturation. However, the activity eluted slightly before the tetraenes (1 and 2) previously isolated from *C. hemipterus*, which occurred in fractions 6.0-7.5 ml after injection.

Pheromone Isolation and Spectra

The $AgNO_3$ HPLC fractions from male and female *C. lugubris* were compared by capillary GC. The highly active 5.5-6.0 ml fraction from males produced a GC peak at 7.87 min (15 m X 0.25 mm ID DB-1 column, 1.0 μm film thickness, 100-200°C at 10°C/min) which was completely lacking from the female sample. This peak had a retention index of 15.15, relative to *n*-alkanes (8), and this peak represented 5% of the total GC peak area in the male-derived fraction; the other peaks corresponded to materials that were diet derived. The male-specific compound was present in the fraction at ca. 100 pg per beetle-day. The 6.0-6.5 ml fraction had a smaller amount of this compound and was still fairly active in the bioassay.

The 5.5-6.0 ml $AgNO_3$ fraction was rechromatographed on a size-exclusion HPLC column (PLGEL 50A, 30 cm X 7.5 mm ID, 10 μm particle size, hexane as solvent). The male-specific compound eluted in a fraction 11.0-11.7 ml after injection, and by GC, appeared uncontaminated by other compounds. A total of ca. 700 ng was obtained. Qualitatively, the beetles responded instantly to 1 ng of this preparation in the bioassay.

The electron impact mass spectrum (Figure 1), obtained on a Finnigan 4535 instrument, indicated a molecular weight of 204, which corresponded to the molecular formula, $C_{15}H_{24}$. This formula implies four degrees of unsaturation. In general appearance, the spectrum was similar to those of the tetraenes **1** and **2** of *C. hemipterus* (5). An ultraviolet spectrum was obtained in hexane, and a broad peak with a maximum at 286 nm was observed. The extinction coefficient was approximately 20,000, based on GC integration relative to an external quantitative standard. Again, the UV spectrum was similar to those of **1** and **2** from *C. hemipterus* (5). The spectral data suggested that the *C. lugubris* pheromone was a conjugated tetraene related to **1** and **2** but with 15 instead of 13 or 14 carbons.

Hydrogenation

The purified compound was hydrogenated (9) and analyzed by mass spectrometry to confirm the number of double bonds and to gain information about the carbon skeleton. The reaction was run on ca. 100 ng of pheromone in 25 μl of CH_2Cl_2; PtO_2 was the catalyst. The sample was introduced into the mass spectrometer through a capillary GC column (DB-1, programed at 10°C per min), and spectra were taken at the rate of 1 per sec. A complex mixture of products resulted (Figure 1, total ion GC trace). (A similar mixture was always observed when **1** or **2** was hydrogenated (5) and was due to the creation of asymmetric centers during hydrogenation as well as to a cyclizing reaction which competed with complete hydrogenation). The highest molecular weight in the products from *C. lugubris* was 212, which corresponded to the formula, $C_{15}H_{32}$, and resulted from the uptake of 8 hydrogen atoms. This implied the original compound was indeed a tetraene and had no rings. Although the molecular ion (m/z 212) was of low intensity for these products, most of their major alkyl fragment ions (m/z 57, 71, ...) were easily detected. The single mass chromatograms for these fragments (e.g., m/z 183, Figure 1) indicated that there were four distinct GC peaks due to acyclic,

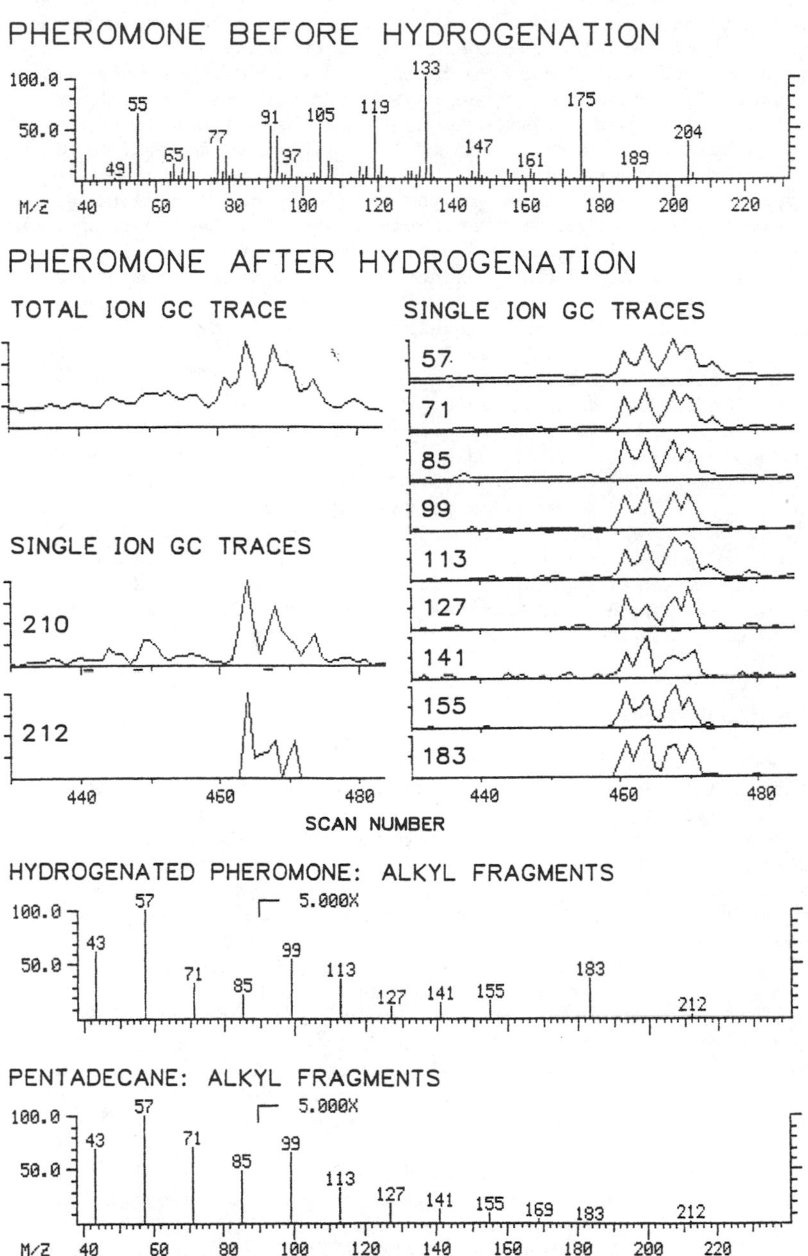

Figure 1. Analysis of pheromone by mass spectrometry, before and after hydrogenation.

saturated derivatives. The cyclic derivatives (molecular ion m/z 210) overlapped with these, by GC, but did not interfere with mass spectral interpretation for the acyclic derivatives.

The generation of four GC peaks from a single pure compound suggested the creation of three asymmetric centers during hydrogenation. With three asymmetric centers, there would be eight possible optical isomers, which would produce, at most, four GC peaks on an achiral column. As with the pheromone components of *C. hemipterus*, the double bonds in the original compound probably involved three branches in the carbon chain, but even more branches were possible if the saturated derivatives did not separate completely by GC or if the branches did not involve asymmetric centers. The GC retention times of the saturated derivatives also indicated a high degree of branching. These 15-carbon alkanes had retention indices of 13.32-13.43.

The intensities of alkyl fragment ions, $C_nH_{2n+1}^+$, provide information about locations of branches in alkanes (10). The spectra of all four acyclic derivatives were nearly identical when only these fragments were considered, and the key feature was the relatively intense peak at m/z 183, compared with the 15-carbon *n*-alkane (Figure 1). This corresponded to the ready loss of an ethyl radical, a feature which was also evident in the mass spectrum of the unsaturated parent compound (m/z 175). Several other fragments were relatively suppressed (m/z 71, 85, 169), but unfortunately, the locations of branches were not as obvious from the fragmentation pattern of the hydrogenated derivatives as they had been for **1** and **2** (5).

Synthesis of Model Tetraenes

Four model compounds were synthesized (Figure 2) to aid the interpretation of spectral and chromatographic features of the unknown compound, particularly the loss of an ethyl group from the carbon skeleton. Because of the close phylogenetic relationship between *C. lugubris* and *C. hemipterus*, model compounds were chosen which were structurally similar to **1** and **2**. Compound **3** was simply the 15-carbon homolog of the *C. hemipterus* compounds. Tetramethyl tetraene **4** was chosen because its carbon skeleton could lose an ethyl group from either end after hydrogenation, perhaps resulting in an enhanced m/z 183 fragment. Compounds **5** and **6** had ethyl groups at the 5 and 7 positions, respectively, instead of methyl groups, so that the loss of an "internal" ethyl group could be studied. No 3-ethyl-5,7-dimethylundecatetraene was synthesized because the carbon skeleton would have only two asymmetric centers (the 5 and 7 carbons) and could not, therefore, generate four GC peaks on an achiral column.

The syntheses of the compounds followed the general procedures outlined earlier (5) with one major modification: ethyl groups instead of methyl groups could be incorporated into the carbon chains by using triethyl 2-phosphonobutyrate (TEPB) in the Wittig-Horner reaction instead of triethyl 2-phosphonopropionate (TEPP). All reactions were monitored by GC and mass spectrometry. The Wittig-Horner reactions produced predominantly *E* isomers, while both *E* and *Z* isomers were obtained from the Wittig reactions. From previous experience (5), the isomers retained the longest by GC were

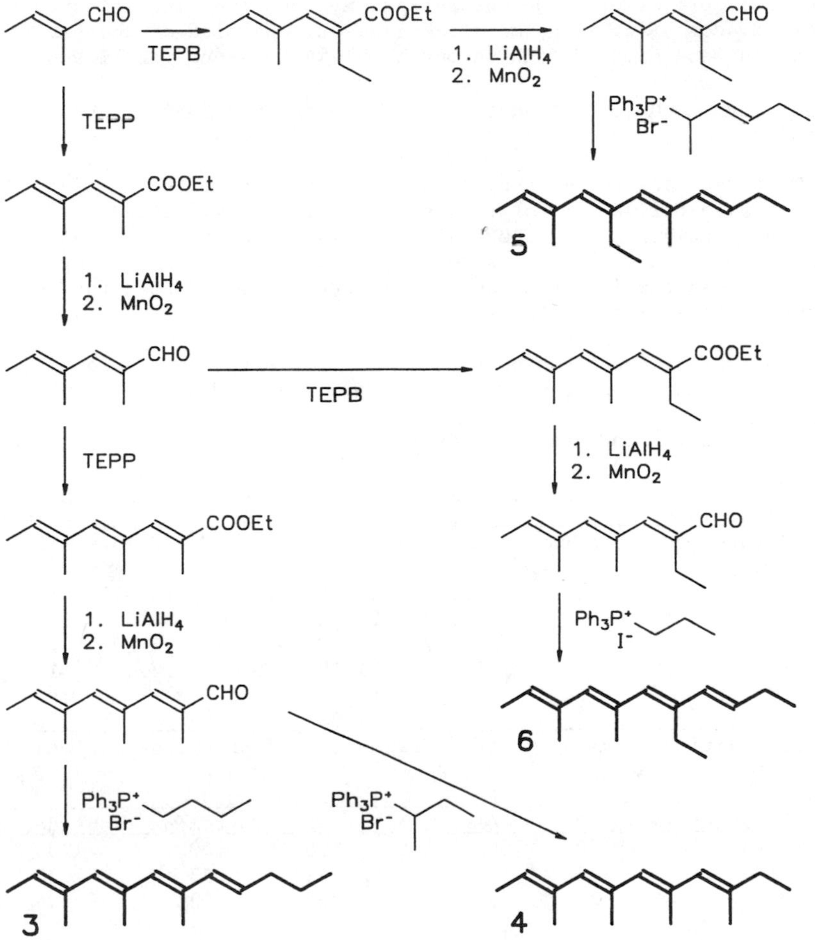

Figure 2. Synthetic scheme for four tetraenes used as model compounds in the structural analysis of the *C. lugubris* pheromone.

assigned the *E,E,E,E* configuration; these represented 50-90% of the synthetic tetraene products, depending on synthetic route and the structure. The synthetic products were purified first on silica gel then by HPLC on the $AgNO_3$ HPLC column. The NMR spectra of the purified synthetic tetraenes always supported the expected branching pattern. By analogy to **1** and **2** (5), the *E,E,E,E* configuration was supported as well. By capillary GC (DB-1), the tetraenes were at least 90% pure after $AgNO_3$ chromatography, and the impurities were mostly traces of other geometrical isomers. Samples of the purified tetraenes were diluted to 1 ng per 10 μl with hexane for bioassays.

Comparison of Chemical Properties between Pheromone and Synthetic Tetraenes

The GC retentions and mass spectral fragmentation patterns of the hydrogenated derivatives of these standards are summarized in Table II. Intensities of mass spectral peaks are relative to the base peak (m/z 57 in all cases) and are rounded to the nearest whole percent; "-" indicates the fragment was not detected, and <1 indicates the fragment was detected but was less than 0.5%.

Table II. Comparison of Four Synthetic Tetraenes, Pheromone, and Pentadecane: GC and Mass Spectral Data

A. GC retention indices (pentadecane = 15.00)

Conditions	Compound 3	4	5	6	Pheromone
Before hydrogenation	15.77	14.98	15.13	15.15	15.15
After hydrogenation	13.45	13.04	13.28	13.31	13.32
	13.48	13.08	13.34	13.34	13.35
	13.57	13.10	13.37	13.39	13.39
	13.59	13.17	13.40	13.42	13.43
		13.22			
		13.24			

B. Intensities of mass spectral fragments (after hydrogenation)

Fragment (m/z)	Compound 3	4	5	6	Pheromone	Pentadecane
57	100	100	100	100	100	100
71	50	46	52	42	33	70
85	42	32	25	24	22	49
99	21	17	11	11	11	13
113	8	6	6	6	7	7
127	2	2	4	3	2	4
141	17	10	2	3	3	3
155	2	1	3	4	3	2
169	<1	-	1	-	-	<1
183	3	1	7	7	7	<1
197	1	1	-	-	-	-
212	1	2	-	<1	1	<1

The derivatives from 3 and 4 were not similar to the derivatized pheromone. Even though the derivative of 4 had two terminal ethyl groups, the M-29 fragment ion (m/z 183) was still of low intensity (1%); instead, m/z 141 was the dominant spectral feature. The derivative of 3 did not produce a large m/z 183 peak either. Furthermore, derivatives of neither 3 nor 4 agreed with those from *C. lugubris* in GC retention. Compound 4, having four asymmetric centers after hydrogenation, generated six GC peaks rather than four. (Symmetry of the carbon skeleton allows only six, instead of eight, diastereoisomers, and all of these were separable by GC). Thus, compounds 3 and 4 had carbon skeletons unrelated to the *C. lugubris* pheromone.

The properties of the hydrogenated derivatives from 5 and 6, however, matched those of the derivatized pheromone very well (Table II). An ethyl branch near the center of the chain generated a more intense M-29 (m/z 183) fragment ion than ethyl groups at the ends of the chain. Furthermore, these alkanes agreed closely in GC retentions with the derivatives from *C. lugubris*. Neither the spectra nor GC retentions provided definitive means for distinguishing between the 5-ethyl and 7-ethyl isomers, but the existence of an internal ethyl branch appeared likely.

The mass spectra of the underivatized standards (Figure 3) provided further information about the location of the ethyl branch. For the tetraenes we have synthesized, there were always fragments at m/z 69, 83, and/or 97. (The next member of this series, m/z 111, was never observed, however). For each compound, the highest mass fragment of this series observed in the spectrum was consistent with cleavage at the 6 double bond, accompanied by a proton transfer (Figure 3). (The whole spectra for 1 and 2 appear in (5)). Compound 5 produced a significant m/z 83 peak (25%), but no 97 peak. However, compound 6 gave a 97 peak (8%) as well as an 83 peak (3%), which was the pattern observed with the *C. lugubris* compound. Thus, the existence of a 7-ethyl group in the pheromone was supported. In fact, the entire mass spectra of 6 and the *C. lugubris* compound agreed very well. Further evidence against a 5-ethyl group was that the 119 and 133 peaks in the mass spectrum of 5 had a distinctly different ratio than that observed for the other two samples.

The agreement of mass spectra does not infer equivalence of double bond configurations because the geometrical isomers of these tetraenes, in our experience, have virtually identical spectra. However, the GC retentions of 6 and the pheromone were identical, and GC retention is very sensitive to double bond configuration. In addition, the samples had identical retentions on the $AgNO_3$ column and UV spectra. Thus, from all available chemical evidence, compound 6 agreed with the unknown pheromone, not only in carbon skelton, but also in the location and configuration of double bonds. We conclude that the male-specific compound from *C. lugubris* is (2*E*,4*E*,6*E*,8*E*)-7-ethyl-3,5-dimethyl-2,4,6,8-undecatetraene, a compound which has not been reported previously.

Nuclear Overhauser experiments on purified synthetic 6 confirmed that it did have the *E*,*E*,*E*,*E* configuration. Irradiation at an olefinic methyl or methylene causes significant enhancement of an olefinic proton that is *cis* to it but not for one that is *trans* (5,11). Proton assignments were made based on double irradiation

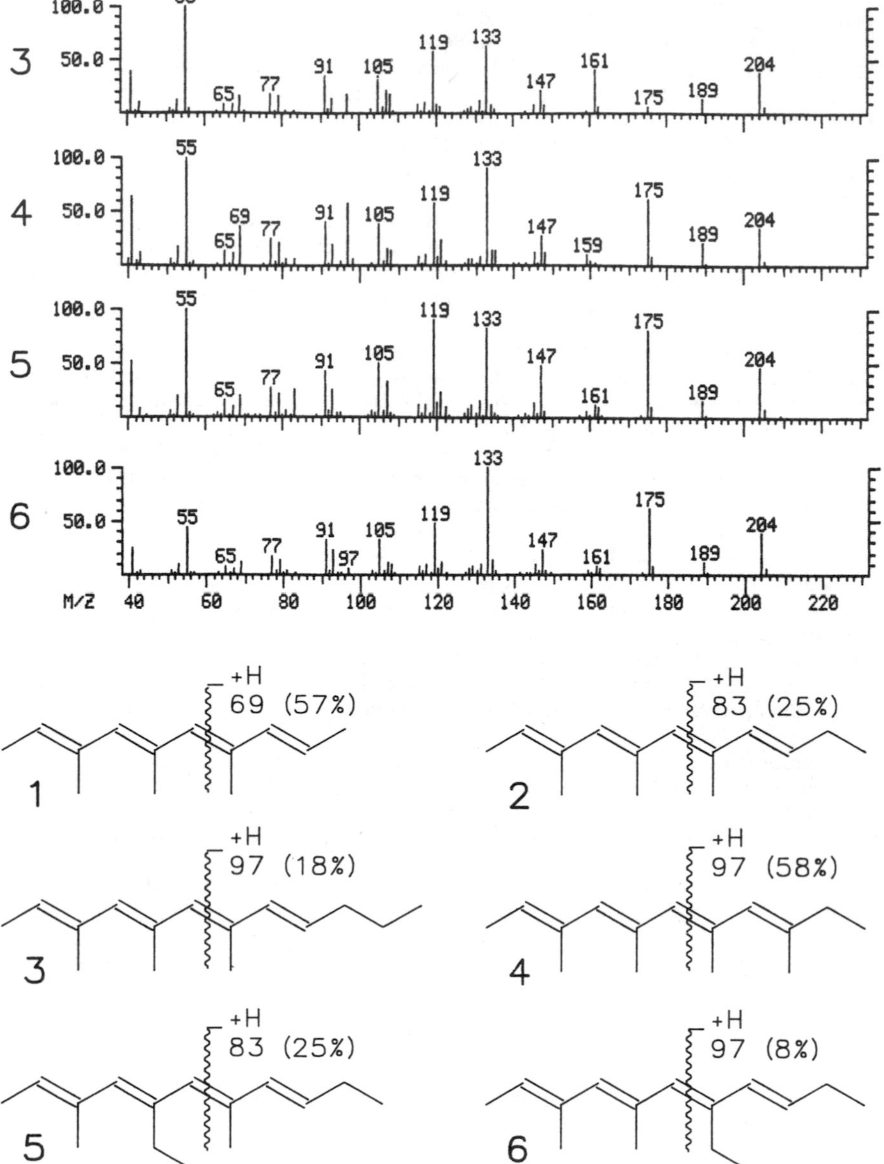

Figure 3. Mass spectra of four synthetic tetraenes and a diagnostic fragmentation.

experiments and by analogy to **1** and **2** as shown below (left). Irradiation at δ 1.75, 2.00, and 2.58 caused no measurable nuclear Overhauser enhancement at δ 5.53, 6.09, or 6.02, respectively, indicating that the trisubstituted double bonds had the *E* configuration. Irradiation at δ 2.11 did cause a 5% enhancement at δ 6.11,

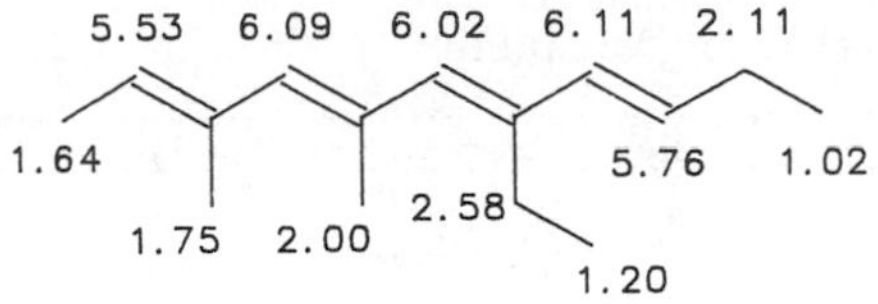

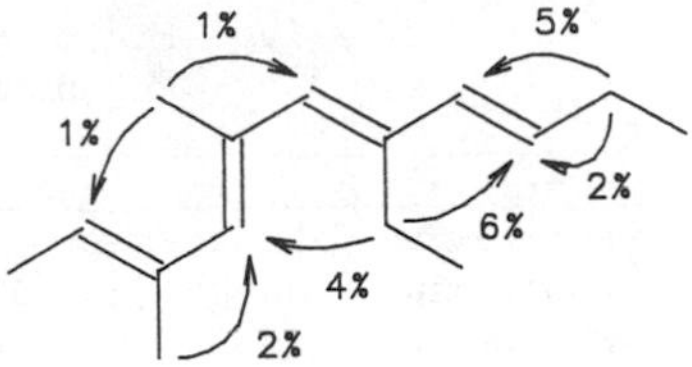

Proton assignments for **6**

Nuclear Overhauser enhancements

which supported the *E* configuration at the disubstituted double bond, as did the coupling constant of 15.7 Hz between the protons at δ 5.76 and 6.11. The observed nuclear Overhauser enhancements did further support the proton assignments and suggested a conformation in solution similar to that shown above (right).

Bioassays with Synthetic Compounds

The synthetic compound **6** was clearly active in the wind tunnel bioassay at doses of 1 ng and even 100 pg (Table III), but the other three model compounds (**3**-**5**) were not. Interestingly, the two pheromone components identified earlier from *C. hemipterus* (**1** and **2**) also elicited significant responses from *C. lugubris*, particularly the 14-carbon component **2**. *C. lugubris* has an extraordinary ability to discriminate among tetraenes with very similar structures, but responses can occur toward more than one compound.

Table III. Wind Tunnel Bioassays of *C. lugubris* with Synthetic Tetraenes

		Mean bioassay count (n=8)	
Tetraene	Test dose	Tetraene	Control
1	1 ng	1.7 *	0.3
2	1 ng	34.8 ***	0.1
3	1 ng	0.5	0.3
4	1 ng	0.1	0.1
5	1 ng	0.2	0.1
6	1 ng	51.0 ***	0.0
6	0.1 ng	19.2 ***	0.1

The crude synthetic pheromone **6** was tested under field conditions. The field traps were made from plastic pipe 6 cm in diameter and wire mesh; they were oriented horizontally and permitted wind to pass through. They had funnels formed from wire mesh on the downwind end, which allowed the attracted beetles to enter but which prevented escape. The traps were baited inside either with the pheromone (formulated at 500 μg per rubber septum), fermenting whole-wheat bread dough (a very effective attractant for *C. lugubris*, 20 g per trap), or a combination of these baits. A piece of screen mesh prevented attracted beetles from feeding on the dough. Unbaited traps served as controls. The traps were suspended ca. 1.5 m above the ground in an oak woods where the beetles were common. There were three traps of each type present during each 5-day trapping period. The whole experiment was replicated three times.

Table IV. Field Bioassay of Synthetic Pheromone **6** in Conjunction with Food Volatiles

Treatment	Mean trap catch (n=9)
Pheromone (500 μg/septum)	0.7 a
Whole wheat dough (20 g)	11.1 b
Pheromone + whole wheat dough	174.2 c
Control	0.0 a

In contrast to wind tunnel studies, the pheromone alone was not effective as a trap bait (Table IV), but it dramatically enhanced the attractiveness of the whole wheat dough, by a factor of over 15. Means followed by the same letter in the table are not significantly different (LSD, 0.05). As in the wind tunnel, both sexes responded. For example, the largest trap catch for the pheromone plus wheat dough treatment consisted of 494 males and 524 females (49% males). The synergistic activity of the pheromone and food volatiles had been previously demonstrated in the wind tunnel for *C. hemipterus* (5). *C. lugubris* did not require a food-derived coattractant to respond to the pheromone in the wind tunnel, but the close association between host-derived and beetle-derived volatiles was nevertheless clearly evident under more natural conditions. It is suggested that nitidulid beetles of the genus, *Carpophilus*, are like bark beetles (12) in using both host-derived odors and pheromones when orienting to new host resources, but the nitidulids differ from the bark beetles in being far less restricted in the range of acceptable hosts. Similar synergistic responses between pheromones and food volatiles have also been reported for *Sitophilus* weevils (13).

Biosynthesis

The pheromone components of *C. lugubris* and *C. hemipterus* are undoubtedly related biosynthetically. Although none of these compounds had been known prior to the nitidulid research, the 13-carbon pheromone component **1** of *C. hemipterus* has the same carbon skeleton as the pheromone of the drugstore beetle, *Stegobium paniceum* (14). A polyketide origin of this pheromone was proposed, in which the polyketide arose from the condensation of one acetate and four propionate units (15). Polyketide biosyntheses have also been proposed for a number of related structures (16, 17). The

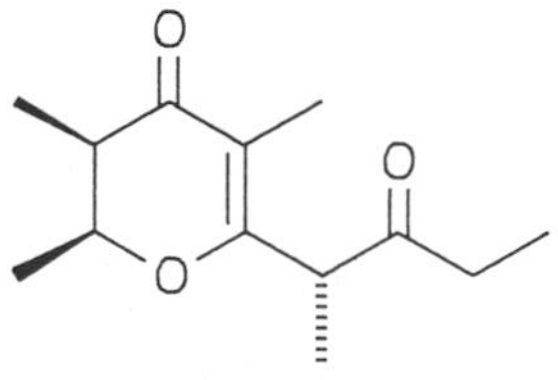

S. paniceum pheromone

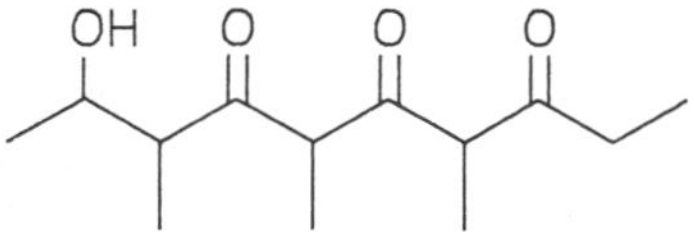

Proposed (15) polyketide precursor

hydrocarbon pheromones of *C. lugubris* and *C. hemiperus* are probably derived from the condensation of small acyl units as well, as suggested in Figure 4. A polyketide intermediate need not be postulated if the reduction and dehydration steps occurred after the acyl additions in a cyclic fashion, as in the biosynthesis of fatty acids. A final decarboxylation would release the hydrocarbon. Reactions analogous to those in this latter scheme are known to occur in insects (18). The methyl branches of the carbon chains would be derived from propionate and the ethyl branch of **6**, from butyrate. Incorporation of butyrate instead of propionate as the final acyl unit would account for **2** and **6** having longer carbon chains than **1**. In Figure 4, [X] represents an acyl carrier such as coenzyme A.

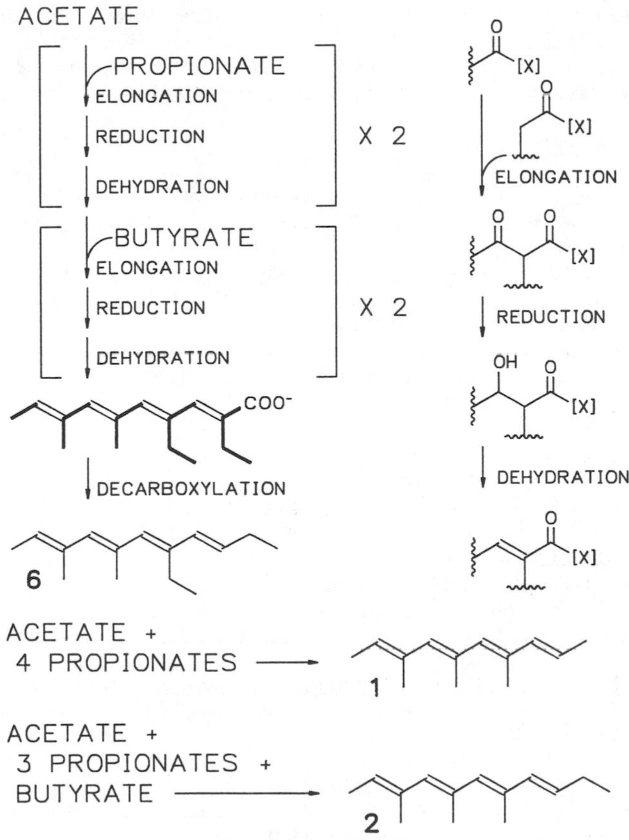

Figure 4. Possible biosynthetic origin of the pheromones of *C. lugubris* and *C. hemipterus*.

Needs for the Future

At present, pheromones are known for only two nitidulid species, and even these may possess additional components of as yet unknown biological importance. Before nitidulid pheromones can realize their full pest management potential, pheromones must be determined for all the species of economic concern. Basic biological research is still needed to explore exactly how the pheromones operate under natural conditions. In addition, chemical syntheses and effective formulation methods for these chemicals must be developed. Finally, applied research must be conducted to develop ways to incorporate these chemicals into pest management programs. Eventually, we would hope to achieve selective, effective control of nitidulid beetles with minimal use of broad spectrum insecticides.

Acknowledgments

We are grateful to Dr. David Weisleder of this laboratory for obtaining the NMR spectral data and to Dr. J. M. Kingsolver of the USDA Systematic Entomology Laboratory for confirming the identity of our beetle species. The field experiments were conducted at the Illinois Valley Sand Field Experiment Station of the University of Illinois; we are grateful to Mr. Stan Sipp for his cooperation there.

Literature Cited

1. Sanford, J. W. Ph.D. Thesis, University of Illinois, 1963.
2. Luckmann, W. H.; Hibbs, E. T. Proc. North Central Br. Entomol. Soc. Am. 1959, 14, 81.
3. Wicklow, D. T. In Phytochemical Ecology: Allelochemicals, Mycotoxins and Insect Pheromones and Allomones; Chou, C. H.; Waller, G. R., Eds.; Institute of Botany, Academia Sinica Monograph Series No. 9: Taipei, ROC. 1989; p. 263.
4. Dorsey, C. K.; Jewell, F. F.; Leach, J. G.; True, R. P. Plant Dis. Rep. 1953, 37, 419.
5. Bartelt, R. J.; Dowd, P. F.; Plattner, R. D.; Weisleder, D. J. Chem. Ecol. 1990, 16, 1015.
6. Dowd, P. F. J. Econ. Entomol. 1987, 80, 1351.
7. Heath, R. R.; Sonnet, P. E. J. Liq. Chromatog. 1980, 3, 1129.
8. Poole, C. F.; Schuette, S. A. Contemporary Practice of Chromatography; Elsevier: Amsterdam, 1984; p 23.
9. Parliment, T. H. Microchem. J. 1973, 18, 613.
10. Nelson, D. R. Adv. Insect Physiol. 1978, 13, 1.
11. Anet, F. A. L.; Bourn, A. J. R. J. Amer. Chem. Soc. 1965, 87, 5250.
12. Birch, M. C. In Chemical Ecology of Insects; Bell, W. J.; Carde, R. T., Eds.; Sinauer Assoc.: Sunderland, Massachusetts, 1984; Chapter 12.
13. Walgenbach, C. A.; Burkholder, W. E.; Curtis, M. J.; Khan, Z. A. J. Econ. Entomol. 1987, 80, 763.
14. Kuwahara, Y.; Fukami, H.; Howard, R.; Ishii, S.; Matsumura, F.; Burkholder, W. E. Tetrahedron 1978, 34, 1769.
15. Chuman, T.; Mochizuki, K.; Mori, M.; Kohno, M.; Kato, K.; Noguchi, M. J. Chem. Ecol. 1985, 11, 417.
16. Vanderwel, D.; Oehlschlager, A. C. In Pheromone Biochemistry; Prestwich, G. D.; Blomquist, G. J., Eds.; Academic Press: Orlando, Florida, 1987; Chapter 6.
17. Phillips, J. K.; Walgenbach, C. A.; Klein, J. A.; Burkholder, W. E.; Schmuff, N. R.; Fales, H. M. J. Chem. Ecol. 1985, 11, 1263.
18. Blomquist, G. J.; Jackson, L. L. Prog. Lipid Res. 1979, 17, 319.

RECEIVED May 16, 1990

Chapter 4

Host-Regulating Factors Associated with Parasitic Hymenoptera

Thomas A. Coudron

Agricultural Research Service, U.S. Department of Agriculture, P.O. Box 7629, Research Park, Columbia, MO 65205–5001

Parasitic Hymenoptera produce natural regulatory factors that influence the physiology, biochemistry, and behavior of the host. Several paralyzing venoms have been characterized. Recently, interest has been expressed in factors that produce nonparalytic effects. These arise from several sites in the reproductive tract of the adult female parasitoid. Endoparasitoids are known to produce teratocytes, viruses, and secretions from glands and other specialized tissues which have effects on the development and behavior of the host. Nonparalytic venoms of endoparasitoids cause delayed development, precocious molt, supernumerary molt, and arrest of embryonic development. Ectoparasitoids rely on venoms to regulate host development. The most studied nonparalyzing venom is produced by the ectoparasitoid *Euplectrus* sp. Its venom has a unique effect on the host and is responsible for arresting larval-larval ecdysis.

There are estimated to be over 100,000 species of parasitic Hymenoptera worldwide (1). Though only a small percentage has been described (2), it is evident that these species have a variety of methods of coping with the physiological ecology of the host. Some parasitoids conform their development with the growth of the host, causing only moderate changes in the host. Lawrence (3) first used the term host "conformers" when referring to parasitoids whose eggs, deposited in the host, hatch but the first instars remained inactive while allowing the host to develop. Later certain host conditions triggered the final development of the parasitoid. Other parasitoids regulate host growth to benefit their development, causing major changes in the host (4). "Conformers" and "regulators" may represent two points of a continuum of interactions between parasitoids and their hosts (5, 6).

Some parasitoids are regulators in certain environments and conformers in other environments. *Microctonus aethiopoids*, the

This chapter not subject to U.S. copyright
Published 1991 American Chemical Society

adult parasitoid of the alfalfa weevil, *Hypera postica*, is a good example. *M. aethiopoids* is an endoparasitoid that rapidly completes its development early in the season in its non-diapausing adult host. Later in the season *M. aethiopoids* enter diapause with the adult *H. postica*. In the spring, when the diapause of the host comes to an end, the parasitoid resumes development and eventually kills the host (Puttler B., University of Missouri, personal communication, 1990.). *Biosteres longicaudatus* is a larval-pupal parasitoid, of the Caribbean fruit fly *Anastrepha suspensa*, that allows the host to develop to the pupal stage before actively developing itself (3). The influence of the host hormones upon parasitoid development may be the primary mechanism of parasitoid-host synchrony. An interesting delayed effect is also seen with several encyrtids belonging to the genus *Copidosoma*. These encyrtids allow their hosts to grow to accommodate their polyembryonic development (7-9). Although some species do act as conformers during part of the interaction with their host, ultimately they spend part of that time as regulators and eventually cause incomplete development and premature death of the host.

Regulation does not imply total domination of the host biology by the parasitoid. Certain host resources and physiological processes place constraints on the parasitoids. Parasitoids that paralyze their hosts, and egg and pupal parasitoids experience finite resources. These parasitoids tend to conform to the limitations of the host, often adjusting their size, number or sex depending on the extent of the nutritional resource (6). Ultimately, the success of a parasitoid-host interaction depends on the ability of the parasitoid to affect certain biochemical processes within the constraints imposed by the host.

Many factors contribute to the interactions of parasitoids and their host insects (10). Table I summarizes data of parasitic Hymenoptera that regulate the biochemistry, development, and behavior of their host. This is an attempt to list those species that have been subjected to thorough studies and is not intended to be a complete list.

The presence of the immature stage of an endoparasitoid can affect the host in several ways. Besides competing for host nutrients, they elicit responses in the host due both to their presence and due to substances they secrete.

The adult female parasitoid may transfer one or more type(s) of regulatory factor(s) to the host during parasitism. Some genera of braconid and scelionid families produce teratocytes. These cells originate from an embryonic membrane in the parasitoid egg and are released into the host hemocoel. Certain parasitoid females of the families Ichneumonidae and Braconidae contain symbiotic viruses that are injected into the host with the egg. Diverse functions have been attributed to the teratocytes and the symbiotic viruses. Both play a major role in the interaction of the parasitoid with their respective hosts.

The ovipositor of most parasitic hymenopteran is also used as a ductus venatus for the injection of material from within glandular structures and specialized epithelium cells attached to the common oviduct (11). Two common glandular structures are the Dufour's gland (= alkaline gland) and the venom gland (= poison gland, =

acidic gland). Originally, the Dufour's gland was postulated to be part of the venom system. That idea is now thought to be inaccurate (12). The oily secretions produced by the Dufour's gland have been shown to function as host marking compounds and sex pheromones (13). The venom gland is usually paired and in some species it is extensively branched. An unpaired segment of the gland, common to both pairs of the gland, can be partly swollen and serves as a reservoir for the venom produced by the gland (11).

The contents of the venom gland, reservoir, and the material secreted by specialized epithelium cells affect the physiology, development, and immune systems of the host. In some instances the venomous material interacts with other injected factors that regulate the host biology (10, 14-16).

A survey of Table I allows for several general comparisons. Many parasitic Hymenoptera have adapted to developing in specific stages of the host. The development of that host stage is critical in determining host suitability. Egg, egg-larval, larval, and larval-pupal parasitoids have different mechanisms of host regulation. Endoparasitoids seldom cause permanent paralysis of their hosts and many allow some development of the host to continue. Larval endoparasitoids must contend with the competent immune system, hormones and metabolic changes of the host. In order to develop in such an environment, often highly specialized tactics are used and a variety of host pathologies result. There are several reviews on the physiological interactions of endoparasitoids with their hosts (3, 17-20). There are two possible ways for endoparasitoids to contend with their hosts' immune system: (a) the egg and larva stages of some parasitoids appear to have surface properties that resist encapsulation or avoid sensitizing the host (21, 22); (b) the adult female parasitoid transfers regulatory factors to the host that control the host response and development. The role of regulatory factors has been established for several endoparasitoids and frequently a combination of factors is used.

Ectoparasitism is an effective strategy for circumventing some problems encountered by the endoparasitoids. Ectoparasitoids appear to rely on venoms as the sole source of host regulatory factors. In these instances the venoms arrest development by causing paralysis or by arresting the molting process.

Regardless of whether a parasitoid is an endo- or ectoparasitoid, or solitary or gregarious, most species use regulatory factors to adapt the host environment to their needs. The following is a review of what we presently know of these factors.

Factors Associated with the Developing Parasitoid

Effect on the Composition of the Host. The primary purpose for the association of the parasitoid and the host is for the parasitoid to use the host as a source of nutrients (for reviews of parasitoid nutrition see 23, 24). However, the effect of parasitism extends beyond the depletion of host nutrients by the feeding parasitoid. Clearly the interaction between the parasitoid and the host is bidirectional, with the parasitoid responding to host substances and the host responding to the parasitoid.

Table I. Parasitic Hymenoptera With Host-Regulating Factors[a]

Family Parasitoid	Host Stage[b]	Endo or Ecto	Teratocytes	Parasitoid Derived Viruses[c]	Venoms & Tissue Secretions	References[d]
Paralyzing Effect[e]						
Braconidae						
Bracon sp.[f]	L	Ec	-	-	+	116
Nonparalyzing Effect						
Braconidae						
Apanteles kariyai	L	En	*	*	+	15, 54
Biosteres longicaudatus	L-P	En	+	+	+	53, 112, 114
Cardiochiles nigriceps	L	En	*	+	+	14, 46, 97
Chelonus near *curvimaculatus*	E-L	En	*	*	+	39, 51, 120
Clinocentrus gracilipes	L	En	*	*	+	119
Cotesia congregata	L	En	*	+	*	29, 33, 52, 69, 108
Cotesia marginiventris	L	En	*	+	*	36, 113
Microplitis croceipes	L	En	+	+	*	55, 85, 111
Microplitis mediator	L	En	*	+	*	105, 106
Encyrtidae						
Copidosoma floridanum	E-L	En	-	*	*	9
Copidosoma truncatellum	E-L	En	-	*	*	7
Eulophidae						
Eulophus larvarum	L	Ec	-	-	+	119
Euplectrus comstockii[g]	L	Ec	-	-	+	
Euplectrus kuwanae	L	Ec	-	-	+	131, 141, 146
Euplectrus plathypenae	L	Ec	-	-	+	132, 142, 143

Ichneumonidae						
Campoletis sonorensis	L	En	–	+	+	47, 58, 88, 96, 100, 101, 107, 109
Hyposoter exiguae	L	En	–	+	*	34, 35, 100, 102
Hyposoter fugitivus	L	En	–	+	*	99
Venturia canescens	L	En	–	+	+	126, 127
Scelionidae						
Telenomus heliothidis	E	En	+	–	+	77
Trichogrammatidae						
Trichogramma pretiosum	E	En	*	*	*	77

[a] This is a partial listing. +, regulatory factor is present in species; –, regulatory factor has not been reported in species; *, host regulation has been reported, but analysis of all regulatory factor(s) has not been completed.

[b] Designations for the host growth stages. L = larval stage, L-P = parasitism of the larval stage and emergence from the pupal stage of the host, E-L = parasitism of the egg and emergence from the larval stage of the host, E = egg stage.

[c] *B. longicaudatus* contains a pox virus. *C. marginiventris* contains a virus that is morphologically distinct from the polydnaviruses. All other species contain polydnaviridae viruses.

[d] See the text for references to material in press.

[e] Piek and Spanjer (23) have compiled an extensive listing of species with paralyzing venoms.

[f] Includes the following species *B. brevicornis*, *B. gelechiae*, *B. hebetor*, and *B. mellitor*.

[g] T. A. Coudron, USDA, ARS, unpublished data.

Relatively little is known of the physiology and metabolism of insect parasitoids. Though it is generally assumed that the maturing parasitic hymenopteran eggs take up host hemolymph components and ions through its thin wall (17, 25) it has been shown that some parasitoids rely on their protein synthetic machinery for growth (25, 26). Much of this synthesis occurs after the parasitoid is associated with the host (27).

Parasitism has resulted in changes in the specific gravity, freezing point depression and dry weight of the hemolymph of the host. This correlates with alterations in the concentration and composition in the host hemolymph (for review see 18). However, the change in host protein concentration or composition is not likely to be due to excretion of waste material by the parasitoid. Parasitic hymenopteran larvae have an imperforate gut, with a midgut that is not joined to the hindgut. Much of the waste generated by the parasitoid is stored internally until pupation when the waste is released as meconium outside the host.

The developing parasitoid may secrete material into the host that affects the composition of the host tissue. Such secretions are likely to originate from the salivary glands of the parasitoid. Phenoloxidase activity is secreted from the salivary gland of *Exeristes roborator* and is thought to help in the preservation of the host tissue (28). *Cotesia* (= Apanteles) *congregata* secreted proteins *in vitro* that arose from *de novo* synthesis by the wasp (29). This is not unique to hymenopteran parasitoids. The dipteran *Blepharipa sericariae*, parasitoid of the silkworm *Philosamia synthia*, secretes a small peptide into its host that inhibits lipid transport by lipophorin (30). A possible biological significance of this peptide may be to divert lipid consumption by the host during diapause. This type of secretion may be directed at providing nutrients for the parasitoid.

Parasitism also causes alterations in the constitutive hemolymph proteins of the host. Hemolymph storage proteins in the cabbage butterfly, *Pieris rapae*, decreased in concentration following parasitism by *Cotesia glomerata* (31). This decrease may have been due to an observed concurrent uptake of the same proteins by the parasitoid (32). A reduced synthesis of arylphorin by the tobacco hornworm, *Manduca sexta*, parasitized by *C. congregata* was hypothesized to be due to inhibitory effects of parasitism on host fat body, food consumption and growth (33). Arylphorin concentrations increased precociously in larvae of the cabbage looper, *Trichoplusia ni*, parasitized by *Chelonus* sp. (Kunkel, J. G.; Grossniklaus, C.; Karpells, S. T.; Lanzrein, C. *Arch. Insect Biochem. Physiol.*, in press). Concentrations of several hemolymph proteins decreased in parasitized *T. ni*, during development of the parasitoid *Hyposoter exiguae* (34, 35). Several high molecular weight host proteins were detected 40 hours earlier in hemolymph from larvae of the fall armyworm, *Spodoptera frugiperda*, parasitized by *Cotesia marginiventris* than in hemolymph of control larvae (36).

Parasitism has also been reported to cause the synthesis of proteins that were not present in unparasitized hosts (18, 37, 38). Two predominant and one minor proteins were found in the hemolymph of *M. sexta* larvae parasitized by larval parasitoid, *C. congregata*. These proteins were apparently produced by the host during the final stages of the parasitoid's development indicating their synthesis was activated in response to the development of the parasitoid (33). In comparison, one minor high molecular weight protein was found in the hemolymph of T. ni parasitized by the egg-larval parasitoid, Chelonus near curvimaculatus (39). It was not determined whether the protein was encoded by the host, the parasitoid, or a factor derived from the parasitoid. However, the protein was only found in hosts containing a developing parasitoid. The cause of these changes in host proteins is unclear and it is uncertain what, if any, regulatory role these changes play.

Host Response to the Developing Parasitoid. Sometimes alterations in the composition of the host tissues are part of the response of the host defense reaction. An initial hemolymph response to foreign material is through the hemocytes (plasmatocytes and granulocytes), produced in hemopoietic organs near the dorsal diaphragm (40). Part of the humoral defense response is the synthesis of several enzymic, lectin, and bactericidal proteins (41-43). It is also probable that some parasitoids elicit unique humoral proteins, other than the bactericidal proteins (29).

Some parasitic Hymenoptera appear equipped to protect themselves against the host defense responses. Early studies by Salt (44, 45) suggested different methods existed for parasitoid resistance to the host immune system. More recent studies indicate in some cases the ability of the parasitoid to overcome the host defenses is associated with factors (e.g., teratocytes, viruses and venoms) that originate from the adult female parasitoid. These factors will be discussed in the subsequent sections of this chapter. The texture and composition of the outer surface of the parasitoid egg can be a passive means of protection against the host defense reactions. Fibrous layers on the surface of Cardiochiles nigriceps (46), Campoletis sonorensis (47) and C. glomerata (48) eggs have been implicated in the protection of the egg from encapsulation by hemocytes of their hosts. Histochemical studies showed the fibrous material was composed of neutral glycoproteins and neutral and acidic mucoproteins. The mechanism by which these complex proteins prevent haemocyte adhesion is unclear. This type of passive defense is proposed to delay encapsulation until more permanent means of host immunosuppression (i.e., teratocytes, parasitoid-derived viruses or venoms) is established (46). It is unlikety that the parasitoid egg secretes haemocyte-repelling substances, since in some cases dormant and dead eggs evade encapsulation. It should be noted that protection from host immune responses is an important issue in larval parasitism. However, protection may not be important in true egg parasitoids since insect eggs apparently lack the ability to encapsulate foreign objects (44).

Effect on the Host Endocrine System. Several publications document an endocrine basis of several parasitoid-host interactions (18, 20, 49, 50). However, most of the mechanisms involving the endocrine interactions remain to be clarified. At best, we can describe the qualitative effects but not the regulatory factors responsible for the interactions. Until the characterization of regulatory factors responsible for these effects, it is perhaps most fitting to discuss these effects in this section.

Several systems have been studied and although each is unique, some common features exist. *Chelonus* sp. stimulate precocious spinning of a cocoon in their lepidopteran hosts (51). Presumably this is a result of a premature decline in the juvenile hormone (JH) level in the host. In contrast, other genera (e.g., gregarious *Cotesia* sp. and *Copidosoma* sp.) cause a delay or suppression of metamorphosis of the host, sometimes causing supernumerary or intercalated molts into immature stages. Presumably this is a result of a decline in the JH specific esterase activity (i.e., causing a higher than normal level of JH) (52), and of a disruption of the ecdysteroid levels in the host (33). *B. longicaudatus* superparasitism of the larvae of *A. suspensa* causes an increase in host JH levels and a decrease in JH esterase activity that apparently is the cause in delayed host larval-pupal metamorphosis (53). These results could be due to the presence of the parasitoid but also could be a result of the extra-embryonic serosa originating from this parasitoid. This is addressed in a subsequent section.

There is increasing evidence that the braconid parasitoids, *Apanteles kariyai* (54) and *Microplitis croceipes* (55), cause a decrease in the levels of ecdysone, via a mechanism other than a direct effect on the prothoracic gland, the site of synthesis of ecdysone. The braconid endoparasitoids complete their development without extensive destruction of host tissues. However, redirection of metabolism in the fat body tissue within the host could account for an ultimate reduction in ecdysone (55, 56). In contrast, ichneumonid parasitoids may regulate the host ecdysteroid levels by affecting the prothoracic gland tissue (57-59). This regulation is accomplished through the symbiotic viruses that are the subject of a subsequent section.

Indirect evidence implies that some parasitoids produce and release hormones into the host that act as regulatory factors that change host events under endocrine control. The homologue JH III was found in larvae of *M. sexta* parasitized by *C. congregata* (52), though only JH I and II are present in unparasitized *M. sexta* (60). In larvae of the large white cabbage butterfly, *Pieris brassicae*, parasitized by *C. glomerata*, the JH titer increased after neck-ligation of the host. Both cases suggest a possible source of the hormone was the parasitoid (61). This hypothesis is further supported by recent findings. Larvae of *T. ni* parasitized by *Chelonus* sp. showed reduced levels of JH II, but contained high levels of JH III, compared to unparasitized larvae (Jones, G.; Hanzlik, T.; Hammock, B. D.; Schooley, D. A.; Miller, C. A.; Tsai, L. W.; Baker, F. C. *J. Insect Physiol*., in press). The presence of JH III, together with JH I and II, was detected in parasitized eggs, but only JH I and II were present in unparasitized eggs (Grossniklaus-Burgin, C.; Lanzrein, B. *Arch. Insect Biochem. Physiol*., in press). These findings suggest that JH III originated from the parasitoid. In addition, a comparison revealed that JH titer fluctuations in the parasitoid were independent of changes in the JH titer in the host, which supports the idea that the parasitoid can produce its own JH (Grossniklaus-Burgin, C.; Lanzrein, B. *Arch. Insect Bioch. Physiol*., in press).

Levels of JH III in A. suspensa superparasitized by B. longicaudatus were considerably higher than in unparasitized A. suspensa (53). JH III was also found in first instars of the parasitoid. These results suggested that the elevated JH III levels in the superparasitized host may result from a decrease in JH esterase activity with a continued or elevated synthesis of the hormone in the host or from a secretion of JH III by the parasitoids.

Parasitoid Response to the Host. Simultaneous with the effects of parasitism on the host, are the effects of the host on the parasitoid. The response of the parasitoid to host parameters influences the way the parasitoid interacts with the host. This creates a continuum between the "conformers" and the "regulators."

Endo- and ectoparasitoids feed in part on the hemolymph of the host and many endoparasitoids develop within the haemocoel of the host. Both ingestion and cross-cuticular transport of host humoral material by the parasitoid are likely to occur. Apparently, parasitoids are particularly sensitive to the endocrine milieu in the host hemolymph (3, 18, 20) and capable of taking up hormones from the host (Grossniklaus-Burgin, C.; Lanzrein, B. Arch. Insect Biochem. Physiol., in press). The "hormonal hypothesis" proposed that the growth of many parasitoids was affected and possibly controlled by the hormones of their hosts (62, 63). Mellini thought this hypothesis was more applicable to dipteran parasitoids. However, several studies have confirmed the application of the hypothesis to parasitic Hymenoptera as well.

Many parasitoids are sensitive to disturbances of the host endocrine milieu caused by application of insect growth regulators (64-71), and by administering exogenous hormone to the host (72-74). The first larval molt of Opius concolor is regulated by endogenous release of ecdysteroids by its host, the Mediterranean fruit fly, Ceratitis capitata, during puparium formation (73, 74). The larval ecdysis of C. congregata at emergence from the tobacco hornworm, M. sexta, is regulated by host hemolymph ecdysteroid levels (52) and affected by topical application of JH or the JH analogue, methoprene (69). The larval-pupal parasitoid B. longicaudatus deposits an egg in the first, second or third instar larvae of A. suspensa. The parasitoid's first instar remains inactive (conformer), and later coordinates its molt with host pupation (75). This obligatory synchrony of the early larval molt of the parasitoid with the host's metamorphosis is in response to the ecdysteroid levels in the host hemolymph (76). In addition to these cases where a direct association or response has been shown between the host hormonal milieu and the development of the parasitoid, there are other cases where the developmental synchrony of both the parasitoid and its host constitutes indirect evidence of this relationship (for a review see 20).

Teratocytes and Serosa. Teratocytes, which originate from the disintegration of an embryonic membrane or serosa (= trophamnion) of some parasitic Hymenoptera, appear to have a significant role in parasitoid-host interactions (8, 77-80, Dahlman, D. L. Arch. Insect Biochem. Physiol., in press). A discussion of teratocytes is included in this section since these cells are derived from the developing parasitoid embryo and are released into the host when the embryo hatches. However, teratocytes can cause certain effects on the host, independent of the developing parasitoid.

Though teratocytes may increase in size when they are released into the host, they do not multiply within the host (18). Secretory organelles have been observed in teratocytes that may be important to the function of these cells (80, 81). It has been proposed that these cells serve nutritional and gaseous exchange functions that benefit the parasitoid (80, 82, 83). In theory nutrients could be sequestered from the host hemolymph or produced by *de novo* synthesis within the cells.

Several studies have investigated the production and secretion of regulator molecules by teratocytes. The presence of rough endoplasmic reticula and other organelles in the teratocytes suggests these cells are actively metabolizing and synthesizing material (8). Substances released by teratocytes have been shown to affect phenoloxidase activity (9, Dahlman, D. L. *Arch. Insect Biochem. Physiol.*, in press).

An endocrine role has also been postulated in some species (84). Injection of teratocytes from the braconid parasitoid *M. croceipes* into larvae of the tobacco budworm, *Heliothis virescens*, caused an elevation of the ecdysteroids, an increase in JH levels and a decrease in the JH esterase activity in the host (85, Dahlman, D. L. *Arch. Insect Biochem. Physiol.*, in press).

Results from recent studies show that the presence of teratocytes causes a disruption of the cellular defense system of the host (Tanaka, T.; Wago, H. *Arch. Insect Biochem. Physiol.*, in press, Kitano, H.; Wago, H.; Arakawa, T. *Arch. Insect Biochem. Physiol.* in press). This supports earlier speculation that encapsulation was suppressed by the teratocytes (80) or substances produced by the teratocytes that inhibit the host immune response, (86-88).

In most parasitic Hymenoptera studied the serosa degenerates after hatching. However, the serosa of *B. longicaudatus* remains intact (Lawrence, P. O. *Arch. Insect Biochem. Physiol.*, in press). Though there are no other reports of serosas that remain intact, it is possible that some species not recorded as having teratocytes (see Table I) may maintain an intact serosa. Cells of the serosa of *B. longicaudatus* are secretory and appear to use large coated vesicles to transport molecules from the host, and microvilli to release materials into the host (Lawrence, P. O. *Arch. Insect Biochem. Physiol.*, in press). A polypeptide, approximately 24 Kd, was found in the hemolymph of the host and apparently was produced in the serosa cells. Injection of the protein into healthy *A. suspensa* prepupae inhibits metamorphosis (Lawrence, P. O., University of Florida, personal communication, 1990.). The effect of *B. longicaudatus* on *A. suspensa* is similar to the effect of teratocytes of *M. croceipes* injected into larvae of *H. virescens* reviewed above (Lawrence, P. O. *Arch. Insect Biochem. Physiol.*, in press). The sequestering of host material through coated vesicles could support a nutrient role of the serosa. However, definitive evidence is still lacking.

The proximity of the serosa and teratocytes with the developing parasitoid, and their simultaneous presence in the host, make it difficult to determine which is the source of the parasitoid regulatory factors. It is also possible that regulatory factors are synthesized in one tissue or location and released to the host at another location via a different tissue. These points remain to be determined in most cases.

Factors Associated with the Adult Female Parasitoid

Parasitoid-Derived Viruses. Symbiotic viruses found in some members of the Ichneumonidae and Braconidae families are assembled in the nuclei of the cells of the calyx epithelium (89). Cells that produce the viruses may form a layer around the calyx lumen near the ovariole end of the calyx (90). The viruses are secreted into the lumina of the lateral oviduct and injected into the host with the egg at the time of oviposition. Once in the host the virus material is expressed (91, 92). At present there is no unequivocal evidence of replication of these viruses in the host (93), though this continues to be an active area of investigation. Viral genomes consist of separate, multiple heterologous, double-stranded, circular DNA of various lengths. Viruses isolated from ichneumonids have a fusiform nucleocapsid surrounded by two unit-membrane envelopes. This is characteristic of the genus Polydnavirus (94). Viruses isolated from braconids have cylindrical nucleocapsids of variable lengths surrounded by a single unit membrane, and morphologically have some resemblance to baculoviruses (89). These symbiotic viruses from both the ichneumonids and the braconids, have been assigned to the new virus family, polydnaviridae (95).

Reports of the secretory nature of the calyx region date back over twenty years. Salt (44) recognized that secretions of the calyx of the ichneumonid *Venturia* (= *Nemeritis*) *canescens* affected the host immune system. Studies on the ichneumonid *C. sonorensis* revealed the presence of nuclear secretory particles associated with the calyx cells (96). Those studies were followed by the confirmation of DNA in virus particles associated with the oocytes of the braconid *C. nigriceps* (97).

Several reports now corroborate the role of the symbiotic viruses in inactivating defense mechanisms of the host. Washed eggs of the ichneumonid parasitoids *C. sonorensis* (98) and *Hyposoter fugitivus* (99) are encapsulated and do not develop when introduced artificially into their hosts *H. virescens* and the forest tent caterpillar, *Malacosoma disstria*, respectively. However, when virus isolated from the calyx of the parasitoid accompanies the egg, viable parasitoid progeny develop, suggesting a role for the virus in protecting the parasitoid egg from the immune response system of the host. In a cross-protection experiment, washed eggs of *C. sonorensis* and *H. exiguae* developed in *H. virescens* larvae when heterologous combinations of eggs and symbiotic virus were used (100). These results suggest that the viruses from the two ichneumonids act in a similar manner to promote successful parasitism.

The mechanism by which the virus of *C. sonorensis* suppresses the host's ability to encapsulate the parasitoid egg remains unknown. Davies et al. (101) demonstrated that within 8 hours of injecting calyx fluid, 75% of the circulating capsule-forming hemocytes (plasmatocytes) were removed from the host hemolymph. Such an alteration in the host plasmatocytes could account for the suppression of encapsulation, though the fate of the lost plasmatocytes remains unknown.

The ability of the virus from the ichneumonid, *H. exiguae* to inhibit phenoloxidase activity in the host *T. ni*, may provide another mechanism of suppression of encapsulation (102). A similar decline in the monophenoloxidase activity in the hemolymph of *M. sexta* occurred after injection of virus from the braconid *C. congregata* (Beckage, N. E.; Metcalf, J. S.; Nesbit, D. J.; Schleifer, K. W.; Zetlan, S. R.; de Buron, I. *Insect Biochem.*, in press). The phenoloxidase-tyrosine enzyme system is also thought to be part of the host immune system (103, 104).

Suppression of the host immune system, observed following parasitism by two braconid parasitoids, has also been attributed, in part, to the presence of a symbiotic virus. Long virus-like filaments from the calyx region of the reproductive tract of the braconid parasitoid *Microplitis mediator* were found attached to the surface of the chorion of the oviposited egg (105). These filaments appeared to prevent encapsulation in half the eggs tested in the host, the common armyworm, *Pseudaletia separata*. There was also evidence that the filopodial elongation of the host hemocytes was strongly suppressed (106). Also, half of the sephadex particles, injected into *P. separata* together with calyx fluid from the braconid endoparasitoid *A. kariyai*, were not encapsulated (15).

It should be noted that a mixture of calyx fluid and venom material for both braconids, was more effective in suppressing encapsulation than the calyx fluid alone. Apparently, calyx fluid (containing symbiotic virus) and venom act synergistically and both were essential for complete evasion of the host defense reactions. There was some evidence suggesting the calyx material affected the capsule-forming cells of the host, while the venom material provided a non-antigenic protective covering over the egg (106). A similar effect was observed with a mixture of calyx fluid and venom material for the braconid *C. nigriceps* (14). Another explanation for the synergism observed between symbiotic viruses and venomous material may involve tissue infection and expression of the virus. Venom promoted uncoating of a braconid virus *in vitro* and, also promoted persistence *in vivo* of DNA from the virus (16).

Injection of virus material has also been associated with host synthesis of new proteins. Parasitism by *C. sonorensis* resulted in the synthesis of a new glycoprotein in several of its habitual hosts (107). The appearance of the glycoprotein was duplicated by the injection of either calyx fluid or purified virus. The glycoprotein has been correlated with suppression of encapsulation by the host. It is interesting that parasitism by *C. sonorensis* of two non-permissive hosts, the velvetbean caterpillar, *Anticarsia gemmatalis*, and the bertha armyworm, *Mamestra configurata*, did not stimulate the production of the glycoprotein by those hosts (107).

Definitive indication of transcription of parasitoid-derived viruses in the host has thus far eluded researchers. Parasitism by *C. congregata* induced synthesis of new hemolymph proteins in the host M. sexta (108). Synthesis of one of the major induced proteins occurred within a few hours of parasitism. Induction was also caused by the injection of ovarian calyx fluid from the parasitoid. Exposure of the calyx fluid to psoralen and UV light destroyed its capacity to induce synthesis of the major new protein. This suggested that the synthesis may be mediated by viral nucleic acid, though it was not determined if it represented a viral gene product or a host protein induced by virus-specific activity in the host.

Another type of interaction of the parasitoid-derived viruses with the host is based on the endocrine regulation of development. Injections of the calyx fluid or isolated virus material of C. sonorensis arrested development of 40% of the H. virescens larvae tested (58, 109). Injections into isolated thoraces were most effective. Ecdysteroid production by the prothoracic gland of the host ceased for ca. 10 days following the injection. Arrested development was reversed by injections of ecdysone and 20-hydroxyecdysone. The prothoracic glands of injected host larvae were partially degenerated. In contrast other tissues associated with the endocrine system of the host appeared normal (59). These observations suggested that the virus material induced tissue degeneration that was specific to the prothoracic gland.

Similarly, parasitism of P. separata, by A. kariyai or injection of the calyx fluid or virus material from the parasitoid, caused a decline in the host ecdysteroid titer and arrested metamorphosis of the host (54). Again, injection of exogenous 20-hydroxyecdysone reversed the developmental arrest. In this case, administration of prothoracicotropic hormone caused a reactivation of the prothoracic glands in the treated host. These results showed that the virus material inhibited the synthesis or secretion of prothoracicotropic hormone and also lowered the ecdysteroid level in the host. Also in this case, a mixture of both venom and calyx fluid were needed to obtain the full prolongation of the larval stage in the host (110).

Other developmental effects caused by a symbiotic virus from a parasitoid may involve the nutritional physiology of the host. An increase in trehalose levels in the hemolymph of H. virescens parasitized by M. croceipes could be duplicated by the injection of the calyx fluid from the parasitoid (111). Though there is no known explanation for the increase in trehalose, it is proposed to result from the release of glucose that is catabolized to trehalose in the host (13). An increase in the host hemolymph trehalose level may have a nutritional benefit for M. croceipes which is a hemolymph feeder.

From this review we see that symbiotic viruses from the braconids and the ichneumonids have some differences in their physical structures and their apparent interaction with the host. Some variations noted in the action of these viruses may eventually be correlated with differences in viral structures and the interaction of the viruses with venom components. Certain host tissues may be susceptible to one virus-type and refractive to the other virus-type. It is possible that ichneumonids, which are commonly known as tissue feeders, contain symbiotic viruses that have a more prominent effect on the endocrine system of the host, thereby preserving the host tissue. In comparison, braconids, which are commonly known as hemolymph feeders, may contain symbiotic viruses that contribute more to alterations in the composition of the host hemolymph that would benefit the parasitoid.

Other types of virus material have been recorded from parasitic hymenopteran species. Viral particles of parasitoid origin were reported in tissues of the _A. suspensa_ superparasitized by the solitary endoparasitoid _B. longicaudatus_ (112). Rhabdoviruses were found in vesicles near the basement membrane of the host epidermal cells and pox viruses were found in hemocytes adjacent to the epidermis (Lawrence, P. O.; Akin, D. _Canad. J. Zool._, in press). The rhabdoviruses were proposed to affect migration of vesicles to the cell apices and the activation of the molting fluid. The function of the pox virus is not yet known but could relate to the changes in the hemolymph levels of ecdysteroids, JH, and JH esterases (53).

A long nonoccluded filamentous virus, morphologically distinct from the polydnaviruses, was characterized from the reproductive tract of the parasitoid _C. marginiventris_. Though the role of the filamentous virus in the parasitoid-host interaction is unknown, the virus was reported to replicate in hypodermal and tracheal matrix cells of the host larvae (113).

Secretions from Glands and Other Specialized Tissues. In general, parasitic Hymenoptera are known to inject secretions from glands and specialized tissues into their hosts. Those secretions, except for the symbiotic viruses and virus material or calyx fluid, are collectively presented in the next two sections as venomous substances.

Paralyzing Venomous Substances. The best characterized regulatory factors at present are a group of substances isolated from the venom of the wasps of the parasitic genus _Bracon_ (= _Haborbracon_, = _Microbracon_), and the predatory sphecids _Philanthus trangulum_ and _Sceliphron caemantarium_ (114), and scoliid _Megascolia flavifrons_ (115). These species cause paralysis in their hosts or prey that varies from complete and permanent paralysis, which causes immediate cessation of feeding, to a transient or an incomplete paralysis. Though these venoms act on the nervous system of the host or prey, the active ingredients vary considerably. The chemical nature of the active ingredients range from large highly labile proteins with molecular weights of 43 kd to 56 kd found in the braconid species, and short chain peptides (kinins) found in the scoliid species, to low molecular weight substances like the polyamines found in the sphecid species (116). Piek and Spanjer (114) list members of the families Braconidae and Ichneumonidae having paralyzing venoms. Current articles are available on this area (12, 115, 117, 118) and no further detail will be given in this report.

Nonparalyzing Venomous Substances. Often the injection of the virus material and nonparalyzing venomous substances occur simultaneously so that the effects are not easily separated. Reference is made in this section to reports where venomous substances were shown to perform a function independent of symbiotic viruses, or no reference was made in the report to the presence of virus material.

A diversity of functions and a variety of developmental abnormalities have been associated with nonparalyzing venomous substances. Of the four sources of regulatory factors reviewed in Table I, venomous substances are common to species in all the families represented, by ecto- and endoparasitoids alike, and the only source of regulatory substances in ectoparasitoids known to affect the development of the host.

Growth of H. virescens was arrested following parasitism by C. sonorensis (88). The mechanism for the inhibition in host weight gain was unknown, but the effect could be reproduced by injections of aqueous extracts of the lateral oviducts of the female parasitoid. These contained protein but no RNA or DNA material and was inactivated by heat and exposure to organic solutions or protease action. This evidence strongly suggests the arrest of weight gain in the host was in response to the injection of a proteinacious substance secreted by the lateral oviducts. Vinson (88) proposed the primary purpose for arrested weight gain was to reverse the flow of nutrients from organs of the host to the hemolymph that aid the nutrition of the developing parasitoid.

The scelionid egg parasitoid Telenomus heliothidis interrupts egg development in its host H. virescens. A regulatory substance, apparently produced and secreted by the exocrine cells of the common oviducts of the parasitoid, was responsible for the rapid arrestment of host embryogensis (77). Necrosis of host tissue is also associated with parasitism. However, if oviposition was interrupted prior to egg deposition, the resulting pseudoparasitized host exhibited arrestment of embryogenesis but no tissue necrosis. Apparently in this case the secreted substance was not associated with restructuring of host tissue for the nutrition of the parasitoid. Also, the mode-of-action and the nature of this regulatory substance is unknown.

Results from an early study by Shaw (119) indicated that stinging by the solitary endoparasitic braconid Clinocentrus gracilipes of its host, Anthophila fabriciana was separable from oviposition. Stinging alone caused temporary paralysis followed by reduced feeding, precocious molting, incomplete pupal formation and death of the host. The effects were attributed to the presence of a nonparalyzing venom from the parasitic wasp.

Proteins from the venom gland of the egg-larval parasitoid, C. near curvimaculatus are injected into the host within the first eight seconds of the oviposition period (120). These proteins remain intact in the host for two days at which time they are rapidly degraded prior to hatch of the parasitoid egg. The exact role of these proteins remains unknown. Interesting is the finding that these proteins did not cause the precocious metamorphosis followed by arrested development observed in parasitized T. ni (120, 121). Another regulatory substance, injected after the venom proteins but before the parasitoid egg, caused the developmental redirection of host tissue. The source and nature of this substance is unknown but it is proposed to be different from calyx fluid and virus material, also associated with this parasitoid, and injected at a time distinct from the injection of other regulatory substances.

In some parasitoid-host systems certain material, which would come under the heading of venomous substances as used in this chapter, may play a role in suppression of the immunological response of the host against the parasitoid by acting as a nonreactive egg coating. Examples of these are the mucinous accessory gland secretions of the ichneumonid, *Pimpla turionella* (122-124), and material from the venom gland of *C. glomerata*. This material protects the parasitoid egg without inhibiting the encapsulation ability of the host (125). The parthenogenetic ichneumonid *V. canescens* secretes a particulate material onto the surface of the egg as it passes through the calyx. The material, which contained antigenic determinants similar to a protein component of the host *Ephestia kuehniella*, formed a physical covering for the egg (126, 127). The observed similarity between proteins of the secreted material and a host protein suggested the secreted protein could simply mask the egg surface and prevent the egg from being recognized as a foreign body. In this way the parasitoid protein provided passive protection from the host immune system similar to the fibrous layers found on eggs of *C. nigriceps*. Though the protein occurred in a particulate structure that gave it the appearance of the virus material it was found to contain glycoprotein but no detectable amount of nucleic acids. In other parasitoid-host systems certain material apparently interferes with the activity of the host hemocytes. Examples of these are the accessory genital gland secretions of *P. turionella*, which inhibits capsule-forming hemocytes (128), and accessory gland secretions of the cynipid *Leptopilina heterotoma*, which causes deterioration of the hemocytes (129).

Careful documenting of the inhibition of molting (i.e., arrest of apolysis and ecdysis) in parasitized lepidopteran hosts is an increasingly important area of study (114, 119, 130-133). Members of the Eulophidae family that are in the genus *Euplectrus* and some *Eulophus* species produce a venom that causes this specific type of response in their hosts. In contrast, most all other hymenopteran ectoparasitoids (e.g., Braconidae, Ichneumonidae, Eulophidae, and Bethylidae) of lepidopteran and coleopteran larvae paralyze their hosts (134-136).

The development of *Orthosia stabilis* was arrested after stinging by the gregarious ectoparasitoid *Eulophus larvarum*. During the period of arrested development, the host was recorded as mobile and continued to feed for a period of time but failed to produce a new cuticle or initiate apolysis (119). This effect was attributed to a venomous substance, presumably transferred from the parasitoid to the host at the time of stinging, and was shown to be independent of oviposition or the developing parasitoid.

Euplectrus sp. are naturally gregarious ectoparasitoids that develop on noctuid and geometrid larvae (137-139). *Euplectrus puttleri* and *Euplectrus kuwanae* are host specific (140, 141). In contrast, *Euplectrus plathypenae* has several lepidopteran host species (142, 143). All three species arrest the development of their hosts. Recent studies have shown that *Euplectrus comstockii* also arrests the development of it's hosts (Coudron, T. A., USDA,

ARS, unpublished data). Euplectrus sp. inhibit host ecdysis once stinging (i.e., puncture with the ovipositor), oviposition (i.e., egg deposition), or parasitism (i.e., stinging and oviposition) has occurred (132). Molt inhibition usually occurs without host paralysis and prevents shedding of the cuticle, which is the site of attachment of the parasitoid's eggs. Contact with the host by parasitoid eggs or developing parasitoid larvae is not necessary for the expression of the molt arrest effect. Instead, Euplectrus sp. transfer an arrestment substance to the host at the time of stinging that acts as a potent bioregulator (132, 133). The recording of superparasitism by the solitary endoparasitoid B. longicaudatus in puparia of A. suspensa (112, 144) is the only other instance of arrestment of apolysis and ecdysis of host larvae following parasitism by an entomophagous insect in a family other than Eulophidae. The presence of viable parasitoids was required for this effect to be manifested and the added stress of the additional parasitoid load (i.e., superparasitism) was thought to cause the observed arrestment and not a venom from the parasitoid alone.

Venom from E. plathypenae arrested molting in 44 species of Lepidoptera, plus 19 species in six other orders. The arrestment was expressed in all natural hosts and in most, but not all, insects outside the natural host range of the parasitoid (132). The active substance has been located in hemolymph of parasitized hosts, in a gland/reservoir complex isolated from the female ectoparasitoid, and found associated with the fluid and crystalline structures within the gland/reservoir complex (133, 145). The developmental arrest could result from disrupting events normally under endocrine control (130, 131, 146, Kelly, T. J.; Coudron, T. A. J. Insect Physiol., in press). However, treatment of parasitized larvae with exogenous JH, 20-hydroxyecdysone, and several hormone analogs did not restore the normal molting process in the parasitized host (133, 145). Molting was arrested even when parasitism occurred after the release of endogenous molting hormone (133). Clearly, this system offers a new and intriguing area for research on the regulation of insect development and parasitoid-host interactions.

Future Exploitation

Classical biological control involves the action of beneficial organisms in the suppression of pest insects. A wider view of biological control has evolved to embrace attempts to develop improved strains of beneficial organisms through controlled selection and genetic manipulation of species. Improved strains are targeted at increasing the suppression phenomenon, widening the host range, and reducing the lapse of time between parasitism, or infection, and cessation of host feeding. This has enhanced our efforts to understand desirable attributes of beneficials and has brought on a search to identify the biological substances (i.e., bioregulators) that are responsible for the regulation of molecular and cellular processes involved in growth, development, and behavior of beneficial and harmful insects. Understanding the mechanisms used by parasitic Hymenoptera in regulating their host physiology may provide valuable information in this direction (147).

The first step is the determination of desired traits controlled by bioregulators. Second is the chemical characterization and elucidation of the mode-of-action of the natural substances that regulate the development of pest insects. Thirdly is the refinement of techniques to modify and regulate the expression of the substance at the cellular level. Over the past decade genetic engineering has developed from the advances made in molecular biology and biotechnology areas. This technology has been applied to the production of transgenic plants that are refractive to extensive insect damage, and to transgenic microbials that are designed to produce desirable biological materials. These powerful tools of the molecular geneticists could make substantial and rapid contributions to biological pest control. Applications of such contributions may include the transfer of desirable bioregulators to other beneficial organisms to enhance effectiveness or improve desirable qualities (i.e., speed of kill) and may lead to the production of new substances that influence unique insect biochemical processes. Such contributions may take the form of insect transformations through the use of P-elements (148) and plasmids (149), or the incorporation of bioregulators into the genome of plants consumed by phytophagous insects. Refer to Kuhlemeier et al. (150), McCabe et al. (151) and Benfey and Chua (152) for recent reviews on the incorporation and regulation of foreign genes in plant tissues. It may be possible to use vectors to produce large quantities of desirable substances that can then be formulated for direct use as naturally derived pesticides. Refer to Luckow and Summers (153) and Maeda (154) for recent reviews on the use of insect baculoviruses as vectors for expression of foreign DNA.

Presently, concerns over insect resistance to synthetic insecticides (155) has grown to include concerns over resistance to transgenic organisms that are designed specifically to replace the use of synthetic chemicals (156). One potential solution for avoiding resistance to transgenic organisms is the use of many gene traits, each of which encodes for a natural bioregulator. The rate-limiting step to this solution is the determination of characters that are desirable for manipulation and the identification of the genes that express for those substances with desirable traits.

The regulatory factors produced by parasitic Hymenoptera have promising applications in this area. Though partial descriptions can be given for the effects of these regulatory factors on the host insect, it is apparent that much of their function remains unclear. Chemical characterization is incomplete for most of these factors and a thorough understanding of their effects, individually and in combination, has yet to be detailed. However, the protein nature and independent action of several venomous substances make them amenable to approaches that involve genetic engineering technology. The synthesis and expression of genes encoding insecticidal proteins derived from scorpion venoms have been reported (157, 158). Enriched genetic material (RNA and DNA) coding for these substances can be isolated from specific tissues known to produce the active substances or cDNA can be constructed that encodes for the active substance. The genomic material can be transferred into microbials (both symbiotic or pathogenic bacteria and viruses) which can be used as vectors to transport and express the substances in pest insects. Transgenic plants also could be constructed to produce the active substances directly within the specific plant tissue that would be consumed by the pest insect. The potential for constructing artificial pathogens to act as delivery systems of these active substances is also under consideration.

These approaches create a basis for the construction of an insecticidal method for the use of naturally occurring substances with desirable attributes. Genes encoding for fast-acting venoms could greatly diminish the lapse of time between parasitism by a biological control agent and the resulting cessation of host feeding. Gene products should be selected that have minimal mammalian toxicity. Certain paralyzing toxins from the venoms of the scoliid wasps are now known to act as blockers of the synthesis of acetylcholine by preventing the access of choline to the presynaptic terminals (115, 116, 118). Unfortunately it is generally accepted that this system is similar in insects and vertebrates. In contrast, an insecticidal neurotoxin, found in the venom of the scorpion *Androctonus australis*, is selectively toxic to several species of insects but harmless to crustaceans, arachnids and mammals (159). The venom from the Eulophid species is also fast-acting (133) and it appears to function on a physiological process (i.e., ecdysis) that is more unique to arthropods. These qualities may make the active toxin in the Eulophid venom an attractive alternative to the paralyzing toxins.

The delicate interplay between parasitoid and host development offers a wealth of opportunities for future exploitation. Avenues for exploitation will expand as we continue to characterize the biochemical parameters contributing to the parasitoid-host relationships. A strategy of examining and exploiting these specific interactions could lead to significant new advances in practical biological control measures. Such advances may include new pest control measures targeted at physiological and metabolic pathways specific to insects. An understanding of these interactions also may aid in the structuring of more effective compounds that mimic or antagonize the effects of the biological compounds produced by beneficial agents and aid in the production of pharmacologically active agents that are environmentally acceptable for use to control pest insects.

Acknowledgments

I sincerely thank Mr. Benjamin Puttler for his encouragement and many helpful comments, and Drs. N. Beckage, D. Dahlman, B. Lanzrein, P. Lawrence and M. Strand for providing material in press and their discussions on this subject. The author gratefully recognizes Drs. B. Hammock, D. Jones, P. Lawrence and D. Stoltz for their review of the text, Karen Ridgwell for her assistance in preparing the manuscript, and USDA, ARS for supporting the preparation of this document.

Literature Cited

1. Askew, R. R. In Parasitic Insects; American Elsevier Press: New York, 1971; pp 316.
2. Townes, H. In The Genera of Ichneumonidae; Mem. Ann. Entomol. Inst. 1969; Part I, No. 11.
3. Lawrence, P. O. J. Insect Physiol. 1986, 32, 295-8.
4. Vinson, S. B. In Evolutionary Strategies of Parasitoids and Their Hosts; Price, P., Ed.; Plenum Press: New York, 1975; pp 14-48.
5. Beckage, N. E. In Molecular and Cellular Biology; Baker, R.; Dunn, P., Eds., UCLA Symposium New Series Vol. 112; Alan R. Liss: New York, 1989; pp 497-515.
6. Vinson, S. B. ISI Atlas Science, Animal and Plant Science 1988, 1, 25-32.
7. Jones, D.; Jones, G.; Van Steenwyk, R. A.; Hammock, B. D. Ann. Ent. Soc. Am. 1982, 75, 7-11.
8. Strand, M. R.; Vinson, S. B.; Nettles, Jr., W. C.; Xie, Z. -N. Entomol. Exp. Appl. 1988, 46, 71-78.
9. Strand, M. R.; Dover, B. A.; Johnson, J. A. Arch. Insect Biochem. Physiol. 1990, 113, 41-51.
10. Stoltz, D. B. J. Insect Physiol. 1986, 32, 347-50.
11. van Marle, J.; Piek, T. In Venoms of the Hymenoptera: Biochemical, Pharmacological and Behavioral Aspects; Piek, T., Ed.; Academic: New York, 1986; Chapter 2.
12. Piek, T. In Venoms of the Hymenoptera: Biochemical, Pharmacological and Behavioral Aspects; Academic Press: New York, 1986; Chapter 1.
13. Vinson, S. B.; Dahlman, D. L. Southwest Entomol. 1989, Suppl. 12, 17-37.
14. Guillot, F. S.; Vinson, S. B. J. Insect Physiol. 1972, 18, 1315-21.
15. Tanaka, T. J. Insect Physiol. 1987, 33, 413-20.
16. Stoltz, D. B.; Guzo, E. R.; Belland, C. J.; Lucarotti, C. J.; MacKinnon, E. J. Gen. Virol. 1988, 69, 903-7.
17. Fisher, R. C. Biol. Rev. 1971, 46, 243-78.
18. Vinson, S. B.; Iwantsch, G. F. Q. Rev. Biol. 1980, 55, 143-65.
19. Thompson, S. N. Comp. Biochem. Physiol. 1983, 74B, 183-211.
20. Beckage, N. E. Ann. Rev. Entomol. 1985, 30, 371-413.
21. Salt, G. Proc. Royal Soc. London 1973, 183, 337-50.
22. Lynn, D. C.; Vinson, S. B. J. Invertebr. Pathol. 1977, 29, 50-5.
23. Thompson, S. N. Ann. Rev. Entomol. 1986, 31, 197-219.
24. Thompson, S. N. J. Insect Physiol. 1986, 32, 421-3.
25. Ferkovich, S. M.; Dillard, C. R. Insect Biochem. 1986, 16, 337-45.
26. Jones, D.; Barlow, J. S.; Thompson, S. N. Experimental Parasitology 1982, 54, 340-51.
27. Tilden, R. L.; Ferkovich, S. M. Insect Biochem. 1987, 17, 783-92.
28. Thompson, S. N. J. Insect Physiol. 1980, 26, 505-9.
29. Beckage, N. E.; Nesbit, D. J.; Nielson, B. D.; Spence, K. D.; Barman, M. A. E. Arch. Insect Biochem. Physiol. 1989, 10, 29-45.

30. Hayakawa, Y. Fed. European Biochem. Soc. 1986, 195, 122-4.
31. Smilowitz, Z.; Smith, C. L. Ann. Entomol. Soc. Amer. 1977, 70, 447-54.
32. Kawai, T.; Maeda, S.; Ono, H.; Kai, H. J. Fac. Agric. Tottori Univ. 1983, 18, 18-26.
33. Beckage, N. E.; Templeton, T. J. J. Insect Physiol. 1986, 32, 299-314.
34. Smilowitz, Z. Ann. Entomol. Soc. Amer. 1973, 66, 93-9.
35. Thompson, S. N. Parasitology 1982, 84, 491-510.
36. Ferkovich, S. M.; Greany, P. D.; Dillard, C. R. J. Insect Physiol. 1983, 29, 933-43.
37. Fisher, R. C.; Ganesalingam, V. K. Nature, 1970, 227, 191-2.
38. King, P. E.; Rafai, J. J. Exp. Biol. 1970, 53, 245-54.
39. Jones, D. Insect Biochem. 1989, 19, 445-55.
40. Gupta, A. P. Insect Hemocytes; Cambridge Univ. Press: New York, 1979; p. 614.
41. Boman, H. G.; Hultmark, D. Trends Biochem. Sci. 1981, 6, 306-9.
42. Dunn, P. E. Ann. Rev. Entomol. 1986, 31, 321-9.
43. Boman, H. G., Hultmark, D. Ann. Rev. Micro. 1987, 41, 103.
44. Salt, G. Biol. Rev. 1968, 43, 200-32.
45. Salt, G. Proc. Royal Soc. London 1980, 207, 351-3.
46. Davis, D. H.; Vinson, S. B. J. Insect Physiol. 1987, 32, 1003-10.
47. Norton, W. M.; Vinson, S. B. J. Invertebr. Pathol. 1977, 30, 55-67.
48. Kitano, H. XVII Int. Congress Entomol. 1984, Abstr. S4.9, 19.
49. Grossniklaus-Burgin, C.; Lanz, F.; Lanzrein, B. In Endocrinological Frontiers in Physiological Insect Ecology; Seknal, F.; Zabza, A.; Denlinger, D. L., Eds.; Wroclaw Tech. Univ. Press: Wroclaw, 1988; pp 437-441.
50 . Lawrence, P. O. In Endrocrinological Frontiers in Physiological Insect Ecology; Sehnal, F.; Zabza, A.; Denlinger, D. L., Eds.; Wroclaw Tech. Univ. Press: Wroclaw, 1988; pp 432-435.
51. Jones, D. Ann. Ent. Soc. Amer. 1985, 78, 141-8.
52. Beckage, N. E.; Riddiford, L. M. Gen. Comp. Endocr. 1982, 47, 308-22.
53. Lawrence, P. O.; Baker, F. C.; Tsai, L. W.; Miller, C. A.; Schooley, D. A.; Geddes, L. G. Arch. Insect Biochem. Physiol. 1990, 13, 53-62.
54. Tanaka, T.; Agui, N.; Hiruma, K. Gen. Comp. Endocrin. 1987, 67, 364-74.
55. Dahlman, D. L.; Coar, D. L.; Koller, C. N.; Neary, T. J. Arch. Insect Biochem. Physiol. 1990, 13, 29-39.
56. Racioppi, J. V.; Dahlman, D. L. Comp. Biochem. Physiol. 1980, 67C, 35-9.
57. Dover, B. A.; Davies, D. H.; Vinson, S. B. J. Invertebr. Pathol. 1988, 51, 80-91.
58. Dover, B. A.; Davies, D. H.; Vinson, S. B. Arch. Insect Biochem. Physiol. 1988, 8, 113-26.
59. Dover, B. A.; Strand, M. R.; Davies, D. H.; Vinson, S. B. Int. J. Insect Morphol. & Embryol. 1989, 18, 47-57.

60. Schooley, D. A.; Judy, K. J.; Bergot, J.; Hall, M. S.; Jennings, R. C. In The Juvenile Hormones; Gilbert, L. I., Ed.; Plenum Press: New York, 1976; p 101-17.
61. Schopf, A. Entomologia Exp. Appl. 1984, 36, 265-72.
62. Mellini, E. Mem. Soc. Ent. Ital. 1969, 48, 324-50.
63. Mellini, E. Boll. Ist. Ent. Univ. Bologna 1975, 31, 165-203.
64. Divakar, B. J. Experentia 1980, 36, 1332-3.
65. McNeil, J. Science 1975, 189, 640-2.
66. Ascerno, M. E.; Smilowitz, Z.; Hower, A. A. Environ. Entomol. 1980, 9, 262-4.
67. Outram, I. Environ. Entomol. 1974, 3, 361-3.
68. Vinson, S. B. J. Econ. Entomol. 1974, 67, 335-7.
69. Beckage, N. E.; Riddiford, L. M. J. Insect Physiol. 1982, 28, 329-34.
70. Smilowitz, Z.; Martinka, C. A.; Jowyk, E. A. Environ. Entomol. 1976, 5, 1178-82.
71. Fashing, N. J.; Sagan, H. Ann. Entomol. Soc. Amer. 1979, 8, 816-8.
72. Webb, B. A.; Dahlman, D. L. J. Insect Physiol. 1986, 32, 339-45.
73. Cals-Usciati, J. C. R. Acad. Sci. (Paris), Ser. D. 1969, 269, 342-4.
74. Cals-Usciati, J. C. R. Acad. Sci. (Paris), Ser. D. 1975, 281, 275-8.
75. Lawrence, P. O. Expl. Parasit. 1982, 53, 48-57.
76. Lawrence, P. O. In Biology of the Parasitic Hymenoptera; Gupta, V.; Brill, E. J., Eds.; Amsterdam, 1988; pp 351-366.
77. Strand, M. R.; Meola, S. M.; Vinson, S. B. J. Insect Physiol. 1986, 32, 389-402.
78. Vinson, S. B.; Lewis, W. J. J. Invertebr. Pathol. 1973, 22, 351-5.
79. Hashimoto, K.; Kitano, H. Zool. Mag. 1971, 80, 323.
80. Vinson, S. B. J. Invertebr. Pathol. 1970, 16, 93-101.
81. Strand, M. R.; Quarles, J. M.; Meola, S. M.; Vinson, S. B. In Vitro Cell and Dev. Biol. 1985, 21, 361-7.
82. Hagen, K. S. In Biological Control of Insect Pests and Weeds; DeBach, P., Ed.; Reinhold: New York, 1964; Chapter 7.
83. Ivanova-Kasas, O. M. In Developmental Systems: Insects; Counce, S. J. and Waddington, C. H., Eds.; Academic Press: New York, 1972; pp 243-271.
84. Joiner, R. L.; Vinson, S. B.; Benskin, J. B. Nature 1973, 245, 120-1.
85. Zhang, D.; Dahlman, D. L. Arch. Insect Biochem. Physiol. 1989, 12, 51-61.
86. Kitano, H. Bull. Tokyo Gakuge Univ. 1969, 21, 95-136.
87. Kitano, H. J. Insect Physiol. 1974, 20, 315-27.
88. Vinson, S. B. J. Insect Physiol. 1972, 18, 1509-14.
89. Stoltz, D. B.; Vinson, S. B. Adv. Virus Res. 1979, 24, 125-71.
90. Krell, P. J. Can. J. Microbial. 1987, 33, 176-83.
91. Fleming, J. G. W.; Blissard, G. W.; Summers, M. D.; Vinson, S. B. J. Virol. 1983, 48, 74-8.
92. Blissard, C. W.; Fleming, J. G. W.; Vinson, S. B.; Summers, M. D. J. Insect Physiol. 1986, 32, 351-9.
93. Theilman, D. A.; Summers, M. J. Gen. Virol. 1986, 67, 1961-9.

94. Federici, B. A. Proc. Natl. Acad. Sci. USA 1983, 80, 7664-8.
95. Stoltz, D. B.; Krell, P.; Summers, M. D.; Vinson, S. B. Intervirology 1984, 21, 1-4.
96. Norton, W. N.; Vinson, S. B.; Thurston, E. L. Proc. Electron Microscopy Soc. Am. 1974, 1984, 140-1.
97. Vinson, S. B.; Scott, J. R. J. Invertebr. Pathol. 1975, 25, 375-8.
98. Edson, K. M.; Vinson, S. B.; Stoltz, D. B.; Summers, M. D. Science 1981, 211, 582-3.
99. Stoltz, D. B.; Guzo, D. J. Insect Physiol. 1986, 32, 377-88.
100. Vinson, S. B.; Stoltz, D. B. Ann. Entomol. Soc. Am. 1986, 79, 216-8.
101. Davies, D. H.; Strand, M. R.; Vinson, S. B. J. Insect Physiol. 1987, 33, 143-53.
102. Stoltz, D. B.; Cook, D. I. Experientia 1983, 39, 1002-4.
103. Soderhall, K. Dev. Comp. Immunol. 1982, 6, 601-11.
104. Ratcliffe, N. A.; Leonard, C.; Fowley, A. F. Science 1984, 226, 557-9.
105. Tanaka, T. Entomophaga 1987, 32, 9-17.
106. Tanaka, T. Developmental and Comparative Immunology 1987, 11, 57-67.
107. Cook, D. I.; Stoltz, D. B.; Vinson, S. B. Insect Biochem. 1984, 14, 45-50.
108. Beckage, N. E.; Templeton, T. J.; Nielsen, B. D.; Cook, D. I.; Stoltz, D. B. Insect Biochem. 1987, 17, 439-55.
109. Dover, B. A.; Davies, D. H.; Strand, M. R.; Gray, R. S.; Keeley, L. L.; Vinson, S. B. J. Insect Physiol. 1987, 33, 333-8.
110. Tanaka, T. Ann. Entomol. Soc. Amer. 1987, 80, 520-3.
111. Dahlman, D. L.; Vinson, S. B. J. Invertebr. Pathol. 1977, 29, 227-9.
112. Lawrence, P. O. Ann. Ent. Soc. Am. 1988, 81, 233-9.
113. Styer, E. L.; Hamm, J. J.; Nordlund, D. A. J. Invertebr. Pathol. 1987, 50, 302-9.
114. Piek, T.; Spanjer, W. In Venoms of the Hymenoptera: Biochemical, Pharmacological and Behavioral Aspects; Piek, T., Ed.; Academic: New York, 1986; Chapter 5.
115. Hue, B.; Piek, T. Comp. Biochem. Physiol. 1989, 93C, 87-9.
116. Piek, T. In Venoms of the Hymenoptera: Biochemical, Pharmacological and Behavioral Aspects; Academic Press: New York, 1986; p 570.
117. Schmidt, J. O. In Ann. Rev. Entomol.; Mittler, T. E., Ed.; Annual Reviews, Inc.: Palo Alto, CA, 1982; Vol. 27, pp 339-68.
118. Piek, T.; Hue, B.; Mony, L.; Nakajima, T.; Pelhate, M.; Yasuhara, T. Comp. Biochem. Physiol. 1987, 87C, 287-95.
119. Shaw, M. R. J. Invertebr. Pathol. 1981, 37, 215-21.
120. Leluk, J.; Jones, D. Arch. Insect Biochem. Physiol. 1989, 10, 1-12.
121. Jones, D. J. Insect Physiol. 1987, 33, 129-34.
122. Fuhrer, E. Naturwissenschaften 1972, 52, 167-8.
123. Fuhrer, E. Z. Parasitenkd 1973, 41, 207-13.
124. Osman, S. E.; Fuhrer, E. Int. J. Invertebr. Reprod. 1979, 1, 323-32.

125. Kitano, H. J. Invertebr. Pathol. 1982, 40, 61-7.
126. Feddersen, L.; Sander, K.; Schmidt, O. Experientia 1986, 42, 1278-81.
127. Berg, R.; Schuchmann-Feddersen, I.; Schmidt, O. J. Insect Physiol. 1988, 34, 473-80.
128. Osman, S. E. Z. Parasitenkd 1978, 57, 89-100.
129. Rizki, R. M.; Rizki, T. M. Proc. Natl. Acad. Sci. U.S.A. 1984, 81, 6154-8.
130. Coudron, T. A.; Kelly, T. J. In New Concepts and Trends in Pesticide Chemistry; Hedin, P. A., Ed.; ACS Symposium Series No. 276, 1985, 319.
131. Uematsu, H.; Sakanoshita, A. Appl. Ent. Zool. 1987, 22, 139-44.
132. Coudron, T. A.; Puttler, B. Ann. Entomol. Soc. Am. 1988, 81, 931-7.
133. Coudron, T. A.; Kelly, T. J.; Puttler, B. Arch. Insect Biochem. Physiol. 1990, 13, 83-94.
134. Beard, R. L. Conn. Agric. Exp. Stn. New Haven Bull. 1952, 562.
135. Puttler, B. Ann. Entomol. Soc. Am. 1963, 56, 857-9.
136. Gordh, G. USDA Technical Bull. 1976, 1594.
137. Noble, N. S. Science Bulletin, Dept. Agric. New South Wales 1938, 63, 5-27.
138. Neser, S. In Entomology Memoir; Dept. Agric. Tech. Ser.: Republic of South Africa, 1973; p 32.
139. Gerling, D.; Limon, S. Entomophaga 1976, 21, 179-87.
140. Puttler, B.; Gordh, G.; Long, S. H. Ann. Entomol. Soc. Am. 1980, 73, 28-35.
141. Uematsu, H. Appl. Entomol. Zool. 1980, 16, 57-9.
142. Wall, R.; Berberet, R. C. Environ. Entomol. 1974, 3, 744-6.
143. Krombein, K. V.; Hurd Jr., P. D.; Smith, D. R.; Burks, B. D. In Catalogue of Hymenoptera in America North of Mexico; Smithsonian Institute Press: Washington, DC, 1979; vol. 1. 1420.
144. Lawrence, P. O. J. Insect Physiol. 1988, 34, 603-8.
145. Coudron, T. A. J. Cell Biochem. 1989, Suppl. 13A, 166.
146. Uematsu, H. Jpn. J. Appl. Ent. Zool. 1986, 30, 55-7.
147. Jones, D. Entomophaga 1986, 31, 153-7.
148. Cooley, L.; Kelly, R.; Spradling, A. Science 1988, 239, 1121-8.
149. Miller, L. H.; Sakai, R. K.; Romans, P.; Guadz, R. W.; Kantoff, P.; Coon, H. G. Science 1987, 237, 779-81.
150. Kuhlemeier, C.; Green, P. J.; Chua, N. -H. Ann. Rev. Plant Physiol. 1987, 38, 221-57.
151. McCabe, D. E.; Swain, W. F.; Martinell, B. J.; Christou, P. Bio/Technology 1988, 6, 923-6.
152. Benfey, P. N.; Chua, N. -H. Science 1989, 244, 174-81.
153. Luckow, V. A.; Summers, M. D. Bio/Technology 1988, 6, 47-55.
154. Maeda, S. Ann. Rev. Entomol. 1989, 34, 351-72.
155. Sparks, T. C.; Hammock, B. D. In Pest Resistance to Pesticides: Challenges and Prospects; Georghiou, G. P.; Saito, T., Eds.; Plenum Press: New York, 1983; pp 615-68.
156. Muller, K. Agrichemical Age 1989, April, 14-21.

157. Carbonell, L. F.; Hodge, M. R.; Tomalski, M. D.; Miller, L. M. Gene 1988, 73, 409-418.
158. Dee, A.; Belagaje, R. M.; Ward, K.; Chio, E.; Lai, M.-H. T. Bio/technology 1990, 8, 339-342.
159. Zlotkin, E. Insect Biochem. 1983, 13, 219-236.

RECEIVED July 2, 1990

Chapter 5

Sex Pheromone Production in Moths

Endogenous Regulation of Initiation and Inhibition

Peter E. A. Teal and James H. Tumlinson

Agricultural Research Service, U.S. Department of Agriculture, Gainesville, FL 32604

Recent studies on a number of moth species have indicated that pheromone production is under endogenous regulation. For many species the diel periodic production period is regulated by a neuropeptide, termed "pheromone biosynthesis activating neuropeptide", which is produced in the brain-subesophageal ganglion complex. This peptide is present in the brain-subesophageal ganglion at all times but is released only during specific periods of the photoperiod. This establishes the diel periodicity of pheromone production. Females of some moth species also inhibit the production of sex pheromone by endogenous means. Inhibition usually follows mating and is either the direct effect of a substance passed from the male to the female during mating or an indirect response to mating, in that the female then produces an inhibitory factor. Virgin females of *Heliothis zea* also produce a pheromone biosynthesis suppression factor as they senesce. The evolutionary significance of this suppression factor may be to ensure that males do not waste reproductive effort on females that will die prior to laying fertile eggs.

Over the past three decades considerable effort has been directed towards the elucidation of the structures and biological effects of sex pheromones produced by moths. These studies have been very fruitful, as is evidenced by the great number of species for which pheromones have been identified (1) and by the significant advances in knowledge about the semiochemical-mediated biology of these insects (2). As a consequence, moth pheromones are now commonplace in population monitoring programs and are being increasingly used as direct control agents (3), which are applied principally for mating disruption. Nonetheless, while so much is known about the chemistry and behavior of moth communication systems, much less is known about the endogenous mechanisms that regulate pheromone production. These mechanisms control when pheromones are produced

This chapter not subject to U.S. copyright
Published 1991 American Chemical Society

and are therefore key components for reproductive success in moth species.

Females of many moth species signal their readiness to mate by producing and releasing sex pheromones during species specific times of the photoperiod (4). At other times little or no pheromone is present in the pheromone gland and none of the behaviors associated with pheromone release are exhibited. Studies on a number of moth species have shown that the induction of pheromone production is under endogenous regulation (5,6,7). The factor responsible for induction of biosynthesis is a neuropeptide produced in the brain-subesophageal ganglion complex. Although present in this complex at all times, its release is apparently under diel periodic control (8,9). The structure of this peptide, termed "pheromone biosynthesis activating neuropeptide" (PBAN), has recently been elucidated for Heliothis zea (10) and for Bombyx mori (11). Although the amino acid sequences of these two forms of PBAN are somewhat different, their activities are the same inasmuch as they both induce the production of pheromone. The exact mechanism by which PBAN stimulates the production of pheromone is unknown in these insects. Furthermore, recent studies (12) have indicated that although some moth species, for example the cabbage looper moth, do produce PBAN, this peptide does not appear to stimulate the insects to produce pheromone. Consequently, the function of PBAN in these species is unknown. Similarly, the reason for the presence of PBAN in the brain-subesophageal ganglia complexes of males of H. virescens is unclear because they maintain equally high titers of pheromone during both the scotophase and photophase (13).

In addition to regulation of the induction of pheromone production, moths and other insect groups (14) inhibit the production of pheromone during periods when mating is not appropriate. Adult females of some species delay reproductive maturity after adult emergence for various reasons and do not engage in reproductive activity until gametes are mature. For example, the spring and fall generations of Pseudaletia unipuncta are migratory and neither produce nor release pheromone until eggs are mature (6). Cusson and McNeil (6) have shown that these insects mediate pheromone production and release via juvenile hormone (JH). Thus, low levels of JH suppress sexual behavior while elevated titers stimulate both pheromone production and release behavior (6). Virgin females, at the peak of their reproductive potential, probably regulate the induction and termination of pheromone production based solely on the diel periodic release of PBAN. However, mated females of many moth species, including the omnivorous leafroller moth, the corn earworm moth, tobacco hornworm moth, and the cecropia moth, switch from a "virgin" condition in which the insects release pheromone, to a "mated" condition indicated by oviposition, cessation of calling, and significant reduction in the amount of pheromone produced (15,16,17,18,19,20). Studies on the omnivorous leafroller have indicated that a topically applied juvenile hormone analogue caused virgin females to oviposit at a high rate and also inhibited pheromone production (17). In the corn earworm moth the titer of pheromone is reduced rapidly after mating and lower titers of pheromone are produced, with respect to virgins, for at least two nights after mating (4). The immediate reduction in titer of pheromone after mating is probably due to the transfer of a substance from the male accessory glands during mating (18). A somewhat different situation occurs in the cecropia moth. Females of this insect release a substance, produced by the bursa copulatrix after fertile mating, which causes a shift from the "virgin" to the "mated" condition (16,19). The behavioral switch by cecropia moth females is mediated through the action of a factor transferred from the male to the female during mating. Stretching of the bursa copulatrix appears to trigger the physiological change and it has been established that continued neural communication between the bursa and corpora allata is required for maintenance of the "mated" condition.

All of the above studies indicate that insects regulate the production of pheromone and mating very precisely by using endogenous factors. Although the many facets of pheromone mediated biology including initiation of production, termination of production and actual release of components are tightly coordinated, it is obvious that no single neural, neuroendocrine, or endocrine factor regulates the complete pheromone production system. Further, while the above studies have laid the foundation for understanding how pheromone production is controlled, they also raise new questions which must be addressed.

Induction of Pheromone Production

The pioneering studies conducted by Raina and co-workers (5,8,10) that led to the discovery of PBAN in *H. zea* have provided a useful tool for studying the physiological and biochemical events that regulate pheromone production as well as for classical pheromone identification studies. Cyclical daily periods during which pheromone is produced by females of many moth species including *Heliothis* correspond with both the calling period, as evidenced by release of volatile pheromone components (21), and the period of response of males (22). During other times little or no pheromone is present in the gland. In these females, periods of production are apparently regulated by the diel periodic release of PBAN. Evidence for this comes from the fact that either neck ligation or decapitation during the photophase inhibits the production of pheromone during subsequent nights (4), but pheromone production can be stimulated in these insects by injection of PBAN during the photophase (7,9). We have found that females of the three species of *Heliothis* currently under study in our laboratory can be induced to produce pheromone by injection of PBAN at times when little or no pheromone is present in the gland, for example, the mid-photophase (Fig. 1). This is of significance because it allows us to conduct our studies during the photophase, precisely timing each experiment after the injection of PBAN.

Our studies on *H. zea*, *H. virescens* and *H. subflexa* have shown that the amount of pheromone present in the gland, as indicated by the titer of the major component, (Z)-11-hexadecenal, increases in a linear fashion for at least the first 90 min after injection of PBAN (Fig. 2). Furthermore, as indicated in Figure 3, the ratio of components present in gland extracts obtained at 15 min. intervals after injection of PBAN remains constant and is not different from the naturally produced ratio. Other studies

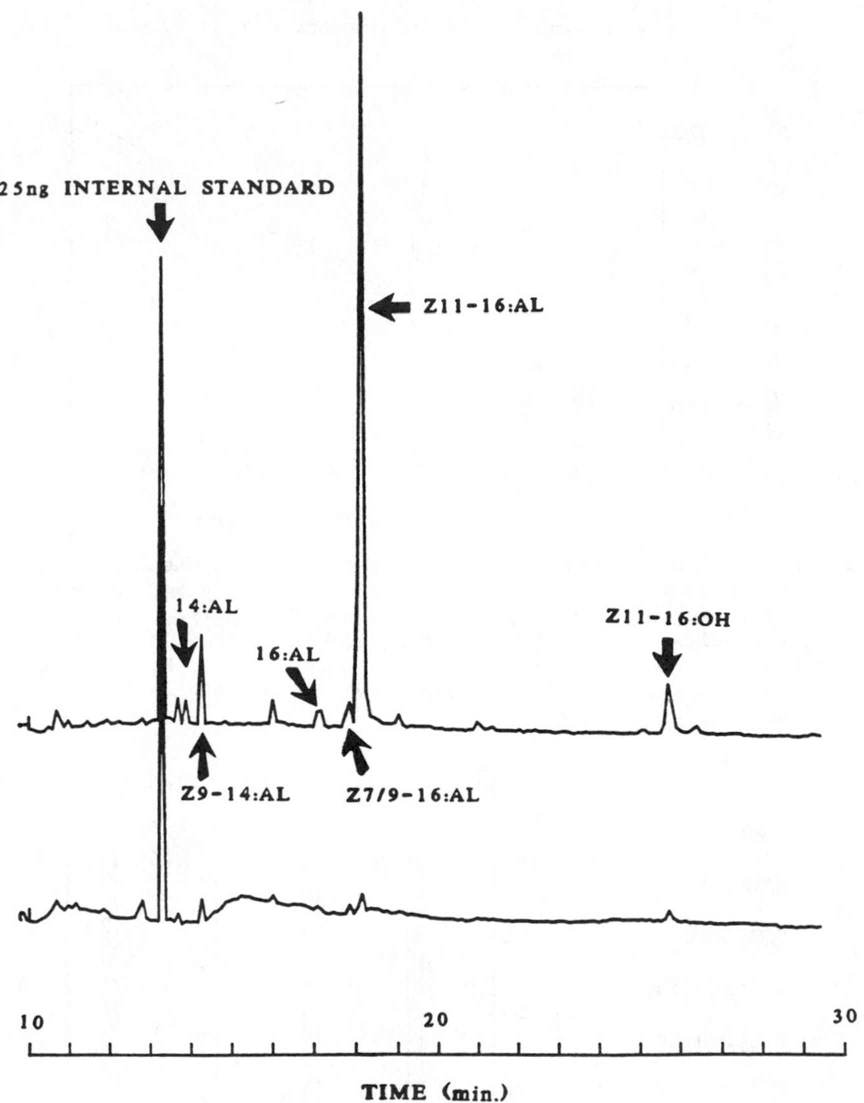

Figure 1. Gas chromatographic analysis (Column = 30m x 0.25mm id Supelcowax 10) of the extract of the pheromone gland of a 3-day-old virgin female of Heliothis virescens obtained 1 h after injection of 0.5 pmoles of synthetic PBAN in 10 µl of water during the photophase (upper), compared to the extract of a gland obtained 1 h after injection of 10 µl of water during the third photophase. Pheromone components are: tetradecanal (14:AL, (Z)-9-tetradecenal (Z9-14:AL), hexadecanal (16:AL), (Z)-7-hexadecenal and (Z)-9-hexadecenal (Z7/Z9-16:AL) (no resolution), (Z)-11-hexadecenal (Z11-16:AL) and (Z)-11-hexadecen-1-ol (Z11-16:OH).

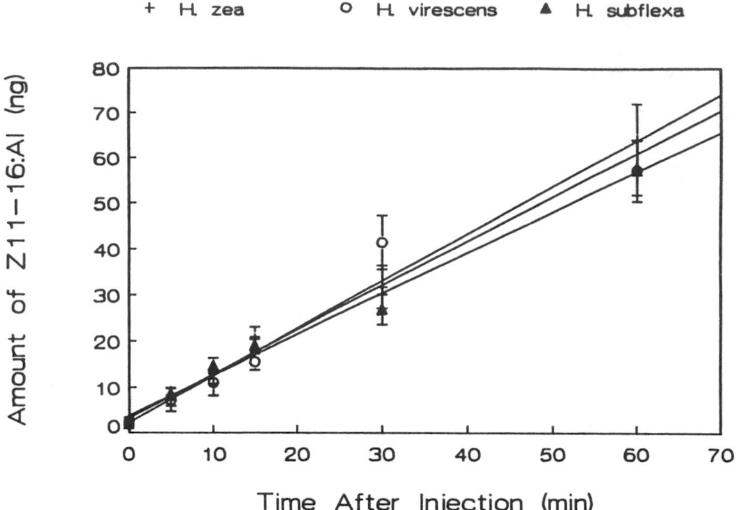

Figure 2. Amounts of Z11-16:AL, the pheromone component present in greatest amount, present in extracts of pheromone glands of virgin females of <u>Heliothis zea</u>, <u>H</u>. <u>virescens</u> and <u>H</u>. <u>subflexa</u> obtained at different times after injection of 0.5 pmole of PBAN in 10 μl of water during the photophase (n=10 each data point). Extracts were analyzed by capillary GC using Supelcowax 10 and SPB1 columns (30m x 0.25mm id).

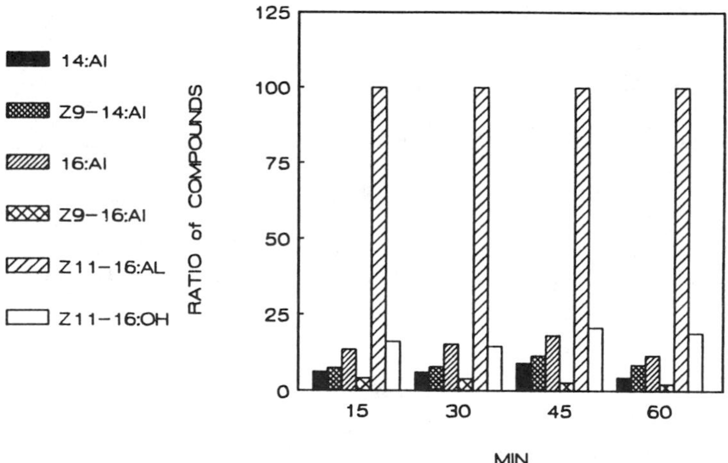

Figure 3. Ratio of pheromone components present in extracts of pheromone glands of virgin females of <u>H</u>. <u>virescens</u> obtained at various intervals after injection of 0.5 pmole of PBAN in 10 μl of water during the photophase (n=10 each data point). Extracts were analyzed as in Figure 2.

conducted in our laboratory have shown that the esterases and oxidases responsible for the terminal steps in pheromone biosynthesis are active at all times providing substrate is available (23). Preliminary work on all three species has shown that (Z)-11-hexadecenoate, the presumed precursor of the major pheromone component, is present in significantly higher titer in gland extracts obtained 30 min. after injection of PBAN than in control insects injected with buffer. Thus, it appears that PBAN acts to increase the availability of substrate, perhaps through mobilization from a storage site (12).

An interesting feature of these artificial induction studies is that females injected with PBAN during the light phase produce the same amount of pheromone in 1 hr. as females sampled at the peak of the natural production period (9). However, it takes several hours after pheromone titer begins to rise for it to peak under natural conditions (21). In nature, the amount of PBAN reaching and then binding with the receptor site may increase slowly which results in a slow rise in pheromone titer. The reason for the rapid rise in pheromone titer after injection of PBAN is not known, but may result from the sudden surge of neuropeptide at the receptor site after injection, thus causing a significant rise in pheromone titer in as little as 15 min. after stimulation. Studies by Tang _et al_. (12) have shown that PBAN does not stimulate the pheromone glands of the redbanded leafroller to produce pheromone. Rather, the neuropeptide appears to stimulate the release of precursors of the pheromones from some tissue other than the pheromone gland. Results of studies conducted in our laboratory parallel those found for the redbanded leafroller inasmuch as we could not stimulate isolated pheromone glands of _H_. _zea_ incubated with PBAN to produce significant amounts of pheromone (24). However, Soroker and Rafaeli (25) have recently reported that _in vitro_ incubation of pheromone glands of the egyptian bollworm with brain extracts resulted in a significant increase in incorporation of [^{14}C] sodium acetate into a compound having the same retention volume on reverse phase HPLC as did the major pheromone component. Thus, in this species, PBAN appears to act directly on the pheromone gland. However, it should be noted that Soroker and Rafaeli (25) used isolated terminal abdominal segments which contain not only the pheromone gland but also hemolymph, nerves, and other tissues, whereas studies conducted on the redbanded leafroller were performed on just the isolated pheromone glands (12). The importance of the presence of other tissues, for example nerves, is indicated by results of studies conducted on _Heliothis zea_ (24,26). These studies have shown that the terminal abdominal ganglion and nerves extending from it to the pheromone gland are required for induction of pheromone production (24). Severing the abdominal nerve cord at the junction of the thorax and abdomen during the light period also inhibits production of pheromone during the subsequent reproductive period (24). Injection of PBAN into these nerve-transected insects restores the ability to produce pheromone. These results indicate that some form of signal, be it neural or neuroendocrine, must be transmitted between the brain and the abdominal nerve cord for induction of pheromone production in this species. This is further supported by the fact that homogenates of the abdominal nerve cords from calling females injected into non-calling, light adapted females (24) induced pheromone production. Females injected with nerve homogenates from other noncalling, light adapted females did not produce pheromone.

Inhibition of Pheromone Production

Studies conducted on mating in the cecropia moth provided the first indication that endogenous factors are involved in causing both the behavioral and physiological changes associated with mating (16,19) in Lepidoptera. These studies demonstrated that females release "bursa factor" into the hemolymph after a fertile mating. The factor then appears to interact with cells of the corpora cardiaca to stimulate release of an oviposition factor. Interestingly, infertile mating does not result in release of "bursa factor". This indicates that the transfer of sperm is required (16,19). Although it is unknown if "bursa factor" inhibits pheromone production, the behaviors and physiological changes that are induced by the factor are the same as those that occur along with reduced pheromone production in other species (17). More recently, Raina *et al*. (4) showed that the pheromone titer of females of H. *zea* dropped significantly almost immediately after mating and remained low thereafter. Subsequent studies indicated that males of this species transfer a substance produced in the accessory glands which appears to cause increased metabolism of pheromone already present in the gland as well as inhibiting further production of pheromone. Based on our knowledge of the enzymatic steps involved in biosynthesis by this species (23) it seems likely that the male accessory gland factor inhibits, or alters, the pathway that regulates the production of the fatty acid precursors and/or the alcohol precursors of the aldehydic pheromone components. It is known that the oxidation of the alcohols to aldehydes is regulated by a cuticular oxidase that is limited only by the availability of substrate (27). An interesting feature associated with the male inhibitory factor produced by *H. zea* is that the effect is cancelled by restimulation with PBAN (15). Consequently, the normal release of PBAN on the night after mating would result in production of pheromone. However, only limited amounts of pheromone are produced. This suggests that some other factor is produced in response to mating and that this second factor is responsible for the maintenance of reduced titers of pheromone in mated females.

Studies on the effect of age on pheromone production (4,28,29) indicated that older virgins produce less pheromone than do females at their reproductive peak. Research conducted in our laboratory (28) indicated that the lower titers of pheromone present in senescing virgins was correlated with increased oviposition. The behavior exhibited by these females was more similar to that of mated females than to that of virgins at their reproductive peak. This suggested that senescing virgins had shifted their behavior and physiology from the "virgin" condition to the "mated" condition as occurs after mating in females of the cecropia and tobacco hornworm moths (16,20). To explore this more fully, we injected females of various ages with brain-subesophageal ganglion homogenates obtained from females during the third photophase (28). We then compared these data to data obtained from oviposition studies where we counted the number of eggs laid per night by individually caged virgins. As indicated in figure 4, there was an inverse relationship between egg laying and titer of pheromone produced. Thus, 5-day-old females laid approximately 60% of the total eggs but only produced 30% of the pheromone produced by females on the third day (28). Three-day-old females injected with brain-subesophageal ganglion homogenates from 5-day-old females produced the same titer of pheromone that females of the same age produced when injected with homogenates from 3-day-old insects. Therefore, the titer of PBAN was the same in 5-day-old and 3-day-old virgins. This indicated that some other factor was responsible for the reduced titer of pheromone present in glands of 5-day-old females (28).

Knowing that the bursa copulatrix of females of the cecropia moth releases a substance after mating which results in inhibition of calling and stimulation of oviposition (16) and that the bursa copulatrix of females of _H_. _zea_ contains secretory cells (30), we hypothesized that the inhibitory factor, which we have termed pheromone biosynthesis suppression factor (PBSF), might be present in the bursa. Injection of 2 female equivalents (FE) of the aqueous homogenate of the bursa obtained from 5-day-old virgins plus 1 FE of PBAN into 3-day-old resulted in complete inhibition of pheromone production (Fig. 5). Studies in which homogenates of other tissues were injected along with PBAN indicated that PBSF activity was associated with the bursa, ovaries and hemolymph (Table 1). These findings indicated that PBSF was not a neural product but did not preclude it from being one of the juvenile (JH) or ecdysteroid hormones. Both of these groups of hormones have been shown to modulate pheromone production in other insects (6,31). However, extraction of aqueous homogenates of the bursa with pentane, hexane or iso-octane did not reduce the inhibitory effect of the aqueous phase (28). This eliminated the JHs from being PBSF candidates because these compounds are extracted from aqueous homogenates with apolar organic solvents. Similarly, ecdysone and 20-hydroxyecdysone were eliminated as candidates because injection of 20-200 ng of either of these compounds along with PBAN did not result in diminished pheromone production. We are currently in the process of isolating and identifying PBSF and are conducting studies to determine the site at which this substance acts and if it has other physiological effects, for example, stimulation of oviposition.

The fact that virgins of _H_. _zea_ produce PBSF as they senesce is probably factitious because in nature the vast majority of moths mate within the first few nights after adult emergence (32). However, we have found that injection of homogenates of the bursae obtained from 3-day-old females mated 12 h earlier along with PBAN resulted in production of only 3.1 ng (±0.9, n=5) of the major pheromone component (28). This was 10% of that produced when homogenates of the bursae from virgin females of the same age plus PBAN were injected. The reduced pheromone titers present in extracts obtained from these studies on mated females parallels data reported by Raina et al. (4) in studies on pheromone titers produced by mated females under natural conditions. As indicated earlier, the effect of the male-produced inhibitor, which causes

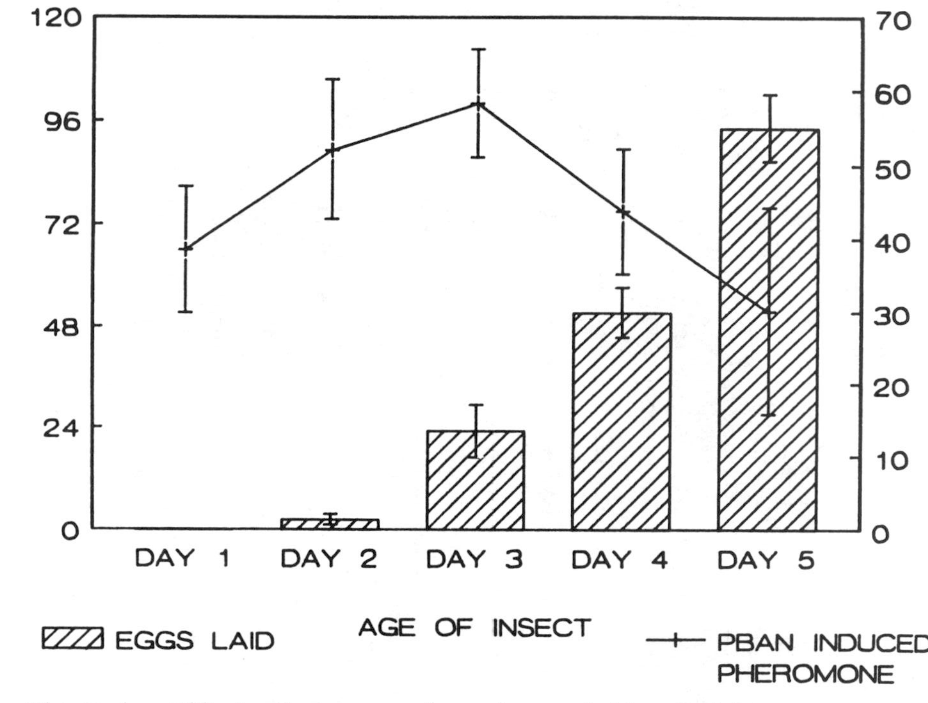

Figure 4. Effect of age on number of eggs laid per night (percentage of total eggs laid) by virgin females of _H. zea_ held in individual cages (n=10) compared to the amount of Z11-16:AL present in extracts of pheromone glands of virgins of the same ages 1 h after injection of 1 female equivalent (FE) of the brain-subesophageal ganglion homogenate obtained from females during the third photophase (n=9 each point). Titer of Z11-16:AL was determined using capillary gas chromatographic analysis as in Figure 1.

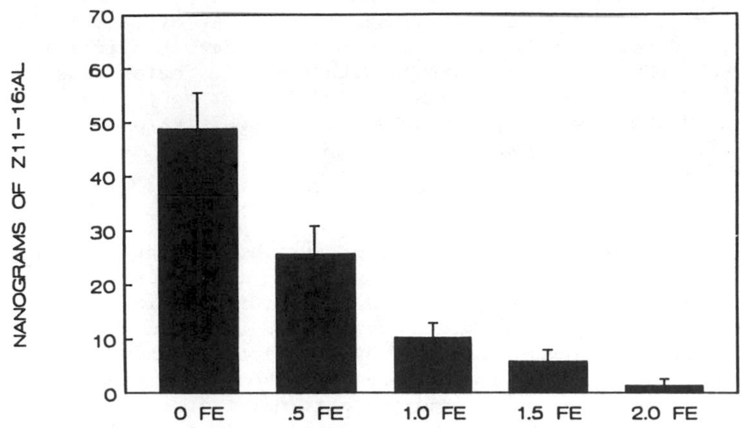

Figure 5. Effect of injection of 1 female equivalent (FE) of brain-subesophageal ganglion homogenate containing PBAN along with various amounts of bursa homogenate (FE) on the amount of Z11-16:AL present in gland extracts of virgin 3-day-old females of _H_. _zea_. Extracts were obtained 1 h after injection of 30 μl of test substances and analyzed for the titer of Z11-16:AL by capillary gas chromatography as in Figure 1 (n=5 each treatment).

Table 1. Effect of injection of PBAN plus 2 FE of tissue extracts obtained from 5-day-old virgin females on pheromone production in 3-day-old virgin females of _H_. _zea_*

Treatment	% Pheromone	
1 FE PBAN + 20 μl buffer (n=10)	100.0%	A
1 FE PBAN + 2 FE washed bursa copulatrix (n=5)	5.1%	D
1 FE PBAN + 2 FE washed ovary (n=5)	12.4%	C
1 FE PBAN + 2 FE washed fat body (n=5)	91.3%	A
1 FE PBAN + 2 FE washed thoracic muscle (n=5)	104.7%	A
1 FE PBAN + 2 FE hemolymph (n=5)	25.3%	B

*Pheromone production as indicated by the titer of Z11-16:AL in gland extracts. Percentages followed by the same letter are not significantly different as indicated by a Duncan's multiple range test at $p = 0.05$ using the amount (ng) of Z11-16:AL present in extracts.

the initial drop in pheromone production, is overcome by restimulation by PBAN and appears to be short-lived (15). Therefore, it is unlikely that the male-produced factor is directly responsible for maintenance of low titers of pheromone for several days after mating, nor is it the same compound that we have found in the bursae of mated and senescing virgins. We hypothesize that the male and female produced factors act in concert to inhibit pheromone production after mating, with the male factor causing the inhibition of continued pheromone production during the hours immediately after mating and PBSF being responsible for maintenance of low titers on subsequent nights.

Conclusions

The precise timing of sex pheromone production to correspond with the peak period of reproductive potential is critical for reproductive success in moth species. As indicated in the preceding discussion, both the induction and inhibition of pheromone production are regulated by a variety of endogenous factors. Consequently, understanding how these endogenous factors function, their chemistries, and the external factors which influence them is fundamental to determining how moths time the production of pheromones. We must also determine what factors regulate the response periods of the receiving sex. This area of research has not been the subject of considerable study but promises to be an exciting area of investigation. Combined results from such studies may lead to development of effective strategies for selectively disturbing the pheromone communication systems of pest insects.

Literature Cited

1. Tamaki, Y. In Comprehensive Insect Physiology, Biochemistry and Pharmacology; Kerkut, G. A., Gilbert, L. I., Eds; Pergamon: New York, 1985; Chap. 9, pp. 145-191.
2. Cardé, R. T.; Baker, T. C. In Chemical Ecology of Insects; Bell, W. J., Cardé, R. T., Eds; Sinauer: Sunderland, M. A., 1984; pp. 355-383.
3. Olkowski, W. IPM Practitioner. 1988, 10, 1-9.
4. Raina, A. K.; Klun, J. A.; Stadelbacher, E. A. Ann. Entomol. Soc Am. 1986, 79, 129-131.
5. Raina, A. K.; Klun, J. A. Science 1984, 225, 531-533.
6. Cusson, M.; McNeil, J. N. Science 1989, 243, 210-212.
7. Martinez, T.; Camps, F. Arch. Insect Biochem. Physiol 1988, 5, 211-227
8. Raina, A. K.; Menn, J. In Pheromone Biochemistry; Prestwich, G. D., Blomquist, G. Eds.; Academic: Orlando, 1987; 159-174.
9. Teal, P. E. A.; Tumlinson, J. H. Can. Entomol. 1989, 121, 43-46.
10. Raina, A. K.; Jaffe, H.; Kempe, T. G.; Keim, P.; Blacher, R. W.; Fales, H. M.; Riley, C. T.; Klun, J. A.; Ridgway, R. L.; Hayes, D. L. Science 1989, 244, 796-798.
11. Kitamura, A.; Nagasawa, H.; Kataoka, H.; Inoue, T.; Matsumoto, S.; Ando, T.; Suzuki, A. Biochem. Biophys. Res. Comm. 1989, 163, 520-525.

12. Tang, J. D.; Charlton, R. E.; Jurenka, R. A.; Wolf, W. A.; Phelan, P. L.; Sreng, L.; Roelofs, W. L. Proc. Nat. Acad. Sci. 1989, 86, 1806-1810.
13. Teal, P. E. A.; Tumlinson, J. H. J. Chem. Ecol. 1989, 15, 413-427.
14. Barth, R. H.; Lester, L.J. Ann. Rev. Entomol. 1973, 18, 445-472.
15. Raina, A. K. J. Chem. Ecol. 1988, 14, 2063-2069.
16. Riddiford, L. M.; Ashenhurst, J. B. Biol. Bull. 1973, 144, 162-171.
17. Webster, R. P.; Cardé, R. T. J. Insect Physiol. 1984 30, 113-118.
18. Bird, T. G.; Raina, A. K. Abs. Am. Chem. Soc. Agro. Chem. Div, Miami, 1989, P. 19, paper #60.
19. Truman, J. W.; Riddiford, L. M. Biol. Bull. 1971, 140, 8-14.
20. Sasaki, M.; Riddiford, L. M. Physiol. Entomol. 1984, 9, 315-327
21. Heath, R. R.; McLaughlin, J. R.; Proshold, F.; Teal, P. E. A. Ann. Entomol. Soc. Am. 1990, in press.
22. Tingle, F. C.; Mitchell, E. R.; Baumhover, A. H. J. Chem. Ecol. 1978, 4, 471-479.
23. Teal, P. E. A.; Tumlinson, J. H. J. Chem Ecol. 1986, 12, 353- 366.
24. Teal, P. E. A.; Tumlinson, J. H.; Oberlander, H. Proc. Nat. Acad. Sci. 1989, 86, 2488-2492.
25. Soroker, V.; Rafaeli, A. Insect Biochem. 1989, 67, 1-6.
26. Christensen, T. A.; Itagaki, H.; Teal, P. E. A.; Hildebrand, J. G. Soc. Neuro. Sci. Abstr. 1989, 15, 365.
27. Teal, P. E. A.; Tumlinson, J. H.; Oostendorp, A. In Biocatalysis in Agricultural Biotechnology; Whitaker, J. R., Sonnet, P. E., Eds; ACS Press: Washington, 1989; pp. 332-343.
28. Tumlinson, J. H.; Teal, P. E. A. In Molecular Insect Science; Hagedorn, H. H., Hildebrand, J. G., Kidwell, M. G., Law, J. H., Eds.; Plenum: New York, 1990; in press.
29. Lawrence, L. A.; Bartell, R.J. Entomol. Exp. Appl. 1972, 15, 455-464.
30. Callahan, P. S.; Cascia, T. Ann. Entomol. Soc. Am. 1963, 56, 535-556.
31. Adams, T. S.; Dillwith, J. W.; Blomquist, G. J. J. Insect Physiol. 1984, 30, 287-294.
32. Hardwick, D. Men. Entomol. Soc. Can. 1965, 40, 247 pp.

RECEIVED September 10, 1990

Chapter 6

Pheromone Biosynthesis and Mating Inhibition Factors in Insects

T. G. Bird, A. K. Raina, and D. K. Hayes

Agricultural Research Service, U.S. Department of Agriculture, Beltsville, MD 20705

The accessory gland of many adult male insects possess biological factors which cause physiological changes in the female when transferred during copulation. We have observed and are isolating two such activities from the accessory ducts of male housefly (Musca domestica) and the accessory glands of corn earworm (Heliothis zea) which affect reproduction in conspecific females. In the housefly virgin females become refractory to mating when a crude homogenate of male accessory duct is injected. Extracts from the accessory glands of the corn earworm injected into females cause rapid, near quantitative reduction in pheromone titer. Isolation of both factors involves solid phase extraction and high performance liquid chromatography.

Recent efforts in our laboratories have addressed the pheromone biosynthesis activating neuropeptide (PBAN) of Heliothis zea (corn earworm, CEW). During those studies, Raina (1) observed that pheromone titers in the CEW declined sharply during copulation. Subsequently, it was shown that a crude homogenate of the male's accessory gland (AG), injected into a pheromone producing female, also caused a rapid drop in titer. The factor responsible for this effect was termed the receptivity termination factor (RTF). An examination of the literature (for reviews, see 2, 3) indicated that this was a previously undescribed function to be added to an extensive list of activities already attributed to factors produced in the male AG, accessory duct (AD), or paragonia (various names for a similarly functioning structure). Other functions included:

1. facilitation of insemination
2. sperm activation
3. formation of oviduct plugs for seminal fluid retention and/or a copulation barrier
4. stimulation of egg maturation and oviposition
5. nutritional source for inseminated female
6. prevention of remating (monogamicity factor, MF)

This chapter not subject to U.S. copyright
Published 1991 American Chemical Society

With the demonstration of the RTF activity, our interest in the MF was piqued due to what appeared to be similarities between the termination of mating and the termination of receptivity. Many of the previous studies addressing inhibition of mating were conducted in one of three dipteran genera: Aedes spp. (family: Culicidae), Drosophila spp. (Drosophilidae), and Musca spp. (Muscidae). In general, those studies involved transplanting the AG or injecting a crude homogenate into virgin females. The resultant female was typically refractory to subsequent mating attempts from males. An extensive series of studies (4-12), begun in 1967 on the housefly, (HF, M. domestica) by USDA-ARS investigators at Fargo, North Dakota, addressed the factor found in the insect's accessory duct (AD). After considerable effort, a single peptide of less than 6,000 molecular weight was isolated from more than 20 AG components. It was found that the factor was composed primarily of basic amino acid residues. No further reports concerning efforts to isolate and sequence the factor were forthcoming. Coincidentally, other groups working with Drosophila spp. and Aedes spp. were also able to identify transferable male factors. Leahy and Craig (13) observed an oviposition stimulation factor as a result of implanting male AG into female A. albopictus and A. aegypti. Subsequenty, Craig (14) demonstrated that in 12 different mosquito species, implants resulted in females which were refractory to mating. Though mating occurred, there were no incidents of sperm transfer to the female following implantation of the gland. The factor(s) responsible for the monogamicity behavior was shown to consist of two distinct peptides, designated the alpha and beta fractions (15-17). Hiss and Fuchs (18) showed that the oviposition behavior required only the alpha fraction while the monogamicity behavior required both peptides. The active factor, termed matrone, was isolated and apparently purified, but never sequenced.

Efforts with the Drosophila peptides have progressed much further than either of the other two studies. In D. funebris Baumann and colleagues (19-21) isolated a 27 amino acid peptide (termed PS-1) and a second, smaller glycine derivative from the paragonia (PS-2). The latter peptide acted as an oviposition stimulant while the former influenced the mating behavior of the females by producing a female refractory to subsequent matings. However, the activity was short lived. More recently, Chen et al (22) isolated a paragonial factor from D. melanogaster which suppressed mating activity in females. The 36 amino acid peptide, when injected into females, produced a monogamicity effect nearly identical to the naturally transferred factor. The investigators, using oligonucleotide cDNA clones, identified nucleic acid sequence coding for the peptide precursor and hydrophobic signal seqence at its N-terminal end.

Our pursuit of mating factors in both CEW and HF stems from a long standing interest in both insects. We report here the current status of research on mating factors in both insects.

Materials and Methods

Insects. Feral HF were collected near the dairy rearing barns of the Beltsville National Agriculture Research Center and reared on fermented CSMA media. This media was innoculated with a smaller amount of fermented media onto which mated females had been allowed to oviposit. After development, pupae were removed and adults allowed to emerge. Flies were segregated by sex within 24 hrs to prevent mating. They were provided a dry mixture of sugar and milk (1:3 w:w) plus water _ad lib_ from a separate container. The insects were held at 27°C in an incubator with a 16:8 L:D photoperiod and RH > 50%.

CEW eggs were supplied by the Southern Field Crop Insect Management Laboratory (USDA-ARS; Stoneville, MS) and rearing media was obtained from Southland Products (Lake Village, AR). The media was prepared according to manufacturer's instructions in one ounce creamer cups. Two to three larvae were added and the sealed cups placed into an incubator at 28°C, >50% rh, with a 16:8 L:D photoperiod. Scotophase began at 0800 hrs daily to allow work with the insects to proceed in the dark as required. Males were segregated from females as pupae. Adults were maintained in the same incubators as for rearing and fed a 10% sucrose solution _ad lib_.

Preparation of Housefly and Corn Earworm Tissue Extracts. All internal reproductive tissues of 4-7 day-old virgin male HF were removed under Milli-Q water, gently dried by touching to a paper towel, and placed in a dry-ice-chilled, 1.5 ml plastic centrifuge tube. Crude extract for bioassay was prepared by homogenizing 1 male equivalent per 30 ul Milli-Q water with a Brinkman Polytron fitted with a 7 mm O.D. diameter generator, and centrifuged at 15,000 x g (4°C). Supernatant was concentrated in a Speed-Vac Concentrator (Savant Inc.; Farmingdale, NY) and resuspended in 1 ul of Milli-Q water per male equivalent for injection. CEW accessory glands were treated exactly as for HF except that accessory glands only were excised and the dissection media was physiological saline (23). Sample for chromatography was homogenized in Bennett's buffer (24).

Purification and Chromatography.
C-18 SEP-PAK. Sample homogenized in Bennett's buffer was passed through C-18 SEP-PAK (Water's Assciates, Milliford MA) solid phase extraction cartridges using the method of Schooley et al (25) at a rate of 25 AG or AD per cartridge. The SEP-PAKs were then washed with 3 ml Milli-Q water, followed by successive 3 ml washes of acetonitrile in 0.1% trifluoroacetic acid (TFA), beginning with 10% ACN and increasing in 10% increments to 50%.

Ion Exchange SEP-PAK. Ion Exchange chromatography was performed essentially by the method of Bennett (24). Either QMA or CM SEP-PAK solid phase extraction cartridges were linked together and washed with 10 mM ammonium acetate (pH 6.0). The linked ion exchangers were then loaded with the fraction of interest in buffer and washed with 6 mls of the loading buffer. The eluant was labeled the "neutral fraction". SEP-PAK's were decoupled and the QMA cartridge washed with 10 mM ammonium acetate buffer plus 1%

triethylammonium. The CM SEP-PAK was washed with 10 mMammonium acetate plus 1% TFA. These eluants were labeled the "Basic" and "Acidic" fractions, respectively.

High Pressure Liquid Chromatography. Size exclusion chromatography was performed isocratically in 0.2 M ammonium acetate (flow=1.5 ml/min) with a Water's Model 840 HPLC with a Toyo-Soda G2000SWXL column (30 cm x 7.8 mm). Eluting components were monitored in the UV at 214 nm and by fluorescence (Ex = 230 nm, Em = 300 nm) and collected in 4 minute fractions for bioassay. Reverse phase chromatography was conducted with a Hewlett-Packard Model 1090 HPLC equipped with a Vydac C-18 column using a linear gradient over one hr from 10% to 50% acetonitrile in 0.1% TFA. Sample was monitored with a diode array detector from 190 to 350 nm in the UV. Fractions were collected in 1 min aliquots.

Bioassays

Housefly. Virgin females, 4-8 days post-emergence, were anesthesized with CO_2. Test fractions were administered to the HF females in 1 ul droplets of Milli-Q water through a small slit made anterior to the left wing. An excess of males was introduced to the cage containing the females 1 hr after treatment and mating was observed at 20 min intervals for a total of 3 hours. Mating response was expressed as:

$$\% \text{ REDUCTION} = \frac{(\text{\# OF TREATED PAIRS MATING}) - (\text{\# OF CONTROL PAIRS MATING})}{(\text{\# OF CONTROL PAIRS MATING})} \times 100$$

Corn Earworm. Bioassay for RTF activity was conducted in three- to four-day-old, adult, virgin CEW females. Neck ligations were performed 12-18 hr prior to bioassay. Five picomoles of synthetic PBAN were injected into each female in 5 ul distilled water at the onset of scotophase. One hour later this was followed with 10 ul of the test fraction. Pheromone glands were excised and pheromone titer determined as previously described (26) using a Shimadzu Model GC14A gas chromatograph equipped with a Hewlett-Packard (Avondale, PA) methyl silicone column.

Results

Housefly. Studies were initiated by comparing the effects of a crude homogenate on mating to those of an injection of deionized water (Table I). The presence of the factor(s) which caused the female to reject mating attempts from the male was confirmed. Initially, we hoped that a size exclusion step would separate a large proportion of non-active material, however, activity was found in all eluting fractions (Table II), rather than in a single distinct fraction. These results suggested that either more than one component is present which is responsible for the activity or that the lack of an organic component in the solvent allows interactions to occur between the factor(s) and the silica based column matrix. We did not test this latter possibility, but subsequently C-18 SEP-PAK's were used in the

Table I. Mating Inhibition in Housefly Following Injection of Crude Homogenate of the Male Accessory Duct

Treatment	Total Number	Pairs Mating	% Reduction	Confidence Interval
Control	24	10	---	22.11 - 63.36
Extract	24	0	100	0.00 - 14.25

The "Total Number" is the number of test insects while the "Pairs Mating" is the observed number of insect pairs mating within the three hrs of the test. "Confidence Interval" determined by Binomial Distribution. Values significantly different at $P \leq 0.05$.

Table II. Mating Inhibition in Housefly Females Following Injection of Size Exclusion Prepared Samples

Treatment	Pairs Mating	Corrected % Mating	Confidence Interval
Control	9	---	66.4 - 100.0 A
5- 9 min	3	33	7.5 - 70.1 A
9-13 min	1	11	0.3 - 48.3 B
13-17 min	2	22	2.8 - 48.3 B
17-21 min	4	44	13.7 - 78.8 A

A total of 20 insect pairs were tested for each treatment. "Pairs Mating" indicates the number of insect pairs observed mating within the three hr of the study. Confidence intervals determined with Binomial Distribution. C.I. followed by same letter not significantly different at $P \leq 0.05$.

purification scheme (Table III) to achieve a rapid reduction in the amount of contaminating materials. Mating bioassays of the SEP-PAK fractions indicated a decline in mating acceptance by the females, however, not at the same level as seen for the crude fractions, above. The most active fractions were eluted with 30, 40, and 50% acetonitrile. Interestingly, the effects of the material were obvious for the first two hours of the study, but after this time, they were not as easily detected.

During the course of this work we noted that in HF females, the duration of the monogamicity effect was short-lived (note Table III). Previously, Adams and Nelson (5) reported a shortened duration of activity with approximately 50% mating refusal in treated females. In general, abdominally injected material required a minimum of 3-6 male equivalents to generate the monogamous behavior. To overcome this limitation, Terranova and Leopold (8) developed a bioassay technique which allowed injection of an aqueous, sonicated extract of AD directly into the vaginal opening of the female HF. Mating refusal exceeded 90% and required less than 0.4 male equivalents introduced per female.

<u>Corn Earworm</u>. Crude extracts of male CEW accessory glands were highly effective in reducing the titers of pheromone produced in

Table III. Mating Inhibition in Housefly Females Following Injection of C-18 SEP-PAK Fractions

Treatment	Pairs Mating	Corrected % Mating	Confidence Interval
ONE HOUR			
Control	7	---	59.0 - 100.0 A
0%	11	157	--- A
10%	7	100	59.0 - 100.0 A
20%	4	57	18.4 - 90.1 AB
30%	1	14	0.1 - 57.8 B
40%	1	14	0.1 - 57.8 B
50%	0	0	00. - 41.0 B
TWO HOURS			
Control	19	---	82.3 - 100.0 A
0%	18	95	74.0 - 99.9 A
10%	17	89	66.7 - 98.7 A
20%	12	63	38.4 - 83.7 AB
30%	6	47	12.6 - 56.6 B
40%	8	42	20.3 - 66.5 B
50%	4	21	6.1 - 45.6 B
THREE HOURS			
Control	21	---	83.9 - 100.0 A
0%	19	90	69.6 - 98.8 A
10%	20	95	76.2 - 99.9 A
20%	17	81	58.1 - 94.6 AB
30%	11	52	29.8 - 74.3 B
40%	13	62	38.4 - 81.9 B
50%	18	86	63.7 - 97.0 AB

A total of 25 females and 30 males were confined to the same cage at the beginning of the study. Mating pairs were removed during the study. Confidence intervals determined with Binomial Distribution. Values not followed by the same letter were significantly different at $P \leq 0.05$.

the female. As little as 0.12 male equivalents reduced the titer by over 50% (1). While not as effective, 0.06 male equivalent was able to elicit a sizeable reduction (38%).

Our initial efforts to separate RTF from CEW accessory glands made use of C-18 SEP-PAKs. The bulk of the RTF eluted from C-18 SEP-PAKs in 30-40% acetonitrile (Table IV), though activity was also found in all other fractions. Subsequent preparations (data not shown) confirmed the presence of the RTF in the 30-40% fraction. The activity, when subjected to ion exchange chromatography, was found concentrated in the acidic fraction (Table V). Separation of this fraction using C-18 reverse phase HPLC (data not shown) and subsequent bioassay localized activity in the 43-44 minute range.

During the course of the studies with the corn earworm receptivity termination factor, it became obvious that asignificant amount of variation was inherent in the bioassay methods. Essentially, each female responds differently to the PBAN titer injected. This, coupled with the variable reduction in the titer due to the presence of the RTF, causes a highly variable response in pheromone levels.

Table IV. Pheromone Biosynthesis in *Heliothis zea* Females after Injection of C-18 SEP-PAK Prepared Fractions

Treatment	Z-11-hexadecenal ng Produced	% Reduction
Control[1] (1)	224.7	---
0% (2)	6.7	97.0
10% (3)	67.1 ± 72.9	70.1
20% (3)	111.5 ± 125.5	50.4
30% (3)	26.7 ± 35.1	88.1
40% (3)	16.6 ± 13.9	92.6
50% (2)	91.1	59.5
60% (1)	161.2	28.3

Fraction prepared by loading a C-18 SEP-PAK cartridge with a crude homogenate of male accessory glands in 0.1% TFA. The fractions were eluted with increasing amounts of acetonitrile in 0.1 % TFA. Bioassay conducted as described in text. A total of 3 insects were tested per treatment (unless otherwise indicated in parenthesis) with 1 male equivalent per female .

Table V. Pheromone Biosynthesis in *Heliothis zea* Females after Injection of Ion Exchange SEP-PAK Prepared Fractions

Treatment	Z-11-hexadecenal ng Produced	% Reduction
Control	169.3 ± 25.7	---
30-40% ACN[1]	4.8 ± 1.5	97.2
Neutral	120.1 ± 44.2	29.1
Basic	49.2 ± 4.8	70.9
Acidic	3.6 ± 1.3	97.9

[1] Fraction prepared from C-18 SEP-PAK cartridge previously eluted with 20% acetonitrile in 0.1% TFA followed by 3 ml 40% acetonitrile in 0.1% TFA. Bioassay conducted as described in

text. Three insects were tested per treatment with 1 male equivalent injected per female.

Studies with the monogamicity factor of HF and the receptivitiy termination factor of CEW have progressed rapidly and, during the course of these studies, we have observed that an additional factor may also be present in the AG of CEW. This oviposition stimulant is also transferred to the female during copulation. Research is continuing on all factors.

Acknowledgments

The authors wish to thank Dr. Charles Woods, Ms. Patricia Thomas, and Ms. Monica Berger for advice and technical assistance during the course of these studies. We also thank Dr. Renee' Wagner, USDA-ARS-LPSI-LIL, Beltsville, MD, and Dr. Charles Woods, University of Maryland, College Park, MD; for critical review of the manuscript.

Literature Cited

1. Raina, A.K. J. Insect Physiol. 1989,35, 821.
2. Leopold, R.A. In Annual Review of Entomology; Annual Reviews, Inc; Palo Alto, CA, 1976; Vol. 21, p 199.
3. Chen,, P.S. In Annual Review of Entomology; Annual Reviews, Inc; Palo Alto, CA, 1984; Vol. 29, p 233.
4. Riemann, J.G.; Moen, D.J.; Thorson, B.J. Insect Physiol. 1967, 13, 407.
5. Adams, T.S.; Nelson, D.R. Ann. Ent. Soc. Am. 1968, 61, 112.
6. Riemann, J.G.; Thorson, B.J. Ann. Ent. Soc. Am. 1969, 62, 828.
7. Nelson, D.R.; Adams, T.S.; Pomonis, J.G. J. Econ. Ent. 1969, 62, 634.
8. Terranova, A.C.; Leopold, R.A. Ann. Ent. Soc. Am. 1971, 64, 263.
9. Leopold. R.A.; Terranova, A.C.; Swilley, E.M. J. Exp. Zool. 1971, 176, 353.
10. Leopold, R.A.; Terranova, A.C.; Thorson, B.J.; DeGrugillier, M.E. J. Insect Physiol. 1971, 17, 987.
11. DeGrugillier, M.E.; Leopold, R.A. Ann. Ent. Soc. Am. 1972, 65, 689.
12. Terranova, A.C.; Leopold, R.A.; DeGrugillier, M.E.; Johnson, J.R. J. Insect Physiol. 1972, 18 1573.
13. Leahy, M.G.; Craig, G.B., Jr. Mosquito News 1965, 25, 448.
14. Craig, G.B., Jr. Science 1967, 156, 1499.
15. Fuchs, M.S.; Craig, G.B., Jr.; Hiss, E.A. Life Sci. 1968, 7, 835.
16. Fuchs, M.S.; Craig, G.B., Jr.; Despommier, D.D. J. Insect Physiol. 1969, 15, 701.
17. Fuchs, M.S.; Hiss, E.A. J. Insect Physiol. 1970, 16, 931.
18. Hiss, E.A.; Fuchs, M.S. J. Insect Physiol. 1972, 21, 2217.
19. Baumann, H. J. Insect Physiol. 1974, 20, 2181.
20. Baumann, H. J. Insect Physiol. 1974, 20, 2347.
21. Baumann, H.; Wilson, K.J.; Chen, P.S.; Humbel, R.E. Eur. J. Biochem. 1975, 52, 521.

22. Chen, P.S.; Stumm-Zollinger, E.; Aigaki, T.; Balmer, J.; Bienz, M.; Bohlen, P. Cell 1988, 54, 291.
23. Meyers, J.; Miller, T. Ann. Ent. Soc. Am. 1969, 62, 725.
24. Bennett, H.P.J. J. Chromatogr. 1986, 359, 383.
25. Schooley, D.A.; Miller, C.A.; Proux, J.P. Arch. Insect Biochem. Physiol. 1987, 5, 157.
26. Raina, A.K.; Klun, J.A. Science, 1984, 225, 531.

RECEIVED July 18, 1990

Chapter 7

Insect Cuticle Tanning

Enzymes and Cross-Link Structure

K. J. Kramer[1], T. D. Morgan[1], T. L. Hopkins[2], Allyson Christensen[3], and Jacob Schaefer[3]

[1]U.S. Grain Marketing Research Laboratory, Agricultural Research Service, U.S. Department of Agriculture, Manhattan, KS 66502
[2]Department of Entomology, Kansas State University, Manhattan, KS 66506
[3]Department of Chemistry, Washington University, St. Louis, MO 63130

Insects periodically secrete and stabilize a cuticular exoskeleton to allow for growth and differentiation during development to the adult stage. Sclerotization or tanning is a vital process in which specific regions of the newly secreted cuticle are stabilized by the formation of cross-links between biopolymers such as protein and chitin. Solid state NMR analysis of pupal cuticle has detected covalent bonding between aromatic or aliphatic carbons of catecholamines and protein nitrogen. Weaker secondary bonding and dehydration may also occur as phenolic content increases. The tanning agents are electrophilic derivatives (*o*-quinones and *p*-quinone methides or free radicals) of catecholamines that may be involved in both sclerotization and pigmentation. Phenoloxidases such as laccases and tyrosinases and other types of enzymes such as isomerases catalyze the formation of reactive tanning agents from N-acylcatecholamines. Understanding the chemistry of insect cuticular sclerotization could lead to the development of new insecticides.

Insects and other arthropods obtain structural support and protection from a cuticular exoskeleton secreted by a single layer of epidermal cells. The stages of growth and development are delineated by deposition, expansion, stabilization and pigmentation of the new cuticle and shedding of the old cuticle. Cuticular sclerotization occurs both before and after ecdysis in genetically determined patterns. The exoskeleton may be expandable to accommodate continued growth, flexible to allow articulation between joints and segments, or rigid to provide mechanical stability and resistance to compression. The physical and chemical properties of this

0097–6156/91/0449–0087$06.00/0
© 1991 American Chemical Society

multifunctional structure depend on the types and amounts of minerals, lipids, chitin, phenols, structural proteins and enzymes that catalyze the assembly and stabilization of the completed exoskeleton. Sclerotization is a complex chemical process involving oxidation of diphenols to quinonoid derivatives and possibly free radicals that form covalent bonds between macromolecules in the cuticular matrix. Cross-links and other types of bonds produce a highly insoluble dehydrated structure resistant to chemical and physical degradation.

Enzymes which catalyze the oxidation and isomerization of phenolic metabolites in the cuticle include phenoloxidases, peroxidases, and isomerases. Dopamine, N-acetyldopamine and N-β-alanyldopamine are important substrates for these enzymes and are transported from the epidermis into newly secreted cuticle prior to sclerotization and pigmentation (1,2). Dopamine is associated with the synthesis of black melanin in cuticle, and N-β-alanyldopamine with the synthesis of brown sclerotin. N-Acetyldopamine is often associated with the synthesis of colorless sclerotin, although the presence of dopamine or N-β-alanyldopamine in structures containing N-acetyldopamine will also result in color development. The oxidized products that can be formed from these catecholamines include o-quinones and their isomeric ρ-quinone methides and α,β-dehydrocatecholamines, as well as semiquinones and their isomeric carbon radicals (3,4). Figure 1 illustrates a possible two-electron oxidative pathway for conversion of diphenols to some of these products and includes metabolites involved in cuticular melanization. The reactions include two electron oxidation to o-quinones (Fig. 1, compound 2), spontaneous cyclization to leucoaminochromes (3), two-electron oxidation to ρ-quinoneimines (5), indolization to 5,6-dihydroxyindole derivatives (6), two-electron oxidation to indole-5,6-quinones (7), and polymerization to melanochromes and melanins.

Insects utilize N-acylated catecholamines for sclerotization because the amino group is blocked from intramolecular cyclization and subsequent melanin formation. The o-quinones are sufficiently long-lived so that they may react with nucleophilic groups in the cuticle or they may tautomerize to reactive ρ-quinone methides (Fig. 1, compound 4 and Fig. 2, compound 5) for side chain cross-link formation. It is possible that the ρ-quinone methide may isomerize to an α,β-dehydrocatecholamine (Fig. 1, compound 8 and Fig. 2, compound 6) which may be subsequently oxidized to an α,β-dehydro-o-quinone or to other products (Fig. 2, compound 9). The resulting electrophilic α and β carbon may be involved in cross-link formation. It should be noted that, in addition to two electron oxidation, one electron oxidative products such as semiquinones (2) may be produced in cuticle either enzymatically by laccases and peroxidases or spontaneously by disproportionation of the o-quinone and diphenol. The semiquinones may isomerize to carbon radicals (3), which may undergo additions, substitutions, abstractions and isomerizations to a large number of possible products. The complex mixture of oxidation products from N-acylcatecholamines will be discussed here, together with the evidence for diphenol-protein adducts and cross-links in sclerotized cuticle.

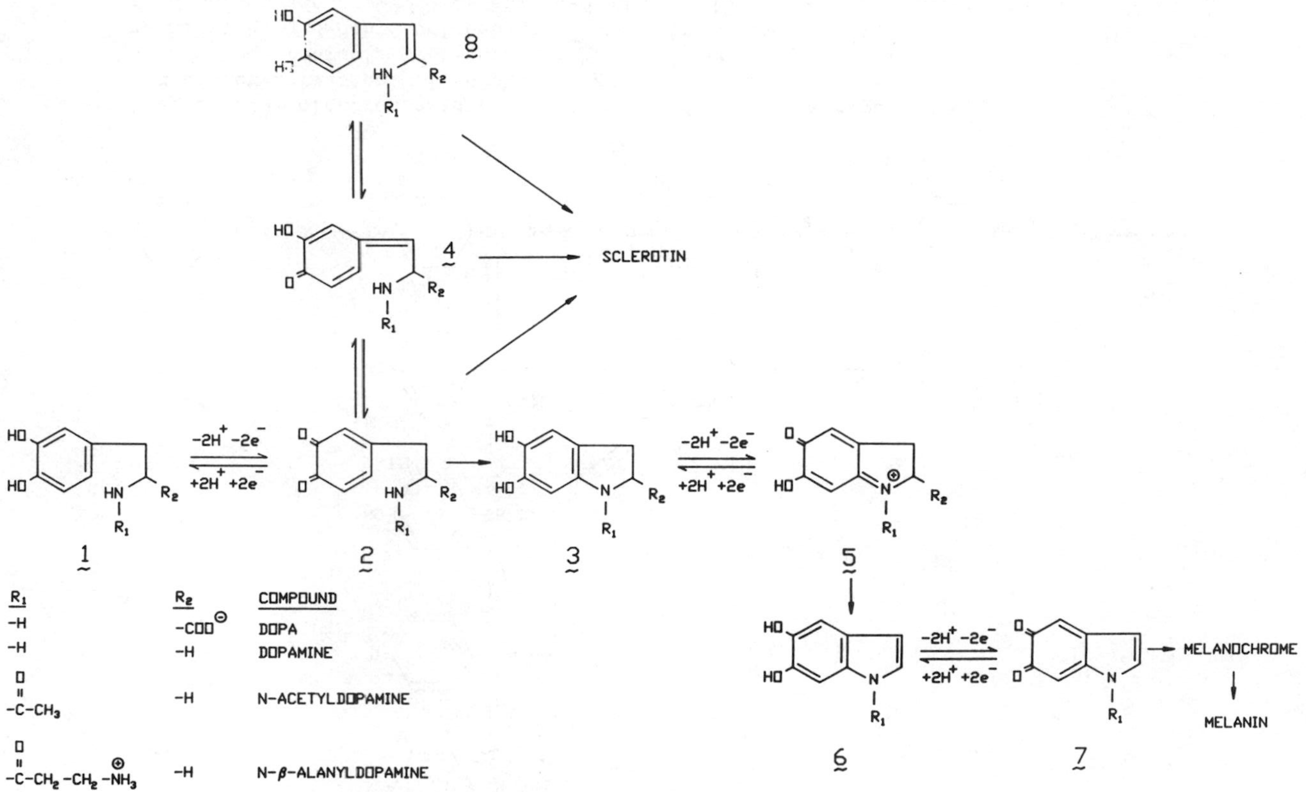

8
4
SCLEROTIN
1
2
3
5
6
7
-2H+ -2e-
+2H+ +2e-
MELANOCHROME
MELANIN
R1 R2 COMPOUND
-H -COO⊖ DOPA
-H -H DOPAMINE
-C(=O)-CH3 -H N-ACETYLDOPAMINE
-C(=O)-CH2-CH2-NH3⊕ -H N-β-ALANYLDOPAMINE

Fig. 1. Two electron oxidative pathway of catecholamines.

1, Catecholamine; 2, *o*-quinone; 3, leucoaminochrome; 4, *ρ*-quinone methide; 5, *ρ*-quinoneimine; 6, 5,6-dihydroxyindole; 7, 5,6-indole quinone; 8, α,β-dehydrocatecholamine.

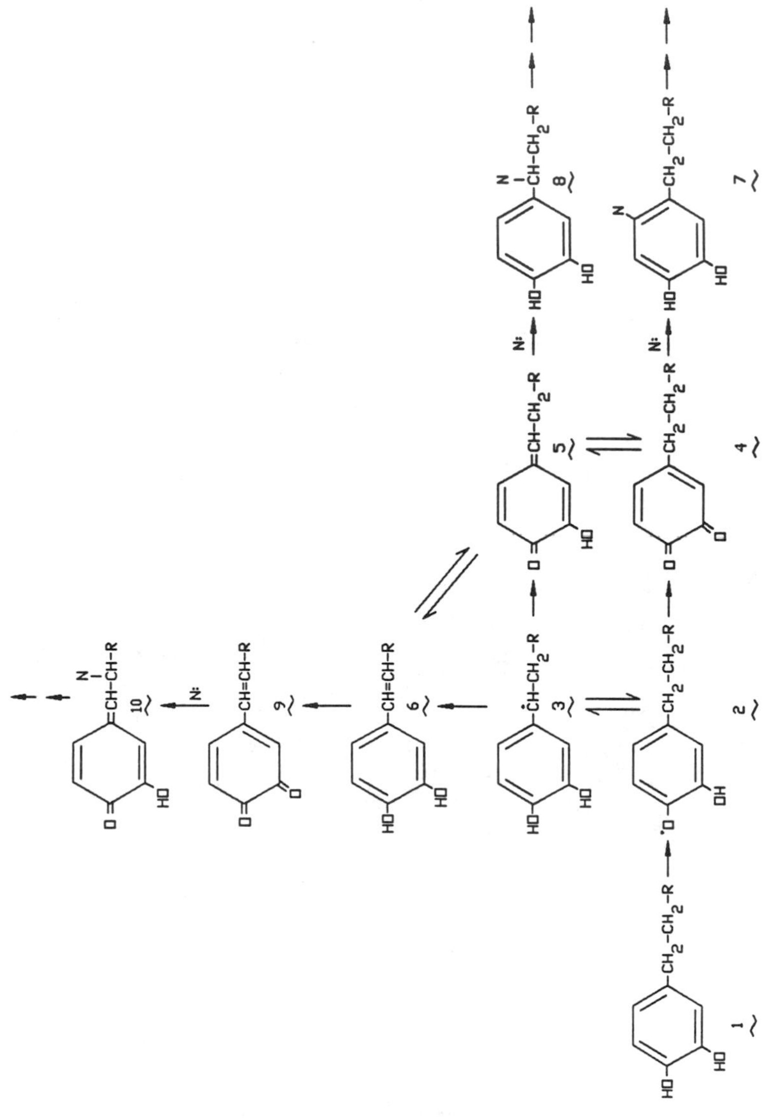

Fig. 2. One electron oxidative pathway of diphenolic compounds and adduct formation. 1, Diphenol; 2, semiquinone; 3, β-carbon radical; 4, *o*-quinone; 5, ρ-quinone methide; 6, α,β-dehydrodiphenol; 7, aromatic carbon adduct; 8, β-carbon adduct; 9, α,β-dehydro-*o*-quinone; 10, α-carbon ρ-quinone methide adduct.

CUTICULAR OXIDASES

The phenoloxidases are a related group of copper containing enzymes that catalyze the oxidation of phenols to quinones in animals and plants (5,6). Two distinct types of phenoloxidases that have substrate specificities and inhibitor sensitivities resembling typical tyrosinases and laccases are found in different types of insect cuticle (7,8). Peroxidases, heme-containing enzymes that also oxidize diphenols to quinones, may be present in cuticle and may play a role in sclerotization (9,10).

Tyrosinase. Tyrosinase is a multi-functional copper oxidase that acts both as a monophenol monooxygenase (EC 1.14.18.1) and as a catechol oxidase (EC 1.10.3.1, o-diphenol O_2 oxidoreductase). Most of the monophenol monooxygenase activity of the pupal integument containing tyrosinase appears to be located in the epidermis instead of the cuticle (11,12, Morgan *et.al.* unpublished data). This localization suggests that the primary role of tyrosinase may be in the epidermis and that some other type of phenoloxidase may be present in the pupal cuticle. Studies with other insect species have shown that tyrosinase is found typically in flexible larval cuticle, where it may be involved in repair of wounds in soft cuticle (8,13,14). Tyrosinase is probably present in flexible, colorless cuticle as an inactive proenzyme that can be activated by a protease following wounding. Active phenoloxidase is present in regions of the larval cuticle that melanize. In the blowfly, *Lucilia cuprina*, active tyrosinase is present in epicuticular filaments and protyrosinase is apparently present in the procuticle of the larvae (15). The phenoloxidase that is responsible for the synthesis of melanin from dopamine in larval cuticle of *M. sexta* has been purified, but it has not been rigorously tested for monophenoloxidase activity (16).

Laccase. Laccase (EC 1.10.3.2, diphenol: O_2-oxidoreductase) is a copper-containing phenoloxidase that has been detected in cuticles of at least 16 species of insects, including the pupal cuticle of *M. sexta* (17,18, Thomas *et.al.* unpublished data). It is present in the inner epicuticle of *L. cuprina* (15). This enzyme is present in the active form during sclerotization and it probably generates the o-quinones that participate in cross-linking reactions. A common problem encountered in studies of cuticular laccase is its insolubility. Soluble laccases have been extracted from larval dipteran cuticle, but endogenous proteases are also known to be present in such cuticle.

Yamazaki (19) was the first to discover laccases in insects. She has recently used sodium dodecyl sulfate to extract laccases from *Bombyx mori* cuticle. The enzymes can be identified by immunoblotting techniques after separation by sodium dodecyl sulfate polyacrylamide gel electrophoresis, but the enzymatic activity has not yet been recovered after the detergent treatment (20). The standard protocol involving proteolysis with exogenous trypsin solubilizes laccase from the pupal cuticle of *M. sexta* (18). Following incubation of pupal cuticle from *M. sexta* with chymotrypsin, we found that laccase activity in the supernatant is elevated after treatment with trypsin (Morgan *et.al.* unpublished data). In addition to its effectiveness in solubilizing laccase, trypsin appears to be an activator of prolaccase. Yamazaki (20,21) has reported a proenzyme from the cuticle of *B. mori* that may be solubilized by treatment with chymotrypsin. However, Andersen (22) and Barrett (23) have not obtained evidence for a prolaccase in their extensive work on cuticular phenoloxidases.

The catalytic properties of laccase differ from tyrosinase in that laccase catalyses the oxidation of diphenols in discrete one-electron steps, whereas a single two-electron step occurs with tyrosinase (5,6). Unlike tyrosinase, laccase has no monophenol o-hydroxylating activity. Both o- and ρ-diphenols are oxidized by laccase, but tyrosinase oxidizes only the former compounds. Methylhydroquinone is a ρ-diphenol that is often used to estimate laccase activity because of the relatively high velocity obtained with this substrate. The high rate of enzymatic oxidation may be due, in part, to the low oxidation potential of methylhydroquinone and this property can easily lead to experimental errors. For example, N-β-alanyldopamine quinone is an effective oxidizing agent for methylhydroquinone; therefore, addition of trace amounts of N-β-alanyldopamine causes rapid oxidation of methylhydroquinone in the presence of tyrosinase (Morgan *et. al.* unpublished data). Nonetheless, methylhydroquinone is a good substrate for the insoluble residue of unsclerotized pupal cuticle and the partially purified trypsin-solubilized laccase from *M. sexta* (18, Morgan *et. al.* unpublished data).

Although the trypsin-solubilized laccase from *M. sexta* oxidizes N-β-alanyldopamine to its o-quinone, the o-quinone does not accumulate to a high concentration (18, Morgan *et. al.* unpublished data). Instead, an enzyme in the preparation, possibly the laccase itself, appears to catalyze the conversion of the o-quinone to a ρ-quinone methide or another intermediate which reacts with water to yield the β-hydroxylated derivative of N-β-alanylnorepinephrine. Further studies of purified *M. sexta* laccase are needed to determine whether it is a bifunctional enzyme which promotes not only o-diphenol oxidation, but also isomerization of the o-quinone to a ρ-quinone methide. Unlike the laccase preparation, tyrosinase from *M. sexta* has little effect on the rate of β-hydroxylation. The functional role of laccase in cuticle sclerotization will, no doubt, receive further attention in the future because it may be the primary phenoloxidase that converts catecholamines into cross-linking agents.

Inhibition of insect laccases was first studied by Yamazaki (19,24), who found that this enzyme is inhibited by cyanide and diethyldithiocarbamate, but is fairly insensitive to thiourea and carbon monoxide. Studies by Andersen (8) and Barrett (23) have shown that laccases from many insect species are less sensitive to phenylthiourea, but more sensitive to azide than are tyrosinases. In our laboratory, we have also found that laccase from *M. sexta* is less sensitive to phenylthiourea than is tyrosinase, but carbon monoxide has not proven useful for distinguishing the two types of phenoloxidases (18). Laccase is reportedly inhibited by monophenols, although it can oxidize monophenols to free radicals at an appreciable rate if other compounds are present that will react with the radicals and prevent inactivation of the enzyme (25). The ability of laccase from *M. sexta* to oxidize monophenols is demonstrated by its oxidation of syringaldazine, a compound with two monophenolic groups (18). One-electron oxidation of each monophenolic group results in rapid rearrangement to a relatively stable quinone methide which has a pink color. Topical application of syringaldazine to the interior scraped surface of *M. sexta* pharate pupal forewing cuticle revealed laccase activity *in situ* (18). Also, N-acetyl-3-methoxy-4-hydroxyphenylethylamine does not inhibit the partially purified laccase from *M. sexta*, but instead is slowly oxidized by the enzyme (Morgan *et. al.*, unpublished data). It appears likely that laccases are critical enzymes involved in sclerotization of insect cuticle. Because of their importance, additional studies are needed to identify effective inhibitors of laccase that can then be tested for potential insecticidal effects.

Peroxidase. Peroxidase (EC 1.11.1.7, phenol: hydrogen peroxide oxidoreductase) is involved in the oxidation of tyrosyl residues of proteins for the production of dityrosine or trityrosine cross-links in the endochorion of eggs of *Drosophila melanogaster* and other species. Tyrosyl cross-links are also present in the highly elastic, rubber-like protein, resilin, which is found in some flexible, articulating regions of insect cuticle (26,27). Bityrosine has also been isolated from larval cuticle of *D. melanogaster* and other dipterans and from larval cuticle of *M. sexta*, but not from the inflexible pupal cuticle of *M. sexta* or fly puparial cuticle (10). Bityrosine has a lower oxidation potential than tyrosine. After one-electron oxidations of both phenolic groups, bityrosine rapidly rearranges to an electrophilic quinonoid structure that can react with nucleophiles in the cuticle. There is some evidence that larval cuticles of dipterans and lepidopterans have peroxidases (9,10), but further work is needed to verify those observations.

ISOMERIZATION OF QUINONOID COMPOUNDS

In addition to phenoloxidases there are other enzymes in insect cuticle which may help to control the formation of intermediates that are involved in sclerotization and melanization. These enzymes include a quinone isomerase that catalyses the conversion of o-quinone to ρ-quinone methide and a quinone methide isomerase that catalyses conversion of ρ-quinone methide to α,β-dehydrocatecholamine.

The ρ-quinone methides of N-β-alanyldopamine and N-acetyldopamine have been proposed as reactive intermediates, but they have not been directly observed either *in vivo* or *in vitro*. However, it is relevant to summarize the indirect evidence for their existence, since the quinone methides are proposed products and substrates of insect cuticular enzymes. All evidence is based on the recovery of catecholamines or other diphenols that have a covalent modification of the benzyl carbon. Whenever N-acetyldopamine or N-β-alanyldopamine is oxidized by cuticular preparations *in vitro*, β-hydroxylation occurs (28-32). The β-hydroxylation product is racemic, suggesting that the hydroxylation step may be nonenzymatic as would occur with a ρ-quinone methide or with a free radical (28,33). The electrophilic intermediate will also react with nucleophiles other than water, such as methanol or kynurenine, as has been demonstrated in several laboratories (1,31,32,34,35, Morgan *et. al.*, unpublished data). If the β-hydroxylated compound is reoxidized and a quinone methide forms, isomerization can be expected to occur with the hydroxyl function being converted to a keto function. This has been observed with some model compounds (29,32,36). There has so far been no spectral or chromatographic evidence for the quinone methides of N-acetyldopamine and N-β-alanyldopamine, although the o-quinones of these catecholamines can be isolated by liquid chromatography (Morgan *et. al.*, unpublished observations). Because of the lack of direct evidence for the quinone methide and its facile reactivity with weak nucleophiles, it is assumed that this intermediate is extremely short lived.

The mechanism of diphenol oxidation may also include free radical formation, either in the enzymatic reaction itself or in subsequent spontaneous disproportionations or cleavages (37,38). The influence of diphenolic radicals on β carbon reactivity has not been well delineated. The metastable radicals are not easily studied in solution or wet solids such as cuticle because they are present at very low concentrations and undergo rapid nonenzymatic reactions.

Quinonoid isomerases. o-Quinone: ρ-quinone methide isomerases have been found in cuticles from four insect species (36,39,40,41, Table I). A quinone tautomerase was also detected in *Sarcophaga bullata* larval hemolymph (42). Although Sugumaran (4) originally proposed that the quinone methides are produced directly by oxidation of the alkyl substituted catechols, later studies (39,41,43) have shown that a strong nucleophile (N-acetylcysteine) does not react with the β carbon, but instead forms an adduct with one of the ring carbons. Therefore, consumption of the quinone by a strong nucleophile was proposed to prevent formation of the quinone methide. A model substrate, 3,4-dihydroxymandelic acid, was also originally believed to be oxidized directly to ρ-quinone methide (44), but data from experiments using short time resolution spectroscopy, kinetics, radiolysis, cyclic voltametry, chronoamperometry and product analysis are inconsistent with the formation of a ρ-quinone methide directly from an o-diphenol (45-47). All of these results indicate that the quinone is the initial oxidation product. Although quinone methides of N-acetyldopamine and N-β-alanyldopamine have been referred to as tautomers of the o-quinones in the literature, their spontaneous isomerization rates are often not reported or assumed to be very low. However, the spontaneous formation of N-β-alanylnorepinephrine from N-β-alanyldopamine-o-quinone is substantial, with about a 14% yield in 10 min in pH 6 phosphate buffer (34, Morgan *et. al.*, unpublished data).

Another possible type of isomerization involves conversion of the ρ-quinone methide to an α,β-dehydrocatecholamine. The latter compound has been identified as a metabolite of N-acetyldopamine during incubation with cuticular preparations and an exogenous nucleophile such as an amino acid (48). Andersen and Roepstorff (48) suggested that the nucleophile competes for a reactive intermediate involved in the formation of catecholamine dimers and that this competition allows significant amounts of unsaturated catecholamine to accumulate. Measurable quantities of a unsaturated catecholamine are not usually present in cuticle. It may not accumulate because the K_m value for this substrate with cuticular laccase is 60 times lower than that for N-acetyldopamine (48). Although Sugumaran *et. al.* (49,50) also demonstrated that rapid consumption of an unsaturated catecholamine can occur, Sugumaran (51) proposed that this compound is never formed in cuticle based on results of radioactive trapping experiments. Recently, an enzyme that apparently converts the ρ-quinone methide to the α,β-dehydro isomer has been detected in cuticular extracts from one species (52, Table I). This result suggests that a direct catecholamine α,β-desaturation step does not occur, at least in puparia of *S. bullata*. Andersen (8) previously suggested that a direct α, β-desaturation step may occur in *Locusta migratoria*, but a desaturating enzyme has not yet been identified. The α,β-dehydrocatecholamine appears to be an important metabolite and its role in sclerotization should receive more study.

The relationship, if any, between phenoloxidases and quinonoid isomerases is poorly understood. Unlike the former, the latter enzymes are not well characterized in terms of chemical, physical or kinetic properties. The reactivity and short lifetime of substrates such as ρ-quinone methides make it difficult to envision simple enzyme interactions with the methides, especially when water and other nucleophiles are available for reaction. It may be that the isomerases simply modulate the activity of phenoloxidases so that tautomerization occurs.

CONVERSION AND BLOCKING FACTORS

For many years the biosynthesis of melanin was thought to result from the spontaneous oxidation and polymerization of dopachrome produced by the tyrosinase-catalyzed hydroxylation of tyrosine to dopa and subsequent oxidation (53). In addition to tyrosinase, however, several enzymatic factors have been recently identified in mammalian tissues that appear to regulate melanogenesis at intermediate steps distal to those involving tyrosine and dopa. The factors include dopachrome conversion factor, dihydroxyindole blocking factor, dihydroxyindole conversion factor and dopachrome oxidoreductase (54-59).

Dopachrome conversion factor catalyzes the decolorization of dopachrome. The mechanism of this conversion apparently involves an isomeric rearrangement of a hydrogen atom from one position of the dopachrome molecule to another, an intramolecular oxidoreduction which results in a tautomeric shift forming 5,6-dihydroxyindole-2-

Table I. Quinonoid isomerizing enzymes in insect tissues.

Order	Species	Tissue	Substrate	Products	Enzyme(s)	Reference
Coleoptera	*Tenebrio molitor*	Pupal & adult cuticles	NADA	NANE	??	75
Dictyoptera	*Periplaneta americana*	Pharate adult cuticle residue	NADA	NANE, NADA quinone, insoluble adducts	Diphenoloxidase, NADA quinone isomerase	41
Diptera	*Calliphora vicina*	Larval & puparial cuticles	NADA, NBAD	NANE, NBANE	??	75,76
	Drosophila melanogaster	Larval & puparial cuticles	NADA, NBAD	NANE, NBANE	??	75,76
	Sarcophaga bullata	Larval cuticle extract	NADA	NANE, NADA quinone, [NADA quinone methide], α,β-dehydro NADA	Diphenoloxidase, NADA quinone isomerase, NADA quinone methide isomerase	36,52,74
		Larval cuticle extract	Dihydroxycaffeiyl methylamide	Caffeiylmethylamide, dihydroxycaffeiyl methylamide quinone, [dihydroxycaffeiyl methylamide quinone methide]	Mushroom tyrosinase, quinone isomerase	77
		Hemolymph	NADA quinone	NANE	Quinone tautomerase, phenoloxidase	42

Lepidoptera	*Dictyoploca japonica*	Pupal cuticle, Silk	N-Acyldopamines	β-Hydroxylated N-acyldopamines	??	32
	Hyalophora cecropia	Larval cuticle extract	NADA	NANE, NADA quinone [NADA quinone methide]	Diphenoloxidase, laccase, quinone isomerase	40
		Wing cuticle	NADA, NBAD	NANE, NBANE	??	75,76
	Manduca sexta	Pharate pupal cuticle residue	NADA	NANE,NADA quinone, [NADA quinone methide], insoluble adducts	Diphenoloxidase, NADA quinone: quinone methide isomerase	30,39,41
		Trypsinized pharate pupal cuticle extract	NBAD	NBANE, [NBANE quinone], [NBAD quinone methide]	Laccase	18, Morgan et. al., unpublished data
Mantodea	*Tenodera sinensis*	Colleterial gland	NADA quinone	NANE	Phenoloxidase, ??	32
Orthoptera	*Locusta migratoria*	Pharate adult cuticle	NADA, NBAD	NANE, NBANE	??	75,76

*Bracket indicates only indirect evidence available for formation of compound. Abbreviations are NADA, N-acetyldopamine; NANE, N-acetylnorepinephrine; NBAD, N-β-alanyldopamine; NBANE, N-β-alanylnorepinephrine.

carboxylic acid (60). Therefore, a more correct nomenclature for dopachrome conversion factor is dopachrome isomerase or tautomerase.

Dihydroxyindole blocking factor blocks the indolization of quinone imine derivatives. Dihydroxyindole conversion factor catalyzes the dehydrogenation of 5,6-dihydroxyindole to indole-5,6-quinone. Dopachrome oxidoreductase converts dopachrome to 5,6-dihydroxyindole and also may block 5,6-dihydroxyindole oxidation and subsequent melanogenic reactions. Relatively little information is available about the physical, chemical and kinetic properties of these proteinaceous factors in mammals. Controversy about melanin-related regulatory factors has focused on whether activity is due to unique individual proteins or is only an expression of activities of a multicatalytic enzyme (61,62). For example, dihydroxyindole conversion activity in mice melanoma is apparently due to tyrosinase, not a unique factor (56).

Although the conversion or blocking factors for dopa metabolism are incompletely understood in mammals, even less is known about these factors in insects (Table II). *Manduca sexta* pharate pupal integument contains a tyrosinase that exhibits dihydroxyindole conversion activity and catalyzes oxidation of dihydroxyindole to indole-5,6-quinone (11,63). The same tissue possesses a dopa quinone imine conversion factor that accelerates the decarboxylation of dopa quinone imine to dihydroxyindole. A similar enzyme that decolarizes dopachrome, presumably dopa quinone imine conversion factor, has been partially purified from hemolymph of *Hyalophora cecropia* diapausing pupae (64). In *Bombyx mori*, dopa quinone imine conversion factors in integument and hemolymph occur in highest concentrations during the latter part of the fifth larval instar (Aso *et. al.*, unpublished data). It is postulated that dopa quinone imine conversion factor in integument participates in wound healing and/or sclerotization, whereas in hemolymph it may facilitate melanization in the humoral immune system.

Table II. Conversion factors for dopa metabolism in insects

Factor	Species	Tissue source	Reference
Dopa quinone imine conversion factor	*M.sexta*	Pharate pupal cuticle	11,63
	B.mori	Hemolymph and cuticle	Aso *et al.*, unpubl. data
	H.cecropia	Hemolymph of diapausing pupae	64
Dihydroxyindole conversion factor (tyrosinase?)	*M.sexta*	Pharate pupal cuticle	11

To summarize, dopa quinone imine conversion factors have been detected in cuticle and/or hemolymph from only three species of Lepidoptera. Other kinds of regulatory factors such as dihydroxy-indole blocking factor have not been detected in insect tissues. The precise physiological roles played by conversion factors that generate indoles is unknown. They may be modulators of reactions associated primarily with melanization.

CROSS-LINK AND PIGMENT STRUCTURES

Progress has been made recently in identifying bonds between catecholamine aromatic and aliphatic carbon to protein nitrogen in insect cuticle by solid-state ^{1}H-^{13}C-^{15}N double-cross polarization (DCPMAS) NMR, (65) and rotational-echo, double resonance (REDOR) ^{13}C and ^{15}N NMR (66,67, Christensen *et.al.*, unpublished data). Tobacco hornworm pupal exuviae were labeled by injection with a combination of [ring-$^{15}N_2$]histidine and either [ring-$^{13}C_6$]dopamine or [β-^{13}C]dopamine. The DCPMAS difference spectrum of [ring-$^{13}C_6$]-labeled cuticle reveals a peak at 135 ppm, upfield from the major oxygenated ring carbon peak at 144 ppm (Fig. 3, middle right spectrum). The same peak was observed in the REDOR difference spectrum, but it was obtained using less than half the number of scans required for the DCPMAS spectrum. The presence of a difference signal at 135 ppm is consistent with the formation of a covalent bond between an aromatic dopamine carbon and a histidyl ring nitrogen.

DCPMAS-NMR of [β-^{13}C]dopamine cuticle was not sensitive enough to observe covalent bond formation between the β carbon of dopamine and histidine due to inherently slow C-N polarization transfer and fast spin-lock relaxation. However, an order-of-magnitude improvement in sensitivity, obtained by using REDOR NMR, facilitated the observation of a sizeable ^{13}C REDOR difference signal at 60 ppm, which is upfield from the major oxygenated aliphalic carbon peak at 75 ppm (Fig. 2, top left spectrum). The difference signal is consistent with direct bonding of the β carbon of dopamine with a ring nitrogen of histidine. We estimate that in *M. sexta* pupal cuticle approximately two-thirds of the bonds to histidine nitrogen are made through catecholamine ring carbons, and about one-third through the β carbon (Christensen *et. al.* unpublished). The two C-N bonds identified by NMR are consistent with an imidazoyl nitrogen attacking either a phenyl carbon of an o-quinone intermediate or a β carbon of a ρ-quinone methide intermediate.

Solid state NMR has been used to demonstrate the presence of melanin-type pigments in beetle cuticles (68). The natural abundance ^{13}C-NMR difference spectrum obtained by subtracting the spectrum of powdered elytra removed from wild-type red flour beetles, *Tribolium castaneum*, from that of powdered elytra from the *black* mutant strain revealed that wild-type and *black* elytra have similar levels of protein, chitin and lipid, but that the *black* elytra have more melanin or other polyphenolic materials. It was estimated that approximately 5% of the total

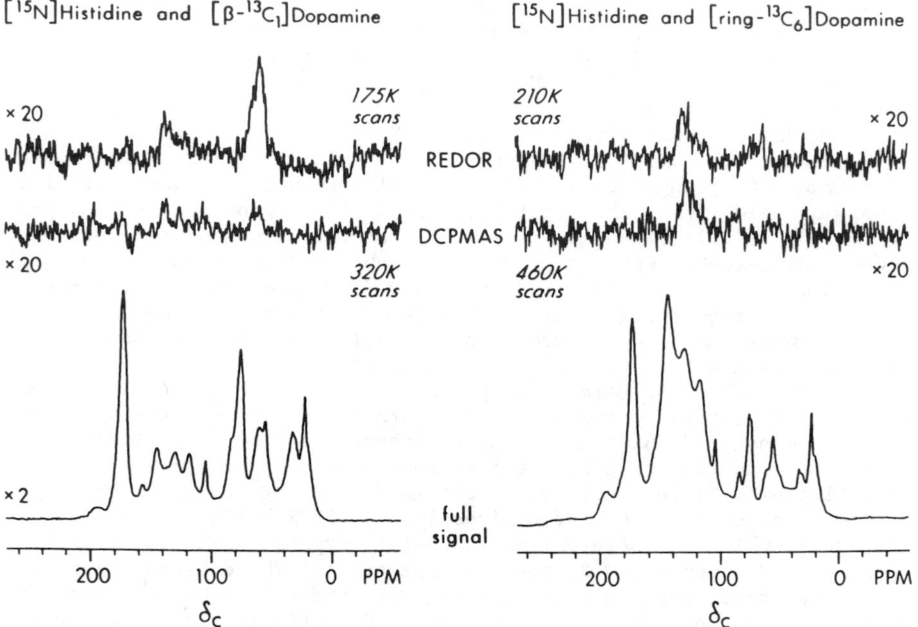

Fig. 3. Single (bottom) and double (middle) cross-polarization magic angle spinning (DCPMAS) and rotational-echo, double resonance (top, REDOR) ^{13}C NMR spectra of tobacco hornworm pupal exuviae double labeled by injection of [β-^{13}C]-dopamine or [ring-$^{13}C_6$]dopamine together with [ring-$^{15}N_2$]histidine. The DCPMAS and REDOR spectra are difference spectra that arise only from those ^{13}C's directly bonded to ^{15}N's.

aromatic carbons in *black* elytra occur as eumelanin or other polyphenols. The spectra of wild-type elytra, on the other hand, had more carbon signals characteristic of β-alanine (both methylene and carbonyl carbons). Apparently, the melanin precursor dopamine is initially directed into the eumelanin pathway in the *black* strain because of a temporary lack of N-acylation with β-alanine. N-β-Alanyldopamine accumulates more rapidly at the expense of dopamine in the wild-type strain such that melanization occurs to a much lesser degree.

The presence of certain low molecular weight catecholamine metabolites in tanned cuticle has been used as indirect evidence for covalent modification of the α and β carbons of catecholamines. Benzodioxin-type dimers of N-acetyldopamine are present in insect cuticle, and the dimeric linkage involves ether bonds between the phenolic oxygens of one monomer and the α and β carbons of the other monomer (8). Mild acid hydrolysis of the dimer produces equal amounts of N-acetyldopamine and 3,4-dihydroxyphenylketoethanol. This ketocatechol is also a major product of mild acid hydrolysis of many insect cuticles. Sugumaran (69) argues that the presence of 3,4-dihydroxyphenylketoethanol does not provide evidence for α-carbon modification of catecholamine. Nevertheless, it is usually accepted that ketocatechols or N-acetyldopamine dimers are evidence for α,β-sclerotization reactions (8). The presence of β-hydroxylated catecholamines such as N-β-alanylnorepinephrine in the pupal cuticle of *M. sexta* provides evidence for quinone methide or β-sclerotization (30). Cold dilute acid extracted substantially more β-hydroxylated catecholamine than did non-acidic solvents, a result suggesting that some of the N-β-alanylnorepinephrine may be a hydrolysis product of N-β-alanyldopamine which has a relatively weak covalent bond to the β carbon.

Model experiments with manducin, which is both a hemolymph and cuticular protein of *M. sexta*, have documented the formation of cross-linked multimers of protein following oxidation of catecholamines by phenoloxidases (70,71, Thomas *et. al.*, unpublished data). N-β-Alanyldopamine, N-acetyldopamine or 1,2-dehydro-N-acetyldopamine caused cross-linking of manducin in the presence of mushroom tyrosinase (70,71). We have also found that these compounds are precursors of cross-linking agents for manducin in the presence of partially purified laccase or tyrosinase from *M. sexta* (Thomas *et. al.*, unpublished data). The cross-linking sites of manducin have not been identified. Solid-state NMR evidence for *M. sexta* cuticle supports bonding of catecholamine aromatic carbons and aliphatic carbons (β) to nitrogen-containing side chains of histidine (65,72, Christensen *et. al.*, unpublished data). Although model experiments and solid-state NMR studies have provided good evidence for cross-linking of cuticular components, noncovalent bonds are also involved in the stabilization of the insect exoskeleton. However, the contribution of these noncovalent interactions to the process of sclerotization is not well understood (73).

CONCLUDING REMARKS

Substantial progress has been made towards the identification of cross-links between polymers in insect exoskeleton and in characterization of enzymes that are required for synthesis of cross-linking agents. These agents may be quinonoid compounds with electrophilic carbons in the ring and in the α and β positions of the side chain. Solid-state NMR has identified bonds between the β and ring carbons of catecholamines and the imidazoyl nitrogen of histidyl residues of protein. In spite of these recent advances, there are still many gaps in our knowledge of the supramolecular structure of insect cuticle and the chemical and enzymatic processes that occur during its formation.

Acknowledgments

This research was a collaborative investigation between the U.S. Grain Marketing Research Laboratory, Agricultural Research Service, U.S. Department of Agriculture, the Department of Entomology, Kansas Agricultural Experiment Station, Manhattan, KS and the Department of Chemistry, Washington University, St. Louis, MO. This is contribution no. 91-68-A from the Kansas Agricultural Experiment Station, Manhattan, KS 66506. Supported in part by National Science Foundation grants DCB 86-09717, DIR-8714035, DIR-8720089 and U.S. Department of Agriculture grant 88-CRCR-3684. We are grateful to Drs. J. Baker, H. Oberlander, R. Howard, S. Andersen and M. Barrett for reviewing the manuscript.

Literature Cited

1. Kramer, K. J.; Hopkins, T. L. Arch. Insect Biochem. Physiol. 1987, 6, 279-301.
2. Hopkins, T. L.; Kramer, K. J. In Physiology of Insect Epidermis; Binnington, K.; Retnakaran, A., Eds.; Inkata Press: Victoria, Australia, 1990; In press.
3. Peter, M. G. Angew. Chem. Int. Ed. Engl. 1989, 28, 555-70.
4. Sugumaran, M. Advances Insect Physiol. 1988a, 21, 179-231.68.
5. Robb, R. In Copper Proteins and Copper Enzymes, Lontie, R., Ed.; CRC Press: Baco Raton, Fl., 1984, Vol. 2, pp 207-41.
6. Reinhammar, B. In Copper Proteins and Copper Enzymes, Lontie, R., Ed.; CRC Press: Baco Raton, Fl., 1984; Vol. 3, pp 1-35.
7. Brunet, P. C. J. Insect Biochem. 1980, 10, 467-500.
8. Andersen, S. O. In Comprehensive Insect Physiol. Biochem. Pharmacol.; Kerkut, G. A.; Gilbert, L. I., Eds.; Pergamon Press: Oxford, 1985; Vol. 3, p 59-74.
9. Locke, M. Tissue Cell 1969, 1, 555-74.
10. Sugumaran, M.; Henzel, W. J.; Mulligan, K.; Lipke, H. Biochem. 1982, 21, 6509-15.
11. Aso, Y.; Kramer, K. J.; Hopkins, T. L.; Whetzel, S. Z. Insect Biochem. 1984, 14, 463-72.
12. Morgan, T. D.; Thomas, B. R.; Yonekura, M.; Czapla, T. H.; Kramer, K. J.; Hopkins, T. L. Insect Biochem. 1990. In press
13. Lai-Fook, J. J. Insect Physiol. 1966, 12, 195-226.

14. Barrett, F. M. Can. J. Zool. 1984, 62, 834-8.
15. Binnington, K. C.; Barrett, F. M. Tissue Cell. 1988, 20, 405-19.
16. Hiruma, K.; Riddiford, L. M. Devel. Biol. 1988, 130, 87-97.
17. Barrett, F. M. Can. J. Zool. 1987, 65, 1158-66.
18. Thomas, B. R.; Yonekura, M.; Morgan, T. D.; Czapla, T. H.; Hopkins, T. L.; Kramer, K. J. Insect Biochem. 1989, 19, 611-22.
19. Yamazaki, H. I. J. Insect Physiol. 1969, 15, 2203-11.
20. Yamazaki, H. I. Zool. Sci. (Tokyo) 1987, 4, 1010.
21. Yamazaki, H. I. Zool. Sci. (Tokyo) 1986, 3, 1015.
22. Andersen, S. O. In Physiology of Insect Epidermis; Binnington, K; Retnakaran, A., Eds.; Inkata Press: Victoria, Australia, 1990; In press.
23. Barrett, F. M. In Physiology of Insect Epidermis: Binnington, K; Retnakaran, A., Eds.; Inkata Press: Victoria, Australia, 1990; In press.
24. Yamazaki, H. I. Insect Biochem. 1972, 2, 431-44.
25. Fahraeus, G.; Ljunggren, H. Biochim. Biophys. Acta. 1961, 46, 22-32.
26. Andersen, S. O. Biochem. Biophys. Acta. 1964, 93, 213-215.
27. Margaritis, L. H. In Comprehensive Insect Physiol. Biochem. Pharmacol.; Kerkut, G. A.; Gilbert, L. I.; Eds; Pergamon Press: Oxford, 1985; Vol. 1, pp 153-230.
28. Peter, M. G. Insect Biochem. 1980, 10, 221-7.
29. Sugumaran, M; Lipke, H. FEBS Letters 1983, 155, 65-8.
30. Morgan, T. D.; Hopkins, T. L.; Kramer, K. J.; Roseland, C. R.; Czapla, T.; Tomer, K. B.; Crow, F. W. Insect Biochem. 1987, 17, 255-63.
31. Andersen, S. O. Insect Biochem. 1989b, 19, 59-67.
32. Yago, M. Insect Biochem. 1989, 19, 673-8.
33. Peter, M. G.; Vaupel, W. J. Chem. Soc., Chem. Commun. 1985, 848-50.
34. Morgan, T. D.; Yonekura, M.; Kramer, K. J.; Hopkins, T. L. Proc. XVIII Internat. Cong. Entomol. Mtg., 1988, p 141.
35. Sugumaran, M.; Saul, S.; Semensi, V. FEBS Letters 1989a, 252, 135-8.
36. Sugumaran, M.; Semensi, V.; Saul, S. J. Arch Insect Biochem. Physiol. 1989d, 10, 13-27.
37. Yamazaki, I. Free Radicals In Biology; Pryor, W. A., Ed.; Academic Press: New York, 1977; Vol. 3, pp 183-217.
38. Peter, M. G.; Stegmann, H. B.; Dao-Ba, H.; Scheffler, K. Z. Naturforsch. 1985, 40C, 535-8.
39. Saul, S.; Sugumaran, M. FEBS Letters 1988, 237, 155-8.
40. Andersen, S. O. Insect Biochem. 1989a, 19, 803-8.
41. Sugumaran, M.; Saul, S. J.; Semamsi, V. Arch. Insect Biochem. Physiol. 1988a, 9, 269-81.
42. Saul, S.; Sugumaran, M. Arch. Insect Biochem. Physiol. 1989c, 12, 157-72.
43. Sugumaran, M.; Dali, H.; Semensi, V. Arch. Insect Biochem. Physiol. 1989a, 11, 127-137.
44. Sugumaran, M. Biochem. 1986, 25, 4489-92.

45. Cabanes, J.; Sanchez-Ferrer, A.; Bru, R.; Garcia-Carmona, F. Biochem. J. 1988, 256, 681-4.
46. Ortiz, F. M.; Serrano, J. T.; Lopez, J. N. R.; Castellanos, R. V.; Teruel, J. A. L.; Garcia-Canovas, F. Biochim. Biophys. Acta. 1988, 957, 158-63.
47. Bouheroum, M.; Bruce, J. M.; Land, E. J. Biochim. Biophys. Acta. 1989, 998, 57-62.
48. Andersen, S. O; Roepstorff, P. Insect Biochem. 1982, 12, 269-76.
49. Sugumaran, M.; Dali, H.; Semensi, V.; Hennigan, B. J. Biol. Chem. 1987, 262, 10546-9.
50. Sugumaran, M.; Hennigan, B.; Semensi, V.; Dali, H. Arch. Insect Biochem. Physiol. 1988b, 8, 89-100.
51. Sugumaran, M. Arch. Insect Biochem. Physiol. 1988b, 8, 73-88.
52. Saul, S.; Sugumaran, M. FEBS Letters 1989b, 255, 340-4.
53. Pawelek, J.; Korner, A. American Scientist. 1982, 70, 136-45.
54. Pawelek, J.; Korner, A.; Berstrom, A.; Bologna, A. Nature 1980, 286, 617-9.
55. Korner, A.; Pawelek, J. J. Invest. Dermatol. 1980, 75, 192-5.
56. Korner, A.; Pawelek, J. Science. 1982, 217, 1163-5.
57. Murray, M.; Pawelek, J. M.; Lamoreux, M. L. Devel. Biol. 1983, 100, 100-20.
58. Barber, J. I.; Townsend, D.; Olds, D. P.; King, R. A. J. Invest. Dermatol. 1984, 83, 145-9.
59. Korner, A. M.; Gettins, P. J. Invest. Dermatol. 1985, 85, 229-31.
60. Pawelek, J. M. Biochem. Biophys. Res. Comms. 1990, 166, 1328-33.
61. Pawelek, J. M. J. Invest. Dermatol. 1985, 84, 234.
62. Barber, J. I.; Townsend, D.; Olds, D. P.; King, R. A. J. Invest. Dermatol. 1985, 84, 234.
63. Aso, Y.; Imamura, Y.; Yamasaki, N. Insect Biochem. 1989, 19, 401-7.
64. Andersson, K.; Sun, S.; Boman, H. G.; Steiner, H. Insect Biochem. 1989, 19, 629-37.
65. Schaefer, J.; Kramer, K. J.; Garbow, J. R.; Jacob, G. S.; Stejskal, E. O.; Hopkins, T. L., Speirs, R. D. Science. 1987, 235, 1200-4.
66. Gullion, T.; Schaefer, J. J. Magn. Reson. 1989, 81, 196-200.
67. Marshall, G. R.; Beusen, D. D.; Kociolek, K; Redlinski, A. S.; Teplawy, M. T.; Pan, Y.; Schaefer, J. J. Amer. Chem. Soc. 1990, 112, 963-6.
68. Kramer, K. J.; Morgan, T. D.; Hopkins, T. L.; Christensen, A. M.; Schaefer, J. Insect Biochem. 1989, 19, 753-7.
69. Sugumaran, M. Bioorg. Chem. 1987, 15, 194-211.
70. Grun, L.; Peter, M. G. The Larval Serum Proteins of Insects: Scheller, K., Ed.; Thieme: Stuttgart, New York, 1983; pp 102-15.
71. Peter, M. G.; Grittke, U.; Grun, L.; Schafer, D. Endocrinological Frontiers in Physiological Insect Ecology, Sehnal, F; Zabza, A.; Denlinger, D. L., Eds.; Wroclaw Technical University Press: Wroclaw, 1988; pp 519-30.

72. Kramer, K. J.; Hopkins, T. L.; Schaefer, J. Adv. Chem. 1988, 379, 160-85.
73. Vincent, J. F. V.; Ablett, S. J. Insect Physiol. 1987, 33, 973-9.
74. Saul, S.; Sugumaran, M. FEBS Letters 1989a, 251, 69-73.
75. Andersen, S. O.; Insect Biochem. 1989, 19, 375-382.
76. Andersen, S. O.; Insect Biochem. 1989, 19, 581-586.
77. Sugumaran, M.; Semesi, V.; Dali, H.; Saul, S. FEBS Letters 1989b, 255, 345-49.

RECEIVED September 10, 1990

Chapter 8

Controlled Release of Insect Sex Pheromones from a Natural Rubber Substrate

Leslie M. McDonough

Agricultural Research Service, U.S. Department of Agriculture, 3706 W. Nob Hill Boulevard, Yakima, WA 98902

A mechanistic model encompassing the factors controlling evaporation rates of pheromones from rubber septa was developed. The model correctly predicted the effect of dose, time, chemical structure, temperature and air speed on evaporation rates. Evaporation rates were first order, and the logarithms of the half-lives were directly proportional to the number of carbon atoms in a homologous series and for a specific compound to the reciprocal of absolute temperature. Also, evaporation rates were directly proportional to air speed. Half-lives depended on thermodynamic relationships and there was an approximate additivity of thermodynamic parameters, making it possible to estimate half-lives and therefore evaporation rates of a large number of compounds at any temperature from a few parameters. There are limitations on uses of natural rubber as a controlled release substrate for conjugated dienes, aldehydes, and certain crystalline compounds. Natural rubber septa are useful for monitoring insect flight activity, for insect behavioral studies, and have potential in control programs based on mating disruption.

A key need for applications of sex pheromones of insect pests are efficacious controlled release substrates (CRS). Applications of pheromones include monitoring insect flight activity as a guide for timing applications of control agents, or direct population control by preventing mating, either by mass trapping the males or permeating the air with pheromone to prevent males from locating females. At the present time there are a number of controlled release materials being studied for the various applications of sex

This chapter not subject to U.S. copyright
Published 1991 American Chemical Society

pheromones and new ones are being introduced. The impetus for developing more effective controlled release substrates is the continuing value of pheromones for monitoring insect flight activity, and the recent commercial success in controlling the oriental fruit moth and the codling moth by permeating the air with their sex pheromones (1, 2).

The requirements for a good controlled release substrate are demanding. Evaporation rates change with chemical structure, dose, time, temperature, and air speed, and one needs to be able to predict or control rates as these variables change. Also, sex pheromones generally consist of mixtures and often the components have different volatilities. Thus, one also needs to be able to predict or control component ratios in the vapor. These are affected by ratios in the substrate, time, and temperature.

This report focuses on natural rubber as the controlled release substrate, and in the physical form of the commercially available sleeve stopper containing a septum and usually referred to as a rubber septum. Interest in this material developed because the rubber septum releases pheromone by only one mechanism (3), and therefore is amenable to systematic study. Also, in many empirical studies the rubber septum had proven to be efficacious for use in traps monitoring insect flight activity, and knowledge of its release characteristics would improve its usefulness. Research that contributes to an understanding of the factors controlling evaporation rates of insect pheromones from natural rubber septa is critically summarized here.

Mechanism of Release from Rubber Septa

In a study evaluating the codling moth sex pheromone [(E,E)-8,10-dodecadien-1-ol] in three different dispensers, Maitlen et al. determined residues of pheromone remaining in the dispensers after a series of time intervals (3). One dispenser displayed a complex loss curve and another allowed extensive chemical decomposition to occur. The rubber septa produced a first order loss curve (logarithm of pheromone content versus time was linear). This experiment was conducted at room temperature at initial doses of 4, 1, and 0.1 mg/septum and the three linear plots were parallel with $t_{1/2}$ = 27 days. In some of these experiments there was an initial fast loss of pheromone (ca. 50% during the first day). Pheromone was impregnated into a septum by adding a solution in an organic solvent to the "cup" of the septum. It was presumed that the initial fast loss was from surface deposited material and that only certain solvents would produce this effect. Subsequently, the surface effect was traced to the quantity rather than the type of solvent: 50 μl or less produced the large first day loss, but with 100 μl or more no fast first day loss occurred (4, 5).

One would expect other organic compounds to behave the same as the codling moth pheromone and also be lost from septa by a first order mechanism. Most studies confirm this. Butler and I showed that several acetates and alcohols were released by this mechanism (6, 7). Also, Leonhardt and Moreno (8), and Heath et al. (9, 10) concurred with this conclusion. Greenway et al. determined the loss of (E)-10-dodecen-1-ol acetate (E10-12:Ac)* from rubber septa outdoors for 131 days when the average daily temperature was 14.5 ± 2.5°C for initial doses of 1, 3, and 10 mg (11). They subjected their data to both single and double exponential curve fitting and concluded that the single exponential curve (i.e., a first order loss curve) best fit the data.

Although a considerable body of information supports a first order mechanism, there are also reports in the literature which appear to be in disagreement with this mechanism. Baker et al. determined evaporation rates of Z8-12:Ac at room temperature by placing a septum with a known dose of pheromone in a 250 ml glass flask, removing the septum after a time interval, and rinsing the compound adsorbed on the walls of the flask (12). For doses of 1,000, 100, and 10 μg/septum they reported evaporation rates of 219, 12, and 1.2 ng/h, respectively. These data are not in agreement with the first order requirement of a direct proportionality between dose and evaporation rate. The reason the direct proportionality was not obtained was because the septa were enclosed in a flask without air movement, and the system would eventually have attained equilibrium. At the beginning of an experiment, the rate of evaporation from a septum would be at a maximum and would continuously decrease to zero at equilibrium. Therefore, the longer the experiment was run, the lower would be the calculated evaporation rates per unit of dose (13). For the 1000 μg dose Baker et al. ran the experiment for 3 h. and for the 100 and 10 μg doses, 16 h. Thus, the shorter run time produced a higher evaporation rate per unit of dose. Also, based on the known evaporation rates of Z7-12:Ac and Z9-12:Ac (14), the evaporation rates of the 100 and 10 μg doses were about 9-fold lower than would be obtained outside the flask. The effect of air speed on evaporation rates is discussed in detail in a later section.

Based on theoretical considerations alone, Zeoli et al. (15) proposed that the evaporation rate of a pheromone from rubber septa is proportional to $t^{-1/2}$ (t is time). This relationship was derived by applying Fick's laws of diffusion to a solute within a polymer matrix and has been successfully applied to the controlled release of therapeutic drugs (16). An assumption in this derivation is that the concentration of solute at the boundary of the polymer matrix is zero. Thus, the removal of the solute as it passes through the boundary must be faster than the diffusion within the matrix, which would be the rate controlling process. Because of this situation, the polymer matrix develops a depletion zone which grows inward from the boundary with time, and the concentration gradient within the matrix as a function of distance from the center becomes more extreme as time progresses (16).

---

* Condensed nomenclature for n-alkyl and n-alkenyl compounds. The letters after the colon indicate the functional type: Ac, acetate; OH, alcohol; Al, aldehyde; H, hydrocarbon. The number between the dash and colon indicate the number of carbon atoms in the longest continuous chain. The letters and numbers before the dash indicate the configuration and position of the double bonds.

The apparent contradiction between the observation of a 1st order mechanism and Zeoli's proposal can be resolved if the rate of evaporation is slower than the rate of diffusion within the matrix. Then the evaporation step is the rate controlling process. Non-kinetic evidence in support of the evaporation step being rate controlling was provided by Greenway et al. (11) who determined the distribution of E10-12:Ac in septa immediately after preparation and after aging outdoors. Before aging, the highest concentration was found in the base of the cups of the septa, but after aging, E10-12:Ac was evenly distributed throughout the septa.

Even through Zeoli's proposal does not apply to rubber septa, one cannot automatically assume all polymeric substrates will release by a first order mechanism. In polymers of greater viscosity or larger volume to surface area ratios, diffusion within the polymeric matrix might become the rate determining step or at least become comparable in magnitude to the evaporation step.

The rubber septum is not the only successful CRS that releases by a first order mechanism. Leonhardt and Moreno (8) reported that some laminated plastic CRS (Hercon Corp.) release first order. Also, Daterman's data (17) on change of release rate with time for Z9-12:Ac in polyvinyl chloride-phthalâte pellets demonstrated a first order loss mechanism.

Methodology for Determining $t_{1/2}$ Values

Because many compounds of interest have very long half-lives, the residue method is very slow. The study of the effect of temperature and other variables required many determinations. Thus a faster method was needed for further progress. My coworkers and I reported a fast method (13, 14) based on determination of the amount of pheromone evaporated over a short period (1-12 h) and the amount remaining in the septum. This method assumes a first order loss mechanism:

$$-\frac{dC_r}{dt} = kC_r$$

Here $-dC_r/dt$ is the instaneous evaporation rate of pheromone, C_r is the amount of pheromone present in a septum at the given instant and k, the proportionality or rate constant, is equal to $(\ln 2)/t_{1/2}$. Of course, instaneous evaporation rates cannot be determined, but if the amount evaporated is small compared to the amount present in the septum then to an excellent approximation C_r does not change and the instantaneous evaporation rate is equal to the amount evaporated per unit time, $-\Delta C_r/\Delta t$. Therefore:

$$E \equiv -\frac{\Delta C_r}{\Delta t} \approx -\frac{dC_r}{dt} = kC_r = \left(\frac{\ln 2}{t_{1/2}}\right)C_r \quad (1)$$

$$\text{and } t_{1/2} = \frac{C_r \ln 2}{E}$$

The half-life is calculated from Equation 1 after experimentally determining $\Delta C_r/\Delta t$ and C_r. Figure 1 shows the apparatus used for these determinations. It consists of two 10-ml hypodermic syringes, A, sealed at the large ends with a rubber gasket, B, and held together by a clamp (not shown). A liquid chromatographic cartridge, C, (Sep-Pak, Waters Associates, Milford Massachusetts) was connected to the small end of the second syringe for collection of evaporated pheromone. A flow meter was connected to the other end of the Sep-Pak cartridge and the flow of purified N_2 over the septum was maintained at 20 cm/sec. After a run, the septum and Sep-Pak cartridge were individually extracted and the pheromone titer was determined by gas chromatography. Sep-Pak cartridges containing octadecyl reverse phase were used for acetates and alcohols (14) and essentially quantitative recoveries were obtained (98 ± 2%). Surprisingly, aldehyde recoveries were erratic and in the range of 70-90% with this phase. A Sep-Pak cartridge containing silica gel produced almost quantitative recoveries for all three compound types (95 ± 4% for aldehydes) (18).

Heath and coworkers also developed a fast method (9, 10) for determining half-lives. Heath and Tumlinson (9) tested the correlation of the logarithm of gas chromatographic retention times on a liquid crystal liquid phase versus the logarithm of half-lives of acetates and alcohols determined by Butler and me (6, 7). For the 13 compounds tested, they reported a correlation (r^2) of 0.995. Subsequently Heath et al. (10) used this correlation to predict the half-lives of acetates, alcohols, and aldehydes. They tested their predictions by determining the vapor ratios of mixtures evaporated from rubber septa. Vapor ratios are equal to the ratios of the evaporation rates, E, as determined by Equation 1. The vapor ratios were determined by collecting the evaporating pheromone components on activated charcoal, extracting the charcoal and determining the titer by gas chromatography. The average of their recoveries for 10 acetates and alcohols was 86 ± 8.7%, and for four aldehydes, 76 ± 11.7%. The titer was corrected for recoveries.

Mechanistic Model of Evaporative Loss

The experimental observation of a first order evaporative loss from septa suggested the following mechanistic model.

$$C_r \underset{k_2}{\overset{k_1}{\rightleftharpoons}} C_v \xrightarrow{k_3}$$

Here C_v is the concentration of pheromone in the vapor and k_1, k_2, and k_3 are, respectively, the rate constants for evaporation from the septum, recondensation into the septum, and movement of evaporated pheromone to a position in space where it cannot recondense into the septum. The rate of evaporation from a septum is the sum of the rates of these three processes (13); therefore

$$-\frac{dC_r}{dt} = k_1C_r - k_2C_v + k_3C_v \qquad (2)$$

Figure 1. Apparatus for determination of half-life by collection of evaporated pheromone.

The k_2 and k_3 processes compete for C_v. The relative rates of these processes determines the mechanism of release. If $k_3 << k_2$, then C_r and C_v are essentially in equilibrium with each other. At equilibrium $k_1C_r = k_2C_v$. Then

$$-\frac{dC_r}{dt} = k_3C_v$$

Also, because of the pseudo-equilibrium,

$$C_v = \frac{k_1C_r}{k_2} = \frac{C_r}{K}$$

where the equilibrium constant, K, is referred to as the partition coefficient. The rate of evaporation can then be expressed by,

$$-\frac{dC_r}{dt} = \frac{k_3C_r}{K} \qquad (3)$$

k_3/K is equal to the experimentally determined rate constant, k, of Equation 1. Thus, the assumption of a pseudo-equilibrium leads to the experimentally observed result of a first order mechanism of evaporation. If k_3 were similar in magnitude to k_2, then an equilibrium would not exist. Therefore k_1C_r would not equal k_2C_v and as Equation 2 shows, the mechanism would not be first order. If $k_3 >> k_2$, recondensation of pheromone vapor into the septum would be essentially completely suppressed (i.e., $k_2C_v \approx 0$). Then the concentration of the pheromone at the boundary of the septum would be zero and the evaporation rate would be proportional to $t^{-1/2}$ as proposed by Zeoli (15). Therefore, only if a pseudo-equilibrium prevails, will the experimentally observed first order evaporation rate be obtained.

Relationship of Half-Lives to Structure of n-alkyl and n-alkenyl Acetates, Alcohols, and Aldehydes

Because $(\ln 2)/t_{1/2} = k = k_3/K$, $t_{1/2}$ is directly proportional to the partition coefficient, K, of the solute. In gas chromatography the retention time, t_r, is likewise directly proportional to the partition coefficient of the solute. Because the values of t_r and $t_{1/2}$ are each directly proportional to the vapor partition coefficients of solutes, the same structure/rate relationships should hold in each case. In gas chromatography the $\ln t_r$ of most homologous series of compounds is directly proportional to the number of carbon atoms. This fact is the basis for the retention index and carbon number systems for organic compounds (19). Correspondingly, in rubber septa the $\ln t_{1/2}$ of most homologous series of compounds should be directly proportional to the number of carbon atoms:

$$\ln t_{1/2} = aN + b, \qquad (4)$$

where a and b are constants and N is the number of carbon atoms in the n-alkyl group.

For the $\ln t_{1/2}$ versus N to be linear, the incremental contribution of the CH_2 group to the $\ln t_{1/2}$ for each added member of a homologous series must be constant. Series such as 10:Ac-16:Ac, 10:OH-18:OH, and 12:Al-18:Al give linear plots for $\ln t_r$ versus N in isothermal gas chromatography and therefore should give linear plots for $\ln t_{1/2}$ versus N.

For n-alkenyl compounds, the position of the double bond will affect half-lives or GC retention times. Its position relative to both the polar and non-polar end of the molecule is significant. How close the double bond can be to the polar end without mutual interaction occurring is uncertain, but because of carbon bond angles, interaction should be negligible for double bonds at Δ^7 or greater. For series such as Z7-12:Ac, Z9-14:Ac, Z11-16:Ac or Z9-14:Al, Z11-16:Al, Z13-18:Al, the double bond is at a constant position relative to the non-polar end of the molecule (ω^5) and is at Δ^7 or greater relative to the polar end and these series show the linear relationship for $\ln t_r$ versus N and therefore should also give a linear plot for $\ln t_{1/2}$ versus N.

Experimental Test of the Proposed Relationship: Acetates and Alcohols

Butler and I first reported an experimental test of the relationship of Equation 4 for series of compounds which are common components of sex pheromones of Lepidoptera (6, 7). Half-lives were determined for saturated and unsaturated acetates and alcohols by determining the residual content of each compound in septa at time intervals after aging indoors at room temperature for up to 387 days. Half-lives were then calculated from the regression of logarithm of residue content versus time. The range of $t_{1/2}$ values of these materials was startling. Thus, the $t_{1/2}$ of 10:OH was about 2 days and the $t_{1/2}$ of the pheromone component, Z3,Z13-18:Ac was over 8000 days. For the compounds 10:Ac-14:Ac, the regression analysis of $\ln t_{1/2}$ versus N (Equation 4) was linear with a slope, a, of 1.041, a constant, b, of -8.825 and $r^2 = 0.997$. For the alcohols, 10:OH-14:OH, the slope was 1.007, the constant was -9.261 and $r^2 = 0.997$. Thus, these compounds followed the expected linear relationship.

In contrast, the higher molecular weight alcohols and acetates produced anamolous results (6, 7): the $t_{1/2}$ of 16:Ac was less than 15:Ac (481 versus 1353 days) and $t_{1/2}$ of 18:OH was less than 17:OH (609 versus 1117 days). These results with the higher molecular weight compounds are now known to be artifacts. These artifacts occurred because of the lack of temperature control and the long half-lives. Temperature was later found to produce a very substantial effect on half-lives and because of the very long half-lives only a small fraction of the total dose was released. Later Heath and Coworkers (9, 10) and studies of my coworkers and I (14) further substantiated the linear plot of $\ln t_{1/2}$ versus N for all members of the tested homologous series. Heath et al. (10) carried out their determinations at room temperature and reported a slope of 1.01 and a constant of -8.48 for the equation of $\ln t_{1/2}$

versus N for the alkyl acetates and alcohols. My coworkers and I determined $t_{1/2}$ values of acetates (Table I) at controlled temperatures (14). The temperature that best agreed with the previous room temperature data was 20°C. At that temperature the slope was 1.0072 and the intercept was -8.4814 for the alkyl acetates.

Table I. Half-lives and heat of evaporation of acetates from rubber septa

Compound	$t_{1/2}$ in days 20°C	$t_{1/2}$ in days 30°C	ΔH, kcal	Compound	$t_{1/2}$ in days 20°C	$t_{1/2}$ in days 30°C	ΔH, kcal
10:Ac	4.91	1.96	16.19	Z7-12:Ac	26.1	9.33	18.13
11:Ac	13.4	5.01	17.36	Z9-14:Ac	196	61.2	20.47
12:Ac	36.8	12.8	18.53	Z11-16:Ac*	1470	402	22.80
13:Ac	101	32.9	19.70	Z7-10:Ac*	4.44	1.78	16.07
14:Ac	276	84.2	20.87	Z9-12:Ac	33.3	11.7	18.41
15:Ac	755	216	22.04	Z11-14:Ac	249	76.7	20.75
16:Ac	2070	553	23.21	Z13-16:Ac*	1870	504	23.08

* Not determined experimentally; values were determined from regression Equations 4 and 6.

Experimental Test of the Proposed Relationship: Aldehydes

Because aldehydes are not entirely chemically stable in rubber septa, half-lives for evaporative loss can not be determined by the residue method. The faster methods previously mentioned are usable. The half-lives so obtained can be used to formulate mixtures that will produce desired ratios in the vapor from freshly prepared septa, but evaporation rates or ratios of mixtures of aldehydes with other types of compounds would not be predictable at significantly later periods unless the rate of chemical degradation could be estimated.

The first data on half-lives of aldehydes were provided by Heath et al. (10). The aldehydes studied and the half-lives in days at room temperature they reported were: Z9-14:Al, 43.4; 14:Al, 63.1; Z11-16:Al, 327.4; 16:Al, 475.8. My coworkers and I determined six saturated (12:Al-17:Al) and 3 homologous unsaturated aldehydes (Z7-14:Al, Z9-16:Al, Z11-18:Al) at different controlled temperatures (Table II) (18). For these compounds at 20°, the regression of $\ln t_{1/2}$ versus N gave a slope of 1.0417 and a constant of -11.0152. In addition, half-lives of the homologous series, Z9-14:Al, Z11-16:Al, and Z13-18:Al, were determined at one temperature (Table II) (18). For the four compounds in common with those of Heath et al., the half-lives at 20°C were: Z9-14:Al, 28.8; 14:Al, 35.5; Z11-16:Al, 231; 16:Al, 285.

Table II. Half-lives and heat of evaporation of aldehydes from rubber septa

Compound	$t_{1/2}$ in days 20°C	$t_{1/2}$ in days 30°C	ΔH, kcal	Compound	$t_{1/2}$ in days 20°C	$t_{1/2}$ in days 30°C	ΔH, kcal
12:Al	4.42	1.73	16.46	Z7-14:Al	27.6	9.71	18.40
13:Al	12.5	4.61	17.57	Z9-16:Al	222	68.8	20.62
14:Al	35.5	12.3	18.67	Z11-18:Al	1784	488	22.83
15:Al	101	32.7	19.78	Z9-14:Al	28.8	ND	ND
16:Al	285	86.9	20.88	Z11-16:Al	231	ND	ND
17:Al	808	232	21.98	Z13-18:Al	1854	ND	ND
18:Al*	2289	616	23.09				

* Not determined experimentally; values were calculated from regression Equations 4 and 6.
ND = not determined.

These values, while similar in magnitude to those of Heath et al., at room temperature are significantly lower. Why the values of my coworkers and myself differ from those of Heath et al. (10), is not known. However, when they determined recoveries they did not aerate the adsorbent (activated charcoal) as would occur during the actual process of collecting evaporated compounds. It may be that aldehydes breakdown or are irreversibly adsorbed on activated charcoal to a greater extent than acetates or alcohols and then this factor would account for the differences in the two studies.

The slope of the regression of $\ln t_{1/2}$ vs N at 20°C is steeper for aldehydes (1.0417) than for the acetates (1.0072). In principal all homologous series should have the same slope because all are incrementally increased by a methylene group, but this principal is only approximately true. Deviations of similar magnitude in slopes for different functional group series are also known in gas chromatography (20).

The Effect of Double Bonds on Half-Life

The half-lives of the unsaturated compounds were smaller than the half-lives of the corresponding saturated compounds. This effect was expected because natural rubber is non-polar, and on non-polar GC liquid phases unsaturated compounds have shorter retention times than saturated. By broadening the definition of N to equivalent carbon number with the saturated compounds as the reference, monoenes can be correlated with the same regression equation parameters (Equation 4) as the saturated compounds. The N values may then possess fractional values. In the acetate series at 20°C, the N values for the monoenes were as follows (14): Z7-12:Ac, Z9-14:Ac, Z11-16:Ac (ω^5 double bonds), N = 11.66, 13.66, 15.66; Z9-12:Ac, Z11-14:Ac, Z13-16:Ac (ω^3 double bonds), N = 11.90, 13.90, 15.90. For the aldehydes (18) Z7-14:Al, Z9-16:Al, Z11-18:Al (ω^7 double bonds), N = 13.76, 15.76, 17.76 and for Z9-14:Al, Z11-16:Al, Z13-18:Al (ω^5 double bonds), N = 13.81, 15.81, 17.81.

Based on the relationship of GC retention times of unsaturated compounds having different positions of unsaturation, a priori one would expect half-lives to be in the order $\omega^7 < \omega^5 < \omega^3 <$ saturated. This effect is borne out for each homologous series. Also a priori one would expect an ω^5 double bond to produce the same effect in any series, but it is different for the aldehydes (0.19 N units less than the saturated) and acetates (0.34 N units less than the saturated). Although this difference might be real, it should be noted that the standard deviations of the N values overlap each other.

The data on alcohols is not as extensive as for acetates and aldehydes. Based on room temperature data (10), the $\ln t_{1/2}$ versus N for the alcohols has the same slope ("a" of Equation 4) as the acetates, and therefore the average ratio of $t_{1/2}$ of saturated and unsaturated acetates to the corresponding alcohols is constant at 2.10 ± 0.01 (n = 9).

Effect of Temperature on Half-Lives

If the half-lives are to have predictive value, the quantitative effect of temperature must be known. Because evaporation rates are controlled by a pseudo-equilibrium, the rates will be determined by thermodynamic relationships. From Equations 3 and 1,

$$-\frac{dC_r}{dt} = \frac{k_3 C_r}{K} = kC_r = \left(\frac{\ln 2}{t_{1/2}}\right) C_r$$

Consequently, as mentioned earlier, $t_{1/2}$ is directly proportional to the partition coefficient, K. K is equal to C_r/C_v and thus is inversely proportional to the vapor pressure, since C_v/C_r is a measure of vapor pressure. Consequently $t_{1/2}$ is inversely proportional to the vapor pressure. The Clausius-Clapeyron equation relates vapor pressure to temperature, and therefore should also relate half-life to temperature, i.e.:

$$\ln\left(\frac{{}^a t_{1/2}}{{}^b t_{1/2}}\right) = \frac{\Delta H (T_b - T_a)}{R T_a T_b} \quad (5)$$

where ${}^a t_{1/2}$ and ${}^b t_{1/2}$ are the half-lives at temperatures T_a and T_b (°K), respectively, ΔH is the heat of evaporation in calories/mole and R is the gas constant, 1.98 calories/mole-degree. For linear regression, a convenient form of Equation 5 is:

$$\ln t_{1/2} = \frac{\Delta H}{RT} + y_o \quad (6)$$

where $\Delta H/R$ is the slope and y_o is a constant. The Clausius-Clapeyron equation is derived from the laws of thermodynamics and the ideal gas law. Pheromones are obviously not ideal gases, but their vapor pressures are low and all gases approach ideal behavior at low pressures.

The first data on the effect of temperature on half-lives of pheromones in rubber septa were reported for n-alkyl and n-alkenyl acetates (<u>14</u>). The correlations with the Equation 6, were high, ($r^2 > 0.98$). Also, at a given temperature the correlations with Equation 4 were high ($r^2 > 0.99$). The final values of $t_{½}$ were obtained by reiteratively calculating values successively at 3 temperatures from Equation 4 and from Equation 6 until all values agreed with both regression line values. The final values were within the 95% confidence limits of the experimentally determined values. The half-lives so obtained at 20°C and 30°C, and the ΔH values are given in Table I. The ΔH and the half-life at one temperature can be used to calculate half-lives at any temperature by Equation 5.

The effect of temperature on $t_{½}$ was striking. When the temperature changed from 30 to 20°C, $t_{½}$ values changed from 1.96 to 4.91 days (a factor of 2.5) for 10:Ac and from 553 to 2,070 days (a factor of 3.7) for 16:Ac. These results show that male moths responding to their synthetic sex pheromone evaporating from rubber septa will experience substantial variation in evaporation rates.

Because the heat of vaporization, ΔH, for the n-alkyl and n-alkenyl acetates increased with molecular weight, the ratio of pheromone components in the vapor, having different molecular weights, will change with temperature. A somewhat common combination of a 2 component pheromone consists of 2 components separated by 2 carbon atoms in a homologous series such as Z9-14:Ac and Z11-16:Ac. From the half-lives (Table I) and Equation 1, a septum containing 61.2 μg of Z9-14:Ac and 402 μg Z11-16:Ac would produce a ratio in the vapor of 50:50 at 30°C and a modest change to 54:46 in favor of Z9-14:Ac at 20°.

Although the ΔH values of Table I are the only ones reported for these compounds in rubber septa, ΔH values of some of these compounds were reported in the pure state by Olsson et al. (<u>21</u>). The 2 sets of ΔH values are not expected to be exactly the same because the condensed states are not the same. The two condensed states, which consist of a pheromone component dissolved in rubber or in itself, are similar enough, however, so that ΔH values should be of similar magnitude. The ΔH values of Olsson et al. and ours for the given compounds are: 10:Ac, 17.27 and 16.19; Z7-12:Ac, 18.27 and 18.13; Z9-14:Ac, 21.37 and 20.47. Our values are from 0.14 to 1.08 kcal/mole smaller than those of Olsson et al. Thus, these two studies agree as well as should be expected.

Greenway et al. (<u>11</u>) reported that the half-life for E10-12:Ac outdoors for 131 days at an average daily temperature of 14.5 ± 2.5°C for initial doses 1, 3, and 10 mg was 65.2 ± 1.9 days. From Table I and Equation 5, one can calculate the half-lives at 14.5°C for other 12 carbon acetates. Thus, at 14.5°C, $t_{½}$ values in days for the given compounds are: Z7-12:Ac, 47.4; Z9-12:Ac, 61.1; 12:Ac, 67.8. Considering the lack of precise temperatures outdoors and possible differences in air speed outdoors and indoors, these data are in satisfactory and perhaps somewhat fortuitous agreement with that of Greenway et al. (11).

The effect of temperature on half-lives of aldehydes was also determined (<u>18</u>). Again, the correlations with Equation 6 were high

($r^2 > 0.98$), and again the final values were obtained by reiterative calculations. Half-lives at 20° and 30°C and ΔH values are reported in Table II.

If the effect of temperature on half-lives is to be used to predict evaporation rates in field tests, the actual temperatures of the test septa must be known. Temperatures of natural rubber septa have not been reported in field studies, but temperatures of natural rubber discs (2.8 cm diameter by 1.2 cm thick) unshaded, shaded, and in Pherocon IC monitoring traps in an apple tree have been determined (22). On sunny afternoons the unshaded discs typically superheated 8°C above ambient, while the shaded discs were only about 1° above ambient. The discs in the traps were always at ambient temperature. Thus, temperatures of septa used in Pherocon IC monitoring traps can be taken as ambient, while those in more exposed positions as may occur with other trap types and for those used for mating disruption by air permeation should be monitored, if meaningful predictions of evaporation rates are to be made. With modern data logers, temperature monitoring is convenient.

Additivity of Evaporation Rate Parameters

If the effect of each chemical group on $\ln t_{1/2}$ or ΔH values were additive, all half-lives of the studied classes of compounds could be calculated from a few parameters. Although additivity does not strictly hold (because "a" values of Equation 4 are slightly different for aldehydes and acetates), rules can be developed that will generate $t_{1/2}$ values that generally deviate by 20% or less from the most accurate values. Because of the large change of $t_{1/2}$ with chemical structure (650% for 2 carbon atoms) and temperature (150-270% for 10°C), deviations of this magnitude are acceptable. The rules and parameters for calculating the $t_{1/2}$ values are summarized in Table III. The value of ΔH of the OH group (Section B2d of Table III) was determined from two compounds (Z7-12:OH and Z11-16:OH) (18). Also, the hydrocarbon rule (Section A3, Table III) is based on 2 comparisons of relative half-lives at 21°C: $t_{1/2}$ ratio of 14:H/10:Ac = 0.98; 20:H/16:Ac = 1.03 (5).

From Table III one can estimate $t_{1/2}$ values for monoenes with double bond positions not listed. For example, for Z8-12:Ac and Z8-12:OH (components of the sex pheromone of the oriental fruit moth), one can average the Table III parameters for the ω^3 and ω^5 positions. Then $t_{1/2}$ at 20° is 30.0 for Z8-12:Ac and 14.3 for Z8-12:OH; and ΔH is 18.15 kcal/mole for Z8-12:Ac and 17.01 kcal/mole for Z8-12:OH.

Effect of Air Speed on Half-Life

There are only a few reports in the literature on the effect of air speed on evaporation rates of pheromone components from rubber septa. Teal et al. reported a linear change of evaporation rate of the sex pheromone of *Heliothis virescens* with air speed over an air speed change of 4-fold (23). Also, in a study of behavioral response of moth species to their sex pheromones, Landolt and Heath (24) reported that the evaporation rates of Z7-12:Ac at two

Table III. Tabulation of rules for calculating $t_{1/2}$ values of Lepidoptera pheromones in rubber septa at an air speed of 20 cm/sec

A. To Determine $t_{1/2}$ Values in Days at 20°C

1. The reference point is $t_{1/2}$ of 14:Ac = 281
2. For each carbon atom added to or subtracted from the longest continuous chain of 14:Ac, multiply or divide the $t_{1/2}$ of 14:Ac by 2.79.
3. $t_{1/2}$ of X:Ac = $t_{1/2}$ of (X + 2):Al = $t_{1/2}$ of (X + 4):H where X, (X + 2), and (X + 4) are the number of carbon atoms in the longest continuous chain of each compound type.
4. For Z-double bonds, multiply: 0.90 x $t_{1/2}$ of saturated for ω^3; 0.76 x $t_{1/2}$ of saturated for ω^5; 0.70 x $t_{1/2}$ of saturated for ω^7.
5. For alcohols, divide the $t_{1/2}$ value for the corresponding acetate by 2.1.

B. To Determine $t_{1/2}$ Values in Days at Another Temperature

1. Use the $t_{1/2}$ values at 20° and the Clausius-Clapeyron equation:

$$\ln\left(\frac{{}^{a}t_{1/2}}{{}^{b}t_{1/2}}\right) = \frac{\Delta H(T_b - T_a)}{RT_aT_b}.$$

2. To obtain ΔH, add:
   a. 1.14 kcal/carbon atom in the alkyl or alkenyl group.
   b. 3.07 kcal for the aldehyde oxygen.
   c. 4.65 kcal for the acetate group.
   d. 3.51 kcal for the alcohol hydroxyl.
   e. For double bonds, subtract: 0.12 kcal for ω^3; 0.25 kcal for ω^5; 0.30 kcal for ω^7.

different concentrations increased linearly with air speed over a speed change of 4-fold. The slopes of the straight lines were different for the two concentrations.

These observations can be explained by the mechanistic model as expressed in Equation 3. k_3 is the rate constant for the movement of evaporated pheromone to a position in space where it cannot recondense into the septum. The value of k_3 will depend on the rate of diffusion of the pheromone components, and on the speed of air movement. One would expect an air speed of 20 cm/sec to significantly exceed the rate of diffusion. Then to a good approximation, k_3 is proportional to air speed:

$$k_3 = mS + n$$

where S is air speed, and m and n are constants. Therefore, from Equation 3:

$$-\frac{dC_r}{dt} = \frac{C_r mS}{K} + \frac{C_r n}{K}, \text{ and}$$

a plot of evaporation rate, $-dC_r/dt$, versus air speed, S, will give a straight line with a slope of $C_r m/K$ and an intercept of $C_r n/K$. If plots of evaporation rate versus air speed are made for one compound at two different values of C_r, the ratio of the slopes and the ratio of the intercepts will be equal to the ratios of the C_r values. The study of Landolt and Heath (24) was conducted at doses of 60 and 400 µg (ratio equal to 6.7); the ratio of the slopes was 5.1, and the ratio of intercepts was 7.2. Thus, the effect was in accordance with that predicted from the mechanistic model and the theoretical and actual ratios are probably within experimental error of each other.

When $t_{½}$ values are determined in a laboratory, the rubber septum will usually be contained in some type of glass container. Condensation and re-evaporation from the inside surfaces of the glass container can affect evaporation rates and calculated $t_{½}$ values (13). The relevant factors can be represented as follows.

$$C_r \underset{k_2}{\overset{k_1}{\rightleftharpoons}} C_v \xrightarrow{k_3} C_g, \quad C_g \xrightarrow{k_4} C_v \tag{11}$$

Here C_g is the concentration of pheromone on the glass surface, and k_4 is the rate constant for evaporation from the glass surface. The rate equation becomes:

$$-\frac{dC_r}{dt} = \frac{k_3 C_r}{K} - k_4 C_g \tag{12}$$

Initially $C_g = 0$ and will increase with time. As a result, the evaporation rate will not be proportional to C_r and the degree of disproportionality will increase with time, i.e., the longer the experimental determination of $t_{½}$ is run, the longer will be the calculated $t_{½}$. Therefore, this situation is similar to that in an enclosed container without air flow even though there is some air flow. For determinations of $t_{½}$ to adequately represent half-lives outside the glass container, air flow must be high enough to maintain C_g at negligible levels.

Limitations on Uses of Natural Rubber Septa

If half-lives are very long, the dose needed to produce desired evaporation rates in monitoring traps may exceed the practical absorption limit of the septa of about 50-70 mg. Also, even if the absorption limit is not exceeded, 50-70 mg is wasteful of an expensive pheromone component. If half-lives are very short, the evaporation rate will be changing so rapidly that the desired range of evaporation rates will be too short lived. Thus, the rubber septum is not an ideal CRS for 18 carbon acetates ($t_{½}$ at 20° = 8,000-15,000 d) nor for 10 carbon acetates and alcohols ($t_{½}$ at 20°, ca. 5 d and 2 d, respectively). Rubber septa generally provide very desirable release rates for compounds with volatilities in the ranges 12:Ac-16:Ac, 12:OH-16:OH, 14:Al-18:Al, and 16:H-20:H.

Heath et al. (10) suggested that some crystalline compounds may give anomolous release rates. Subsequently, my coworkers and I found that 16:OH crystallizes from septa leaving a visible surface deposit and the release rate was faster than would be expected from the regression line of Equation 4 based on the lower members of this series (18). Tetradecanol did not show this effect. Since many E isomers of hexadecen-1-ol are crystalline, these compounds may also show this effect. Fortunately most Lepidopteran sex pheromones of 16 carbon alcohols have the Z configuration and do not crystallize from septa.

Saturated and monoene alcohols and acetates are quite stable in rubber septa. In general in traps in field applications, they neither hydrolyse nor oxidize (5, 25). However, several studies showed that conjugated dienes isomerize in rubber to ultimately form an equilibrium mixture (25-28). At equilibrium the content is 65-70% EE, 2-5% ZZ and the percentages of EZ and ZE are comparable to each other (27, 28). The exact values depend on the position of the double bonds. Because non-pheromone isomers may decrease or prevent trap catch, the useful life of a lure may be determined by this factor. Complete equilibration only requires about a month (27, 28). The reason for the rapid isomerization appears to be a combination of sunlight and catalysis by the sulfur used to cure the rubber (26-28). An alternate synthetic-elastomeric septum is available which reduces the rate of isomerization by about 8-fold (28).

Aldehydes are not entirely stable in rubber septa. Although chemical breakdown of aldehydes in rubber septa has not been reported, loss of aldehydes from septa as determined by the residue method is faster than loss as determined by the rate of evaporation, thereby indicating some form of chemical degradative loss (5). The known tendency of aldehydes to trimerize has been documented in polyethylene vials (29). Also aldehydes oxidize in air to carboxylic acids and other degradation products. Certain aldehydic pheromone components have been shown to form carboxylic acids and other oxidation products in hexane under fluorescent light (30). Aldehydes have been reported to persist for as long as 10 weeks in rubber septa in traps based on insect response (31), but significant degradative loss had probably occurred. Aldehydes react with amines and at least one manufacturer of rubber septa adds amines as stabilizers of the rubber. The loss of lure effectiveness from this factor is well documented (32, 33). The amines can be removed by pre-extracting the septa with an organic solvent (32).

Applications of Rubber Septa

For some insects, rubber septa have potential use for population control by permeating the air with pheromone. For many crops it is economically practical to place the CRS on the foliage by hand labor. Rubber septa could be used for such applications if the half-lives of the pheromone components were suitable for the intended use. An equation relating the concentration of pheromone (i.e., the amount applied per unit area), C, needed for control as a function of the first order rate constant, k, has been reported (34). When the minimum evaporation rate per unit area required for mating disruption is E_m, and the required period of control from one application is t, the equation is:

$$C = \frac{E_m \exp(kt)}{k}$$

$$\text{or since } k = \frac{\ln 2}{t_{1/2}},\ C = \frac{E_m t_{1/2} \exp(t t_{1/2}^{-1} \ln 2)}{\ln 2} \qquad (13)$$

In Equation 13, the value of C as a function of $t_{1/2}$ has a minimum. At the minimum:

$$t_{1/2} = t\ln 2 \approx 0.7t$$

$$C = eE_m t \approx 2.7\ E_m t$$

where e is the base of natural logarithms. Thus, C is minimized when the half-life is 0.7t. For CRS providing half-lives shorter or longer than 0.7t more pheromone is required, but the increased requirement is greater for slightly shorter than slightly longer half-lives. For t values of 30-100 days, $t_{1/2}$ values of 21-70 days would minimize pheromone useage and somewhat longer $t_{1/2}$ values would also be practical. It is obvious from Tables I, II, and III that many pheromone components in rubber septa, even at the temperatures expected from direct exposure to sunlight, could be efficiently used in mating disruption programs.

The principal practical application of rubber septa as a CRS for sex pheromones has been in traps used to monitor flight activity of insect pests. Rubber septa have also been used as a research tool to study behavioral responses of insects to pheromones (35, 36). With the information now available, it is possible to specify blends of pheromone components which will produce desired evaporation rates and vapor ratios, and with known changes of these rates and ratios with time, temperature and air speed for a large number of insects. Thus, more effective use of septa in developing monitoring lures and in behavioral studies is possible. Also, in the future, it is to be expected that applications of septa for mating disruption will be found.

Literature Cited

1. Charmillot, P. J. In Practical Use of Insect Pheromones and Other Attractants; Ridgway, R.; Silverstein, R. M.; Inscoe, M., Ed.; Marcel Dekker, In press, Chapter 11.
2. Audemard, P. H.; Leblon, C.; Neumann, U.; Marboutie, G.; J. Appl. Entomol. 1989, 108, 191-207.
3. Maitlen, J. C.; McDonough, L. M.; Moffitt, H. R.; George, D. A.; Environ. Entomol. 1976, 5, 199-202.

4. Golub, M.; Weatherston, J.; Benn, M. A. J. Chem. Ecol. 1983, 9, 323-33.
5. Butler, L. I.; McDonough, L. M. Unpublished.
6. Butler, L. I.; McDonough, L. M. J. Chem. Ecol. 1979, 5, 825-37.
7. Butler, L. I.; McDonough, L. M. J. Chem. Ecol. 1981, 7, 627-33.
8. Leonhardt, B. A.; Moreno, D. S. In Insect Pheromone Technology: Chemistry and Applications; Leonhardt, B. A.; Beroza, M., Ed.; ACS Symposium Series No. 190: Washington D.C., 1982; Chapter 8.
9. Heath, R. R.; Tumlinson, J. H. J. Chem. Ecol. 1986, 12, 2081-88.
10. Heath, R. R.; Teal, P. E. A.; Tumlinson, J. H.; Mengelkoch, L. J. J. Chem. Ecol. 1986, 12, 2133-43.
11. Greenway, A. R.; Davis, S. A.; Smith, M. C. J. Chem. Ecol. 1981, 7, 1049-56.
12. Baker, T. C.; Carde, R. T.; Miller, J. R. J. Chem. Ecol. 1980, 6, 749-58.
13. McDonough, L. M.; Butler, L. I. J. Chem. Ecol. 1983, 9, 1491-1502.
14. McDonough, L. M.; Brown, D. F.; Aller, W. C. J. Chem. Ecol. 1989, 15, 779-90.
15. Zeoli, L. T.; Kydonieus, A. F.; Quisumbing, A. R. In Insect Suppression with Controlled Release Pheromone Systems; Kydonieus, A. F.; Beroza, M., Ed.; CRC Press: Boca Raton, Florida, 1982; Chapter 3.
16. Roseman, T. J.; Cardarelli, N. F. In Controlled Release Technologies: Methods, Theory, and Applications; A. F. Kydonieus, Ed.; CRC Press: Boca Raton, Florida, 1980; Chapter 2.
17. Daterman, G. E. USDA Forest Service Research Paper 1974, PNW-180.
18. McDonough, L. M.; Brown, D. F.; Aller, W. C. Unpublished.
19. Dal Nogare, S.; Juvet, Jr., R. S. Gas-Liquid Chromatography: Theory and Practice; Interscience Publishers: New York, New York, 1962; Chapter 3.
20. McReynolds, W. O. Gas Chromatographic Retention Data; Preston Technical Abstracts: Evanston, Illinois, 1966.
21. Olsson, A.; Jonsson, J. A.; Thelin, B.; Liljefors, T. J. Chem. Ecol. 1983, 9, 375-85.
22. Davis, H. G.; Brown, D. F.; McDonough, L. M. Unpublished.
23. Teal, P. E. A.; Tumlinson, J. H.; Heath, R. R. J. Chem. Ecol. 1986, 12, 107-26.
24. Landolt, P. J.; Heath, R. R. J. Chem. Ecol. 1987, 13, 1005-18.
25. Shani, A.; Klug, J. T. J. Chem. Ecol. 1980, 6, 875-81.
26. Fugiwara, H.; Sato, Y.; Nishi, K. J. Takeda Res. Lab. 1976, 35, 52-59.
27. Davis, H. G.; McDonough, L. M.; Burditt Jr., A. K.; Bierl-Leonhardt, B. A. J. Chem. Ecol. 1984, 10, 53-61.
28. Brown, D. F.; McDonough, L. M. J. Econ. Entomol. 1986, 79, 922-7.
29. Dunkeblum, E.; Kehat, M.; Klug, J. T.; Shani, A. J. Chem. Ecol. 1984, 10, 421-8.

30. Shaver, T. N.; Ivie, G. W. J. Agric. Food Chem. 1982, 30, 367-71.
31. Flint, H. M.; Butler, L. I.; McDonough, L. M.; Smith, R. L.; Forey, D. E. Environ. Entomol. 1978. 7, 57-61.
32. Steck, W. F.; Bailey, B. K.; Chisholm, M. D.; Underhill, E. W. Environ. Entomol. 1979, 8, 732-3.
33. Kamm, J. A.; McDonough, L. M. Environ. Entomol. 1979, 8, 773-5.
34. McDonough, L. M. U.S. Dept. of Agriculture/Science and Education Administration/Agricultural Research Results 1978, ARR-W-1/May.
35. Grant, A. L.; Mankin, R. W.; Mayer, M. S. Chem. Senses. 1989, 14, 449-62.
36. Baker, T. C.; Hansson, B. S.; Lofstedt, C.; Lofquist, J. Chem. Senses. 1989, 14, 439-48.

RECEIVED May 16, 1990

MECHANISMS OF PLANT RESISTANCE TO INSECTS

Chapter 9

Potato Glandular Trichomes

Defensive Activity Against Insect Attack

Ward M. Tingey

Department of Entomology, Cornell University, Ithaca, NY 14853–0999

Insect-resistant cultivars offer a realistic and practical foundation upon which to build an economical and environmentally sound crop protection system. Insect resistance of a wild Bolivian potato species is conditioned by the presence of glandular trichomes on its foliage. Resistance is expressed both mechanically and behaviorally. Mechanical resistance involves release of a viscous exudate from the trichome gland upon contact by an insect. The trichome secretions accumulate initially on tarsi, first impeding movement and later entrapping the insect. In addition to immobilization, entrapment and resultant mortality, glandular trichomes dramatically alter normal insect behaviors, particularly those involving host acceptance and feeding. Pest species adversely affected by glandular trichomes include the Colorado potato beetle, green peach aphid, potato leafhopper, potato flea beetle, and spider mites. The mechanical and behavioral defensive properties of glandular trichomes are conditioned by the presence of specialized chemistry in trichome exudate. The Cornell University potato breeding program has introgressed the genetic information for expression of glandular trichomes into germplasm of the cultivated potato. Commercially acceptable levels of resistance have been maintained in 8 successive breeding cycles under development for horticultural adaptation. Elite hybrids can experience greater than 85% reduction in populations of insect pests, compared to those on susceptible commercial cultivars. Insecticide use on resistant hybrids can be reduced by at least 40% of that presently needed on susceptible cultivars.

To-date, over 180 wild, tuber-bearing species of *Solanum* are known (1) and while many of these have been screened for insect resistance over the past ten years, the first large-scale systematic efforts to examine wild *Solanum* germplasm for insect resistance were initiated by E. B. Radcliffe at the University of Minnesota and reported in a series of papers first appearing in 1968 (2). These studies led to the identification of several excellent sources of resistance to aphids and leafhoppers. Prior to this time, very few species were known to resist attack by

0097–6156/91/0449–0126$06.00/0
© 1991 American Chemical Society

major potato insects and in most cases, the underlying plant phenomena conferring resistance were unknown although the presence of steroid glycosides in the foliage of some species had been associated with resistance as early as 1950 by European workers (3-7).

One of the wild potato species Radcliffe reported resistant to the potato leafhopper had been described in 1944 from a collecting expedition in Bolivia. J. G. Hawkes, the British potato systematist, described this species, *Solanum berthaultii*, as follows (8):

> "Very glandular species with pale violet-blue pentagonal sub-stellate corolla. Almost certainly formed as a natural hybrid between *S. tarijense* (Commersoniana) and a blue-flowered mountain species, possibly *S. sparsipilum*. Distribution: Bolivia, eastern slopes of Andes in rather dry valleys amongst bushes and in waste places from 2,100 - 2,700 m."

In 1971, R. W. Gibson at the Rothamsted Experimental Station reported that *S. berthaultii* was resistant to aphids and furthermore that resistance was associated with the presence of glandular trichomes (9). Subsequent work in Britain, the U.S., and Peru has expanded the list of pest species against which *S. berthaultii* is defended. Currently, this species is known to resist herbivory by at least 10 major groups or species of pests including leaf miner flies, the potato tubermoth complex, aphids, leafhoppers, fleabeetles, mites , and the Colorado potato beetle (10-13). In 1977, research was initiated at Cornell University to systematically explore the nature of insect resistance in this wild potato species and to examine its potential value in management of insect pests.

Glandular Trichomes of *Solanum berthaultii*

Resistance of *S. berthaultii* to insects and mites is associated with the presence of two types of glandular trichomes on foliage, fruiting organs, and stolons. Type A trichomes are about 120 to 210 mu in length and bear a membrane-bound tetralobulate gland about 50 to 70 mu in diameter at their apices (14). Release of secretory material occurs upon mechanical breakage of the gland at its junction with the stalk. The tetralobulate gland is not renewed following rupture.

Type B trichomes, in contrast, are longer (ca. 600 to 950 mu in length) and bear an ovoid droplet of exudate (20 to 60 mu in diameter) at their tips (14). The naked droplets of exudate are extremely adhesive and transfer readily upon contact by an insect. Following mechanical or solvent removal, droplets of type B exudate are renewed to their original dimensions within 8 days (15).

Defensive Biology of *Solanum berthaultii*

Many species of small-bodied pests experience classical responses to host resistance phenomena during an encounter with foliage of *S. berthaultii*. These include host avoidance and restlessness, reduced feeding, delayed development, reduced adult weight and reproductive performance, increased mortality, and diminished longevity. In several cases, these responses are associated with entrapment or immobilization by trichome exudate.

Entrapment/Immobilization. Aphids and leafhoppers experience significant mortality by entrapment in trichome exudate. Initial contact with the foliage leads to an encounter with the tall Type B hairs which confer a high degree of adhesiveness to the tarsi (Fig. 1). Tarsi coated with this viscous material are very effective in

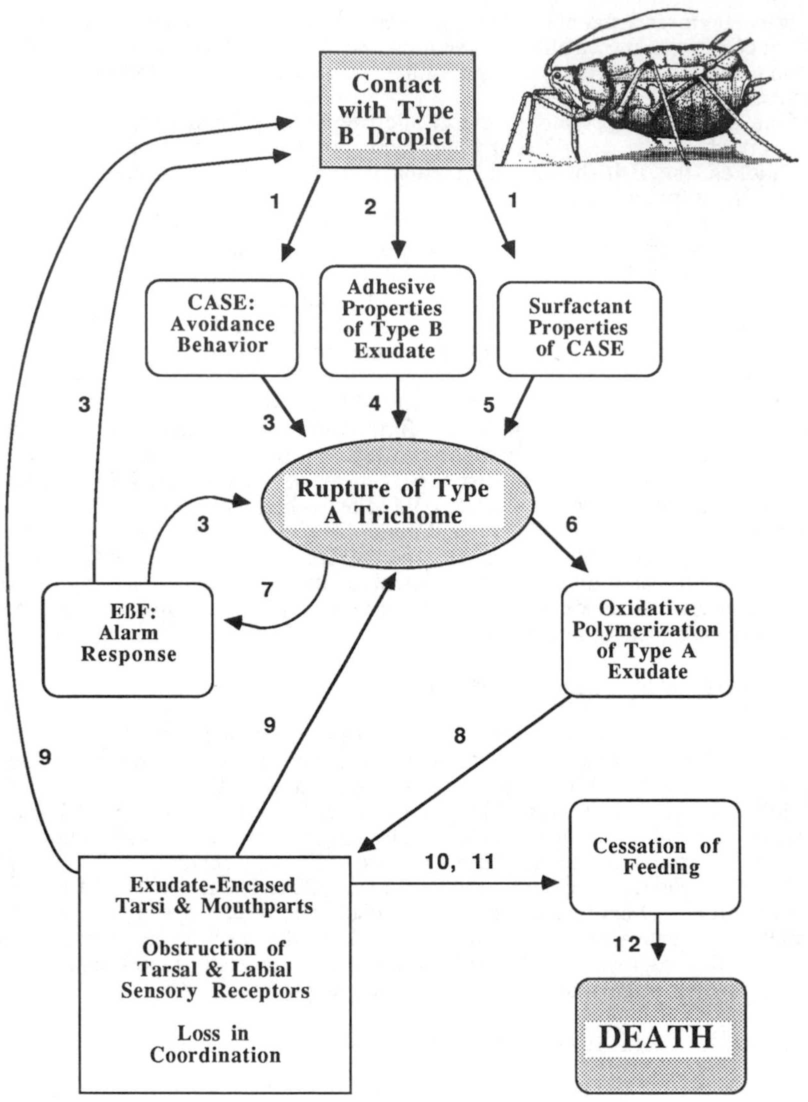

Figure. 1. Nature of glandular trichome-mediated aphid resistance in *S. berthaultii*. Key: (1) Carboxylic acid sucrose esters (CASE), (2) Viscous type B trichome exudate, (3) Increased aphid movement and attempts to escape, (4) Adhesive aphid tarsi, (5) Enhanced rupture of type A trichome membrane?, (6) Polyphenoloxidase + O2 + substrate, (7) Aphid alarm pheromone, E-(ß)=farnesene, (8) Encasement of tarsi by trichome exudate, (9) Greater effective tarsal size, (10) Decreased aphid mobility, (11) Occlusion of mouthparts by trichome exudate, (12) Starvation and death. Adapted from (38).

breaking the junction between the head and stalk of a Type A trichome thus resulting in release of Type A exudate onto the insect's body (16). Within several hours, the exudate darkens and hardens by virtue of its phenolic oxidation chemistry, immobilizing these insects or severely restricting their movements. The dispersal rate of aphids with exudate-encased tarsi on *S. berthaultii* and on hybrids is only 1/3 that of aphids with unencased tarsi on nonglandular plants (17). Trichome exudate also accumulates on the mouthparts and may totally occlude the stylets of insects with sucking mouthparts, thus preventing feeding (18). The defensive activity of *S. berthaultii* foliage increases along with the densities of both types of trichomes and with increased volume of Type A trichome glands. Young aphids and leafhoppers experience greater mortality from an encounter with glandular trichomes than do adults, probably because of a more limited ability to escape the viscous and adhesive trichome barrier (16).

Behavioral and Sensory Disturbance. In addition to entrapment and immobilization, the glandular trichomes of *S. berthaultii* interfere with host acceptance and condition an avoidance for oviposition and feeding. Agitation and avoidance by aphids is conferred by a complex of sesquiterpenes located primarily in Type A trichomes and released when the gland is broken (19, 20). The aphid-immobilizing potential of trichome exudate is magnified by the presence of these compounds because the resulting behavioral excitation and increased locomotion promote a greater frequency of encounters with undisturbed Type A trichomes. Interestingly, Type A trichomes of some commercial potato cultivars contain high levels of the aphid alarm pheromone, E-ß-farnesene (20). However, aphid acceptance and feeding behavior on these plants are not adversely affected because the tetralobulate glands on foliage of the cultivated potato do not readily rupture upon contact.

Type B trichome exudate of *S. berthaultii* also conditions abnormal behaviors in aphids and leafhoppers, particularly delay in host acceptance as measured by an increase in time to the first probe and a decrease in feeding time (Table I) (21, 22). The avoidance/deterrance responses of aphids are conditioned by the presence of sucrose esters of short-chain branched carboxylic acids in the type B exudate (Table II) (23, 24). Sensory receptors on the tarsi and/or antennae of aphids are the likely target sites of sucrose esters in type B exudate (24).

Table I. Effect of type B trichome removal on feeding behavior of fourth-instar potato leafhopper nymphs on resistant and susceptible potato clones. Adapted from (22)

Clone	Type B trichomes	Preprobe time (min)	Total feeding time (min)	% of nymphs feeding
S. tuberosum X *S. berthaultii* (F_3)				
	Intact	14.2 a	5.7 b	25 b
	Removed	11.2 b	5.3 b	95 a
S. tuberosum				
	Intact	1.2 c	21.0 a	100 a
	Removed [a]	0.6 c	16.9 a	100 a

[a] A glass microscope slide was gently pressed to the leaf surface to simulate the conditions used to remove type B exudate from the F_3 hybrid.

Coupled with the consequences of entrapment by Type A exudate, (i.e. high mortality, limited population development, and the disabling influence of trichome exudates on locomotion), the avoidance behaviors conditioned by sesquiterpenes in type A trichomes and by sucrose esters in type B trichomes contribute to the reduced vector efficiency of the green peach aphid for potato virus Y as measured by impaired acquisition and transmission ability (Table III) (25, 26). This interplay of several trichome defensive phenomena adds further value to the plant breeding attributes of *S. berthaultii* because it provides an additional barrier of disease protection for hybrid germplasm carrying genetic resistance to the potato virus Y pathogen.

Table II. Deterrence of green peach aphid feeding by sucrose esters from type B glandular trichomes of *S. berthaultii* (PI 473331) applied to diet membranes. Adapted from (24)

Concentration $\mu g/cm^2$	No. feeding sheaths/mm^2 Treated	Untreated	*P* [a]
100	0.21	2.24	0.0001
33	0.38	3.36	0.0001
10	1.31	1.77	0.06
3	1.39	1.17	0.33
1	1.66	1.30	0.15

[a] Probability that sucrose esters have no effect on the distribution of feeding sheaths by G-test, goodness of fit.

Table III. Final potato virus Y (PVY) incidence as a percent of exposed target plants for four source plant/target plant combinations of *S. tuberosum* and *S. berthaultii* (PI 310927). Target plants exposed to green peach aphids for 4 weeks. Adapted from (26)

Source plant/target plant	% PVY incidence	S.E.	N
Tuberosum/Tuberosum	39	3.67	3
Tuberosum/Berthaultii	30	9.15	2
Berthaultii/Tuberosum	17	4.16	2
Berthaultii/Berthaultii	6	2.78	3

Nature of Resistance to the Colorado Potato Beetle. Over the past two decades, the Colorado potato beetle, *Leptinotarsa decemlineata* (Say), has become the major limiting factor in potato production in the northeastern and mid-Atlantic regions of the U.S. (27) and is an increasingly serious constraint in Europe. The magnitude of the problem has stimulated considerable interest in *S. berthaultii* as a source of resistance to this pest. When confined on *S. berthaultii*, both adults and larvae accumulate type A and B exudate on their tarsal pads and claws but they do not become entrapped nor does their mobility appear affected (28, 29). Rather, the expression of resistance against the Colorado potato beetle is more subtle and characterized initially by abnormal behaviors. Females display a marked reluctance to accept, feed, and oviposit on *S. berthaultii*. Numbers of eggs per egg mass and, indeed, total fecundity on *S. berthaultii* are typically less than 1/2 that on commercial cultivars of potato (13, 30-33). Neonates experience significant mortality within 72 h of confinement on *S. berthaultii* because of starvation (29). This reluctance to accept *S. berthaultii* foliage for food can be largely eliminated by mechanical and/or solvent removal of glandular trichomes (Table IV). Our most recent findings indicate that the presence of type A trichomes is a fundamental requirement for expression of *S. berthaultii* resistance to neonate *L. decemlineata*. Type B droplets containing sucrose esters increase the expression of resistance in the presence of defensively-active type A trichomes (29). However, we have been unable to demonstrate a density-dependent relationship between the two types of glandular trichomes and levels of resistance.

Table IV. Effect of removing type A (wipe) and type B (methanol dip) glandular trichomes of *S. berthaultii* (PI 310927) on feeding, growth, and mortality of neonate Colorado potato beetle. Adapted from (29)

Treatment	N	% Feeding	% Mortality	Weight (mg)
None	65	40 a	63 a	1.6 a
Wipe	50	60 ab	48 ab	1.9 a
MeOH dip	57	72 ab	32 bc	1.2 a
MeOH dip and wipe	39	90 cd	15 cd	1.4 a

abcd. Values followed by the same letter are not significantly different ($p > 0.05$) by R X C test of independence using the G statistic for feeding and mortality and analysis of variance, LSD for weight.

Such subtle effects of the trichome barrier might explain our recent finding that feeding is a much smaller portion of the activity budget of Colorado potato beetle larvae on *S. berthaultii* than on non-glandular susceptible cultivars (Table V) (34) and thus provide an explanation for several of the major impacts of resistance on this pest, i.e. decreased food consumption leads to the reduced growth rates, retarded development, and their cumulative suppressive effects on survival, fecundity, and population dynamics.

Genetic Manipulation of *S. berthaultii* and Variety Development

The breeding program at Cornell University to introgress genetic information for insect resistance of *S. berthaultii* was initiated in 1977 by crossing selected clones of *S. berthaultii* (as males) with several tetraploid (2n=48) clones of *S. tuberosum* (35-37). The successful production of tetraploid hybrids between these two species results from unreduced male gamete production in diploid (2n=24) *S. berthaultii*. All subsequent breeding has been with tetraploids. The F2 generation was produced by random intermating of the F1 generation. The F3 and F4 generations were produced by intercrossing progenies selected for presence of type B droplets and for horticultural adaptation. The F3 generation also included outcrossing to selected F2's of *S. andigena* x *S. berthaultii* to introduce genes for resistance to potato virus Y. In subsequent generations, the breeding method has involved a system in which back-crossing to *S. tuberosum* has been alternated with intercrossing within the hybrid population. Parents used in these latter generations had high densities of both types of trichomes, large Type B droplets and the presence of sucrose esters in type B exudate, high levels of trichome phenolic oxidation activity, insect, disease and nematode resistance, and horticultural adaptation. We are currently 8 generations advanced beyond the F1.

Table V. Activity of neonate Colorado potato beetle larvae on excised leaflets of either *S. tuberosum*, *S. berthaultii* PI 310927, or PI 310927 from which most of the glandular trichome exudate had been removed by wiping between tissue papers. Adapted from (34)

Host plant	n	% of larvae in contact with leaflet	No. of larvae feeding as % of: Total no. larvae	No. larvae on leaflet
S. tuberosum	16	96 a	12.5 a	13.1 a
S. berthaultii				
PI 310927 (wiped)	16	79 b	8.5 b	11.1 a
PI 310927	16	74 b	1.0 c	1.7 b

Means within the same column followed by the same letter not significantly different in multiple paired t tests at $\alpha = 0.017$ (experimentwise $\alpha = 0.05$).

The inheritance of glandular trichomes in *S. berthaultii* and in crosses with *S. tuberosum* has been studied by several groups (11, 35-37). Type B trichome density, droplet size and presumably the presence of sucrose esters appear to be controlled by relatively few recessive genes. Heritability estimates ranged from 20-30% for density of Type A and B trichomes and 60% for Type B droplet size. The presence of polyphenoloxidase in type A trichomes is controlled by a single dominant gene (38). The inheritance of sesquiterpenes in type A trichomes has not been studied.

Recent studies at Cornell indicate that with the exception of Type B trichomes, none of the trichome insect-resistance traits being selected in our breeding program has any deleterious associations with horticultural adaptation. In the case of type B trichomes, their presence in hybrid clones is associated with reduced yielding ability, late maturity, and other characteristics unique to the wild parent (39). Alternate breeding schemes and somaclonal variation are being explored to determine whether gene linkage is responsible for this undesirable association (39, 40).

Potential Pest Management Applications

Clones selected from these hybrid populations have demonstrated excellent levels of resistance to aphids, leafhoppers, and the Colorado potato beetle in field and laboratory studies. We have demonstrated season-long reduction in aphid populations on hybrid clones of up to 60%, compared with populations on commercial cultivars (17, 41). We have also demonstrated population reductions of leafhopper adults and nymphs of over 80% on hybrid clones, compared to those on commercial susceptible potato cultivars (22). This level of resistance eliminates the need for insecticides in management of leafhoppers. As for the Colorado potato beetle, we have documented the following negative impacts by resistant clones compared to commercial cultivars: 30% delay in time to 1st egg laying, 80% reduction in total egg production, 20% increase in larval development time, and 25% reduction in adult weight (Table VI). Under field conditions, these impacts translate into a nearly 90% reduction in densities of second generation larvae with accompanying protection from defoliation. Insecticide use on resistant hybrids can be reduced by at least 40% of that presently needed on susceptible cultivars ($80-160 per acre per growing season) (32).

Although glandular trichomes can interfere with predators and parasitoids of aphids and the Colorado potato beetle, the inhibitory effect is largely associated with entrapment by type B exudate and is minimized on plants bearing moderate densities of these trichomes (42, 43).

Table VI. Vital Statistics of Colorado Potato Beetle on *S. berthaultii and S. tuberosum*. Adapted from (31)

	S. berthaultii	*S. tuberosum*
Age at 1st Egg Laying (days)	46	31
Total Eggs per Female	351	2063
Larval Development Time (days)	14.3	11.2
Female Pupal Weight (mg)	104	138
Female Adult Weight (mg)	88	105

Future Outlook

Our experience with *S. berthaultii* indicates that it is a useful source of insect resistance for genetic improvement of the potato. Major attributes of this resistance include its broad-spectrum nature and the diversity of its mechano-chemical impact across a range of pest species. However, insects as a group have demonstrated a remarkable ability to adapt to stress. And, in fact, the durability of some types of host resistance is not appreciably greater than that of insecticides. We believe, however, that adaptation by pests to the multiplicity of negative impacts conditioned by glandular trichomes is likely to be a lengthy process requiring substantial genetic changes involving behavior and morphology. For this reason, glandular trichome-mediated host resistance to such pest species is likely to be more stable than that conditioned by the presence of a single toxic or deterrent allelochemical.

Acknowledgments

I am indebted to the many present and former Cornell University academic personnel contributing to the research efforts described above and with whom I've been fortunate to interact: Dick A. Avé, Pierre Y. Bouthyette, Michael B. Dimock, Felix H. França, Joseph J. Goffreda, Peter Gregory, Robert Hoopes, Julio C. Kalazich, Stanley P. Kowalski, Stephen L. Lapointe, Zaida Lentini-Gil, Shawn A. Mehlenbacher, Jonathan J. Neal, John J. Obrycki, Robert L. Plaisted, John R. Ruberson, James D. Ryan, John C. Steffens, Maurice J. & Catherine A. Tauber, Robert G. Wright, and G. Craig Yencho. This research was supported by the International Potato Center and the United States Department of Agriculture under grant number 7800454 and subsequent grants from the Competitive Research Grants Office.

Literature Cited

1. Ochoa, C.; Schmiediche, P. In Research for the Potato in the Year 2000 Hooker, W.J., Ed.; Int. Potato Ctr., Lima, Peru. 1983, 199 p.
2. Radcliffe, E. B.; Lauer, F. I. Minn. Agric. Exp. Stn. Tech. Bull. 1968, 259.
3. Buhr, H.; Toball, R.; Schrieber, K. Entomol. Exp. Appl. 1958, 1, 209-224.
4. Kuhn, R.; Löw, I. In Origins of Resistance to Toxic Agents Sevag, M.G.; Reid, R.D.; Reynolds, D. E., Eds. Academic Press: New York, 1955; p. 122-132.
5. Schreiber, K. Züchter 1957, 27, 289-299.
6. Schreiber, K. Entomol Exp. & Appl. 1958, 1, 28-37.
7. Tingey, W. M. Am. Potato J. 1984, 61, 157-167.
8. Hawkes, J. G. Bull. Imp. Bur. Plant Breed and Genet. Cambridge, 1944; pp 142.
9. Gibson, R. W. Ann. Appl. Biol. 1971, 68, 113-119.
10. Gibson, R. W.; Turner, R.H. PANS 1977, 23, 272-277.
11. Gibson, R. W. Potato Res. 1979, 22, 223-236.
12. Tingey, W. M.; Sinden, S.L. Am. Potato J. 1982, 59, 95-106.
13. Casagrande, R. J. Econ. Entomol. 1982, 75, 368-372.
14. Tingey, W. M. In The Leafhoppers and Planthoppers Nault, L.R.; Rodriguez, J.G., Eds. 1985; pp. 217-234.
15. Lapointe, S. L.; Tingey, W. M. J. Econ. Entomol. 1986, 79, 1264-1268.
16. Tingey, W. M.; Laubengayer, J. E. J. Econ. Entomol. 1981, 74, 721-725.
17. Tingey, W. M.; Plaisted, R. L.; Laubengayer, J. E.; Mehlenbacher, S. A. Am. Potato J. 1982, 59, 241-251.
18. Tingey, W. M.; Gibson, R. W. J. Econ. Entomol. 1978, 71, 856-858.
19. Gibson, R. W.; Pickett, J. A. Nature 1983, 302, 608-609.
20. Avé, D. A.; Gregory; P.; Tingey, W. M. Entomol. Exp. Appl. 1987, 44, 131-138.
21. Lapointe, S. L.; Tingey, W. M. J. Econ. Entomol. 1984, 77, 386-389.
22. Tingey, W. M.; Laubengayer, J. E. J. Econ. Entomol. 1986, 79, 1230-1234.
23. King, R. R.; Pelletier, Y.; Singh, R. P.; Calhoun, L. A. Chem. Comm. 1986, 14, 1078-1079.
24. Neal, J. J.; Tingey, W. M.; Steffens, J. C. J. Chem. Ecol. 1990, 16, 487-497.
25. Gunenc, Y.; Gibson, R. W. Potato Res. 1980, 23, 345-351.
26. Lapointe, S. L.; Tingey, W. M.; Zitter, T. A. Phytopathol. 1987, 77, 819-822.
27. Radcliffe, E. B. Ann. Rev. Entomol. 1982, 27, 173-204.
28. Dimock, M. B.; Tingey, W. M. Am. Potato J. 1987, 64, 507-515.
29. Neal, J. J.; Steffens, J. C.; Tingey, W. M. Entomol. Exp. Appl. 1989, 51, 133-140.
30. Dimock, M. B. Ph. D. dissertation. Cornell University, 1985.
31. Dimock, M. B.; Tingey, W. M. 1985. In: Proc. Symp. Colorado potato beetle, XVIIth Intern. Congress Entomol. Ferro. D. N; Voss, R. H., Eds. Mass. Agric. Exp. Stn. Bull. 704, 1-144.
32. Wright, R. J.; Dimock, M. B.; Tingey, W. M.; Plaisted, R. L. J. Econ. Entomol. 1985, 78, 576-582.
33. Groden, E.; Casagrande, R. A. J. Econ. Entomol. 1986, 79, 91-97.
34. Dimock, M. B.; Tingey, W. M. Physiol. Entomol. 1988, 13, 399-406.
35. Mehlenbacher, S. A.; Plaisted, R. L. Proc. Int. Congress: Research for the Potato in the Year 2000., 1983, pp. 1228-130.

36. Mehlenbacher, S. A.; Plaisted, R. L.; Tingey, W. M. Am. Potato J. 1983. 60, 699-708.
37. Mehlenbacher, S. A.; Plaisted, R. L.; Tingey, W. M. Crop Sci. 1984, 224, 320-322.
38. Kowalski, S. P. Ph.D. dissertation. Cornell University, Ithaca, NY 1989.
39. Kalazich, J. C. Ph.D. dissertation. Cornell University, Ithaca, NY 1989.
40. Lentini-Gil, Z. Ph.D. dissertation. Cornell University, Ithaca, NY 1989.
41. Xia, J.; Tingey, W. M. J. Econ. Entomol. 1986, 79, 71-75.
42. Obrycki, J. J.; Tauber, M. J.; Tingey, W. M. J. Econ. Entomol. 1983, 76, 456-462.
43. Ruberson, J. R.; Tauber, M. J.; Tauber, C. A.; Tingey, W. M. Can. Ent. 1989, 121, 841-851.

RECEIVED May 16, 1990

Chapter 10

Biochemical Aspects of Glandular Trichome-Mediated Insect Resistance in the *Solanaceae*

John C. Steffens and Donald S. Walters

Department of Plant Breeding and Biometry, Cornell University, Ithaca, NY 14853

Glandular trichomes of wild *Solanaceae* produce an array of biochemical defenses against insects. This chapter reviews our knowledge of two of these mechanisms in the wild potato, *Solanum berthaultii*, and the wild tomato, *Lycopersicon pennellii*. Ruptured trichomes of *S. berthaultii* release a highly expressed, epidermis-specific polyphenol oxidase that polymerizes trichome exudate and entraps insects. Glandular trichomes of *L. pennellii* continuously secrete a viscous mixture of glucose esters which act as insect feeding deterrents. Glucose ester acyl constituents are composed of branched short and medium chain length fatty acids. Biosynthetic investigations reported here show that the branched amino acid biosynthetic pathway is used to form 4 and 5 carbon branched acyl groups which are subsequently elongated by acetate to form the medium chain acyl groups. It appears that, in both species, primary metabolic enzymes have been recruited and modified for trichome-specific insect resistance mechanisms.

Development of plants for use as food crops has gradually stripped these species of their natural resistance to insects and pathogens. Consequently, most modern cultivars rely upon inputs of pesticides to produce an acceptable yield. Insecticide resistance and the increasing social and economic costs of pesticide application have prompted efforts to reduce the requirement of crop plants for these inputs. Breeding for host plant resistance is one approach to reducing

0097–6156/91/0449–0136$06.00/0
© 1991 American Chemical Society

insecticide use. Wild relatives of crop plants frequently possess useful insect or pathogen resistance traits, and are often crossable with crop plants.

Efforts to transfer these resistance traits to crop plants have thus far proceeded by conventional plant breeding approaches. The prospect of producing transgenic crop plants possessing resistance traits derived from wild species provides a new avenue for utilizing resistance traits of wild species and removes the barrier to their utilization imposed by incompatibility with crop species. Knowledge of the basic biochemical processes that contribute to resistance is critical to the success of molecular biological approaches since it is, at present, necessary to characterize specific products of resistance genes in order to identify and transfer these genes to other species. Although knowledge of resistance mechanisms may not be necessary to employ traditional breeding methods, information on the biochemical mechanism of resistance may permit direct screening for the mechanism in the absence of pest populations, and can be far less time consuming than traditional screens for insect or pathogen susceptibility.

We have been involved in mechanistic studies aimed at understanding the basis of glandular trichome-based insect resistance in wild *Solanum* (potato) and *Lycopersicon* (tomato) species. Much effort has focused on identification of wild members of the *Solanaceae* with potentially useful resistance traits for introgression into *Solanum tuberosum* and *Lycopersicon esculentum*. In many cases resistance has been shown to be conferred by glandular trichomes, modified epidermal cells (1) which function as physical and/or chemical barriers against insect attack (2-10, Tingey, this volume).

The wild tomato, *Lycopersicon pennellii,* and the wild potato, *Solanum berthaultii,* are two species which exhibit insect resistance conferred by glandular trichomes. *S. berthaultii* and *L. pennellii* have been the focus of efforts at Cornell University to transfer trichome-based insect resistance traits. This chapter reviews our knowledge of the biochemistry of glandular trichome-based insect resistance in these species.

Type A (VI) Trichomes and Polyphenol Oxidase-Mediated Insect Entrapment

There are primarily two classes of foliar glandular trichomes that contribute to the insect resistance of *S. berthaultii* and *L. pennellii*. Type A trichomes of *S. berthaultii* (referred to as Type VI in the genus *Lycopersicon* (11)) condition resistance to small-bodied insect pests such as aphids and leafhoppers by an adhesive entrapment mechanism. The type A trichome is stalked with a tetralobulate head, approximately 50-65 microns in diameter, and found at high densities on the leaf surface of many solanaceous species. Contact with this trichome by an insect causes the membrane-enclosed head of the trichome to rupture, coating the legs and mouthparts of the insect with the trichome contents. Immediately this coating begins to brown and harden; as the insect ruptures additional trichomes, the dark masses on the legs and mouthparts increase dramatically. These accretions disrupt insect feeding by restricting movement, by occluding the mouthparts, or by entrapment of the insect on the leaf (12).

Browning and hardening of exudate on the insect's mouthparts and tarsi is due to the enzymatic activity of an O_2- requiring oxidase (13), and has been associated with both peroxidase and polyphenol oxidase activities (6,14,15). We recently showed that these enzymatic activities are due to a 59 kD polyphenol oxidase (PPO) which is present in the type A trichomes of *S. berthaultii* at approximately 1 ng/trichome head (a concentration of approximately 0.3 mM) (16). The specialization of the glandular trichome is reflected in the fact that PPO comprises approximately 60% of the total soluble protein of the organ.

Trichome PPO appears to be an example of an evolutionary modification of an existing plant enzyme for use as an insect defense mechanism. PPO enzymes, of molecular weight 45,000, are localized in the thylakoid and are nearly ubiquitous in tissues of plants (17,18). Unlike other known nuclear encoded chloroplast proteins, the 45 kD thylakoid PPO, whose function is unknown, is translated at its mature size of 45 kD, and does not possess a transit peptide sequence (19). In contrast, the 59 kD trichome PPO is translated as a 67 kD precursor and localized in the leucoplasts of trichome and outer epidermal cells. Immunological and primary sequence similarities between the 59 kD trichome PPO and the 45 kD thylakoid PPO underscore the close evolutionary relationship between these two proteins. The apparent advantage of the very high concentration of PPO in the trichome is the high initial rate of catalysis which results upon trichome rupture, facilitating the entrapment of mobile insects.

We have cloned the 59 kD trichome PPO from a cDNA library constructed from epidermal mRNA (H. Yu and J. Steffens, unpublished data) and are currently using this clone as a probe to isolate the genomic DNA encoding the *S. berthaultii* trichome PPO (S. Newman and J. Steffens, unpublished data). These studies may allow us to increase the insect entrapping abilities of cultivated potato, which has retained low Type A trichome densities and no longer possesses the biochemical machinery necessary for insect entrapment. Study of the regulatory sequences controlling expression of this enzyme may provide an understanding of how glandular trichomes are able to synthesize the high levels of specific products which frequently characterizes these organs in plants. Alternatively, the regulatory sequences of glandular trichome genes may be exploited to express other insecticidal proteins in the trichome and epidermis as a first line of defense against insect pests.

Secretion of Sugar Esters by Glandular Trichomes

A similar degree of biochemical specialization exists in glandular trichomes which secrete sugar esters. Glucose and sucrose esters are produced by several genera within the *Solanaceae* (20-25). In contrast to the PPO-containing trichome, these sugar esters are continuously secreted from the apex of a tall (750μm), tapered trichome designated as the Type B or Type IV trichome in *Solanum* and *Lycopersicon*, respectively (11). Described initially as epicuticular lipids by Fobes et al. (26), the glucose and sucrose esters of *Solanum* and *Lycopersicon* species have been shown to physically entrap small pests such as spider mites and aphids (27-29; Tingey, this volume), and to act as aphid feeding deterrents. As little as 33 μg sucrose ester/cm^2 significantly deters feeding by green peach and potato aphids in artificial feeding chamber bioassays (8,30). Levels of sucrose esters present on leaflets of *S. berthaultii* range above 100 mg/ cm^2 (unpublished), and in *L. pennellii* are frequently above 300 mg/cm^2 (26). In addition to directly reducing aphid feeding, sugar esters indirectly contribute to reduction in virus transmission by aphids (31). Other results suggest that sugar esters may also inhibit bacterial and fungal growth (7,32).

The sugar esters produced by solanaceous trichomes are primarily composed of short to medium chain length fatty acids (C_4-C_{12}) esterified to glucose or sucrose. Both branched and straight chain acyl components are present in these esters. Each species produces a characteristic combination of fatty acids esterified at 2 to 4 sugar hydroxyls (Table I). Among the known species producing sucrose esters, acylation of every possible hydroxyl has been described except the 1', 4', and 6' positions (20-25). The diversity of fatty acid composition as well as the acylation positions are puzzling phenomena. In structure/activity studies we are unable to detect differences in aphid feeding deterrence when comparing sucrose and glucose esters. Esters bearing different chain lengths of fatty acid also do not exhibit significant activity differences (11). The origin of sugar ester structural diversity

Table I. Sugar Ester Composition of Trichome Exudates in the *Solanaceae*

	Solanum berthaultii	*Lycopersicon pennellii*	*Solanum neocardensii*	*Solanum aethiopsicum*	*Datura metel*	*Nicotiana tabacum*
	3,3′,4,6 sucrose	2,3,4 glucose	2,3,3′,4 sucrose	2,3,4 glucose	1,2,3 glucose	2,3,4,6 sucrose
<u>Acyl Group</u>						
acetate			+	+		+
isobutyrate	+	+	+	+		*
2-methylbutyrate	*	+		+		+
3-methylbutyrate		*		+		+
valerate					+	
caproate			+		+	
3-methylvalerate						+
heptanoate					+	
5-methylcaproate						*
octanoate					+	
nonanoate					*	
decanoate	+	+	*			
8-methylnonanoate	*	+		+		
undecanoate					*	
9-methyldecanoate		*				
dodecanoate		*				
10-methylundecanoate	*					

+ Major acyl groups, * Minor acyl groups (<5% of total)

and the biological function of this diversity are unknown.

The metabolic investment required to support the level of sugar ester biosynthesis shown by some species is another striking aspect of these compounds. *Lycopersicon pennellii* accumulates glucose esters up to 25% of leaf dry weight (26), an energy commitment equivalent to that expended by the plant for cell wall biosynthesis (J. Steffens and D. Walters, unpublished). Elegant studies with chimeric plants composed of the epidermis of *L. pennellii* and mesophyll and vascular tissues of *L. esculentum* (which does not synthesize sugar esters) show that the epidermal layer of cells is sufficient to specify production of glucose esters. Chimeric plants possessing only the epidermis of *L. pennellii* secrete glucose esters at the same level as normal *L. pennellii* plants (J. Goffreda, unpublished data). Like *L. pennellii,* the esters of these chimeras are composed of 2,3,4-tri-O-acylglucose, and possess qualitatively similar fatty acid compositions (J. Steffens and D. Walters, unpublished data). It seems likely therefore, that the production of sugar esters is confined to epidermal cells (possibly the trichomes themselves) which are highly specialized and capable of extremely high rates of biosynthesis.

Sugar Ester Biosynthesis. Because of the massive production of glucose esters by the relatively few cells of the epidermis we have been interested to understand how sugar esters are synthesized. As a necessary first step to isolating the enzymes and genes involved in sugar ester biosynthesis, we have studied the synthesis of the branched chain acyl groups of *L. pennellii* glucose esters. Branched chain fatty acids are presumed to arise from carbon skeletons derived from branched chain amino acid biosynthesis (33). Accordingly, we have proposed a biosynthetic pathway for *L. pennellii* glucose esters (Fig. 1). The majority of the reactions proposed to function in this hypothetical pathway are essentially those found in the pathway of branched chain amino acid biosynthesis. The difference in the two pathways lies in the fate of the α-keto acids 2-oxo-3-methylbutanoate, 2-oxo-4-methylpentanoate, and 2-oxo-3-methylpentanoate. In branched chain amino acid biosynthesis these structures are reductively aminated to form the amino acids Val, Leu, and Ile, respectively. In the proposed pathway for glucose ester biosynthesis, the α-keto acids are oxidatively decarboxylated to form acyl CoA intermediates. These intermediates are then either directly esterified to the glucose moiety, or undergo elongation via a fatty acid synthetase system to form branched medium chain acyl groups which are then esterified to glucose. In contrast, straight medium

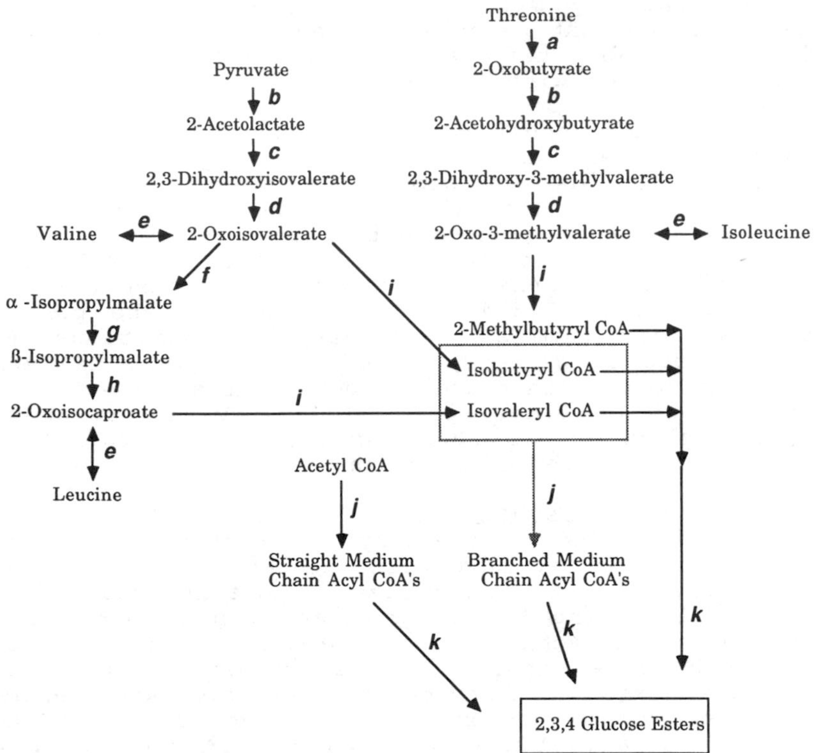

Figure 1. Proposed pathway of glucose ester biosynthesis in *Lycopersicon pennellii*. Letters represent the following enzymatic activities: a) Threonine Dehydratase b) Acetohydroxyacid Synthase c) Acetohydroxyacid Reductoisomerase d) Dihydroxyacid Dehydratase e) Aminotransferase f) α-Isopropylmalate Synthase g) Isopropylmalate Isomerase h) β-Isopropylmalate Dehydrogenase i) Branched-Chain Oxoacid Dehydrogenase j) Fatty Acid Synthetase k) Glucose Acyltransferase(s)

chain acyl groups arise by *de novo* synthesis from acetate. The experiments described below were designed to test the validity of this proposed pathway of glucose ester biosynthesis.

Material and Methods

Stable Isotope Administration. The biosynthesis of *L. pennellii* glucose esters was investigated using deuterium labeled precursors. Compound leaves of *L. pennellii* were removed and briefly rinsed in 100% EtOH before precursor administration. The preincubation sample of the trichome exudate was saved to establish fatty acyl composition prior to precursor feeding; its removal served to increase the proportion of labeled glucose esters after precursor incubations.

Following exudate removal, the leaf petiole was cut diagonally and the leaf immediately placed in deionized water with a 5 mM concentration of a potential precursor. Leaves were illuminated continuously with a 100W incandescent lamp 12 inches from the leaf surface. Chlorsulfuron and sulfometuron methyl were administered similarly as 10 μM solutions. Incubations proceeded for 48 hr with deionized water added to replenish evapotranspirational losses. Following incubation the newly secreted glucose esters were removed by rinsing the leaf with 100% EtOH.

Determination of acyl composition and analysis of stable isotope incorporation were accomplished by GLC and GC/MS of the ethyl esters formed from Na^+EtO^- transesterification of the glucose esters. A polyethylene glycol phase capillary column was used in conjunction with an FID detector. The chromatograph was operated with a temperature program from 36° to 220°C. GC/MS analysis was performed using similar chromatographic conditions. The deuterium-labeled fatty acids possess different retention times from their unlabeled analogs, simplifying GC/MS analyses and permitting analysis of isotopic incorporation using the FID detector.

Results

Analysis of Labeled Acyl Groups. Deuterated branched chain amino acids were all incorporated at extremely high rates into sugar ester acyl groups, an indication of the active nature of this pathway. In some instances the amount of deuterium labeled fatty acids exceeded (146%) that of the unlabeled material remaining on the leaf surface (10-20% of the original amount of glucose ester). GC/MS data for labeled 3-methylbutyrate from the incubation of [d_{10}] Leu are shown in Fig. 2. As

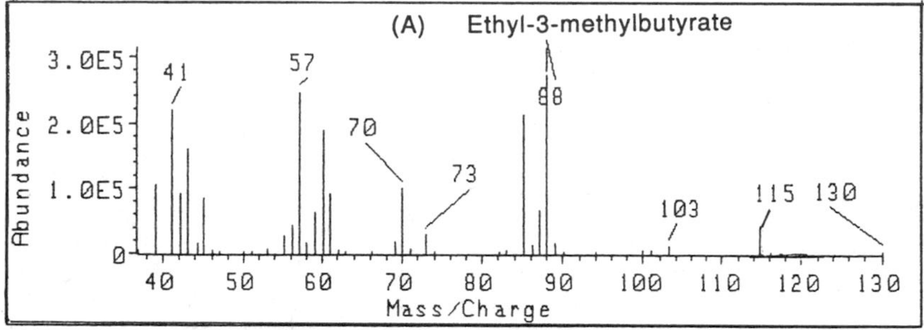

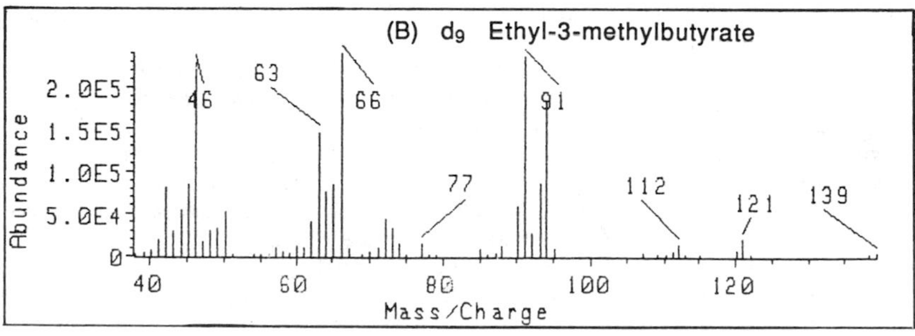

Figure 2. Mass Spectra of A) unlabeled ethyl-3-methylbutyrate B) deuterated (d_9) ethyl-3-methylbutyrate from glucose esters following incubation of *L. pennellii* leaves with d_{10}- Leucine.

predicted by the biosynthetic outline d_{10}- Leu was incorporated intact into [d_9] 3-methylbutyrate and its elongation products. Ethyl [d_9] 3-methylbutyrate exhibits an M^+ ion at *m/z* 139 as well as abundant ions at *m/z* 121, 94, 91, 66, and *m/z* 50 that correspond to ions at *m/z* 115, 88, 85, 57, and 43 in the spectrum of unlabeled ethyl 3-methylbutyrate. The ion at *m/z* 115 in the unlabeled 3-methylbutyrate spectrum is produced by loss of a methyl group (M^+-15) and is a common fragment in the other esters analyzed. The *m/z* 121 ion of the deuterated product is similarly produced by loss of a perdeuterated methyl group (M^+-18). The shift of the base peak from *m/z* 88 in the unlabeled compound to *m/z* 91 in ethyl 3-methylbutyrate results from the substitution of 2 deuterons on C-2 and one deuteron on C-3 such that McLafferty rearrangement produces a transfer of a deuteron rather than the proton transfer observed in the unlabeled compound. The M^+-45 (ethoxy loss) yields an ion of *m/z* 85 in unlabeled isobutyrate. The d_9- compound produces the same ion but at *m/z* 94 as expected. Both the d_9 and the unlabeled compound exhibit an isobutyryl ion at *m/z* 66 and *m/z* 57, respectively. Thus, incorporation of d_{10}- Leu is consistent with a pathway involving transamination and oxidative decarboxylation prior to acylation and secretion of the sugar ester.

d_{10}- Leu is also efficiently incorporated into the longer chain fatty acids of *L. pennellii* glucose esters. The predominant elongated compound resulting from d_{10}- Leu incorporation was [d_9] 9-methyldecanoate (ethyl ester, M^+=223). A detectable amount of [d_9] 7-methyloctanoate (ethyl ester, M^+=195) was also present, suggesting the sequential elongation of the short chain precursor by 2 carbon units. No other deuterated products were detected from d_{10}- Leu incubation. Both elongated d_9 compounds produce the same series of hydrocarbon clusters at *m/z* 101, 115, 143 and *m/z* 157 (*m/z* 157 does not occur in 7-methyl octanoate for obvious reasons) as the unlabeled compounds and also exhibit the analogous M^+-29, M^+-43 and M^+-45 ions indicative of their isotopic composition. The substitution of the *m/z* 43 (isopropyl ion) with *m/z* 50 in the fatty acids incorporating d_{10}- Leu indicates unscrambled isotopic incorporation of the deuterated carbons at the alkyl end of the carbon chain. These results are consistent with a branched pathway with two possible fates of the branched short chain acyl groups: 1) direct acylation, and 2) elongation by several cycles of acetate addition in a modified fatty acid synthetase system, prior to acylation and secretion. Deuterated Val was also incorporated into sugar ester acyl groups in a manner entirely consistent with the proposed pathway (data not shown).

The hypothesis that branched amino acids are metabolized to branched fatty acids via α-keto acid intermediates is further supported by incubations with deuterated 2-oxo-3-methylpentanoate and 2-oxo-4-methylpentanoate, the transamination products of Ile and Leu respectively. GC/MS of acyl groups resulting from incubation with d_2 -2-oxo-4-methylpentanoate (deuterated at C-3) revealed deuterium incorporation into glucose esters containing d_2-3-methylbutyrate and d_2-9-methyldecanoate. Trace amounts of d_2-7-methyloctanoate were also observed, supporting the hypothesis that branched chain precursors are elongated by the addition of acetate.

Biosynthesis of Novel Glucose Esters. The facile incorporation of amino acid and α-keto acid precursors into glucose esters led us to test the extent to which unusual amino acids would be accepted, metabolized, and incorporated into glucose esters by *L. pennellii*. Administration of unusual amino acids in the same manner described for deuterated Leu produced dramatic changes in the acyl group composition of *L. pennellii* glucose esters. *L. pennellii* leaves administered norvaline and norleucine secreted glucose esters containing *n*-butyrate and *n*-pentanoate, respectively, as major acyl constituents, despite the absence of these compounds in *L. pennellii* under normal conditions. Glucose esters containing *n*-nonanoate and *n*-undecanoate are also not found in *L. pennellii* trichome exudate, but accumulate to significant levels following norleucine administration. Again, this indicates that newly synthesized short chain acyl groups can either be esterified directly, or elongated with acetate for two to three cycles prior to esterification as the medium-chain length fatty acid. Administration of methionine resulted in the production of novel glucose esters containing methylthiopropionate (although in lesser amounts than the other novel acyl groups). Methylthiopropionate, like the other acyl groups, results from transamination and oxidative decarboxylation of methionine. In contrast to other precursors, no evidence was seen for incorporation of methylthiopropionate into longer acyl substituents. These experiments indicate that the acyl composition of *L. pennellii* glucose esters is regulated mainly by the composition of acyl substrates available to the acyltransferase(s). The fact that norleucine is incorporated into *n*-nonanoate and *n*-undecanoate also demonstrates that a variety of acyl groups are accepted as primers and elongated by the fatty acid synthetase system.

Acetolactate Synthetase Inhibition. Acetolactate synthetase (ALS; acetohydroxy acid synthetase) is a thiamine pyrophosphate- requiring enzyme involved in biosynthesis of the branched amino acids Val, Leu, and Ile. This enzyme has been investigated more thoroughly than most of the branched chain amino acid biosynthetic enzymes of plants because it is the target of sulfonylurea, imadazolinone and triazolo pyrimidine herbicides (34). Sulfometuron methyl is a sulfonylurea herbicide that has been shown to cause an accumulation of 2-oxo-butyrate in bacteria through inhibition of ALS activity (34,35). In the preceding section we summarized results indicating that *L. pennellii* is able to incorporate a number of α-amino acids and α-keto acids into the acyl substituents of glucose esters. Thus, it might be expected that 2-oxo-butyrate accumulating as a result of ALS inhibition would also be metabolized by branched chain oxoacid dehydrogenase (BCOAD) to form propionyl CoA. Two different ALS inhibitors (chlorsulfuron and sulfometuron methyl) induced accumulation of significant amounts of propionyl glucose esters (up to 8% of the total short chain acids). In the absence of ALS inhibitors, propanoic acid is not detectable in *L. pennellii* glucose esters. ALS inhibition also produced an 800% increase in the amounts of C_9 and C_{11} straight chain fatty acids. This is consistent with elongation of the propionyl-CoA formed in response to ALS inhibition. Similar to the elongation of the four- and five- carbon branched chain precursors, propionyl-CoA appears to undergo three to four cycles of extension reactions with acetate prior to release from the fatty acid synthetase and subsequent acylation. These results provide strong evidence that synthesis of the acyl groups of glucose esters proceeds by the branched chain amino acid biosynthetic pathway, and also that a specialized fatty acid synthetase system is responsible for elongating the acyl primers. It also further demonstrates the low substrate specificity of both the fatty acid synthetase system and glucose acylating enzymes.

Evolutionary Perspective

The evolutionary modification of enzyme systems of primary metabolism to produce secondary compounds appears to be a common theme in plants. Apart from the sugar esters in the *Solanaceae*, the anacardic acids of geranium and the acyl nornicotines of *Nicotiana* have both been suggested to result from a modified fatty acid synthetase system (36-38). The fatty acid synthetase system responsible for synthesis of *L. pennellii* glucose esters is modified to accept branched chain primers in addition to the usual straight chain fatty acids, and elongates them to a length of 10 to 12 carbons rather than the C-16 and C-18 chains more typical of lipid biosynthesis in primary metabolism. Similarly modified biosynthetic systems may be responsible for the production of the wide variety of sugar esters isolated from other members of the *Solanaceae* (Table I).

Conclusions

In addition to their specialized function in plant resistance to insects, glandular trichomes are highly specialized organs and cells which may uniquely possess the complete pathways of the novel secondary metabolism that distinguishes trichomes from other plant organs. Further exploration of pubescent wild germplasm for other biochemically interesting or economically important chemicals may reveal additional instances in which the biochemistry of these organs can be exploited to obtain the enzymes and genes for the production of these chemicals, or for their transfer to crop plants to enhance productivity while decreasing the use of applied pesticides.

Literature Cited

1. Unzelman, J.M.; Healey, P.L. Protoplasma 1974, 80, 285-303.
2. Gibson, R.W.; Turner, R.H. PANS 1977, 23, 272-277.
3. Schnepf, E. In Dynamic Aspects of Plant Ultrastructure; Robards, A.W., Ed.; McGraw-Hill: New York, 1974; Chapter 9; pp 331-357.
4. Georgiev, K.; Satirova V. Genet. Sel. 1978, 11, 214.
5. Dimock, M.B.; Kennedy, G.G. Ent. Exp. Appl. 1983, 33, 263-268.
6. Duffey, S.S. In Insects and the Plant Surface; Juniper, B.E.; Southwood, T.R., Eds.; Edward Arnold Ltd.: Baltimore, 1986; pp 151-172.
7. Cutler, H.G.; Severson, R.F.; Cole, P.D.; Jackson, D.M.; Johnson, A.W. In Natural Resistance of Plants to Pests: Role of Allelochemicals; Green, M.B.; Hedin, P.A., Eds.; ACS Symposium Series No. 296, American Chemical Society: Washington, DC, 1986; pp 178-196.
8. Goffreda, J.C.; Mutschler, M.A.; Avé, D.A.; Tingey, W.M. Steffens, J.C. J. Chem. Ecol. 1989, 15, 2135-2147.
9. Gregory, P.; Avé, D.A.; Bouthyette, P.Y.; Tingey, W.M. In Insects and the Plant Surface; Juniper, B.E.; Southwood, T.R., Eds.; Edward Arnold Ltd.; Baltimore, 1986; pp 173-183.
10. Walters, D.S.; Craig, R.; Mumma, R.O. J. Chem. Ecol. 1990, in press.
11. Goffreda, J.C. Ph.D. Thesis, Cornell University, Ithaca, NY, 1988.
12. Tingey, W.M.; Gibson, R.W.; J. Econ. Entomol. 1978, 71, 856-868.

13. Gibson, R.W.; Ann. Appl. Biol. 1971, 68, 113-119.
14. Ryan, J.D.; Gregory, P.; Tingey, W.M. Phytochemistry 1982, 21, 1885-1887.
15. Bouthyette, P.Y.; Eannetta, N.T.; Hannigan, K.J.; Gregory, P. Phytochemistry 1987, 26, 2949-2954.
16. Kowalski, S.P. Ph.D. Thesis, Cornell University, Ithaca, NY, 1989.
17. Mayer, A.M. Phytochemistry 1987, 26, 11-20.
18. Vaughn, K.C.; Lax, A.R.; Duke, S.O. Physiol. Plantarum 1988, 72, 669-665.
19. Flurkey, W.H. Plant Physiology 1985, 79, 564-567.
20. Burke, B.A.; Goldsby, G.; Mudd J.B. Phytochemistry 1987, 26, 2567-2571.
21. King, R.R.; Singh, R.P.; Calhoun, L.A. Carbohydr. Res. 1987, 166, 113-121.
22. King, R.R.; Calhoun, L.A. Phytochemistry 1988, 27, 3761-3763.
23. King, R.R.; Calhoun, L.A.; Singh, R.P. Phytochemistry 1988, 27, 3765-3768.
24. King, R.R.; Singh, R.P.; Calhoun, L.A. Carbohydr. Res. 1988, 173, 235-241.
25. Severson, R.F.; Arrendale, R.F.; Chortyk, O.T.; Green, C.R.; Thome, F.A.; Stewart, J.L.; Johnson, A.W. J. Agric. Food Chem. 1985, 33, 870-875.
26. Fobes, J.F.; Mudd, J.B.; Marsden, P.F. Plant Physiology 1985, 77, 567-570.
27. Gentile, A.G.; Stoner, A.K. J. Econ. Entomol. 1968, 61, 1152-1154.
28. Gentile, A.G.; Webb, R.E.; Stoner, A.K. J. Econ. Entomol. 1969, 61, 834-836.
29. Juvik, J.A.; Berlinger, M.J.; Ben-David, T.; Rudich, J. Phytoparasitica 1982, 10, 145-156.
30. Neal, J.J.; Tingey, W.M.; Steffens, J.C. J. Chem. Ecol. 1990, in press.
31. Lapointe, S.L.; Tingey, W.M.; Zitter, T.A. Phytopathol. 1987, 77, 819-822.
32. Holley, J.D.; King, R.R.; Singh, R.P. Can. J. Plant Pathol. 1987, 9, 291-294.
33. Kolattakudy, P.E. In Annual Review of Plant Physiology; Machlis, L., Briggs, W.R., Park, R.B., Eds.; Annual Reviews: Palo Alto, CA, 1970; Vol. 21, pp 163-192.
34. LaRossa, R.A.; Van Dyk, T.K. In Methods in Enzymology; Harris, R.A.; Sokatch, J.R., Eds.; Academic: San Diego, 1988; Vol. 166, pp 97-107.
35. LaRossa, R.A.; VanDyk, T.K.; Smulski, D.R. J. Bacteriol. 1987, 169, 1372-1378.
36. Gerhold, D.L.; Craig, R.; Mumma, R.O. J. Chem. Ecol. 1984, 10, 713-722.
37. Huesing, J.; Jones, D.; Deverna, J.; Myers, J.; Collins, G.; Severson, R.; Sisson, V. J. Chem. Ecol. 1989, 15, 1203-1217.
38. Severson, R.F.; Arrendale, R.F.; Snook, M.E.; Sisson, V.A. Ga. J. Sci. 1985, 43, 21.

RECEIVED May 16, 1990

Chapter 11

2-Tridecanone–Glandular Trichome-Mediated Insect Resistance in Tomato

Effect on Parasitoids and Predators of *Heliothis zea*

George G. Kennedy[1], Robert R. Farrar, Jr.[1], and R. K. Kashyap[2]

[1]Department of Entomology, North Carolina State University, Raleigh, NC 27695–7630
[2]Department of Entomology, Haryana Agricultural University, HISAR 125004, India

2-Tridecanone/glandular trichome-mediated resistance to the tobacco hornworm, *Manduca sexta* (L.) and the Colorado potato beetle, *Leptinotarsa decemlineata* (Say) in a wild tomato, *Lycopersicon hirsutum* f. *glabratum* C. H. Mull, PI 134417, adversely affects several species of parasitoids and predators of the tomato fruitworm, *Heliothis zea* (Boddie). Rates of parasitism or predation, and parasitoid survival, were lower on PI 134417 foliage than rates on susceptible foliage. Removal of trichomes from PI 134417 foliage greatly reduced or eliminated these effects. Filter paper treated with 2-tridecanone at levels comparable to those in PI 134417 foliage had similar effects. Intermediate effects were found on a hybrid line with intermediate densities of trichomes but no 2-tridecanone, indicating at least a partial role of trichome density.

Research on the biochemical mechanisms of plant defenses has expanded greatly in recent years and has revealed the rich diversity and complex nature of plant defenses. It has provided information important to the efficient development and utilization of arthropod resistant crop varieties for pest management (1) and has contributed new insight into the role played by plant defensive chemistry in the structuring of arthropod communities associated with particular plant species or varieties (2).

0097–6156/91/0449–0150$06.00/0
© 1991 American Chemical Society

The suitability of arthropods as prey or as hosts of insect predators and parasitoids has been repeatedly shown to be affected by host plant chemistry. The mechanisms by which plant chemical factors mediate suitability of herbivores as hosts (prey) for their natural enemies fall into two broadly overlapping categories. The first involves host plant compounds ingested, and in some cases sequestered, by the herbivores that are toxic or distasteful to parasitoids or predators (4-6). Such compounds may be active as repellents or as acute or chronic toxins (5, 7-10). The second category involves the nutritional quality of the host plant mediating herbivore utilization by parasitoids and predators (11-14). Changes in herbivore size due to host plant effects, for example, have been associated with differences in size and sex ratio of parasitoids (15, 12), and differences in functional responses of parasitoids and predators (16).

In addition to plant chemistry, morphological features of the host plant have also been shown to affect parasitoids and predators. However, with the exception of trichomes, these effects have been studied less than chemical effects. Both glandular and nonglandular trichomes have been widely reported to influence the effectiveness of natural enemies (17-18). Although nonglandular trichomes have been shown, in some cases, to impede searching by some parasitoids (19-20) and predaceous hemiptera and coleoptera (21-23), the effect is by no means universal (24-25).

Glandular trichomes have also been associated with reduced natural enemy effectiveness. They have been shown to entrap several species of hymenopterous parasitoids (26-28) and to reduce the mobility of some species of coccinellids and chrysopids, but not others (25, 29-31).

In the use of plant resistance for crop protection, it is important to recognize potential incompatibilities that may exist between particular resistance mechanisms and the natural enemies of target or nontarget pests of resistant crops. The integration of host plant resistance with biological control, ideally, should combine the cumulative, density independent, pest population suppressive features of the resistant cultivar with the density dependent regulation of pest populations by natural enemies of the crop pests. Although, as described above, there are numerous documented cases of adverse effects of plant resistance factors on herbivore natural enemies, there are also many examples demonstrating the successful integration of plant resistance and biological control (24, 32-37, but see 38).

In general, where resistance alone will not provide acceptable control of the target pest species, it is important to consider the nature of the interaction between plant resistance and biological control in developing strategies for deploying resistance to achieve the desired level of pest control. In some instances, intermediate levels of plant resistance may permit more efficient biological control with the combination resulting in a higher level of pest control than either one alone. A high level of resistance (or other plant characters) may result in a loss or diminution of biological control with a reduction in the overall level of pest control (23, 39).

The situation is further confounded by the reality that most crops are attacked by a complex of arthropod pest species. Plant resistance mechanisms that provide an acceptable level of control of the resisted pest species may also interfere with the effectiveness of natural enemies attacking other pest species that are otherwise unaffected by the resistance mechanisms. Knowledge of such relationships and their impact on the overall level of pest control for the entire pest complex attacking the crop is essential for making intelligent decisions regarding whether or not, and how best, to use a particular resistance mechanism for pest management.

Some of the complex ecological interactions mediated by plant defenses are illustrated by the resistance of a wild tomato, *Lycopersicon hirsutum* f. *glabratum* C. H. Mull, accession PI 134417, to *Manduca sexta* (L.), the tobacco hornworm, and its effects on several predators and parasitoids of *Heliothis zea* (Boddie), the tomato fruitworm.

Resistance in L. hirsutum f. glabratum

Accession PI 134417 is highly resistant to *M. sexta* due to the acute toxicity of its foliage to young larvae (40-41). This resistance is conditioned by the presence of the 13-carbon methyl ketone, 2-tridecanone, in the tips of type VI (42) glandular trichomes (43-46).

2-Tridecanone also conditions resistance to the Colorado potato beetle, *Leptinotarsa decemlineata* (Say), causing mortality of young larvae which disrupt the glandular trichomes (47-51). PI 134417 also possesses factors associated with leaf lamellae that kill *L. decemlineata* larvae during the late instars. These factors are mechanistically distinct from 2-tridecanone-mediated resistance (51-52). Unlike 2-tridecanone-mediated resistance, which is inherited recessively (45) and kills early instar larvae, the lamellar-based resistance is a dominant or incompletely dominant character. Its presence is observed when larvae are fed PI 134417 foliage from which glandular trichomes have been removed, or F_1 hybrid (*L. esculentum* x PI 134417) foliage, which does not possess 2-tridecanone-mediated resistance (52).

PI 134417 is also highly resistant to *H. zea*, but the major portion of this resistance is associated with the leaf lamellae, not the trichomes (50, 53). 2-Tridecanone associated with PI 134417 occurs in quantities potentially lethal to *H. zea*. However, few larvae (13-20%) experience lethal doses of 2-tridecanone. Most receive sublethal exposures that apparently induce elevated levels of detoxifying enzymes (cytochrome P-450 isozymes) and allow them to recover within a few hours. Recovered larvae show increased tolerance to 2-tridecanone in subsequent exposures and are able to feed on the foliage for several days before they succumb to other resistance factors associated with the foliar lamellae. These foliar factors are independent of 2-tridecanone (53-55). Because of the 2-tridecanone-mediated induction of cytochrome P-450, the recovered larvae are also more tolerant to some insecticides (55-58).

2-Undecanone, the 11-carbon methyl ketone, also occurs in the type VI trichomes of PI 134417 and is active against *H. zea* and several other pest species (59-60). It is less acutely toxic than 2-tridecanone against neonate *H. zea* (59) and is less abundant ($\bar{x}$ = 1.1 ug vs. 6.2 ug/trichome tip for 2-tridecanone in PI 134417). However, in combination, 2-tridecanone and 2-undecanone are synergistic in their toxicity to several insect species, including H. zea (60-61). At levels frequently found in PI 134417, 2-undecanone also causes extensive mortality of H. zea pupae when ingested by fifth instar larvae (61-62).

Because 2-tridecanone-mediated resistance to M. sexta and L. decemlineata is genetically distinct from the lamellar-based resistance to H. zea, it would be possible to develop tomato cultivars resistant to M. sexta and L. decemlineata but susceptible to H. zea. On such cultivars, the 2-tridecanone-mediated induction of elevated tolerance to some insecticides might make control of H. zea more difficult (55, 58). The severity of H. zea as a pest problem on such cultivars would be further exacerbated if 2-tridecanone associated with the resistant foliage seriously interfered with insect parasitoids and predators important in suppressing natural populations of H. zea. Because of its occurrence in the foliar glandular trichomes, and because of its broad spectrum toxicity to insects, we considered it likely that parasitoids and predaceous insects searching the resistant foliage for prey would be exposed to potentially toxic levels of 2-tridecanone.

Effects of PI 134417 and 2-Tridecanone on an Egg Parasitoid of H. zea

Parasitism of eggs by Trichogramma spp. is a major source of mortality among H. zea populations on tomato in the Coastal Plain of N. Carolina. In field studies involving tomato lines varying in the level of 2-tridecanone-mediated resistance to M. sexta, parasitism of H. zea eggs by Trichogramma spp. (primarily T. pretiosum Riley) was greatly reduced on plants expressing high levels of 2-tridecanone and high densities of glandular trichomes (Table I, 63). In this study, there was a significant negative relationship between percent parasitism by Trichogramma and glandular trichome density (% parasitism = -2.6 (trichome density) + 82.1, $R^2 = 0.64$, $P \leq 0.0001$), but not between percent parasitism and density of host eggs ($R^2 = 0.07$, $P > 0.16$). However, because 2-tridecanone is contained within the glandular trichome tips of the more resistant lines, it was not possible to separate the effects of 2-tridecanone from those of the trichome density alone (63). High densities of foliar trichomes have been shown to adversely affect Trichogramma spp. in other plant species (28), and it is clear that they have an effect in this system. This is shown by the reduction of parasitization on the F_1 line, which has a much higher trichome density than L. esculentum, but no 2-tridecanone.

In a series of studies to determine if the methyl ketones 2-tridecanone and 2-undecanone were contributing to the reduced parasitization by Trichogramma on resistant plant lines, we found that adult T. pretiosum contacting PI 134417 foliage or exposed to foliar volatiles of PI 134417 (rich in 2-tridecanone vapors (54) suffered elevated levels of mortality. When foliage from which the glandular trichomes had been removed was used, this mortality was eliminated among insects exposed to foliar volatiles, and greatly reduced among insects held in contact with foliage (Table II). In addition, fewer parasitized eggs incubated on PI 134417 foliage with trichomes intact produced parasitoid adults than was the case for parasitized eggs incubated on PI 134417 from which the glandular trichomes were removed or on L. esculentum foliage (Table III). This difference is apparently attributable to mortality of immature parasitoids within host eggs.

Table I. Parasitism of H. zea eggs by Trichogramma spp. on tomato lines selected for varying levels of 2-tridecanone-mediated resistance to M. sexta - Clinton, NC - 1986[1/]

Plant line	Mean Trichome Density no./mm^2	Mean 2-tridecanone ng/mm^2	Mean Percent Parasitism July 29	Aug 8	Aug 13
L. esculentum	1.2	0	82	83	96
F_1	10.2	0	42	78	82
BC_2	15.2	80.2	0	51	70
PI 134417	23.3	352.8	8	19	55

1/ Data from Kauffman and Kennedy (63)

Table II. Percent mortality of T. pretiosum adults following confinement on PI 134417 or L. esculentum foliage or exposure to foliar volatiles for 4 hours

Treatment	Type of Foliage: PI 134417 Trichomes Present[1/]	PI 134417 Trichomes Absent[1/]	L. esculentum
Confined on foliage	35 a	10 b	5.1 c
Exposed to foliar volatiles	16 a	6	5 b

1/ Mean separation horizontal by LSD at $P \leq 0.05$; values are means for 10 replicates with 25 T. pretiosum adults per replicate.

Table III. Emergence of T. pretiosum adults from parasitized H. zea eggs incubated on PI 134417 or L. esculentum foliage[1/]

Foliage Type	Percent parasitized eggs yielding adults[2/]
PI 134417	
trichomes present	56 a
trichomes removed	84 b
L. esculentum	86 b

1/ Eggs exposed, on a neutral substrate, to parsitization by T. pretiosum over a 24 h period prior to transfer to the foliage treatments for incubation.

2/ Values are means of 10 replicates with 25 parasitized eggs per treatment. Mean separation by LSD at $P \leq 0.05$.

Similar but more dramatic effects were observed when T. pretiosum adults were confined on filter paper disks treated with 2-tridecanone at concentrations comparable to those found in PI 134417 foliage and when parasitized H. zea eggs were incubated on 2-tridecanone treated filter paper disks (Table IV). In similar tests, 2-undecanone, at rates comparable to those associated with PI 134417 foliage, had no effect on mortality of adult T. pretiosum but caused a significant reduction in the percentage of parasitized eggs yielding adult parasitoids (Table V).

Table IV. Mortality of T. pretiosum adults on, and emergence of T. pretiosum from, parasitized H. zea eggs incubated on filter paper treated with 2-tridecanone,

2-Tridecanone1/ (ug/cm^2)	% Mortality of T. pretiosium adults2/	% Parasitized eggs yielding adult parasitoid3/
0	26 a	93 a
45	85 b	10 b
90	96 b	0 b

1/ PI 134417 foliage contains a mean of 44.6 ug 2-tridecanone/cm^2 of foliage (50).
2/ Mortality after 4 hours of confinement in 5.5 cm Petri dish containing 5 cm filter paper disk treated with 2-tridecanone at the indicated concentration. Values are means of 10 replicates with 50 adults/replicate. Mean separation vertical by LSD at $P \leq 0.05$.
3/ Values are percent parasitized H. zea eggs producing one or more adult T. pretiosum when incubated on 2-tridecanone treated filter paper. Values are means of 10 replicates with 25 parasitized eggs per replicate. Mean separation vertical by LSD at $P \leq 0.05$.

Table V. Effect of 2-undecanone on emergence of T. pretiosum adults from parasitized H. zea eggs incubated on 2-undecanone treated filter paper

2-undecanone (ug/cm^2)1/	Percent parasitized eggs yielding wasps2/	Percent eggs containing dead parasitoid pupae2/
0	81 a	1 a
4.47	7 b	82 b
8.94 3/	38 b	53 b
17.88	11 b	80 b

1/ Mean concentrations associated with PI 134417 foliage range from 1.9 to 9.6 ug/cm^2 leaflet surface (63).
2/ Values are means of 5 replicates with 25 parasitized eggs per replicate. Mean separation vertical by LSD, $P \leq 0.05$.
3/ Approximate concentration found in PI 134417 foliage.

Although the level of exposure to the methyl ketones experienced by T. pretiosum in these "treated filter paper tests" is likely to be higher than those experienced by parasitoids on PI 134417 foliage, the types of effects caused by 2-tridecanone and 2-undecanone are the same as those caused by normal PI 134417 foliage but not by foliage from which the glandular trichomes have been removed. Therefore, it is likely that 2-tridecanone and to a lesser extent 2-undecanone contribute to the adverse effects of PI 134417 on *Trichogramma*.

Effects of PI 134417 and 2-Tridecanone on Larval Parasitoids of *H. zea*

Although parasitism of *H. zea* larvae on tomato is variable and often low in North Carolina, we have consistently observed lower levels of larval parasitism by an array of parasitoids on plant lines selected for elevated levels of 2-tridecanone-mediated resistance to *M. sexta* (Table VI) (64). To investigate the causes of reduced parasitism, we focused on two species of larval parasitoids: *Campoletis sonorensis* (Cameron), an ichneumonid, and *Archytas marmoratus* (Townsend), a tachinid.

Table VI. Percent parasitism of *H. zea* larvae collected from PI 134417, *L. esculentum* or F_1 plants, Clinton, NC, 1987

Plant Line	Mean Percent Parasitism 1/
L. esculentum	15.5 a
F_1	12.6 a
PI 134417	3.4 b

1/ Values are means of 6 replicates; mean separation by LSD at $P \leq 0.05$. They reflect total parasitism by a complex of larval parasitoids consisting of *Cotesia marginiventris*, *Campoletis sonorensis*, *Microplitis croceipes* and *Meteorus autographae*.

In field cage studies in which *H. zea* larvae on several plant lines were exposed to adult parasitoids (one plant line per cage), percent parasitization by both species was significantly lower on BC_2, (a backcross ((*L. esculentum* x PI 134417) x PI 134417) line with intermediate levels of 2-tridecanone and glandular trichome density), and PI 134417 than on *L. esculentum* and the F_1 hybrid (*L. esculentum* x PI 134417) (Table VII). This pattern reflects levels of 2-tridecanone-mediated resistance to *M. sexta*. These data indicate that fewer *H. zea* larvae are parasitized on the more resistant plant lines but do not indicate whether this is due to elevated densities of glandular trichomes or methyl ketones, or both.

Because of their life histories, larvae of both parasitoids directly contact the foliage of the food plants of their host larvae. Thus, potential exists for both to be affected directly by the methyl ketones present in the glandular trichomes of the resistant plants.

C. sonorensis oviposits in second and early third instar *H. zea* larvae. The parasitoid larvae kill and emerge from late third and early fourth instar host larvae, crawl onto the foliage, and spin a cocoon in which pupation occurs.

Close observation of *C. sonorensis* larvae emerging from their hosts on PI 134417 foliage indicated that they disrupt a large number of glandular trichomes during the cocoon spinning process and contact the methyl ketones contained therein. In studies in which parasitized *H. zea* larvae were reared on artificial diet and transferred to foliage of *L. esculentum*, BC_2 or PI 134417, significantly higher levels of parasitoid mortality were observed on foliage from plant lines possessing elevated levels of 2-tridecanone-mediated resistance (Table VIII). It is noteworthy that on PI 134417 foliage, which contained the highest levels of methyl ketones (Table I), virtually all parasitoid mortality occurred among larvae during the cocoon spinning process. In contrast, on BC_2 foliage, which has lower levels of methyl ketones (Table I), a significant portion of total parasitoid mortality occurred among pupae and adults within the cocoon. This difference may be attributable to reduced dosages of methyl ketones experienced on BC_2 (*64*).

Table VII. Percent parasitization of *H. zea* larvae by *C. sonorensis* and *A. marmoratus* in nonchoice field cage studies

Plant Line	Percent Parasitization	
	C. sonorensis 1/	A. marmoratus 1/
L. esculentum	66.6 a	76 a
F_1	66.2 a	73 a
BC_2	47.5 b	54 b
PI 134417	38.9 c	57 b

1/ Mean separation vertical by LSD at $P \leq 0.05$. See Table 1 for trichome densities and methyl ketone levels associated with each plant line.

Table VIII. Mortality of *C. sonorensis* on foliage for cocoon spinning following development of parasitoids in *H. zea* larvae reared on artificial diet 1/

Plant Line	Percent Mortality 2/		
	larvae 3/	in cocoon 4/	Total
L. esculentum	0.7 a	9.5 a	10.2 a
BC_2	43.6 b	16.3 b	59.9 b
PI 134417	98 c	0.7 c	98.7 c

1/ Data from Kauffman and Kennedy (*63*).
2/ Mean separation vertical $P \leq 0.05$ by Duncan's multiple range test.
3/ Data include all parasitoid larvae that died prior to cocoon construction.
4/ Data include mortality of larvae, pupae and adults within the cocoon.

In a related study, Kauffman and Kennedy (*65*) demonstrated that the toxicity of PI 134417 and BC_2 foliage to *C. sonorensis* larvae during cocoon spinning was eliminated when the glandular trichomes were removed. They further demonstrated the acute toxicity of 2-tridecanone and 2-undecanone to *C. sonorensis* larvae on treated filter paper disks (2-tridecanone LC_{50} = 13.0, 95% fiducial limits =

10.88, 17.95); 2-undecanone LC_{50} = 38.9 (32.34, 47.79 ug/cm^2) and showed levels of parasitoid mortality on methyl ketone treated filter paper disks that were quantitatively similar to those observed on resistant foliage (65).

Although the toxicity of 2-tridecanone to C. sonorensis and H. zea is similar (LC_{50} = 13.9 and 17.1 ug/cm^2, treated surface, respectively (59, 65) in treated filter paper bioassays, C. sonorensis is affected to a much greater degree than its host. Unlike fifth instar C. sonorensis larvae, which discharge numerous glandular trichomes and directly contact substantial quantities of 2-tridecanone, neonate H. zea are very small and rarely discharge trichome tips before 2-tridecanone vapors surrounding the foliage induce elevated levels of cytochrome P-450 which allows the larvae to tolerate subsequent exposures to 2-tridecanone (58).

Archytas marmoratus larviposits first instar maggots (planidia) on the host's food plant. Planidia remain on the foliage until they contact and attach themselves to second or subsequent instars of their host. Larval development takes place within host pupae (66). Since H. zea pupate in the soil, the only immature stage of A. marmoratus to contact plant foliage is the planidium.

On F_1 and PI 134417 foliage, A. marmoratus planidia suffer extensive mortality as compared to that observed on L. esculentum foliage. Since elevated mortality is not observed among planidia on F_1 and PI 134417 foliage from which the glandular trichomes have been removed (Table IX), the trichomes are an important cause of the mortality. Trichome factors other than the methyl ketones must be largely responsible, however, because extensive mortality occurred on F_1 hybrid foliage, which contains neither 2-tridecanone nor 2-undecanone (Table I).

Table IX. Mortality of A. marmoratus planidia on foliage of three tomato lines with and without trichomes

Plant Line	% Mortality[1/]	
	With Trichomes	Without Trichomes
Better Boy	48.9 a	36.8 a
F_1	83.0 b	35.1 a**
PI 134417	95.8 c	36.8 a**

1/ Means separation vertical by LSD; means with the same letter are not significantly different (P>0.05).

** Means for foliage with and without trichomes are significantly different by F-test (P<0.05).

H. zea larvae ingesting PI 134417 foliage during the fifth instar suffer extensive mortality during the pupal stage as a result of the toxic effects of 2-undecanone in the glandular trichome tips (62). Because A. marmoratus complete their development within host pupae, we reared parasitized H. zea larvae through pupation on artificial diet containing 2-tridecanone, 2-undecanone or the combination at rates (% wt/wt) comparable to those associated with PI 134417 foliage. The results of this experiment show clearly that the presence of 2-undecanone in the host's diet caused a significant reduction in the percentage of parasitized pupae producing A. marmoratus adults (Table X). 2-Tridecanone, which does not kill H. zea pupae, had no such effect.

Table X. Effects of methyl ketones in host diet on emergence of A. marmoratus adults.

Chemical	Rate, (% wt/wt)	% of Hosts Yielding Flies[1/]
Control	--	71 a
2-Tridecanone	0.3	64 a
2-Undecanone	0.05	33 b
2-Tridecanone + 2-Undecanone	0.3 + 0.05	31 b

1/ Means separation vertical by Chi-square; means with the same letter are not significantly different (P>0.05).

Effects of PI 134417 and Glandular Trichomes on Generalist Predators of H. zea

The big-eyed bug, Geocoris punctipes (Say), a lygaeid, and the lady beetle Coleomegilla maculata (DeGeer), a coccinellid, are important predators of Heliothis spp. eggs and early instar larvae in a number of crops including tomato. In preliminary experiments, there were no differences in survival of adults of either predaceous species when confined on foliage of the various plant lines for 24 h. However, consumption of H. zea eggs over a 24 h period by adults of both species was significantly reduced on PI 134417 foliage relative to that on L. esculentum foliage and intermediate on F_1 and BC_2 foliage (Table XI). These differences among plant lines were not observed when the predators were confined on foliage from which the glandular trichomes had been removed (Barbour and Kennedy, unpublished). The significant reductions in consumption on F_1 foliage indicate that trichome-associated factors other than the methyl ketones account for at least a portion of the reduced consumption of prey. It is quite possible, even likely, that high densities of glandular trichomes associated with F_1 and BC_2 foliage simply impede movement and therefore searching efficiency of those predators (25, 30, 67-68). Work is currently underway to further define the causes of effects seen in Table XI.

Conclusion

2-Tridecanone/glandular trichome-mediated resistance of L hirsutum f. glabratum PI 134417 to M. sexta and L. decemlineata adversely affects an array of parasitoids and predaceous insects that are important natural enemies of the tomato fruitworm, H. zea. These natural enemies represent a diversity of life histories and are affected by the defenses of PI 134417 in different ways. Because the resistance of PI 134417 decreases the parasitization rates and increases mortality of immatures of the three parasitoid species studied, it interferes with both their functional and numerical responses to changes in host density. The data presented for *G. punctipes* and *C. maculata* indicate that the functional responses of both predators are also adversely affected.

Table XI. Consumption of *H. zea* eggs by adult *Geocoris punctipes* and *Coleomegilla maculata* on normal foliage of several tomato plant lines or foliage divested of glandular trichomes

	No. eggs consumed in 24 h[1/]			
	G. punctipes		C. maculata	
	Trichomes		Trichomes	
Plant Line	present	absent	present	absent
L. esculentum	2.9 a	4.1 a	4.1 a	4.4 a
F_1	1.4 b	4.5 a	1.8 b	4.6 a
BC_2	1.5 b	4.6 a	1.0 b	4.6 a
PI 134417	0.3 c	3.8 a	0.3 c	4.6 a

1/ Vertical mean separation by LSD, $P \leq 0.05$. Differences between trichomes present and trichomes absent were significantly different for all plant lines for both predators F-test, $P \leq 0.05$.

The diversity and magnitude of the negative tritrophic level effects of 2-tridecanone/glandular trichome-mediated resistance suggest that the deployment of tomato cultivars possessing this resistance, but no additional resistance to other important arthropod pests, could have undesirable pest management consequences.

The negative tritrophic level effects associated with PI 134417 foliage are likely to be extreme relative to those associated with other plant defenses, because of the high acute toxicity of 2-tridecanone and 2-undecanone and their presence at biologically active concentrations in the tips of glandular trichomes on the leaf surfaces. Despite this, the tritrophic level interactions associated with PI 134417 provide some appreciation for the diversity and complexity of ecological effects of even relatively simple plant defenses.

Acknowledgments

The technical assistance of Vann Covington, Dana Bumgarner and Diana Afra is greatfully acknowledged as are the valuable comments, ideas and suggestions of J. D. Barbour, F. P. Hain, J. R. Meyer, and G. C. Rock. This research was supported by USDA Competitive Grants 87 CRCR-1-2505 and 88-37252-4017 to G. G. Kennedy and the North Carolina Agricultural Research Service.

"Literature Cited"

1. Smith, C. M. 1989. Plant resistance to insects: a fundamental approach. J. Wiley & Sons, NY. 282 p.
2. Barbosa, P. and D. K. Letourneau (eds.). 1989. Novel aspects of insect-plant interactions. J. Wiley &Sons, NY. 362 p.

3. Duffey, S. S. 1980. Sequestration of plant natural products by insects. Ann. Rev. Entomol. 25: 447-477.
4. Duffey, S. S., K. A. Bloem and B. C. Campbell. 1986. Consequences of sequestration of plant natural products in plant-insect-parasitoid interactions. pp. 31-60. In D. J. Boethel and R. D. Eikenbary (eds.). Interactions of Plant Resistance and Parasitoids and Predators of Insects. Ellis Horword Ltd., Chichester.
5. Roeske, C. N., J. N. Seiber, L. P. Brower and C. M. Moffit. 1976. Milkweed cardenolides and their comparative processing by monarch butterflies. pp. 93-167. In J. W. Wallace and R. L. Mansell (eds.). Biochemical Interactions Between Plants and Insects. Rec. Adv. Phytochem. Plenum Press, NY.
6. Eisner, T. 1970. Chemical defense against predation in arthropods. pp. 157-217. In E. Sondheimer and J. B. Simeone (eds). Chemical Ecology. Academic Press, NY.
7. Aplin, R. T., R. Ward d'A. and M. Rothachild. 1975. Examination of the large white and small white butterflies (Pieris spp.) for the presence of mustard oils and mustard oil glycosides. J. Entomol. Ser. A. 50: 73-78.
8. Bernays, E. A. and R. F. Chapman. 1978. Plant chemistry and acridoid feeding behavior. pp. 99-142. In J. B. Harbourne (ed.). Biochemical Aspects of Plant and Animal Coevolution. Academic Press, London.
9. Campbell, B. C. and S. S. Duffey. 1979. Tomatine and parasitic wasps: potential incompatibility of plant antibiosis with biological control. Science (Wash). 205: 700-702.
10. Campbell, B. C. and S. S. Duffey. 1981. Alleviation of alpha-tomatine-induced toxicity to the parasitoid Hyposoter exiguae by phytosterols in the diet of the host Heliothis zea. J. Chem. Ecol. 7: 927-946.
11. Price, P. W. 1986. Ecological aspects of host plant resistance and biological control: interactions among three trophic levels. pp. 11-30. In D. J. Boethel and R. D. Eikenbary (eds.). Interactions of Plant Resistance and Parasitoids and Predators of Insects. Ellis Horwood Ltd., Chichester.
12. Vinson, S. B. and G. F. Iwantsch. 1980. Host suitability for insect parasitoids. Ann. Rev. Entomol. 25: 397-419.
13. Pimentel, D. 1966. Wasp parasite (Nasaria vitripennis) survival on its fly host (Musca domestica) reared on various foods. Ann. Entomol. Soc. Amer. 59: 1031-1038.
14. Zohdy, H. M. 1976. On the effect of the food of Myzus persicae Sulz. on the hymenopterous parasite Aphelinus asychis Walker. Oecologia (Berlin) 26: 185-191.
15. Smith, J. M. 1957. Effects of the food plant of the California red scale Aonidiella aurantii (Mask.) on reproduction of its hymenopterous parasites. Can. Entomol. 89: 219-230.
16. Luck, R. F. and H. Podoler. 1985. Competitive displacement of Aphytis lignanesis by A. melinus: the role of host size and female progeny production. Ecology 66: 904-913.
17. Levin, D. A. 1973. The role of trichomes in plant defense. Quart. Rev. Biol. 48: 3-15.
18. Stipanovic, R. D. 1983. Function and chemistry of plant trichomes and glands in insect resistance: protective chemicals in play epidermal glands and appendages. pp. 69-100. In P. A.

Hedin (ed.). Plant Resistance to Insects. ACS Symposium Series 208. ACS, Washington, D.C.

19. Hulspas-Jordaan, P. M. and J. C. van Lenteren. 1978. The relationship between host-plant leaf structure and parasitization efficiency of the parasitic wasp Encarsia formosa Graham (Hymenoptera: Aphelinidae). Med. Fac. Landbouww. Rijksuniv. Gent. 43: 431-440.

20. Kumar, A., C.P.M. Triphati, R. Singh and R. K. Pandey. 1983. Bionomics of Trioxys (Binodoxys) indicas, an aphid parasitoid of Aphis craccivora. 17. Effects of host plants on the activities of the parasitoid. Z. Agnew. Ent. 96: 304-307.

21. Banks, C. J. 1957. The behavior of individual coccinellid larvae on plants. Brit. J. Anim. Behav. 5: 12-24.

22. Johnson, B. 1953. The injurious effects of the hooked epidermal hairs of the French bean (Phaseolus vulgaris L.) on Aphis craccivora Koch. Bull. Ent. Res. 44: 779-788.

23. Obrycki, J. J. 1986. The influence of foliar pubescence on entomophagous species. pp. 61-83. In D. J. Boethel and R. D. Eikenbary (eds.). Interactions of plant resistance and parasitoids and predators of insects. Ellis Horwood Ltd., Chichester.

24. Lampert, E. P., D. L. Haynes, A. J. Sawyer, D. P. Jokinen, S. G. Wellso, R. L. Gallum and J. J. Roberts. 1983. Effects of regional releases of resistant wheats on the population dynamics of the cereal leaf beetle (Coleoptera: Chrysomelidae). Ann. Entomol. Soc. Amer. 76: 972-980.

25. Shah, M. A. 1982. The influence of plant surfaces on the searching behavior of coccinellid larvae. Entomol. Exp. Appl. 31: 377-380.

26. Kantanyukul, W. and R. Thurston. 1973. Seasonal parasitism and predation of eggs in the tobacco hornworm on various host plants in Kentucky. Environ. Entomol. 2: 939-945.

27. Elsey, K. D. and J. F. Chaplin. 1978. Resistance of tobacco introduction 1112 to the tobacco budworm and green peach aphid. J. Econ. Entomol. 71: 723-725.

28. Rabb, R. L. and J. R. Bradley, Jr. 1968. The influence of host plants on parasitism of eggs of the tobacco hornworm. J. Econ. Entomol. 61: 1249-1252.

29. Gurney, B. and N. W. Hussey. 1970. Evaluation of some coccinellid species for the biological control of aphids in protected cropping. Ann. Appl. Biol. 68: 451-458.

30. Elsey, K. D. 1974. Influence of plant host on searching speed of two predators. Entomophaga 19: 3-6.

31. Rao, R.S.N. and I. J. Chandra. 1984. Brinkochrysa scelestes (Neur.: Chrysopidae) as a predator of Myzus persicae (Hom.: Aphididae) on tobacco. Entomophaga 29: 283-285.

32. Casagrande, R. A. and D. L. Haynes. 1976. The impact of pubescent wheat on the population dynamics of the cereal leaf beetle. Environ. Entomol. 5: 153-159.

33. Schuster, D. J. and K. J. Starks. 1975. Preference of Lysiphlebus testaceipes for greenbug resistant and susceptible small grain species. Environ. Entomol. 4: 887-888.

34. Schuster, M. F., D. G. Holder, E. T. Cherry and F. G. Maxwell. 1976a. Plant bugs and natural enemy insect populations on frego bract and smooth-leaf cottons. Miss. Agric. For. Exp. Stn. Tech. Bull. 75.

35. Schuster, M. F., M. J. Lukafahr and F. G. Maxwell. 1976b. Impact of nectariless cotton on plant bugs and natural enemies. J. Econ. Entomol. 69: 400-402.
36. Starks, K. J., R. Muniappan and R. D. Eikenbary. 1972. Interaction between plant resistance and parasitism against greenbug on barley and sorghum. Ann. Entomol. Soc. Amer. 65: 650-655.
37. Pimentel, D. and A. G. Wheeler, Jr. 1973. Influence of alfalfa resistance on a pea aphid population and its associated parasites, predators and competitors. Environ. Entomol. 2: 1-11.
38. Hare, J. D. 1983. Manipulation of host suitability for herbivore management. pp. 655-680. In R. F. Denno and M. S. McClure (eds.). Variable Plants and Herbivores in Natural and Managed Systems. Academic Press, NY.
39. Van Emden, H. F. 1986. The interaction of plant resistance and natural enemies: effects on populations of sucking insects. pp. 138-150. In D. J. Boethel and R. D. Eikenbary (eds.). Interactions of Plant Resistance and Parasitoids and Predators of Insects. Ellis Horwood Ltd., Chichester.
40. Kennedy, G. G. and W. R. Henderson. 1978. A laboratory assay for resistance to the tobacco hornworm in Lycopersicon and Solanum spp. J. Amer. Soc. Hort. Sci. 103: 334-336.
41. Kennedy, G. G. and R. T. Yamamoto. 1979. A toxic factor causing resistance in a wild tomato to the tobacco hornworm and some other insects. Entomol. Exp. Appl. 26: 121-126.
42. Luckwill, L. C. 1943. The genus Lycopersicon: an historical, biological and taxonomic survey of the wild and cultivated tomatoes. Aberdeen Univ. Studies No. 120.
43. Williams, W. G., G. G. Kennedy, R.T. Yamamoto, J. D. Thacker and J. Bordner. 1980. 2-Tridecanone: a naturally occurring insecticide from the wild tomato species Lycopersicon hirsutum f. glabratum. Science 207: 888-889.
44. Kennedy, G. G., R. T. Yamamoto, M. B. Dimock, W. G. Williams and J. Bordner. 1981. Effect of daylength and light intensity on 2-tridecanone levels and resistance levels in Lycopersicon hirsutum f. glabratum to Manduca sexta. J. Chem. Ecol. 7: 707-716.
45. Fery, R. L. and G. G. Kennedy. 1987. Genetic analysis of 2-tridecanone concentration, leaf trichome characteristics, and tobacco hornworm resistance in tomato. J. Amer. Soc. Hort. Sci. 112: 886-891.
46. Schwartz, R. F. and J. C. Snyder. 1983. Characterization of resistance to tobacco hornworm in Lycopersicon leaflets. HortScience 18: 170.
47. Kennedy, G. G. and M. B. Dimock. 1983. 2-Tridecanone: a natural toxicant in a wild tomato responsible for insect resistance. pp. 123-128. In: J. Miyamoto and P. C. Kearney (eds.). Pesticide Chemistry, Human Welfare and The Environment. Vol. 2. Pergamon Press, Tokyo.
48. Kennedy, G. G. and C. E. Sorenson. 1985. Role of glandular trichomes in the resistance of Lycopersicon hirsutum f. glabratum to Colorado potato beetle (Coleoptera: Chrysomelidae). J. Econ. Entomol. 78: 547-551.
49. Kennedy, G. G., C. E. Sorenson and R. L. Fery. 1985. Mechanisms of resistance to Colorado potato beetle in tomato.

pp. 107-116. In D. N. Ferro and R. H. Voss (eds.). Proceedings of the Symposium on the Colorado Potato Beetle. XVII Intern. Congr. Entomol. Mass. Agr. Exp. Stn. Bull. 704.

50. Kennedy, G. G. 1986. Consequences of modifying biochemically mediated insect resistance in Lycopersicon species. pp. 130-141. In M. R. Green and P. A. Hedin (eds.). Natural Resistance of Plants to Pests - Roles of Allelochemicals. ACS Symposium Series 296, Washington, D.C. 243 p.

51. Kennedy, G. G. and R. R. Farrar, Jr. 1987. Response of insecticide-resistant and susceptible Colorado potato beetles, Leptinotarsa decemlineata to 2-tridecanone and resistant foliage: the absence of cross resistance. Entomol. Exp. Appl. 45: 187-192.

52. Sorenson, C. E., R. L. Fery and G. G. Kennedy. 1989. Relationship between Colorado potato beetle (Coleoptera: Chrysomelidae) and tobacco hornworm (Lepidoptera: Sphingidae) resistance in Lycopersicon hirsutum f. glabratum. J. Econ. Entomol. 82: 1143-1148.

53. Farrar, R. R., Jr. and G. G. Kennedy. 1987a. Growth, food consumption and mortality of Heliothis zea larvae on the foliage of the wild tomato Lycopersicon hirsutum f. glabratum and the cultivated tomato, L. esculentum. Entomol. Exp. Appl. 44: 213-219.

54. Dimock, M. B. and G. G. Kennedy. 1983. The role of glandular trichomes in the resistance of Lycopersicon hirsutum f. glabratum to Heliothis zea. Entomol. Exp. Appl. 33: 263-268.

55. Kennedy, G. G. 1984. 2-Tridecanone, tomatoes and Heliothis zea: potential incompatibility of plant antibiosis with insecticidal control. Entomol. Exp. Appl. 35: 305-311.

56. Riskallah, M. R., W. C. Dauterman and E. Hodgson. 1986a. Nutritional effects on the induction of cytochrome P-450 and glutathion transferase in larvae of the tobacco budworm, Heliothis virescens. Insect Biochem. 16: 491-499.

57. Riskallah, M. R., W. C. Dauterman and E. Hodgson. 1986b. Host plant induction of microsomal monooxygenase activity in relation to diazinon metabolism and toxicity in larvae of the tobacco budworm Heliothis virescens (F.). Pest. Biochem. & Physiol. 25: 233-247.

58. Kennedy, G. G., R. R. Farrar, Jr. and M. R. Riskallah. 1987. Induced tolerance in Heliothis zea neonates to host plant allelochemicals and carbaryl following incubation of eggs on foliage of Lycopersicon hirsutum f. glabratum. Oecologia 73: 615-620.

59. Dimock, M. B., G. G. Kennedy and W. G. Williams. 1982. Toxicity studies of analogs of 2-tridecanone, a naturally-occurring toxicant from a wild tomato. J. Chem. Ecol. 8: 837-842.

60. Lin, S.Y.H., J. Trumble and J. Kumamoto. 1987. Activity of volatile compounds in glandular trichomes of Lycopersicon species against two insect herbivores. J. Chem. ecol. 13: 837-850.

61. Farrar, R. R., Jr. and G. G. Kennedy. 1987b. 2-Undecanone, a constituent of the glandular trichomes of Lycopersicon hirsutum f. glabratum: effects on Heliothis zea and Manduca sexta growth and survival. Entomol. Exp. Appl. 43: 17-23.

62. Farrar, R. R. and G. G. Kennedy. 1988. 2-Undecanone, a pupal mortality factor in Heliothis zea: sensitive larval stage and in planta activity in Lycopersicon hirsutum f. glabratum. Entomol. Exp. Appl. 47: 205-210.
63. Kauffman, W. C. and G. G. Kennedy. 1989a. Relationship between trichome density in tomato and parasitism of Heliothis spp. (Lepidoptera: Noctuidae) eggs by Trichogramma spp. (Hymenoptera: Trichogrammatidae). Environ. Entomol. 18: 698-704.
64. Kauffman, W. C. and G. G. Kennedy. 1989b. Inhibition of Campoletis sonorensis parasitism of Heliothis zea and of parasitoid development by 2-tridecanone-mediated insect resistance of wild tomato. J. Chem. Ecol. 15: 1919-1930.
65. Kauffman, W. C. and G. G. Kennedy. 1989c. Toxicity of allelochemicals from wild insect-resistant tomato Lycopersicon hirsutum f. glabratum to Campoletis sonorensis, a parasitoid of Heliothis zea. J. Chem. Ecol. 15: 2051-2060.
66. Hughes, P. S. 1975. The biology of Archytas marmoratus (Townsend). Ann. Entomol. Soc. Am. 68: 757-767.
67. Obrycki, J. J. and M. J. Tauber. 1984. Natural enemy activity on glandular pubescent potato plants in the greenhouse: an unreliable predictor of effects in the field. Environ. Entomol. 13: 679-683.
68. Belcher, D. W. and R. Thurston. 1982. Inhibition of movement of larvae of the convergent lady beetle by leaf trichomes of tobacco. Environ. Entomol. 11: 91-94.

RECEIVED July 2, 1990

Chapter 12

Enzymatic Antinutritive Defenses of the Tomato Plant Against Insects

Sean S. Duffey and Gary W. Felton

Department of Entomology, University of California, Davis, CA 95616

The tomato plant, Lycopersicon esculentum, contains constitutive and inducible chemical defenses that play a significant role in protection from attack by a variety of insects and pathogens. Most studies on chemical bases of resistance of the tomato plant against insects have focused on constitutive, single-factor components which exert their effect upon the insect by directly poisoning it. We discuss an approach to resistance which relies upon the simultaneous action of multi-component constitutive and inducible defenses which indirectly retard insect growth by depriving the insect of essential nutrients. The driving force of this resistance is predominantly derived from oxidative enzymes which, upon damage to the plant, activate certain constitutive components to highly reactive alkylating agents (electrophiles), which in turn render dietary protein and other essential or limiting nutrients unutilizable. The production of these electrophiles, in conjunction with reduced protein quality (e.g., lowered sulphur amino acid intake), may also place a strain on the insect's ability to generate reducing and conjugating agents (e.g., glutathione). Hence, not only is nutrient intake by the insect compromised, but so may be the insect's ability to detoxify natural and synthetic toxins. The benefits and detractions of this approach in terms of compatibility with biological control agents and biotechnological approaches to host plant resistance are discussed.

During the last decade our knowledge of the bases and genetics of resistance of solanaceous crop plants to insects (see other chapters in this volume) has increased markedly. Resistance has been primarily based on the enhancement of the levels of one or two constitutive chemical and/or physical resistance-conferring characteristics in commercial varieties via breeding with exotic germplasm (see other chapters in the volume). The predominant mode

0097-6156/91/0449-0166$09.00/0
© 1991 American Chemical Society

of host plant resistance is antibiosis, whereby the insect species is directly poisoned and/or physically impeded by these characteristics. In contrast, we describe an approach to host plant resistance of the tomato plant against noctuid larvae that attempts to derive resistance through antinutritive resistance. This resistance results from the chemical interaction of a multitude of plant factors that irreversibly limit the bioavailability of essential or limiting dietary nutrients during the early stages of ingestion and digestion of plant material. Antinutritive resistance derives from the action of several plant oxidative enzymes that activate a variety of chemical defenses and nutrients to nutrient-destroying agents.

Rationale for Antinutritive Resistance

Great theoretical importance has been given to the role of plant nitrogen in regulating insect population dynamics (1-7). General ecological theory also proposes that acquisition of plant nitrogen, a major hurdle for phytophagous insects, can also be hampered by the coingestion of several types of natural products (e.g., phenolics, tannins, lignins, and proteinase inhibitors) (4,8-14) whose putative mode of action involves interference with digestive processes. It is surprising that host-plant resistance programs do not include more concerted efforts to employ resistance aimed at impeding utilization of nitrogen by insects. The majority of research on chemical bases of resistance of crop plants to insects has focused on natural products that are directly toxic, form physical barriers, and/or are modifiers of behavior (15-18).

For sake of argument, these antibiotic factors can be viewed as "positive" traits of resistance because their bioavailability in the plant permits them to exert their direct effects after contact with the insect. In contrast, it is also possible to have "negative" resistance, which results from the absence or deficit of plant chemical traits that are essential or limiting for the insect. In principle, the putative antidigestive properties of tannins and gossypol represent negative resistance because they are thought to reduce the digestion and utilization of dietary protein (19-21), although such antidigestive action is questioned (22-26). In theory, such resistance results from a chemically reduced bioavailability of nutrients essential for growth an development. Thus, the insect is not poisoned directly (or its behavior modified) by an overload of allelochemicals, but rather indirectly by a deficit of essential nutrients.

"Negative" resistance has been documented in a pea cultivar against an aphid and in a rice cultivar against a leaf-hopper because of reduced quantities of amino acids (18). Generally, the detrimental effect upon aphids of absence or limitation of specific amino acids has been comparatively well studied (27-29). Also, the resistance of maize to the European corn borer has been related to insufficient levels of ascorbic acid to permit proper growth (30). A variety of other plant natural products are known to adversely affect the insect's ability to utilize food. Saponins have been implicated as antitryptic agents (31) and were also shown to interfere with cholesterol absorption (31,32). α-Amylase inhibitors are potent inhibitors of amylase digestion in beetles (33,34).

Lipoxygenase has been shown to degrade essential fatty acids (e.g., linoleic) as well as produce toxic metabolites (35). The ability to utilize such factors as negative forms of resistance has not been explored fully.

Using the tomato plant *Lycopersicon esculentum*, we have been exploring the possibility of utilizing plant enzymes to activate constitutive defenses and nutrients to chemically reactive products that reduce the utilization of dietary plant nitrogen, and hence, confer negative resistance against the tomato fruitworm *Heliothis zea* and the beet armyworm *Spodoptera exigua* (Figures 1 & 2). The detrimental effects of these enzymatic reactions occur during the early stages of feeding upon foliage. These enzymes are primarily oxidative (polyphenol oxidase, peroxidase, lipoxygenase), although a variety of enzymes have potential roles (e.g., catalase, superoxide dismutase, glutathione reductase, and phenylalanine and tyrosine ammonia lyases) in modulating the antinutritive effect.

Biochemical Nature of Response of Plant to Wounding. One aim of our program is to utilize some of the immediate defenses of plants to attack by microorganisms coincidentally against insects. Plants have a generalized defensive response to wounding that can be arbitrarily divided into two phases -- activation and induction. Activation represents the immediate response to cellular damage wherein cell integrity is lost and a variety of hydrolytic (e.g., acyl lipases, chitinases, and glucanases) and oxidative (e.g., lipoxygenases, polyphenol oxidases, and peroxidases) enzymes are released from compartmentalization. This release results in the generation of chemical signals that trigger the systemic and/or local induction of defenses (e.g., lignification, isoflavonoids, phenolics, and proteinase inhibitors) (36-46), and in the generation of chemically reactive products that lead to cell death through destruction of membranes and polymerization of cellular components (38,42,43,47-50). This polymerization is primarily mediated by polyphenol oxidases, peroxidases, and lipoxygenases. Such polymerization leads to an insoluble matrix that is thought to present a physical barrier to the progression of disease (51-57).

These enzymes (e.g., lipoxygenases, polyphenol oxidases, peroxidases) also occur in the tomato plant and are locally and/or systemically inducible, as a result of infection by pathogens (49,53,55,58-62). It should follow that they are also inducible by insect-feeding damage such as that inflicted by *H. zea* or *S. exigua*, and amplify the antinutritive defense.

A brief discussion of the chemical reactivity of the products of these enzymes is central to our proposed use of these enzymes as antinutritive bases of resistance. Polyphenol oxidase (PPO) and peroxidase (POD) oxidize phenolics to quinones, which are strong electrophiles that alkylate nucleophilic functional groups of protein, peptides, and amino acids (e.g., -SH, $-NH_2$, -HN-, and -OH)(Figure 1)(53,63-65). This alkylation renders the derivatized amino acids nutritionally inert, often reduces the digestibility of protein by tryptic and chymotryptic enzymes, and furthermore can lead to loss of nutritional value of protein via polymerization and subsequent denaturation and precipitation (63,66-69). POD is also capable of decarboxylating and deaminating free and bound amino acids to aldehydes (e.g., lysine, valine, phenylalanine,

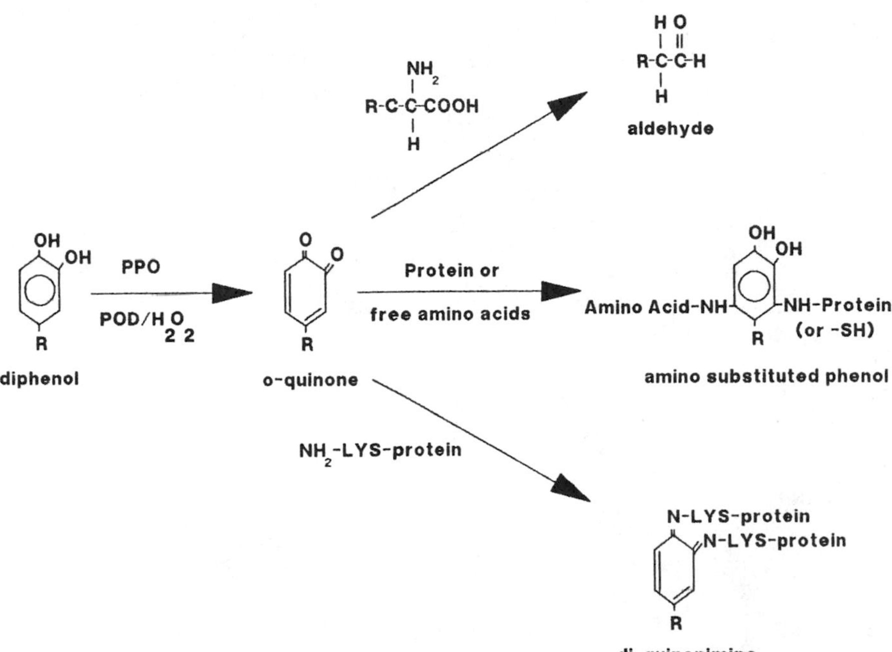

Figure 1. A Simplified Version of Some Chemical Reactions Mediated by Polyphenol Oxidase and Peroxidase that Contribute to Impairment of Protein Quality. Details of reactions and end-products are not specified. diphenol = generalized o-dihydroxy-phenolic; aldehyde = product derived from generalized amino acid.

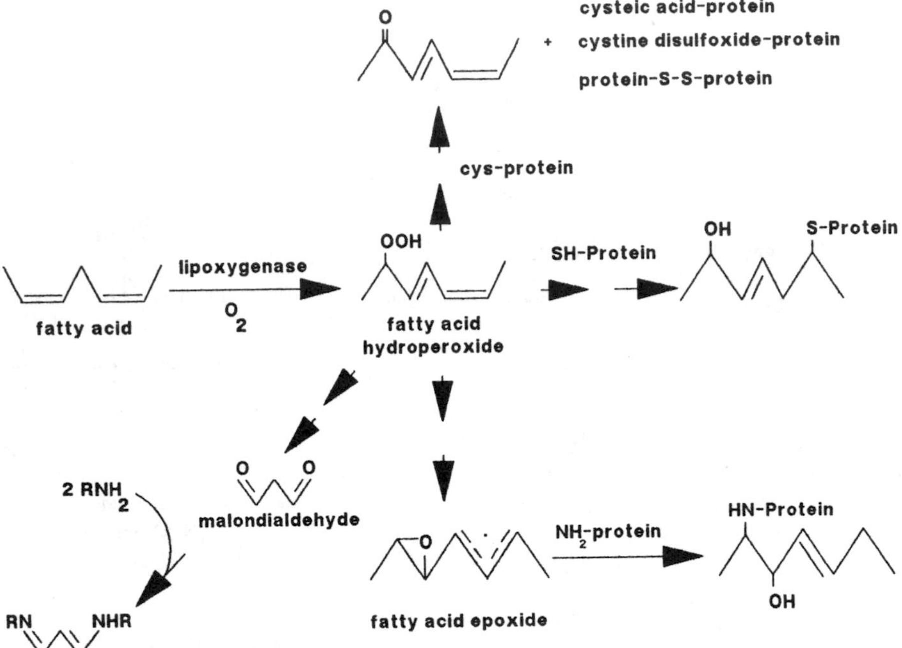

Figure 2. A Simplified Version of Some Chemical Reactions Mediated by Lipoxygenase that Contribute to Impairment of Protein Quality. Only the cis,cis-1,4-pentadiene portion of a fatty acid is indicated: cys-protein = cysteine in protein. Details of reactions and end-products are not specified.

methionine, and leucine), further resulting in nutritional loss (Figure 1)(57,70,71). These aldehydes (in protein) facilitate polymerization by forming Schiff bases with $-NH_2$ functions of other protein molecules. POD can also initiate free radical formation on -SH and tyrosinyl functions of protein, which leads to polymerization of protein and possible nutritive loss (63,65,69,72).

Lipoxygenase (LOX) converts polyunsaturated fatty acids, such as linoleic and linolenic acids, to lipid hydroperoxides (Figure 2)(52,73,74). The lipid hydroperoxides then form hydroperoxide radicals, epoxides, and/or are degraded to form malondialdehyde. These products are also strongly electrophilic, and can destroy individual amino acids by decarboxylative deamination (e.g., lysine, cysteine, histidine, tyrosine, and tryptophan); cause free radical mediated cross-linking of protein at thiol, histidinyl, and tyrosinyl groups; and cause Schiff base formation (e.g., malondialdehyde and lysine aldehyde) (39,49,50,74-78).

Hence, these oxidative enzymes have the potential to chemically destroy a variety of essential or limiting amino acids (e.g. arginine, cysteine, histidine, leucine, lysine, methionine, serine, tryptophan, and tyrosine) (27). In addition, linoleic acid, another essential dietary component (27) is destroyed. These reactions (Figures 1 & 2) represent the activation of constitutive defenses and nutrients to potent antinutritive agents.

PPO, POD, and LOX activities are immediately released when noctuid larvae masticate foliage. Depending upon substrate availability, these activities persist in the gut during digestion, and often persist in the faeces (79: unpubl. data). The pH-activity profiles of these enzymes permit their action in the basic environment of the digestive fluid. PPO and POD are resistant to inactivation by a variety of proteases (e.g., insect and bovine trypsin and chymotrypsin, cathepsin, pepsin, and pronase)(79; unpubl. data). This extensive action provides the appropriate time frame for significant depreciation of nutritional quality by the above mechanisms.

Nutritional Consequences. These oxidative enzymes can have a major destructive impact upon dietary protein and free amino acids (e.g., arginine, cysteine, histidine, lysine, methionine, tryptophan, and tyrosine)(Figures 1 & 2). Our projected use of these defensive reactions as an antinutritive form of resistance is well justified by studies on nutrition of vertebrates. The negative impact of PPO and POD activity (with chlorogenic or caffeic acids as substrates) on protein quality for animals is well established (63,66,69,80-87). The primary negative effect is exerted by reducing digestibility of protein and assimilation of digested protein through precipitation, destruction of sites of attack of tryptic (e.g., at lysine and arginine) and chymotryptic (at tyrosine and tryptophan) enzymes, as well as destruction of other essential or limiting amino acids such as histidine, proline, cysteine, methionine (63,65,84,87-90). Note that many of these amino acids are essential amino acids for insects (27,28). Some depreciation of protein quality can also result from inhibition of digestive enzymes by reactive products (91,92).

The nutritional requirements of these insects have been studied to a very limited degree. A variety of L-amino acids are indispensable for growth of H. zea (valine, leucine, lysine, arginine, histidine, isoleucine, tryptophan, tyrosine, phenylalanine, and methionine)(93), the very amino acids destroyed by these oxidative enzymes. Additionally, we have demonstrated that the growth of both H. zea and S. exigua is strongly inhibited by low-quality dietary protein regimes which are particularly poor in lysine and sulfur amino acids (79,94-97). Lysine and sulfur amino acids are among the most sensitive to such oxidative/alkylative reactions.

The proteolytic gut enzymes of H. zea and S. exigua have been shown to be in majority trypsin and in minority chymotrypsin (94), which require basic (lysine and arginine) and aromatic (tyrosine and tryptophan) amino acids, respectively, as sites to hydrolyze protein. Hence, derivatization of these protein-bound or free amino acids by any of the reactive enzyme-products should lead to reduced utilizability of dietary nitrogen.

The detrimental effects of these enzyme-mediated chemical reactions upon dietary nitrogen should be manifested in insects such as H. zea and S. exigua initially as reduced growth rate and subsequently as potentially detrimental effects on life-history traits (e.g., longevity, fecundity, and survivorship)(97,98).

Evidence for Antinutritional Resistance

The Tomato Plant. Does the tomato plant provide the appropriate chemical milieu necessary to create "negative resistance"? High levels of constitutive PPO and POD exist in foliage (53,54,59,60, 79,99-101). Also, the levels of foliar PPO and POD increase after attack by microorganisms (55,59,62,100), as well as by insects (61). Catalase has been identified in the tomato plant (60; Felton and Duffey, unpubl.).

The substrates for PPO and POD activity are ample in foliage and green fruit of the tomato plant. Chlorogenic acid, rutin, quercetin, caffeic acid, and caffeoyl glutaric acid, and other caffeoyl derivatives occur in the tomato plant with chlorogenic acid and rutin predominating (99,102-104). Most of these phenolics have been reported as substrates for PPO and POD activity (66,69,81,83,84,87,105). Our assays with tomato plant PPO and POD show that chlorogenic acid and caffeic acids are good substrates for these enzymes; whereas, rutin is poor. Moreover, catecholic phenolics (e.g., chlorogenic acid and rutin) are also inducible above constitutive levels after attack by microorganisms (106-110). In tobacco, PPO and POD are induced both locally and systemically by attack from Fusarium (49,57).

Likewise, LOX occurs constitutively in tomato fruit and leaves (49,58,60,111-113). Both Type I and II LOX's are present in the tomato fruit which produce both lipid hydroperoxides (of linoleic and linolenic acids) and carbonyls (e.g., malondialdehyde)(49,58). The tomato variety "Castlemart" shows low levels of constitutive LOX type I activity, but much higher levels have been found in wild species of Lycopersicon, particularly in L. hirsutum f. glabratum (99; unpubl. data). It has been suggested that induced LOX activity contributes to resistance of soybean to mites (114,115); this enzyme is induced by aphid feeding (116).

For purposes of breeding, our survey of the genus Lycopersicon points to accessions of L. hirsutum f. glabratum as the best sources of phenolics, PPO, POD, and LOX (99; unpubl data).

Proteinase Inhibitors in Resistance. Proteinase inhibitors (PI's) have attracted considerable attention because their inducible nature, in conjunction with their antiproteolytic properties, offer the plant the possibility of effective "post-infectional" defense against certain pest insects (15,46,117-119). The promise of this form of resistance has led several research groups to transfer and/or amplify the genes for proteinase inhibitor production in(to) several crop plants (119-122).

In the tomato plant, L. esculentum, damage to foliage of young plants by the feeding of larval S. exigua and H. zea induces the de novo production of PI's I and II (123-125). In the case of S. exigua, PI production was doubled by feeding damage; growth was negatively correlated with feeding damage (79,123,124). Reduction of larval growth results from several causes. The PI, as well as inhibiting alimentary proteolytic activity, induces an hypertrophic synthesis of trypsin in the insect's gut. This induction of trypsin increases the insect's demand for dietary sulphur amino acids; the net uptake of sulphur amino acids is simultaneously being reduced because of the antiproteolytic action of the PI's. Hence, the insect enters a pernicious state of limitation of sulphur amino acids where the hypertrophic synthesis of trypsin detracts from rapid growth. The higher demand for sulphur amino acids can be met by the dietary protein rich in sulphur amino acids, and/or by a surplus of free sulphur amino acids in the diet (94,95). In fact, the degree of growth inhibitory effects of PI's is inversely related to the nutritional quality of the dietary protein. Proteins relatively rich in lysine, arginine, and sulphur amino acids reduce the toxicity of PI substantially compared to proteins low in these amino acids (94-96).

In theory, the antidigestive properties of PI's combined with the above enzyme-based defenses should provide "double-barrel" negative resistance, since the first increases the need for sulphur amino acids, and the second has the potential to decrease the availability of these and other amino acids. However, in order to jointly utilize these two defenses effectively, a deeper knowledge of their chemical interrelationships is necessary. Potential complications in the use of proteinase inhibitors will be discussed below.

Polyphenol Oxidase, Peroxidase, and Lipoxygenase in Resistance. Polyphenol oxidase in conjunction with chlorogenic acid as a substrate has the potential to reduce the ability of larval H. zea and S. exigua to utilize dietary protein. For example, alkylation of casein in artificial diet (at 1.0% wwt) by PPO (from mushroom or tomato plant) and chlorogenic acid, at levels commensurate with that found in tomato foliage, inhibits the growth of both larval species by up to 70% (Table I). Rutin is a very poor substrate for mushroom tyrosinase and tomato PPO (79; unpubl data), and hence, has negligible effect upon protein utilization by alkylation; however, it is an antibiotic in its own right (97).

Table I. Influence of CHA and PPO on Relative Growth of _Heliothis zea_ and _Spodoptera exigua_

Treatment	_Heliothis zea_	_Spodoptera exigua_
Control	100 a	100a
7.0 mM CHA	48.6b	58b
7.0 mM CHA + PPO	29.7c	42c

CHA = chlorogenic acid at 7.0 mM/kg diet wwt.;PPO = 0.100 O.D./min/gm diet:Dietary protein is casein at 1.0 % wwt.; larvae were grown on diet from neonate stage to 10 days old. Significant differences between means within a column, based on 95% confidence intervals from ANOVA, are shown by different letters.

The depreciation of protein quality is not restricted to casein. The growth of larval _H_. _zea_ depends on the quality and quantity of dietary protein (96,97) which is highly correlatable with the relative levels of certain amino acids (lysine. cysteine, histidine, and methionine)(Table II). Proteins with low amounts of these amino acids support growth poorly. These amino acids are the most susceptible to alkylation by _o_-quinones particularly in the basic conditions of the gut. Therefore, it follows that the severity of reduction of the nutritive quality of a protein is dependent upon its quantity and quality (amount of the above amino acids as alkylatable amino acids = AAA; Table II). Casein, which is the most nutritive protein tested and has the highest relative quantities of AAA, is the least affected by alkylation. If a variety of proteins, at similar dietary concentrations and similar exposures to the combination of PPO + chlorogenic acid (to produce chlorogenoquinone), are compared for their ability to support growth of either insect, the least nutritious proteins (e.g., glutein and tomato protein) are the most affected by alkylation compared to soy protein or casein. Amino acid analyses of treated proteins show losses of up to 30% of essential amino acids such as lysine, methionine, histidine, and cysteine (79; unpubl. data).

Table II. The Relationship Between the Nutritive Value of Various Proteins, as Indexed by Relative Growth Rate, to Larval *Spodoptera exigua* and *Heliothis zea* as a Function of Alkylatable Amino Acids

Dietary protein	Total AAA (μmoles/100 gm of diet)	% Reduction of Rel. Growth Rate (mg/day/mg larva)	
		H. zea	S. exigua
Casein	895	10	10
Soy	565	20	50
Tomato	482	35	73
Glutein	292	50	80
Zein	238	90	99

%Relative growth based on untreated casein as 100%: Total AAA= total μmoles of alkylatable amino acids (lysine, histidine, methionine, and cysteine) per 100 gm of diet: Dietary proteins at 1.0%; protein alkylated with 3.5 mM chlorogenic acid/kg diet wwt.

The ability of chlorogenoquinone to render protein less nutritious is not peculiar to casein, but generalizable to a variety of plant proteins. Hence, the efficacy of such an oxidative defense will in part be dependent upon the quality of plant protein as indexed by the quantity of alkylatable groups. Also, the efficacy of this defense will also be contingent upon the absolute levels of protein. The higher the dietary level of protein (e.g., casein), the greater the alleviation of the antinutritive effect of chlorogenoquinone (Table III). This effect is not restricted to casein; similar results are obtained with soy protein and tomato protein in proportion to the number of AAA'a (unpubl. data). The alleviation occurs because the relative number of alkylatable groups exceeds that destroyed by the equivalents of chlorogenoquinone. The most nutritious proteins are the best alleviators of this antinutritive effect (124). Therefore, the efficacy of such a defense will depend both upon quality and quantity of protein, as well as the relative quantity of oxidizable phenolics. Plant phenology may have a great bearing on the magnitude of the antinutritive effect (79,124,125).

Table III. Effect of Quantity of Protein on Quinone-mediated Depreciation of Nutritive Quality of Dietary Proteir Quality for larval *S. exigua*

Protein	% in Diet (wet wt.)	Total AAA (μmoles/100 gm diet)	% Relative Growth (mg/day/mg larva)
Casein	0.5	448	75
	1.0	895	82
	2.0	1790	90
	4.0	3580	90

See legend of Table II.

The levels of PPO in foliage (at immature green fruit stage) of the tomato plant are strongly negatively correlated ($r = -0.72$, $p > 0.001$, $n = 75$) with the ability of _H. zea_ to grow on foliage. Also, foliar PPO activity increases with age, which is also strongly negatively correlated with a reduced ability of larval _H. zea_ to grow (79; unpubl. data). In addition, it has been shown that plant PPO activity persists in the insect's gut hours after ingestion, and in fact the insect's digestive processes actually activate PPO by 50% within minutes after ingestion (79). Because up to 50% of ingested chlorogenic acid is covalently bound to plant protein (the cause of a significant loss of amino acids; e.g., up to 38% of lysine and cysteine, and 10-20% of histidine) when the insect feeds on foliage, it is concluded that PPO is operating as a strong antinutritive defense against these insects.

Although PPO is the major phenolase of the tomato plant, peroxidase (POD) also is present (79). A major limitation of the use of POD as a defense is that, unlike PPO, it requires a continuous source of H_2O_2 in order to oxidize phenolics. One possible advantage of POD is that it is able to oxidize a greater variety of phenolics. Foliar PPO has significant activity with caffeic acid and chlorogenic acid; whereas, POD can oxidize a much greater variety of substrates (e.g., chlorogenic acid, caffeic acid, rutin, esculetin, ferulic acid, tyrosine, and coumaric acid). Moreover, POD can attack protein by oxidizing tyrosinyl and sulphydryl moieties, as well as deaminating and decarboxylating amino acids such as lysine. As pointed out before, these reactions are very detrimental to protein integrity. Thus, this greater spectrum of activity may make POD a strong or indispensible component of enzyme-based defense.

To this point our work with POD is limited. Addition of POD with chlorogenic acid or rutin to artificial diet strongly reduces the nutritive value of dietary casein for larval _H. zea_ (Table IV). This again is in part the result of covalent binding of the oxidized phenolic to the protein.

Our major evidence for the antinutritive action of POD comes from feeding experiments with foliage in which catalase was added to macerated foliage. Catalase converts H_2O_2 (needed by POD) to water; hence, the presence of high catalase activity in foliage reduced the impact of POD on food quality. The addition of catalase to foliage not only reduced the level of POD activity in freshly crushed tissue 36-fold but correspondingly improved the relative growth of larval _H. zea_ by 0.042 mg/day/mg larva (Table V). This evidence implicates POD as a potentially important factor in reducing the growth of _H. zea_. However, because the insect's gut content contains high levels of catalase, the antibiotic action of POD may be limited. In contrast, PPO does not require H_2O_2 and remains highly active in the insect's gut during the digestion of food and even remains active in the faeces.

Table IV. Effect of Oxidative Enzyme Activity on the Growth of larval <u>Heliothis zea</u>

Treatment	% Relative Growth
Protein alone (tomato foliar protein at 0.5%)	100a
Protein + 3.5 mM linoleic acid	100a
Protein + 3.5 mM chlorogenic acid	102a
Protein + 10mM H_2O_2	105a
Protein + lipoxygenase + 3.5 mM linoleic acid	48.1b
Protein + peroxidase + H_2O_2 (10mM) + 7.0 mM chlorogenic acid	27.1c
Protein + peroxidase + H_2O_2 (10mM) + 7.0 mM rutin	78.0d
Protein + polyphenol oxidase + 7.0 mM chlorogenic acid	32.0c

Relative growth = mg/day/mg of larva: lipoxygenase, peroxidase, and polyphenoloxidase were added to diet at activities corresponding to that found in tomato foliage: rutin and chlorogenic acid at 3.5 mM/kg diet wwt. Significant differences between means within a column, based 95% confidence intervals from ANOVA, are shown by different letters. PPO = 0.100 O.D./min/gm diet wwt.; POD = 27.0 O.D./min/gm diet wwt.

Table V. Effect of Exogenous Catalase on Larval Growth and Foliar Oxidative Activities

Treatment	CAT[1]	POD[2]	RGR[3]
Catalase added	3570a	1.85a	0.296
No catalase added	11b	65.20b	0.254

[1] CAT = catalase activity in units/min/gm foliage
[2] POD = peroxidase activity in OD470/min/gm foliage
[3] RGR = relative larval growth rate (mg/day/mg larva).
Significant differences between means within a column, based 95% confidence intervals from ANOVA, are shown by different letters.

We have the least information about the antinutritive effects of lipoxygenase (LOX). Experiments (Table IV) show that the nutritive value of protein to _H. zea_ is reduced by treatment with LOX and linoleic acid (52% reduction in growth). Tomato foliage contains significant quantities of LOX activity and linoleic acid has been shown to be covalently bound to protein both _in vitro_ and _in planta_ (unpubl. data). Studies are underway to determine which amino acids are preferentially destroyed. Also, preliminary studies with foliage demonstrate that copious quantities of malondialdehyde are generated in crushed foliage; the antinutritional effects of this Schiff base former are as yet undetermined.

Currently we are determining if the joint antinutritional effects of POP, POD, and LOX activity are additive or synergistic. If these activities are found to differentially destroy amino acids, then their battery of effects may be synergistic. If their effects are equal (same amino acids destroyed), then perhaps only one enzyme, say PPO (with high phenolic levels) requires amplification to prove a sufficient antinutritive defense. We are also determining if various oxidized phenolics (e.g., caffeic acid, chlorogenic acid, coumaric acid, and rutin) are equivalent in their ability to alkylate and impair protein quality.

A breeding program to enhance resistance would benefit from the knowledge of which enzymes and/or substrates to enhance. Our surveys of wild species of _hycopersicon_ show that certain genotypes (particularly _L. hirsutum_ f. _glabratum_) not only have higher constitutive levels of catecholic phenolics than found in _L. esculentum_ but also higher levels of PPO, POD, and LOX (99; unpubl. data).

Induction of Enzymes by Feeding-Damage. Although the tomato plant has constitutive levels of PPO, POD, and LOX, the defensive ability of the plant may be enhanced if these enzymes were induced by insect feeding damage. The inducible nature of PPO, POD, and phenolics as a result of infection by microorganisms has already been discussed. Similarly, the induction of PI's by noctuid larvae has been pointed out. Indeed, feeding damage by larval _H. zea_ is able to systemically induce very high levels of PPO activity in tomato foliage 24 hours after damage (Table VI). Induction of POD and LOX by H. zea was not observed. In contrast, the Tomato Russet mite Aculops lycopersici dramatically induced LOX and POD levels (Table VI). We have not yet related the induction of these enzymes to the levels of phenols, protein, and/or insect performance on induced versus uninduced foliage.

Table VI. Induction of Tomato Plant Foliar Oxidative Enzymes by Arthropods

	%Relative Activity[1]	
Enzyme	H. zea[2]	Russet Mite[3]
Lipoxygenase	105	2367
Peroxidase	100	264
Polyphenol Oxidase	980	96

[1] undamaged plants represent 100% activity;
[2] represents systemic induction (other leaves);
[3] represents localized induction (within a leaf).

Impact of Oxidative Enzymes on Aphids. Tomato plants are also attacked by a variety of other arthropods such as whiteflies, mites, and aphids (126,127). Trichomes have been studied as a basis of resistance against many of these arthropods (99). A consistent feature of resistance has been the allusion to the sticky entrapping properties of trichomes (type VI). We have shown that these entrapping properties result from the presence of compartmentalized catecholic phenolics and PPO/POD in the tips of type VI trichomes (99: unpubl. data), which upon breakage, for example by aphids, lead to quinone mediated polymerization of trichomal protein. This polymerized protein forms the classical blackened boots on aphid's feet. Utilizing a series of crosses between PI 134417 and Walter [the parents in Kennedy's studies of 2-tridecanone resistance (128)], we have shown a significant negative correlation between the number of aphids, Macrosiphon euphorbiae, per leaflet and the density of type VI trichomes and their phenolase activity. These findings parallel the studies of Tingey's group on resistance in potato (see chapter in this volume). As mentioned above, certain accessions of L. hirsutum f. glabratum are sources of high levels of foliar phenolics and oxidative enzymes. Some of the same accessions of L. hirsutum f. glabratum have been shown to be highest in type VI trichomal densities with commensurately high trichomal PPO/POD activity. Hence, it should be possible, by employing oxidative enzymes, to breed for simultaneous resistance against pests such as aphids, H. zea and S. exigua.

Potential Incompatibility with Proteinase Inhibitors

It was suggested earlier that PI's and PPO might serve as complementary defenses because of the demand they place upon the insect for sulphur amino acids. Unfortunately, the two types of defenses may be incompatible. PI's are polypeptides (8,000 - 40,000 mw) which have a variety of alkylatable amino acids (e.g., cysteine-cysteine and lysine). PI's often have lysine near the active site, and many PI's have multiple disulphide bonds which are integral for activity (129-131). Derivatization of these amino acids by quinones should render PI's less active. Indeed, this has been shown to occur both in vitro and in planta (125). Treatment of soybean trypsin inhibitor (II), tomato PI (I and II), and lima bean inhibitor with PPO and chlorogenic acid in vitro caused a loss of ability to inhibit bovine trypsin: this loss of activity corresponded to a loss of up to 30% of detectable amino acids such as lysine and cysteine as a result of alkylation. Furthermore, it was shown the action of PPO plus chlorogenic acid against PI's I and II in tomato foliage resulted in up to 70% loss in their detectability by immunological assay with a corresponding 50% loss in the ability of foliage to inhibit the growth of S. exigua. The loss of PI identity and biological activity was magnified by wounding (124,125).

Crop plants contain a variety of potential alkylating or Schiff base forming agents (e.g., gossypol, isothiocyanate from mustard oils, $^{-}$CN from cyanogensis, DIMBOA, epoxides, sesquiterpene lactones, oxidized tannins, and aldehydes) which are implicated as defenses against insects and pathogens. These alkylating agents also have the ability to significantly reduce (generally 30-70%) the inhibitory properties of a variety of PI's (e.g., Kunitz and Bowman-Birk inhibitors, tomato PI I and II, potato inhibitor I and II). Generally, PPO and POD with a variety of substrates (e.g., caffeic acid, chlorogenic acid, coumaric acid, and esculetin) were the most effective at inactivating PI's compared to epoxides and Schiff base formers. Some selected results are shown in Table VII. Tomato PI II was more resistant to inactivation than several other PI's (unpubl. data), which suggests that chemical incompatibilities in transgenic plants may be partially avoided by the correct choice of PI. In view of the recent emphasis on transgenic alteration of crop plants with PI genes (119-122,132), our results suggest that unless the chemical milieu of the receiving plant is properly accounted for the efficacy of PI's as a basis of resistance may be compromised.

Potential Incompatibilites with Biological Control Agents

Another potential constraint upon the use of oxidative enzymes as bases of host plant resistance may be their potential incompatibility with biological control agents. Tomato plant phenolics (rutin and chlorogenic acid) have been shown to be

Table VII. Impact of Selected Plant Natural Products upon Proteinase Inhibitor Activity against Bovine Trypsin[1]

Chemical Treatment	Inhibitor	% Inactivation
Allylisothiocyanate	STI	32
	Tomato PI II	0
DIMBOA	STI	49
Gossypol	STI	5
	Tomato PI I	64
Tannic acid	STI	50
Tannic acid + H_2O_2 + POD	STI	74
Lipoxygenase + linoleic acid	STI	33
	Tomato PI I	47
	Tomato PI II	20
POD + H_2O_2 + coumaric acid	STI	53
	Tomato PI I	75
	Tomato PI II	12

[1] Data derived from Workman, Felton, and Duffey, unpubl. data: % Inactivation is determined by comparing ability of untreated PI versus pretreated proteinase inhibitor to inhibit bovine trypsin hydrolysis of TAME *in vitro*; for treatment all chemicals were used at 3.5 mM: STI = soy trypsin inhibitor I; POD = peroxidase:

moderately safe compared to tomatine in their detrimental effects upon the ichneumonid parasitoid *Hyposoter exiguae* (32,97). The effects of phenolics in conjunction with PPO and POD upon this parasitoid are unknown. Their action may be incompatible if the growth of the host larvae were sufficiently restricted so as to impair the growth of the parasitoid (97). The use of high trichomal density in conjunction with high trichomal PPO activity to control aphids may compromise the efficacy of such parasitoids. It is not known if high trichomal density/high trichome PPO activity is closely genetically linked with the production of 2-tridecanone in *Lycopersicon hirsutum* f. *glabratum*. Kennedy's group has found that the presence of 2-tridecanone in trichomes to be incompatible with the action of the parasitoid *Campoletis sonorensis* against *H. zea* (98).

In the absence of PPO, rutin and chlorogenic acid inhibit the replication of a nuclear polyhedrosis AcMNPV in insect tissue culture (*Tricoplusia ni*). These chemicals in artificial diet also strongly reduced the infectivity of HzSNPV in *H. zea* (133). In the presence of PPO and chlorogenic acid, the solubility of the occlusion body of HzSNPV is markedly reduced, and correspondingly, infectivity in larval *H. zea* is reduced up to 90% (124,134). Such a negative impact upon the virus arises from alkylation of polyhedron proteins by chlorogenoquinone. Hence, resistance based on PPO and POD may severely compromise the efficacy of viral control agents in the field. Furthermore, if insect viruses are to be used as vectors of alien genes (e.g., neuropeptides or enzymes; 132,135-137), consideration must be given to the potential of such oxidative enzymes to detract from control. Suprizingly, lipid hydroperoxides produced from the action of LOX on linoleic acid enhanced the infectivity of HzSNPV in H. zea (unpubl. data; Table VIII). In addition, it has been found that the toxicity of purified toxin from Bacillus thuringiensis kurstaki is enhanced up to 50% by alkylation with chlorogenoquinone (Ludlum, Felton, and Duffey, unpubl. data) (Table. VIII). Other chemicals such as PI from soybean and tomatine also enhanced the activity of both NPV and BTk against H. zea (Table VIII).

Table VIII. Effect of Phytochemicals on the Infectivity of NPV and BTK in Heliothis zea

	Mortality Ratio[1]	
Treatment	NPV[2]	BTK[3]
STI	1.34a	1.21a
CHA	0.74a	1.68b
CHA + PPO	0.50b	2.76c
LOX + linolenic acid	2.60c	-----
Rutin	0.53b	1.22a
Tomatine	1.61d	------

STI = soy trypsin inhibitor I in diet at 0.18% wwt; CHA = chlorogenic acid in diet at 3.5 mM/kg diet wwt; rutin at 3.5 mM/kg diet wwt; tomatine at 0.9 mM/kg diet wwt.:
NPN = HzSnPV; LOX = lipoxygenase; PPO = polyphenol oxidase:
[1] mortality ratio = %mortality in treatment diet/mortality in control diet. Ratio <1.0 shows inactivation of infectivity; ratio > 1.0 shows enhancement of infectivity:
[2,3] Fixed dose of pathogen such that larvae ingesting control diet suffered 40-50% mortality. Significant differences between means within a column, based on 95% confidence intervals from ANOVA, are shown by different letters.

Critical Complications in the Use of Oxidative Enzymes as Bases of Resistance against Noctuid Larvae

Our knowledge of how to most effectively utilize PPO, POD, and LOX as antinutritive bases of resistance against noctuid larvae is insufficient. A number of other critical enzymatic and chemical reactions occur in both wounded plant tissue and the insects' digestive fluids that might strongly modulate the overall effect.

PPO has high activity over a broad pH range (5.5 - 10.0) and efficiently oxidizes a variety of caffeic acid derivatives to quinones without requiring a cofactor. PPO is operational through all phases of digestion. In comparison to POD, the use of PPO for resistance is relatively straightforward. Although POD also has high activity over a broad pH range, its full activity is compromised by the fact that it requires H_2O_2 as a cofactor. Although plants are known to produce a localized burst of H_2O_2 at sites of damage (36), it is not known in the tomato system if this burst provides sufficient H_2O_2 to facilitate degradation of protein quality during the early stages of feeding. Both foliage, insect regurgitate, midgut tissues and lumen contents all contain catalase. In fact, the regurgitate alone of larval H. zea significantly impedes plant POD activity because of the presence of high catalase activity (Figure 3). We have already provided evidence that addition of catalase to crushed foliage enhances the ability of H. zea to grow on that foliage (Table IV), presumably through reduction of endogenous levels of plant H_2O_2 with consequent diminution of POD activity. However, in order to formally establish the defensive role of plant POD, one must understand its relationship not only with plant and insect derived catalase, but also with other chemical and enzymatic systems that degrade or generate H_2O_2.

For example, both foliage and the insect's gut fluid contain superoxide dismutase (SOD), an enzyme which converts superoxide ions ($O_2^{\cdot-}$) to H_2O_2 (Figure 3). The formation of superoxide ion is favored in basic conditions (52,138,139), and is also a by-product of semiquinone action on O_2 (73) and of autoxidative reactions (e.g., catecholic phenolics, thiols, and leukoflavans) (139). Also, certain enzymes produce $O_2^{\cdot-}$ or H_2O_2 as end products (Table IX). Hence, the counter-balancing activities of catalase and SOD in the generation of H_2O_2 in the plant and the insect must also be accounted for in order to assess the efficacy of POD as a plant defense.

H_2O_2 can also be generated non-enzymatically from co-oxidative processes (Figure 3) such as via the oxidation of a catecholic phenolic by an o-quinone (139-143). Quinone formation can be the result of PPO activity, but also can arise from spontaneous oxidation in basic media (142,143). Although we have established in vitro that chlorogenic acid and caffeic acid can generate significant quantities of H_2O_2, we have not established whether this process occurs in crushed foliage and/or the insect's gut and simultaneously furnishes sufficient H_2O_2 to permit POD to operate during the digestion of food. It also remains to be determined whether certain phenolics (e.g., rutin, vs chlorogenic acid, vs. caffeic acid) are more efficient than others at generating H_2O_2. Also, certain enzymes present in plant foliage (e.g., catalase, and ascorbic acid peroxidase; 144-146) counteract the production of H_2O_2 by reducing it to water (Figure 3).

It may be possible to breed for the appropriate enzymatic and chemical milieu which will favor a higher pre-injury level of H_2O_2 production, a more rapid post-damage burst, and/or its maintenance during feeding by the insect, thereby permitting the

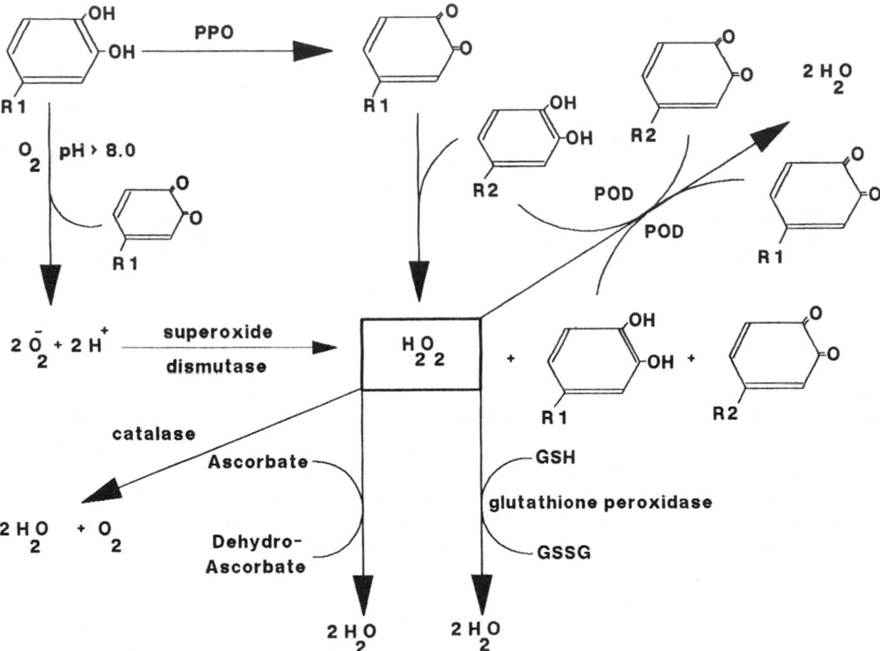

Figure 3. Interrelationship of Oxidative and Reductive Processes Linked to the Production of o-Quinones. POD = peroxidase, PPO = polyphenol oxidase, GSG = glutathione, GSSG = oxidized glutathione.

joint antinutritive action of PPO and POD. The simultaneous use of LOX may be complicated by the fact that catechols such as rutin are known to inhibit LOX activity (147) as well as scavenge free radicals generated during these oxidative processes (148).

Other interactive factors which will determine the ability of oxidized phenolics (quinones and semiquinones) to damage protein are the levels of ascorbic acid, glutathione, and other reductants in foliage and the insect's digestive system. The maintenance of high levels of H_2O_2, to run POD, may be compromised by high levels of ascorbic acid which can reduce H_2O_2 to water (144-146)(Figure 3). Furthermore, high levels of ascorbic acid are also capable of reducing quinones to phenolics, which then can counteract the antinutritive effects of quinone production (Figures 1,2, & 3). *In vitro*, the presence of ascorbic acid impairs the production of chlorogenoquinone by PPO. However, this counteractive effect may be overcome by the enzyme ascorbic acid oxidase (AOX). which oxidizes ascorbic acid to dehydroascorbic acid. Addition of AOX to artificial diets containing ascorbic acid, chlorogenic acid, and PPO causes a greater reduction in larval growth (unpubl. data). Tomato foliage contains both high levels of ascorbic acid and AOX (60; unpubl. data). Thus, if one were to breed for high levels of AOX, quinone production would occur more rapidly because of lowered levels of the reductant/nutrient ascorbic acid (Figures 3 & 4).

This multi-enzyme approach has several potential advantages for controlling noctuid larvae. First, ascorbic acid is a nutritional requirement for such lepidopterans (27,30,149); its oxidation to dehydroascorbic acid demands that the insect use reducing power in the form of glutathione and NAD(P)H to reclaim ascorbic acid. The simultaneous production of quinones may also place a drain on reducing power because glutathione is readily alkylated by *o*-quinones (140,141; unpubl data) rendering it unreclaimable. Glutathione may also reduce *o*-quinones directly or with an intervening step involving ascorbic acid, thereby placing a further drain on reducing power (Figure 4). If H_2O_2 is detoxified by glutathione peroxidase a further drain is placed on reducing power. Glutathione peroxidase and glutathione may also be involved in detoxication of products from lipid peroxidation resulting from LOX activity (Figure 3). Reduced sulphur amino acid intake as a result of the action of PI's or quinones may may further exacerbate the requirement for reducing power and glutathione. Hence, this multiple drain may be of significance in controlling the insect if high levels of reducing power are required to simulataneously detoxify other ingested toxins such insecticides or other natural products.

Since PI's, PPO, POD, and LOX have the potential to impair sulphur amino acid intake and utilization by these insects, and such intake is important for detoxication (e.g., glutathione), it may be possible to control *H. zea* and *S. exigua* not just through antinutritive effects but also through impairment of detoxicative abilities. However, a deeper knowledge of these insects' detoxicative abilities is essential if these enzymes are to be used successfully. Futher complications arise, for like the foliage they ingest (146,148,150-153), their guts also contain inherent catalase, superoxide dismutase, glutathione peroxidase,

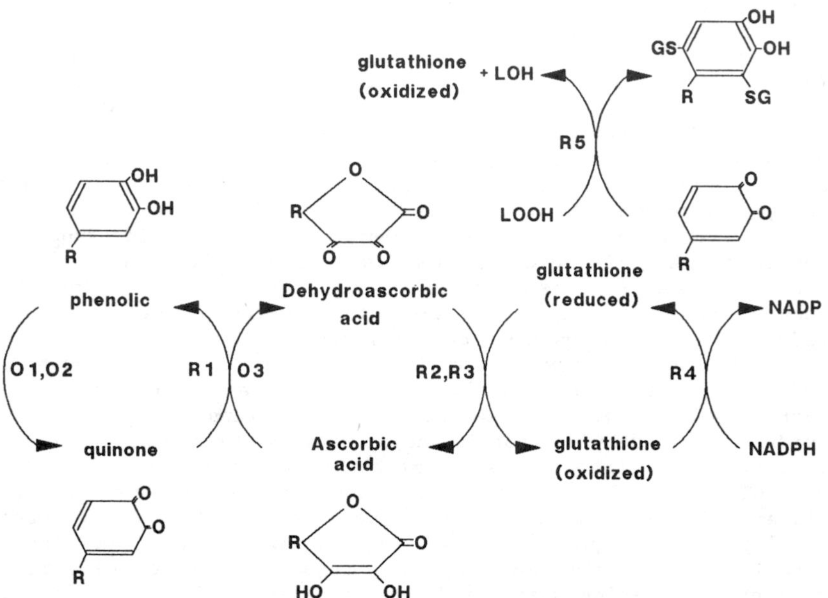

Figure 4. Some Interrelationships Between Quinones, Ascorbic acid, Glutathione, and Reducing Power. The cycle of redox events can occur from left to right without intervention of enzymes. Certain enzymes can facilitate these reactions: O1 = polyphenol oxidase, O2 = peroxidase, R1 = quinone reductase without ascorbic as a substrate, O3 = ascorbic acid oxidase, R2 = ascorbic acid free radical reductase, R3 = dehydroascorbic acid reductase, R4 = glutathione reductase, R5 = glutathione peroxidase, LOOH = lipid hydroperoxide, and GS = covalently bound glutathione.

peroxidase, glutathione reductase, and other enzymes and reductants (147,154-157). The consequences of the balance between the plant driven chemical and/or enzymatic reactions and those of the insect are poorly understood. A variety of plant enzymes (Table IX) generate toxic by-products. It certainly may be possible to simultaneously challenge the insect's acquisition of nutrients, its balance of detoxicative reducing power, and its ability to handle the dietary and bodily generation of superoxide ions, hydroxyl ions, free radicals and H_2O_2 (52,139,147,154,155,158). The degree to which these plant factors can be genetically manipulated without impairing the plant's productivity, is undetermined.

Table IX. Some Oxidants and Reductants/Antioxidants in Plants

Oxidants and By-products	Reductants/Antioxidants
Aldehyde oxidase (superoxide ion)	glutathione
Glucose oxidase (superoxide ion)	phenolics
Xanthine oxidase (superoxide ion)	polyamines
Lipoxygenase (hydroperoxides, epoxides, free radicals)	carotene
	superoxide dismutase
Polyphenoloxidase (quinones)	glutathione peroxidase and reductase
Peroxidase (quinones, semiquinones free radicals)	catalase
Chlorophyll (in light)	chlorophyll (in dark)
Flavan dehydrogenase (superoxide ion)	vitamin E
Galactose oxidase (superoxide ion)	ascorbic acid
H_2O_2 generators	
Superoxide dismutase	
Amine oxidase	
Polyamine oxidase	
Glycolate oxidase	
Uric oxidase	
Autooxidation and cooxidation of/by phenolics	

See references 52,139,146,148,151,152,159.

Multiple Onslaughts against Acquisition of Nutrients

We have described an approach to host plant resistance that involves the use of plant oxidative enzymes to irrevocably deprive the insect of nutrients. We have emphasized that the chemical reactions catalyzed by POD and PPO have the potential to destroy a variety of essential or limiting amino acids (Table X). In particular, these reactions are adept at destroying lysine and cysteine. Integral lysine is necessary for proper enzymatic hydrolysis of protein. Cysteine and methionine, amongst other uses, are required to synthesize trypsin. The action of PPO and POD in conjunction with PI's are proposed to place a severe strain on the insect for high sulphur amino acid intake. This strain may be further exacerbated by the complementary action of quinones depleting available glutathione via formation of conjugates. Furthermore, simultaneous action of AOX may limit the quantity of essential ascorbic acid and place a further strain on chemical and enzymatic reducing power. These oxidative conditions also have the potential to destroy essential nutrients such as tocopherol (151) and thiamine (160).

Table X. Some Nutrients Destroyed or Rendered Less Available by Plant Chemical or Enzymatic Interactions

Agent	Nutrient
Ascorbic acid oxidase	ascorbic acid
Lipoxygenase	linolenic and linoleic acids, ß-carotene amino acids (lysine, cysteine, histidine)
Peroxidase	protein (lysine,cysteine, methionine, histidine, tyrosine, tryptophan), above free amino acids, ascorbic acid, thiamine
Phenylalanine ammonia lyase	phenylalanine
Polyphenol oxidase	protein (lysine, cysteine methionine, histidine, tyrosine), above free amino acids, ascorbic acid, thiamine
Proteinase inhibitors	digestion and utilization of protein (sulphur amino acids)
Tomatine	cholesterol, sitosterol and related phyto-sterols
Tyrosine ammonia lyase	tyrosine

LOX destroys linoleic and linolenic acids via their oxidation to lipid hydroperoxides. These lipid hydroperoxides subsequently form hydroperoxides, hydroperoxide free radicals, epoxides and malondialdehyde which can impair the nutritive quality of protein via mechanisms similar to those mediated by POD and PPO. These unsaturated fatty acids are essential for normal larval growth and maturation.

We have presented evidence (124) that phenylalanine and tyrosine ammonia lyases, two enzymes induced during wounding, have the potential to limit the insects' intake of free phenylalanine and tyrosine because they are converted to nutritionally inert cinnamic acid derivatives.

Insects also require an exogenous source of phytosterols (27). The glycoalkaloid tomatine is an effective precipitator of certain phytosterols such as sitosterol and cholesterol, and as such may provide a means of reducing sterol intake for noctuid larvae (97).

The feasibility of using these antinutritive plant systems as multiple-factor/multiple-mechanism resistance against noctuid larvae remains to be determined. It is possible that such a multiple onslaught against nutrient acquisition is redundant. In other words, perhaps merely the use of PPO and chlorogenic acid is sufficient. It also remains to be determined whether this proposed multiple-factor/multiple-mechanism of resistance renders the insects' detoxicative systems more susceptible to traditional control tactics, and whether the evolution of resistance to such multiple antinutritive factors is more difficult than to insecticides.

Advantages of the Use of Enzymatic Defenses

Our evidence supports the contention that plant oxidative enzymes can be used as constitutive and/or inducible antinutritive bases of resistance against insects. This resistance is based on the irreversible chemical degradation of multiple essential or limiting nutrients, which may be more difficult for the insect species to evolve biochemical resistance against than against classical "toxins".

Other advantages also accrue from their use. Because they are enzymes, only catalytic amounts are required to drive the reactions, provided substrates are not limiting. If one is concerned about the cost of defense rendering the plant less agronomically efficient (15), when employing a battery of secondary gene products as the bases of resistance (e.g., tomatine, phenolics, and 2-tridecanone), perhaps the utilization of PPO and POD in conjunction with phenolics is more efficient. The synthesis of catalytic amounts of enzyme should place less metabolic drain on the plant than synthesizing one or several secondary products that usually require levels of 0.1% and above to be active.

In terms of breeding programs, it should be easier to manipulate the expression of primary gene products through classical (e.g., introducing exotic genes from related species) or modern biotechnological procedures (e.g., amplifying gene expression) than to manipulate the expression of secondary gene products. Considering the increasingly severe constraints upon registering genetically modified organisms for commercial use, it might be easier to register plants that contain only catalytic quantities of enzymes, derived from related species, than to register plants that must express high levels of transgenic gene products (e.g., BT toxin, lectins, and PI's).

Other advantages bear repeating. These enzymes have been implicated in resistance to pathogens, and hence, their directed utilization against insects may complement resistance to pathogens. Three enzymes (PPO, POD, and LOX) have broad pH profiles which permit their operation in the acidic conditions of immediately crushed foliage (pH = 5.5) and in the basic conditions of the insect's gut (pH 8.5). Furthermore, PPO and POD are almost completely resistant to digestion by tryptic and chymotryptic enzymes and thus remain active in the insect's gut during digestion of food. PPO is actually activated by the gut proteases. These plant enzymes are inducible by various kinds of feeding damage, which should exacerbate their effects. And, finally, the phenology of the tomato, like most plants (151,161-164), favors the action of these enzymes. As tomato plants age they become more oxidative by having levels of PPO, POD, and LOX increase, while levels of plant nitrogen and catalase fall (79; unpubl. data). Even in the green fruit stage, tomato plants are oxidative. It may be possible to accentuate the oxidative state of the plant throughout its life to facilitate resistance. However, this effort might be at odds with others'attempts to breed less "oxidative" plants in order to decrease susceptibility to herbicides (165).

This form of resistance is targeted against insects with basic guts (i.e., lepidopteran larvae). The basic gut environment favours oxidative conditions and the types of reactions proposed (e.g., Schiff base formation, alkylation, co-oxidation, and autooxidation). However, with insects (e.g., the Colorado Potato beetle <u>Leptinotarsa</u> <u>decemlineata</u>) having acidic gut fluid (pH 6.0 - 6.5), the proposed mechanisms may be of little value. In acidic conditions, Schiff base formation and alkylation of nucleophilies are not favored because the nucleophilic groups are protonated. We have shown in larval <u>L</u>. <u>decemlineata</u> that plant protein is not significantly alkylated by PPO and chlorogenic acid (unpubl. data). Alternate tactics might be necessary for simultaneous control of larval beetles and noctuids.

<u>Conclusion</u>

We have proposed the use of several plant enzymes as a polygenic basis of resistance against noctuid larvae through activation of both "toxins" and "nutrients" to forms that chemically reduce the nutritional value of the tomato plant. Although this approach may be useful in developing resistance that is durable, the use of such enymes is complicated by their mutual interactions and the chemical context of the plant, by the detoxicative abilities of the insect, and by their unpredictable effects upon biological control agents. Such an approach may be warranted in view of public disdain for pesticides; such resistance may offer not only simultaneous resistance against pathogens and insects, but also offer resistance that is environmentally safe, thus, lessening reliance on conventional control tactics.

On a less optimistic note, the forthright utilization of the above antinutritive defenses may be pragmatically difficult from the standpoint of breeding. The expression of phenolics, and likely many of the other characters, are under quantitative genetic control. Hence, deriving predictable resistance, as prescribed above, without the highly modulating effects of environment and potential gene interactions may present its own set of problems. Furthermore, many of the candidate enzymes are mutually and intimately involved in the plant's defense against microbes, in detoxication and general metabolism, and in processes of maturation and senescence. The effects of manipulating the expression of such enzymes on the general agronomic value of the plant is unknown.

Acknowledgments

This work was supported by the USDA Competitive Grants Program through grants USDA 89-37250-4639 to S.S.D and G W.F., and USDA 87-CRCR-1-2371 to S.S.D.

Literature Cited

1. Hare, J. D. In Variable Plants and Herbivores in Natural and Managed Systems; Denno, R. F.; McClure, M. S., Eds; Academic Press: New York, 1983; Chapter 18.
2. Mattson, W. J. Ann. Rev. Ecol. Syst. 1980, 11, 119-61.
3. Mattson, W. J.; Haack, R. A. BioScience 1987, 37, 110-18.
4. Price, P. W. Insect Ecology; John Wiley & Sons: New York, 1984; p 606.
5. Scriber, J. M. In Chemical Ecology of Insects; Bell, W. J.; Carde, R. T., Eds.; Sinauer Associates Inc.: Sunderland, Massachusetts, 1984b; pp 159-202.
6. Strong, D. R.; Lawton, J. H.; Southwood, Sir R. Insects On Plants. Community Patterns and Mechanisms; Harvard University Press: Cambridge, Massachusetts; 1984.
7. White, T. C. R. Oecologia 1984, 63, 90-105.
8. Barbosa, P.; Schultz, J. C., Eds. Insect Outbreaks; Academic Press: New York, 1987; p 578.
9. Coley, P. D.; Bryant, J. P.; Chapin, F. S. Science 1985, 230, 895-99.
10. Denno, R. F.; McClure, M. S., Eds. Variable Plants and Herbivores in Natural and Managed Systems; Academic Press: New York; 1983, p 717.
11. Mattson, W. J.; Levieux, J.; Bernard-Dagan, C. Eds. Mechanisms of Woody Plant Defenses Against Insects. Search for Pattern; Springer-Verlag: Berlin, 1988; p 416.
12. McClure, M. S. In Variable Plants and Herbivores in Natural and Managed Systems; Denno, R.F.; McClure, M. S., Eds.; Acadmeic Press: New York, 1983, pp 135-54.
13. Rhoades, D. F. In Variable Plants and Herbivores in Natural and Managed Systems; Denno, R.F.; McClure, M. S., Eds.; Acadmeic Press: New York, 1983, pp 155-220.
14. Rhoades, D. Amer. Nat. 1985, 125, 205-38.
15. Kogan, M. In IPM and Ecological Practise; Kogan, M., Ed.; John Wiley and Sons: New York, 1986; pp 83-134.
16. Maxwell, F. G.; Jennings, P. R., Eds. Breeding Plants Resistant to Insects; Wiley Interscience: New York, 1980; p 683.
17. Panda, N. Principles of Host-Plant Resistance to Insect Pests; Allanheld/Universe: New York, 1979; p 386.
18. Smith, C. M. Plant Resistance to Insects. A Fundamental Approach; John Wiley and Sons: New York, 1989; p 286.

19. Feeny, P. P. Coevolution of Animals and Plants; Gilbert, L. E.; Raven, P. H., Eds.; University of Texas Press: Austin, 1975, pp 1-19.
20. Moore, R. F. J. Econ. Entomol. 1983, 76, 697-99.
21. Rhoades, D. F. In Herbivores: Their Interaction with Secondary Plant Metabolites; Rosenthal, G. A.; Janzen, D. H., Eds.; Academic Press: New York, 1979; pp 1-54.
22. Bernays, E. A. Ecol. Entomol. 1981, 6, 353-60.
23. Martin, J. S.; Martin, M. M. Oecologia 1982, 54, 205-311.
24. Martin, J. S.; Martin, M. M.; Bernays, E. A. J. Chem. Ecol. 1987, 13, 605-21.
25. Reese, J. C.; Chan, B. G.; Waiss, A. C., Jr. J. Chem. Ecol. 1982, 8, 1429-36.
26. Zucker, W. V. Amer. Natur. 1983, 121, 335-65.
27. Dadd, R. H. Ann. Rev. Entomol. 1973, 18, 381-419.
28. Brodbeck, B.; Strong, D. In Insect Outbreaks; Barbosa, P.; Schultz, J. C., Eds.; Academic Press: New York, 1987; pp 347-64.
29. van Emden, H. F. In Phytochemical Ecology; Harborne, J.,Ed.; Academic Press: London, 1972; pp 25-43.
30. Penny, L. H.; Scott, G. E.; Guthrie, W. D. Crop Sci. 1967, 7, 407-9.
31. Birk, Y.; Peri, I. In Toxic Constituents of Plant Foodstuffs; Liener, I. E., Ed.; Academic Press: New York, 1980; pp 161-82.
32. Bloem, K. A.; Kelley, K. C.; Duffey, S. S. J. Chem. Ecol. 1989, 15, 387-98.
33. Silano, V. Cereal Chem. 1978, 55, 722-31.
34. Silano, V.; Zahnley, J. C. Biochim. Biophys. Acta. 1978, 533, 181-5.
35. Shukle, R. H.; Murdock, L. L. Environ. Entomol. 1983, 12, 787-91.
36. Apostol, I.; Heinstein, P. F.; Low, P. S. Plant Physiol. 1989, 90, 109-16.
37. Boller, T. In Cellular and Molecular Biology of Plant Stress; Key, L.; Kosuge,T., Eds; Alan R. Liss, Inc.: New York, 1985; pp 247-62.
38. Boller, T. In Plant-Microbe Interactions. Molecular and Genetic Perspective; Kosuge, T.; Nester, E. W., Eds.; MacMillan Publ. Company: New York, 1987; pp 385-413.
39. Davies, E. The Biochemistry of Plants. A Comprehensive Treatise. Physiology of Metabolism; Davies, E., Ed.; Academic Press: New York, 1987; Vol. 12, pp 243-64.
40. Esquerre-Tuagaye, M. T.; Mazau, D.; Pelissier, B.; Roby, D.; Runeau, D.; Toppan, A. In Cellular and Molecular Biology of Plant Stress; Key, J. L.; Kosuge, T., Eds.; Alan R. Liss Inc.: New York, 1985; pp 459-73.
41. Gianinazzi, S. Plant-Microbe Interactions. Molecular and Genetic Perspective; Kosuge, T.; Nester, E. W., Eds.; MacMillan Publ. Company: New York, 1984; pp 321-42.
42. Hahlbrock, K.; Scheel, D. In Innovative Approaches to Plant Disease Control; Chet, I.,Ed; Wiley and Sons: New York, 1987; pp 228-54.

43. Kahl, G. In The New Frontiers in Plant Biochemistry; Akazawa, T.; Asahia, T; Imaseki, H., Eds; Martinus Nijhoff/Dr. W. Junk Publishers: The Hague, 1983; pp 193-216.
44. Kuc, J.; Lisker, N. In Biochemistry of Wounded Plant Tissues. Kahl, G., Ed.; Walter de Gruyter: Berlin, 1987; pp 203-42.
45. Rhodes, J. M.; Wooltorton, L. S. C. In Biochemistry of Wounded Plant Tissues; Kahl, G., Ed.; Walter de Gruyter: Berlin, 1978; pp 243-86.
46. Ryan, C. A.; Bishop, P. D.; Graham, J. S.; Meyer-Broadway R.; Duffey, S. S. J. Chem. Ecol. 1985, 12, 1025-36.
47. Butt, V. S. In The Biochemistry of Plants. A Comprehensive Treatise. Metabolism and Respiration; Stumpf, P. K.; Conn,.E. E., Eds; Academic Press: New York, 1981; Vol. 2, pp 81-123.
48. Fry, S. C. Ann. Rev. Plant. Physiol 1986, 37, 165-86.
49. Galliard, T. In Biochemistry of Wounded Plant Tissues; Kahl, G., Ed.; Walter de Gruyter: Berlin, 1978; pp 155-201.
50. Gardner, H. W. J. Agric. Food Chem. 1979, 27, 220-29.
51. Ampomah, Y. A.; Friend, J. Phytochemistry 1988, 27, 2533-41.
52. Elstner, E. F. In The Biochemstry of Plants. A Comprehensive Treatise. Biochemistry and Metabolism; Davies, D. D., Ed.; Academic Press: New York, 1980; Vol. 11, pp 253-315.
53. Mayer, A. M. Phytochemistry 1987, 26, 11-20.
54. Mayer, A. M.; Harel, E. Phytochemistry 1979, 18, 193-215.
55. Mohan, R.; Kolattukudy, P. Plant Physiol. 1990, 921, 276-80.
56. Pierpoint, W. S.; Ireland, R. J.; Carpenter, J. M. Phytochemistry 1977, 16, 29-34.
57. Stahmann, M. A. In The New Frontiers in Plant Biochemistry; Akazawa, T.; Asahia, T.; Imaseki, H., Eds; Martinus Nijhoff/Dr. W. Junk Publishers: The Hague, 1983; pp 237-50.
58. Galliard, T.; Matthew, J. A. Phytochemistry 1977, 16, 339-43.
59. Gentile, I. A.; Ferraris, L.; Matta, A. J. Phytopathol. 1988, 122, 45-53.
60. Hobson, G. E.; Davies, J. N. In The Biochemistry of Fruits and Their Products.; Hulme, A. C., Ed.; Academic Press: New York, 1971; Vol. 2, pp 437-82.
61. Raman. K.; Sanjayan, K. P. Proc. Indian Acad. Sci. (Animal Sci.), 1984, 93(6), 543-7.
62. Reuveni, R.; Ferreira, J. F. Phytopath. Z. 1985, 112, 193-7.
63. Hurrell, R. F.; Finot, P. A.; Cuq, J. L. J. Nutrition 1982, 47, 191-211.
64. Vaughn, K. C.; Duke, S. O. Physiol. Plant 1984, 60, 106-12.
65. Hurrell, R. F.; Finot, P. A. In Digestibility and Amino Acid Availability in Cereals and Oilseeds; Finley, J. W.; Hopkins, D. T., Eds.; American Association of Cereal Chemists, Inc.: St. Paul, 1982; pp 233-46.
66. Davies, A. M. C.; Newby, V. K.; Synge, R. L. M. J. Sci. Food Agric. 1978, 29, 33-41.

67. Leatham, G. F.; King, V.; Stahmann, M. A. Phytopathology 1980, 70, 1134-40.
68. Matheis, G.; Whitaker, J. R. J. Food Biochem. 1987, 11, 309-27.
69. Pierpoint, W. S. In Leaf Protein Concentrates; Telek, L.; Graham, H. D., Eds.; Avi Publ. Comp., Inc.: Wesport, Connecticut, 1983; pp 235-67.
70. Motoda, S. J. Ferment. Technol. 1979, 57, 395-9.
71. Stahmann, M. A.; Spencer, A. K.; Honold, G. R. Biopolymers 1977, 16, 1307-18.
72. Whitmore, F. W. Plant Sci. Letters 1978, 13, 241-5.
73. Kanner, J.; German, J. B.; Kinsella, J. E. In Critical Reviews in Food Science and Nutrition; Furia, T. E., Ed.; CRC Press: Boca Raton, 1987; pp 317-65.
74. Vick, B. A.; Zimmerman, D. C. In The Biochemistry of Plants. A Comprehensive Treatise. Lipids: Structure and Function; Stumpf, P. K., Ed.; Academic Press: New York, 1987; Vol. 9, pp 53-90.
75. Belitz, H.-D.; Grosch, W., Eds. Food Chemistry. (translated from German by D. Hadziyev); Springer Verlag: Berlin, 1987; p 774.
76. Matsushita, S. J. Agric. Food Chem. 1975, 23, 150-5.
77. Ory, R. L.; St. Angelo, A. J. In Food Protein Deterioration, Mechanisms and Functionality; Cherry, J. P., Ed.; American Chemical Society: Washington, DC, 1982; pp 55-66.
78. Wills, E. D. In Biochemical Toxicology, A Practical Approach; Snell, K.; Mullock, B., Eds.; IRL Press: Washington, DC, 1985; pp 127-52.
79. Felton, G. W.; Donato, K.; Del Vecchio, R. J.; Duffey, S. S. J. Chem. Ecol. 1989, 15, 2667-94.
80. Dryden, M. J.; Saterlee, L. D. J. Food Science 1978, 43, 650-1.
81. Eklund, A. Nutr. Metab. 1975, 18, 258-264.
82. Friedman, M. Protein Nutritional Quality of Foods and Feeds; Marcel Dekker, Inc.: New York, 1975.
83. Horigome, T.; Kandatsu, T. Agric. Biol. Chem. 1968, 32, 1093-1102.
84. Hurrell, R. F.; Finot, P. In Nutritional and Toxicological Aspects of Food Safety; Friedman, M., Ed.; Plenum Press: New York, 1984; pp 423-36.
85. Jung, H.-J. G.; Fahey, G. C., Jr. J. Agric. Food Chem. 1981, 29, 817-20.
86. Lahiry, N. L.; Satterlee, H. W.; Wallace, G. W. J. Food Science 1977, 42, 83-85.
87. Lee, H. S.; Louriminia, S. S.; Clifford, A. J.; Whitaker, J. R.; Feeny, R. A. J. Nutrition 1978, 108, 687-97.
88. Barbeau, W. E.; Kinsella, J. E. J. Agric. Food. Chem. 1983, 31, 993-98.
89. Barbeau, W. E.; Kinsella, J. E. J. Food Sci. 1985, 50, 1083-1100.
90. Finot, P. A. Qual. Plant Foods Hum. Nutr. 1983, 32, 439-53.
91. Anderson, P. A. In Digestibility and Amino Acid Availability in Cereals and Oilseeds; Finley, J. W.;

Hopkins, D. T., Eds.; Amer. Assoc. Cereal Chemists, Inc.: St. Paul, 1985; pp 31-44.
92. Igarashi, K.; Tsunekuni, T.; Yasui, T. J. Nutrition Sci. Vitaminol. 1983, 29, 227-32.
93. Rock, G. C.; Hogdson, E. J. Insect Physiol. 1971, 17, 1087-97.
94. Broadway, R. M.; Duffey, S. S. J. Insect Physiol. 1986, 32, 673-80.
95. Broadway, R. M.; Duffey, S. S. J. Insect Physiol. 1986, 32, 827-33.
96. Broadway, R. M.; Duffey, S. S. J. Insect Physiol. 1988, 34, 1111-17.
97. Duffey, S. S.; Bloem, K. A. In Ecological Theory and IPM Practice; Kogan, M., Ed.; John Wiley & Sons: London, 1986; pp 135-83.
98. Kauffman, W. C.; Kennedy, G. G. J. Chem. Ecol. 1989, 15, 2051-60.
99. Duffey, S. S. In Insects and Plant Surfaces; Southwood, Sir R.; Juniper, B., Eds.; Edward Arnold: London, 1986; pp 151-72.
100. Hall, C. B.; Knapp, F. W.; Stall, R. E. Phytopathology 1969, 59, 267-8.
101. Rick, C. M.; Fobes, J. F. Proc. Nat. Acad. Sci. 1976, 73, 900-4.
102. Duffey, S. S.; Isman, M. B. Experientia 1981, 37, 574-6.
103. Elliger, C. A.; Wong, Y.; Chan, B. G.; Waiss, A. C., Jr. J. Chem. Ecol. 1981, 7, 753-8.
104. Strack, D.; Gross, W. Plant Physiol. 1990, 92, 41-7.
105. Ryan, J. D.; Gregory, P.; Tingey, W. M. Phytochem. 1982, 8, 1185-7.
106. Carrasco, A.; Boudet, A. M.; Marigo, S. Physiol. Plant Pathol. 1978, 12, 225-32.
107. Chadha, K. C.; Brown, S. A. Can. J. Bot. 1974, 52, 2041-7.
108. Matta, A.; Gentile, I.; Giai, I. Phytopathol. 1969, 59, 512-13.
109. Mendez, J.; Brown, S. A. Can. J. Bot. 1971b, 49, 2101-05.
110. Pollock, C. J.; Drysdale, R. B. Phytopath. Z. 1976, 86, 56-66.
111. Buttery, R. G.; Teranishi, R.; Ling, L. C. J. Agric. Food Chem. 1987, 35,540-546.
112. Buttery, R. G.; Ling, L. C.; Light, D. M. J. Agric. Food Chem. 1987, 35, 1039-42.
113. Zamora, R.; Olias, J. M.; Mesias, J. L. Phytochemistry 1987, 26, 345-7.
114. Hildebrand, D. F. Physiol. Plant. 1989, 76, 249-53.
115. Hildebrand, D. F.; Rodriquez, J. G.; Brown, G. C.; Luu, K. T.; Volden, C. S. J. Econ. Entomol. 1986, 79, 1459-65.
116. Neese, P. A. Poster STD0017; Entomol. Soc. Amer. Natl. Conf. Exh., Dec., 1988, Louisville, Kentucky.
117. Ryan, C. A. In Advances in Plant Gene Research; Verma, O. S. P.; Hohns, T., Eds.; Springer-Verlag: Berlin, 1984; pp 321-32.
118. Weiel, J.; Hapner, K. D. Phytochemistry 1976, 15, 1885-7.
119. Thornburg, R. W.; Kernan, A.; Molin, L. Plant Physiol. 1990, 92, 500-5.

120. Foard, D. E.; Murdock, L. L.; Dunn, P. E. In Plant Molecular Biology; Alan R. Liss, Inc.: New York, 1983; pp 223-33.
121. Hilder, V. A.; Angharad, A. M. R.; Gatehouse, Sheerman S. E.; Barker, R. E.; Boulter, D. Nature 1987, 300, 160-3.
122. Thornburg, R. W.; Cleveland, A. G.; Johnson, R.; Ryan, C. A. Proc. Natl. Acad. Sci. 1987, 84, 744-8.
123. Broadway, R.; Duffey, S. S.; Pearce, G.; Ryan, C. A. Entomol. Exp. Appl. 1986, 41, 33-8.
124. Duffey, S. S.; Felton, G. W. In Biocatalysis in Agricultural Biotechnology; Amer Chem. Soc. Symp. Ser. 389, 1989; pp 289-313.
125. Felton, G. W.; Broadway, R. M.; Duffey, S. S. J. Insect Physiol. 1989, 35, 981-90.
126. Lange, W. H.; Bronson, L. Ann. Rev. Entomol. 1981, 26, 345-71.
127. Kennedy, G. G. Entomol. Soc. Amer. Bull. 1978, 24, 375-84.
128. Dimock, M .A.; Kennedy, G. G. Ent. Exp. Appl. 1983, 33, 263-8.
129. Lee, J. S.; Brown, W. E.; Graham, J. S.; Pearce, G.; Fox, E. A.; Dreher, T. W.; Ahren, K. G.; Pearson, G. D.; Ryan, C. A. Proc. Natl. Acad. Sci. 1986, 83, 7277-81.
130. Liener, I. E. Ed. Toxic Constituents of Plant Foodstuffs; Academic Press: New York, 1980; p 500.
131. Plunkett, G.; Senear, D. F.; Zuroske, G.; Ryan, C. A. Arch. Biohem. Biophys. 1982, 213, 463-72.
132. Meeusen, R. L.; Warren, G. Ann. Rev. Entomol. 1989, 34, 373-81.
133. Felton, G. W.; Duffey, S. S.; Vail, P. V.; Kaya, H. K.; Manning, J. J. Chem. Ecol. 1987, 13, 947-57.
134. Felton, G. W.; Duffey, S. S. J. Chem. Ecol. 1989, 36.
135. Barton, K. A.; Whiteley, H. R.; Yang, N.-S. Plant Physiol. 1987, 85, 1103-09.
136. Hammock, B. D.; Harshman, L. G.; Philpott, M. L.; Szekacs, A.; Ottea, J. A.; Newitt, R. A.; Wroblewski, V. J.; Halarnakar, P. P.; Hanzlik, T. N. In Biomechanisms Regulating Growth and Development; Steffens, G. L.; Rumsey, T. S., Eds; Kluwer Academic Publishers: The Netherlands, 1988; pp 137-73.
137. Maeda, S. Ann. Rev. Entomol. 1989, 34, 351-72.
138. Cadenas, E. Ann. Rev. Biochem. 1989, 58, 79-110.
139. Fridovich, I. Science 1978, 201, 875-80.
140. Cheynier, V. F.; Van Hulst, M. W. J. J. Agric. Food Chem. 1988, 36, 10-14.
141. Cheynier, V.; Osse, C.; Rigaud, J. J. Agric. Food Science 1988, 53, 1729-33.
142. Cohen, G.; Heikkila, R. E. J. Biol. Chem. 1974, 249, 2447-52.
143. Hanham, A. F.; Dunn, B. P.; Stich, H. F. Mutation Research 1983, 116, 333-9.
144. Dalton, D. A.; Russell, S. A.; Hanus, F. J.; Pascoe, G. A.; Evans, H. J. Proc. Natl. Acda. Sci. 1986, 83, 3811-15.
145. Dalton, D. A.; Hanus, F. J.; Russell, S. A.; Evans, H. J. Plant Physiol. 1987, 83, 789-94.
146. Rennenberg, H. Phytochemistry 1982, 21, 2771-81.

147. Pritsos, C. A.; Ahmad, S.; Bowen, S. M.; Elliot, A. J.; Blomquist, G. J.; Pardini, R. S. Arch. Insect Biochem. Physiol. 1988, 8, 101-12.
148. Torel, J.; Cillard, J.; Cillard, P. Phytochemistry 1986, 25, 383-5.
149. Kramer, K. J.; Hendricks, L. H.; Liang, Y. T.; Seib, P. A. J. Agric. Food Chem. 1978, 26, 874-8.
150. Barraccino, G.; Dipierro, S.; Arrigoni, O. Phytochemistry 1989, 28, 715-17.
151. Kunert, K. J.; Edcere, M. Physiol. Planta 1985, 65, 85-8.
152. Larson, R. A. Phytochemistry 1988, 27, 969-78.
153. Meister, A. Science 1982, 220, 472-8
154. Ahmad, S.; Pritsos, C. A.; Bowen, S. M.; Kirkland, K. E.; Blomquist, G. J.; Pardini, R. S. Arch. Insect Biochem. Physiol. 1987, 6, 85-96.
155. Ahmad, S.; Pritsos, C. A.; Bowen, S. M.; Heisler, C. R.; Blomquist, G. J.; Pardini, R. S. Free Rad. Res. Comms. 1988, 4, 403-8.
156. Ahmad, S.; Pritsos, C. A.; Bowen, S. M.; Heisler, C. R.; Blomquist, G. J.; Pardini, R. S. Arch. Insect Biochem. Physiol. 1988, 7, 173-86.
157. Pritsos, C. A.; Ahmad, S.; Bowen, S. M.; Blomquist, G. J.; Pardini, R. S. Comp. Biochem. Physiol. 1988, 90C, 423-7.
158. Fridovich, I. J. Biol. Chem. 1989, 264, 7761-4.
159. Khan, V. Phytochemistry 1983, 22, 2155-9.
160. Panijpan, B.; Ratanaubolchai, K. Internatl. J. Vit. Nutr. Res. 1980, 50, 247-53.
161. Choudhuri, M. A. Plant. Physiol. Biochem. 1988, 15, 18-29.
162. Dhindsa, R. S.; Plumb-Dhindsa, P.; Thorpe, T. A. J. Exp. Botany 1981, 32, 93-101.
163. Stewert, R. R. C.; Bewley, J. D. Plant Physiol. 1980, 65, 245-8.
164. Thompson, J. E. In Senescence and Aging in Plants; Nooden, L. D.; Leopold, A. C., Eds.; Academic Press: New York, 1988; pp 51-83.
165. Schmidt, A.; Kunert, K. J. Molecular Strategies for Crop Protection; Arntzen, C. J.; Ryan, C.; Eds.; Alan R. Liss: New York, 1987; p 443.

RECEIVED July 31, 1990

Chapter 13

Phytoalexins and Their Potential Role in Control of Insect Pests

Jack D. Paxton

Department of Plant Pathology, University of Illinois, 1102 S. Goodwin, Urbana, IL 61801

Plant phytoalexins [natural plant antibiotics] [1] have the potential of becoming a new class of useful compounds in the control of insect pests. Some phytoalexins have been demonstrated as deterrents to insect feeding. Considerable progress has been made to characterize them chemically and to extend the study of their function in plant disease resistance, but exploration of their role in the control of insect pests is just beginning.

Plant Defenses Against Stresses

Plants have developed many responses and ways of defending themselves against attacks by pathogens and insects. Some of these defenses are preformed such as the plant cuticle, which provides a barrier against desiccation and which possibly even creates an inhospitable environment for various pathogens. Another preformed defense against stress is lignification of cell walls. This creates a physical barrier to mechanical penetration by fungi or insect mouthparts, however, the most common preformed defense that has been studied to date is the accumulation of chemical deterrents (allomones) [2].

In preformed defenses the plant invests considerable energy and other metabolic resources in a generalized defense against some stresses that may never materialize.

0097–6156/91/0449–0198$06.00/0
© 1991 American Chemical Society

These defenses have the advantage of being in place when the insect arrives at the plant thereby preventing appreciable damage to the plant. A possible disadvantage is that a considerable amount of the plant's resources are devoted to a defense that may not be necessary and which might ultimately reduce it's ability to reproduce and compete for an ecological niche and resources.

Between preformed (constitutive) and inducible defenses exists a wide range of compounds that are essentially preformed, but stored in an inactive form until damage occurs to the cell. An example of this is juglone, a 5-hydroxy naphthoquinone that is accumulated as its 4-glucoside in members of the Juglandaceae [3]. Upon injury of the cell, b-glucosidases in the cell release the aglucone hydrojuglone, which is then rapidly oxidized to juglone by air and plant cell oxidases. Juglone, for example, is a potent antifeedant to *Scolytus multistriatus* [4]. Hydrojuglone is relatively non-toxic compared to juglone [5].

Another compound bordering between preformed and induced defenses is gossypol. This compound accumulates within epidermal glands on the surface of cotton plants and is very toxic to some insects [6]. Gossypol, and a series of structurally related compounds on the same biosynthetic pathway, accumulate to substantially increased levels in plants that have been challenged by potential pathogens. For this reason they have been considered phytoalexins by some plant pathologists. Phytoalexins are inducible antibiotics and will be discussed later.

Inducible defenses have the advantage of diverting plant metabolic resources to defense only when an insect or pathogen attack actually occurs. While some tissue may be damaged and lost, a response to the insect or pathogen can prevent significant damage to the plant, without diverting the same amount of resources that would be required for preformed defenses. Furthermore, some of the preformed compounds that seem to play a significant role in defense against feeding of some insects, actually are used as 'keys' to feeding on the same plant by other insects that attack that particular plant [7].

Plant Defenses Against Insects

Plants have evolved a number of different defenses against insects. One prominent defense is hairs or other structures which cover the surface of the plant and deter insect feeding. Another is accumulation of allomones (compounds that serve to deter feeding or oviposition or are otherwise toxic to the insect), especially in critical plant parts such as the reproductive structures.

Immature seeds and their seed coats are often a good source of such compounds. As an example, walnut trees contain the richest source of hydrojuglone glucoside in young growing parts and in the pericarp surrounding the seed [3].

Preformed Compounds

A wide range of compounds are now known to accumulate in plants and serve as allomones to some insects and kairomones (feeding attractants or excitants) to other insects [2]. An example of such a compound is glaucolide A, a germacranolide type sesquiterpene lactone found in *Vernonia* spp. Glaucolide A has antifeedant action against some Lepidoptera. But other Lepidoptera, such as the cabbage looper [*Trichoplusia ni*] and the yellow woollybear [*Spilosoma virginica*] actually prefer *Vernonia* spp. with this compound [7].

Inducible Compounds

While inducible compounds that deter insect feeding have not been widely studied, some interesting examples of these compounds are known. A proteinase inhibitor inducing factor (PIIF), discovered by Ryan in wounded tomatoes, has been extensively studied in insect resistance [8]. PIIF activity has been isolated from tomato leaves as a single, broad Sephadex G-50 peak with a Mr range of 5,000-10,000 and is primarily carbohydrate. PIIF can elicit the accumulation of two proteinase inhibitors in wounded tomato leaves and these inhibitors interfere with protein digestion by insects. Inhibitor I has a molecular ratio of 41,000 and is a pentamer. Each subunit has an active site, specific for inhibiting chymotrypsin, with a K_i of about 10^{-9}M. Inhibitor II is a dimer with a molecular ratio of 23,000 and strongly inhibits both trypsin and chymotrypsin with K_i values of about 10^{-8} and 10^{-7}M respectively [9]. Furthermore, plant and fungal cell wall fragments activate expression of proteinase inhibitor genes for plant defense just like they activate phytoalexin accumulation [10].

Carroll and Hoffman [11] presented evidence that a feeding stimulant for *Acalymma vittata* is rapidly mobilized into damaged *Cucurbita moschata* leaves. These same leaves rapidly accumulate a feeding inhibitor of *Epilachna tredecimnotata*. This plant response is circumvented by *Epilachna tredecimnotata* by trenching around the leaf area to be eaten, prior to feeding.

Unfortunately the active compounds have not been chemically characterized.

Another group of inducible compounds that probably have an important role in plant insect resistance is phytoalexins.

Phytoalexins

Phytoalexins are low molecular weight, antimicrobial compounds that are both synthesized by and accumulated in plants after exposure to microorganisms [1]. Several lines of evidence suggest that these compounds have an important role in plant disease and pest resistance [12].

Phytoalexins from many different plants have been chemically characterized. The phytoalexins include isoflavanoid-derived pterocarpan compounds characteristic of the Leguminosae, sesquiterpenoid compounds characteristic of the Solanaceae, phenanthrene compounds characteristic of the Orchidaceae and acetylenic compounds characteristic of the Compositae.

Glyceollin, a phytoalexin of soybeans, accumulates as a series of isomers first identified by Lyne et al.[13]; the three most common isomers are shown in Figure 1. Plants frequently produce a series of active isomers of related phytoalexins, and soybeans are a useful example [14].

Phytoalexins as Insect Feeding Deterrents

In the first work to implicate phytoalexins in feeding-deterrent activity of plants, Russell et al. [15] found that 3R-(-)-vestitol and sativan from *Lotus penduculatus* leaf extracts were major deterrents for *Costelytra zealandica* larvae. Furthermore pastures containing as little as 20% *Lotus penduculatus* were relatively free of this insect pest.

Vestitol had an ED_{50} with *Costelytra zealandica* larvae of about 8.5 µg/g of medium and feeding was completely inhibited at 100 µg/g of medium. Vestitol however could not account for all of the feeding deterrency of the *Lotus* extracts, and sativan was implicated in this additional activity. The dual function of vestitol as a phytoalexin and insect feeding deterrent is of ecological interest. It must still be determined whether vestitol is induced by insect feeding or is present in uninfected field plants.

Subsequent research found that *Costelytra zealandica* and *Heteronychus arator* larvae are sensitive to seven phytoalexins including pisatin, maackiain and medicarpin [16]. Related compounds naringenin, apigenin, morin,

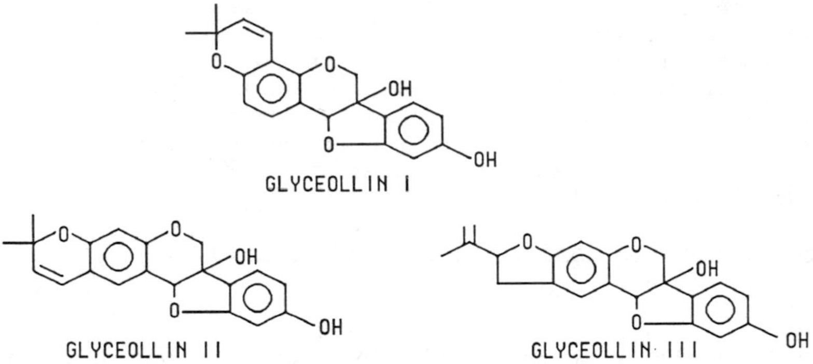

GLYCEOLLIN I

GLYCEOLLIN II

GLYCEOLLIN III

Figure 1. Major soybean phytoalexins.

coumestrol, formononetin, biochanin A and genistein did not significantly affect feeding activity. Phaseollin, vestitol, phaseollinisoflavan, and 2'-methoxyphaseollin-isoflavan were the most active, showing significant feeding reduction at 10μg/ml or less. *Heteronychus arator* seems more sensitive than *Costelytra zealandica* to these compounds, except for pisatin. Phytoalexins may therefore serve two different and perhaps independant roles in the plant.

Research by Hart et al. [17] on the effect of soybean phytoalexins on Mexican bean beetle *Epilachna varivestis* and the soybean looper *Pseudoplusia includens* showed that tissues in which these isomers had been elicited [by exposure to ultraviolet irradiation] strongly deterred the Mexican bean beetle but not the soybean looper. At the induced, physiological concentration of 1% of dry weight, glyceollin had some effect on survival of 1-day-old soybean looper larvae but not on subsequent survival and development. Since this insect readily feeds on soybeans, as its name implies, this ability to tolerate glyceollin is not surprising. In addition, testing single compounds in an artificial diet ignores the possible synergistic effects of the multitude of phenolic compounds ingested from the plant during normal herbivory. Additionally a compound distributed in a diet rather uniformly may not approximate the localized, and sometimes higher concentrations in plants.

On the other hand the oligophagous Mexican bean beetle clearly was deterred from feeding on the phytoalexin-rich tissue. Since this insect is not easily reared on artificial diets, the direct toxicity of glyceollin to the Mexican bean beetle could not be determined in these experiments. This is a common experimental problem when determining the effect of a given compound in vivo.

Some of these problems have been addressed by Fischer, D.C.; Kogan, M.; Paxton, J.D., (Environ. Entomol. submitted). They purified glyceollin, applied it to common bean (*Phaseolus vulgaris*) leaves at rates of 0.22, 1.1, 2.2, 11.0 or 22.0 μg/mg dry weight of leaf tissue. These tissues were then subjected to feeding preference tests in petri plates with the southern corn rootworm (*Diabrotica undecimpunctata howardi*), the Mexican bean beetle (*Epilachna varivestis*) and the bean leaf beetle (*Cerotoma trifurcata*). Glyceollin treated leaves were consumed less by the southern corn rootworm and the Mexican bean beetle, even at levels below that found in elicited plants. On the other hand the bean leaf beetle fed freely, even at very high [22μg/mg] doses of glyceollin. This suggests that phytoalexins may represent a convergence of defenses against both microorganisms and insect herbivores.

Elicitation of Phytoalexins by Stresses

Many environmental factors such as heat, drought, and frost may affect herbivore/plant interactions by increasing or decreasing the level of resistance of the plant to the herbivore [19]. This suggests that the plant is actively responding to the feeding by the herbivore and that phytoalexins might therefore be involved [20]. This also suggests that selective elicitation of phytoalexins in plants might be a desirable strategy for protecting plants against insects [21].

A number of different stresses can elicit inducible compounds important in insect interactions with plants. Some of these compounds appear to attract insects to a plant 'in distress', such as the southern pine beetle, *Dendroctonus frontalis*, to *Pinus* spp. Other compounds appear to prevent attack [22]. Ozone decreases soybean resistance to insect herbivory and overrides induced resistance [Lin, H.C.; Kogan, M.; Endress, A.G., Environ. Ento. 1990, in press.]. This might be due to a diversion of phenolics into coumestrol instead of more toxic compounds such as glyceollin.

Pathogens Elicit Phytoalexins

The most studied aspect of phytoalexin elicitation is that generated by fungal pathogens of plants. Karban et al.[23] found induced resistance and interspecific competition between spider mites and a vascular wilt fungus. McIntyre et al. [24] demonstrated that 7 days after tobacco plants were infected with tobacco mosaic virus, reproduction of the green peach aphid, *Myzus persicae,* was significantly reduced on these infected plants. Inoculation of plants with tobacco mosaic virus induced resistance to several pathogens, however the mechanism for induced resistance was not characterized.

This suggests that a different type of protection of plants against insect attack is possible. Few cases exist where such protective interactions between pathogens and insects on plants have been studied. I believe such interactions are common and deserve more attention than received to date.

Insect Feeding Elicits Plant Responses

Decreased feeding on soybean can be induced by previous insect herbivory as well as by mechanical injury. Lin [25] found mechanical injury and prior herbivory induced resistance in soybean cv Williams 82 to the soybean looper and the Mexican bean beetle. The level of induced resistance depended on the number of injured host cells in contact with healthy cells, and not on the amount of leaf area lost. The resistance induced by the soybean looper resulted from a combination of mechanical injury and compounds in the larval regurgitant. No communication of the induction signals between plants, and no subsequent protection, was found. Induced resistance had a significant retardant effect on development and growth of both insects, but it had no significant effect on total food consumed by either insect.

Lin [25] separated, by high pressure liquid chromatography, from soybean leaf extracts, twelve peaks representing an unknown number of compounds. Feeding by the Mexican bean beetle increased the areas of three peaks and feeding by the soybean looper increased the areas of two peaks. These compounds were significantly correlated with antifeeding activity of the leaf samples. Karban [26] found that cotton resistance to beet armyworms, *Spodoptera exigua,* could be induced by prior exposure to spider mites, *Tetranychus turkestani.* In these cases the compounds responsible for resistance were not identified.

Use of Phytoalexins and Their Elicitors

Based on the previous examples, the elicitation of phytoalexins and other inducible compounds in plants by various elicitors might be a successful tactic for controlling insect pests on important crops. This elicitation would need to occur only when the insect feeds, to prevent unnecessary damage to the plant by accumulation of compounds toxic to plant cells as well. Based on presently known carbohydrate elicitors, it seems plausible that plant surfaces could be treated in such a way that the elicitor is carried into the plant upon insect feeding [21]. These carbohydrate elicitors should be stable and should pose no environmental threat since they are most likely non-toxic themselves.

I suggest that phytoalexins play an important role in insect resistance by plants. This role of phytoalexins and phytoalexin elicitation now should be studied further and tested under field conditions.

Acknowledgments

Certain aspects of the research presented in this chapter were supported in part by the Illinois Agricultural

Experiment Station, The American Soybean Association, and U.S.D.A. Competitive Grants Program.

Literature Cited

1. Paxton, J.D. Phytopath. Zeit. 1981, 101, 106-109.
2. Norris, D.M., In Chemistry of Plant Protection; Haug, G.; Hoffmann, H. Eds.; Springer-Verlag: Berlin, 1986; pp 99-146.
3. Daglish, C. Biochem. J. 1950, 47, 452-462.
4. Gilbert, B.L.; Norris, D.M. Insect Physiol. 1968, 14, 1063-1068.
5. Paxton, J.D.; Wilson, E.E. Phytopathology 1965, 55, 21-26.
6. Stipanovic, R.D.; Bell, A.A.; Lukefahr, M.J. In Host Plant Resistance to Pests; Hedin, P.A., Ed.; ACS Symposium Series No.62; American Chemical Society: Washington, DC, 1977; Chapter 14.
7. Mabry, T.J.; Gill, J.E.; Burnett, W.C.; Jones, S.B. In Host Plant Resistance to Pests; Hedin, P.A., Ed.; ACS Symposium Series No.62; American Chemical Society: Washington, DC, 1977; Chapter 12.
8. Ryan, C.A. Trends Biochem. Sci. 1978, 3, 148-150.
9. Nelson, C.E.; Walker-Simmons, M.; Makus, D.; Zuroske, G.; Graham, J.; Ryan, C.A. In Plant Resistance to Insects; Hedin, P.A., Ed.; ACS Symposium Series No.208; American Chemical Society: Washington, DC, 1983; Chapter 6.
10. Ryan, C.A.; Bishop, P.D.; Graham, J.S.; Droadway, R.M.; Duffey, S.S. J. Chem. Ecol. 1986, 3, 1025-1036.
11. Carroll, C.R.; Hoffman, C.A. Science 1980, 209, 414-416.
12. Hedin, P.A., Ed. Host Plant Resistance to Pests; ACS Symposium Series No.62; American Chemical Society: Washington, DC, 1977; pp 286.
13. Lyne, R.L.; Mulheirn, L.J.; Leworthy, D.P. JCS Chem.Comm. 1976, 1976, 497-498.
14. Ingham, J.L.; Keen, N.T.; Mulheirn, L.J.; Lyne, R.L. Phytochem. 1981, 20, 795-8.
15. Russell, G.B.; Sutherland, O.W.R.; Hutchins, R.F.N.; Christmas, P.E. J.Chem.Ecol. 1978, 4, 571-579.
16. Sutherland, O.W.R.; Russell, G.B.; Biggs, D.R.; Lane, G.A. Biochem. Syst. Ecol. 1980, 8, 73-75.
17. Hart, S.V.; Kogan, M.; Paxton, J. J. Chem. Ecol. 1983, 9, 657-72.
18. Fukami, H.; Nakajima, M. In Naturally Occurring Insecticides; Jacobsen, M.; Crosby, D.G., Eds.; Marcel Dekker, New York, 1971; pp 71-97.

19. Kogan, M., In The Role of Chemical Factors in Insect/Plant Relationships; Proc. XV Int. Congr. Entomol., Washington, D.C. 1977; pp 211-227.
20. Kogan, M.; Paxton, J.D. In Plant Resistance to Insects; Hedin, P.A., Ed.; ACS Symposium Series No.208; American Chemical Society: Washington, DC, 1983; Chapter 9.
21. Paxton, J.D. In Biologically Active Natural Products, Potential Use in Agriculture; Cutler, H.G., Ed.; ACS Symposium Series No.380; American Chemical Society: Washington, DC, 1988; Chapter 8.
22. Schoonhoven, L.M. In Semiochemicals: Their Role in Pest Control; Nordlund, D.A. Ed.; John Wiley and Sons, New York, 1981, pp 31-50.
23. Karban, R.; Adamchak, R.; Schnathorst, W.C., Science 1987, 235, 678-680.
24. McIntyre, J.L.; Kodds, J.A.; Hare, J.D. Phytopathology 1981, 71, 297-301.
25. Lin, H. Ph. D. Thesis, University of Illinois, Urbana-Champaign, Illinois, 1989.
26. Karban, R., Am. Mid. Nat. 1988, 119, 77-82.

RECEIVED May 16, 1990

ALLELOCHEMICALS FOR CONTROL OF INSECTS AND OTHER ANIMALS

Chapter 14

Insect Resistance Factors in Petunia

Structure and Activity

Carl A. Elliger and Anthony C. Waiss, Jr.

Western Regional Research Center, U.S. Department of Agriculture, 800 Buchanan Street, Albany, CA 94710

Petunia species contain an array of more than three dozen steroidal materials, formally related to ergostane, that are involved in the resistance of the plant toward attack by certain lepidopteran larvae. Among these compounds are A-ring dienones and 1-acetoxy-4-en-3-ones. Epoxy functionalities may be found on the side chain in certain cases and on the A- and D-rings. Hydroxy and acetoxy substituents may be located at various positions on the steroid nucleus and the side chain. In a large number of examples, the side chain possesses a bicyclic orthoester moiety which, in certain cases, may also have a thiolester functionality attached. Alteration of ring A from the usual steroid pattern has given rise to a set of compounds which have a spirolactone at position-5. The biological activity of these substances is dependent upon various structural features, most notably the orthoester, which is essential for toxicity of the compounds toward insects.

Foliage of petunia plants is poisonous to various insects. Early reports by Fraenkel, *et al* (1-3) indicated that although tobacco hornworm (*Manduca sexta* Johannson) larvae found *Petunia hybrida* Vilm. plant material to be highly acceptable (ie attractive, or at least non-repellent) as food, growth was not supported and premature death occurred. It was also stated that petunia was toxic to the Colorado potato beetle, *Leptinotarsa decemlineata* Say, and that these insects found the plant to be attractive as well. Subsequently, other investigators (4,5) confirmed that several species of *Petunia* were extremely toxic to first and second instar *M. sexta* although the third and fourth instar larvae were killed less rapidly. Even so, after feeding on *P. inflata* Fries and *P. violacea* Lindl., the larger larvae exhibited convulsive symptoms within four hours and stopped eating. All larvae were dead two days later. These workers found that washing leaves of *P. axillaris* Lam., *P. inflata* and *P. violacea* with water, ethanol-water mixtures, or 95% ethanol did not reduce the toxicity of the leaves toward *M. sexta*. From this they concluded that exudates from foliar trichomes, which they considered to probably contain alkaloids responsible for activity, were not removed by this treatment. No evidence for the presence of alkaloids, other than convulsions of the larvae during toxicosis, was given.

The toxic effect of *P. hybrida* has been used to good account by Dethier (6,7) in conducting an interesting series of experiments on food-aversion learning in certain

This chapter not subject to U.S. copyright
Published 1991 American Chemical Society

caterpillars. Larvae of two species of woolly bear caterpillars, *Spilosoma (Diacrisia) virginica* Fabr. and *Estigmene congrua* Walk. as well as *M. sexta* were allowed to feed on petunia leaves until acute symptoms were observed. Most animals could recover from this effect within twelve hours after they were removed from the leaves under the conditions of the experiment. They were then subjected to preference tests involving *Petunia* and two other plants, one more acceptable, and one less acceptable than *Petunia* to control larvae. The results of these experiments suggested that aversive learning was possible for the woolly bears but not for the hornworm. Dethier mentioned that the symptoms of insect intoxication on *Petunia* included convulsions, partial paralysis, and excessive regurgitation, but that *S. virginica* and *E. congrua* were less affected than was *M. sexta.* Another wooly bear, *Pyrrharctia (Isia) isabella* Smith, was also subject to poisoning when fed *P. violacea* (8). From the results shown above and from similar toxicity data obtained in our laboratory using the corn earworm and fall armyworm, we believed that *P. hybrida* and various species of *Petunia* contained a substance or substances injurious to insects, and that these plants might provide a valuable genetic source of host plant resistance factors.

Isolation of Biologically Active Materials from *Petunia*

Initial experiments were conducted on freeze-dried leaf material of *P. hybrida* by serial extraction with increasingly more polar solvents (9). Fractions obtained were assayed by incorporation into artificial diet (10) upon which newly hatched larvae of the corn earworm (*Heliothis zea* Boddie) were placed. This insect had previously been shown to be adversely affected by petunia foliage, and inhibition of larval growth on these artificial diets (ten-day weights) indicated which fractions contained active material. Insect-inhibitory substances of *Petunia* are relatively nonpolar and some of these are extracted from plant material even with hexane, but we have found that it is more convenient to effect complete extraction of all of the compounds with chloroform followed by chromatographic separation. After evaporation of chloroform from the initial extract, the mixture of lipids thus obtained was stirred with boiling acetonitrile and allowed to cool. Waxy materials, which are not very soluble in acetonitrile, were removed by filtration, after which the acetonitrile solution was passed through a column of preparative C-18 packing to yield an insect-inhibitory, enriched mixture of steroidal materials in the eluate, free from very nonpolar lipids and chlorophyll. Further fractionation was carried out by preparative HPLC on silica, C-18, polar amino-cyano (PAC) and cyanopropyl columns as required by the properties of individual compounds. In many cases, detection of components by UV at 254 nm was satisfactory; however, a variable wavelength detector was used when increased sensitivity for a specific compound was desired (ie 232nm for certain α,β-olefinic ketones), and a refractive index detector was used when substances lacked sufficiently intense UV absorption at any wavelength. Chromatographic profiles of crude extracts could be obtained either on C-18 or silica analytical HPLC columns, but not all components were resolved on even the most efficient analytical columns. However, it was possible to follow the progress of preliminary chromatographic workups by analytical HPLC and also to examine extracts of various plant parts conveniently by this means.

Petunia leaves are typically coated with a sticky exudate, and it was of interest to determine if this exudate contained any of the insect-inhibitory steroids. We subjected fresh leaves to an initial five minute soak in water followed by a one minute dip in chloroform with gentle agitation. This chloroform extract contained nonpolar surface lipids and substances tentatively identified as carbohydrate esters (ca 1 % of fresh wt.), but no biologically active materials were shown to be present by HPLC analysis. Further workup of the solvent-washed leaves by grinding with additional chloroform then yielded the usual quantities of active materials, showing that the insect-inhibitory components were definitely neither part of the exudate nor of the leaf surface coating.

Structures of the *Petunia* Steroids

Petuniasterones. At the present time, thirty seven steroidal ketones, all of which are based on the ergostane skeleton, have been isolated from several species of *Petunia* and from *P. hybrida* (11-16). The structures of these petuniasterones vary considerably, depending upon the plant source, and their biological activity toward insects is structure dependent (*vide infra*). Many of these structures have been described in an earlier

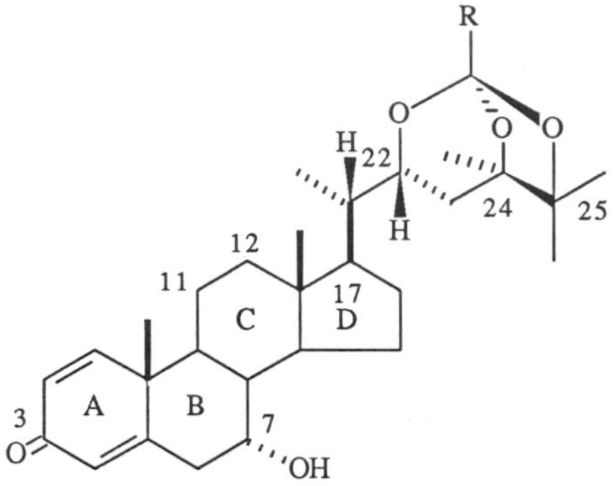

I R = CH_2COSMe
II R = Me

Figure 1. Petuniasterones A (I) and D (II) showing two different orthoester types.

symposium (9); consequently only the most significant types will be mentioned here. Petuniasterone A (I) shows many typical structural features (Figure 1). An A-ring dienone is present in about two thirds of the compounds. More unusual is the pecular bicyclic orthoester system of the side chain, which in the case of I (PS-A) is derived from the uncommon [(methylthio)carbonyl]acetate or from acetate for PS-D (II). About half of the petuniasterones have orthoesters of this sort, approximately evenly divided between the two types above. Two derivatives of PS-A bear hydroxy groups α- to the thiolester group, and a few orthopropionates (only from *P. integrifolia* Hook) have been found. All but one of the petuniasterones have either a 7α-hydroxy (or acetoxy) group, and further hydroxylation or acetoxylation can also occur at positions -11, -12 and -17 of the steroid nucleus.

Petuniasterones with side chains bearing functionalities other than orthoesters also occur. The important PS-B and PS-C series (Figure 2) have epoxy groups at the 24,25-position and hydroxy or acyloxy groups at position-22. We have shown (12) that 22-acyloxy epoxides easily rearrange to the bicyclic orthoesters, and also that a 22-hydroxy epoxide yields the isomeric 5-membered cyclic ether as well as a mixture of side chain triols under mild acid treatment.

Epoxy groups are also found at positions in rings A, B and D (Figure 3). The A-ring epoxy petuniasterones, I (III) and J (IV) are hydroxylated at position -17 and acetoxylated at position -12 respectively, a pattern that is also observed for ring-A dienone 7-acetates derived from I and II. Petuniasterones K (V) and L (VI), bearing epoxy functionality at position -16β,17β, retain the A-ring dienone system of I and II. The substitution pattern of petuniasterone O (VII), with its keto group at position-l and 5α-hydroxy group is not typical of the usual petuniasterone structures, more nearly resembling the substitution pattern found in many withanolides (17) which also may have 6α,7α-epoxides. However, the characteristic orthoacetate side chain reveals the affinity of VII with petuniasterones already discussed.

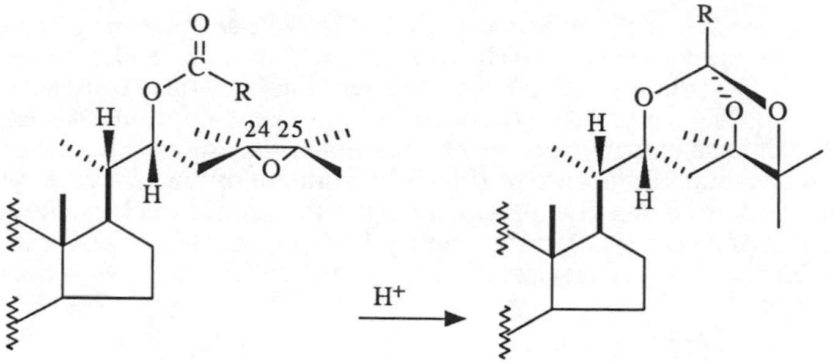

H^+

R = CH_2COSMe, Me

Figure 2. Rearrangement of epoxy side chain petuniasterones.

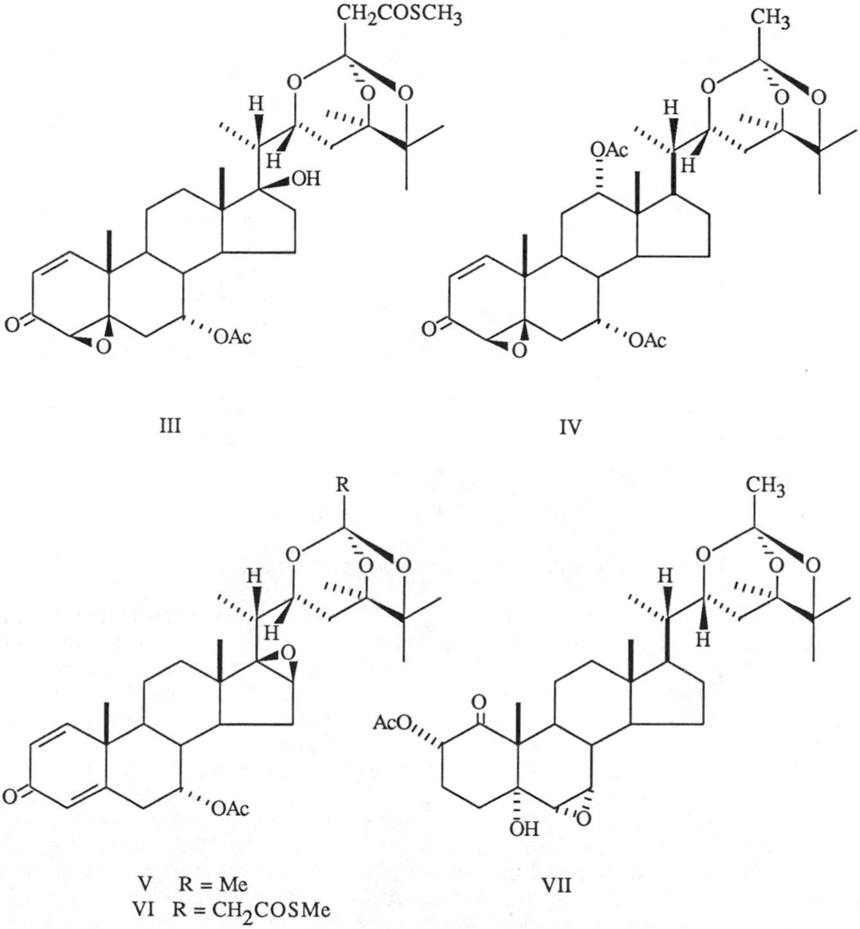

III

IV

V R = Me
VI R = CH_2COSMe

VII

Figure 3. Ring epoxides.

In two examples of petuniasterones, oxidative cleavage of ring D has occurred (Figure 4). Petuniasterone Q (VIII) has a keto group on an extended side chain which still possesses the orthoacetate moiety. The remainder of this seco-steroid bears the familiar A-ring dienone and the 7α-acetoxy group. Petuniasterone N (IX), which has undergone additional oxidative processes, still retains the original number of carbons of the ergostane system. Reclosure of ring D by addition of the 13-hydroxyl group to a ketone at position-16 has given a structure for the steroid nucleus similar in many respects to that of more usual steroids, but the C,D-ring junction is now *cis*. Again, the orthoester functionality has been preserved. Compound IX was always obtained as

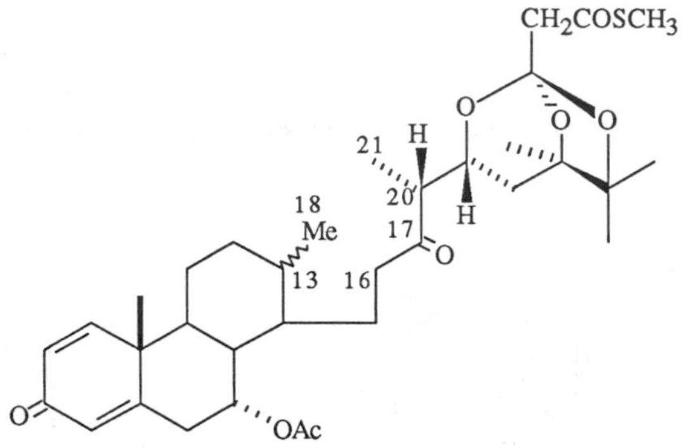

CH_2COSCH_3

VIII

CH_2COSCH_3

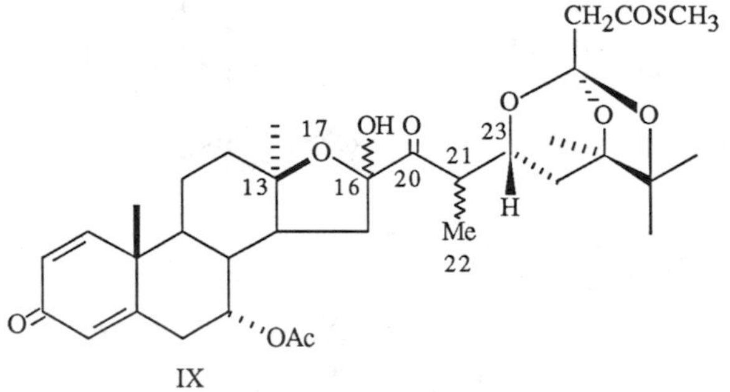

IX

Figure 4. Ring D seco-petuniasterones.

a mixture of isomers that could be separated by HPLC. Reversion to the original epimeric proportions took place in deuteromethanol much faster than did incorporation of deuterium into position-21, thereby indicating that enolization of the C-20 keto group did not occur rapidly, and that epimerization of methyl-22 was not involved in the process. Isomerization under these conditions must occur by very facile ring opening and reclosure at C-16 to give a mixture of epimers at that position.

Recently, several new petuniasterones derived from the epoxides V and VI have been identified (Elliger, et al. unpublished). We found that mild acid treatment (Figure 5) of these epoxides gave 16-ketopetuniasterone D (X) and 16-ketopetuniasterone A (XI) along with a set of side chain esters (XII-XV) having the 17,22-oxido- functionality. Acid catalyzed opening of the oxirane ring of V and VI can occur with carbocation formation at C-17. Interception of this cation by the oxygen attached to position -22 followed by ring opening of the bicyclic orthoester system between this oxygen and the orthoester center (with addition of water) gives the new oxetane system. The monocyclic species remaining on the side chain, with oxygens linked to carbons -24 and -25, may open in either of two directions, to yield the mixture of 24,25-hydroxy esters XII, XIII and XIV, XV. The individual hydroxy esters can be isolated, but they revert to an approximately equal mixture of 24,25-isomers on storage, presumably through the same cyclic intermediate.

Ketone formation can occur by loss of a proton at C-16 from the same cationic intermediate mentioned above to yield the enol forms which then may equilibrate to the 16-ketones (X and XI). It is known that the β-configuration of the side chain at C-17 is preferred for 16-ketosteroids under conditions that effect enolization (18). Although

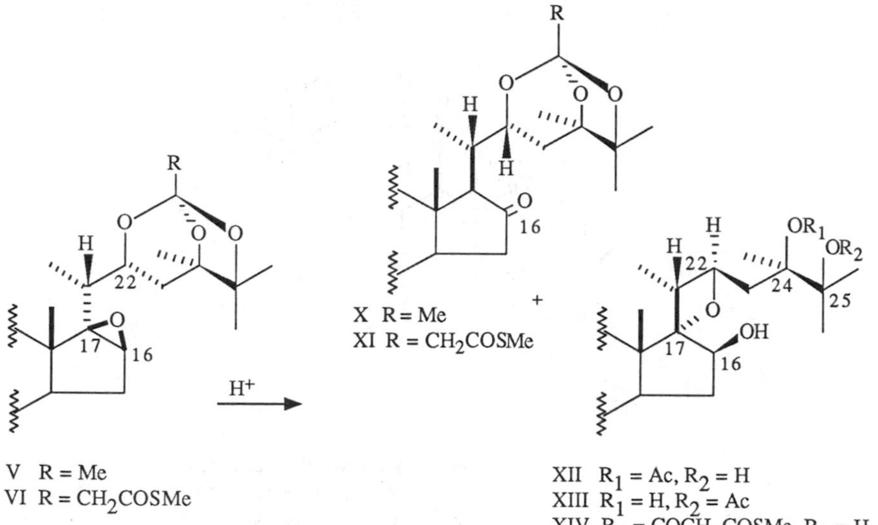

V R = Me
VI R = CH_2COSMe

H^+

X R = Me
XI R = CH_2COSMe

XII R_1 = Ac, R_2 = H
XIII R_1 = H, R_2 = Ac
XIV R_1 = $COCH_2COSMe$, R_2 = H
XV R_1 = H, R_2 = $COCH_2COSMe$

Figure 5. Rearrangement of D-ring epoxides.

all of these substances have been isolated from chromatographic fractions of *P. parodii* Steere, it is likely that they are artifactual, and result from rearrangement during workup of the D-ring epoxides originally present in the plant. The rearrangement products were always found in those fractions that contained their epoxy precursors even though the chromatographic mobilities of the individual products were typically quite different. If their formation had occurred naturally then the products should have appeared in different chromatographic zones from the plant extracts.

Petuniolides. In certain species of *Petunia* there is considerably more insect toxicity than can be accounted for by the presence of known petuniasterones. From these varieties we have isolated a number of related steroidal lactones, some of which are much more toxic than are the petuniasterones, and have named these petuniolides (19) in analogy to the well known withanolides (20). The petuniolides are based on a modified ergostane system in which ring-A has undergone rearrangement with loss of a carbon and formation of a spirolactone (Figure 6). All petuniolides isolated have a 6α,7α-epoxy group on ring-B and either orthoacetate or orthopropionate side chains. No thiolester derivatives analogous to those of the petuniasterone orthoesters have been found at the present time. Petuniolides A-E (XVI-XX) have a 9,11-double bond and differ from each other at position-12 in having allylic acetates, allylic keto groups or a methylene group at that position. Petuniolides A (XVI), B (XVII), and D (XIX) were found in *P. integrifolia* whereas petuniolide C is the major petuniolide of *P. parodii*. Three other petuniolides have been found to occur in *P. parodii* in smaller concentrations (Elliger, et al., unpublished) and include petuniolides E (XX), F (XXI) and G (XXII). Both XXI and XXII have *cis*-stereochemistry at the B,C-ring junction, with petuniolide G being notable in possessing not only a 9β,19-cyclopropane ring but also an 11-hydroxy-12-ketone in ring C. X-ray crystal structure determination of XXII showed that considerable steric interaction occurs between the 11-hydroxy and the *endo* hydrogen of methylene-19 as well as between H-19-*exo* and one of the methylene hydrogens at position-4.

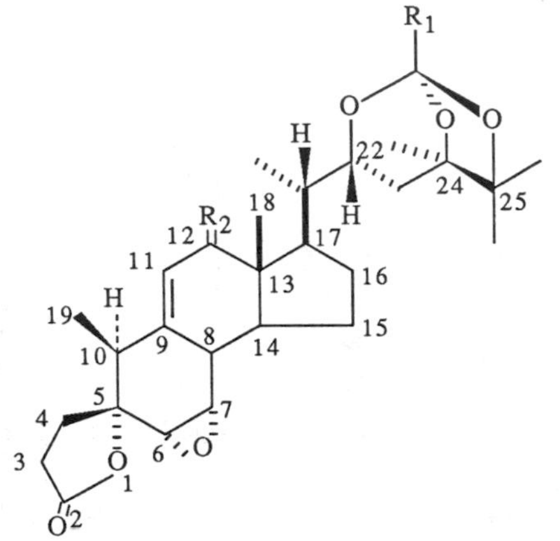

XVI R_1 = Me, R_2 = H, α-OAc
XVII R_1 = Et, R_2 = H, α-OAc
XVIII R_1 = Me, R_2 = O
XIX R_1 = Et, R_2 = O
XX R_1 = Me, R_2 = H_2

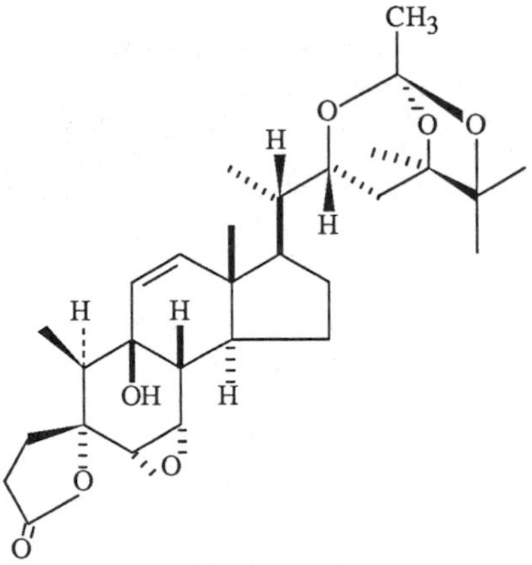

XXI

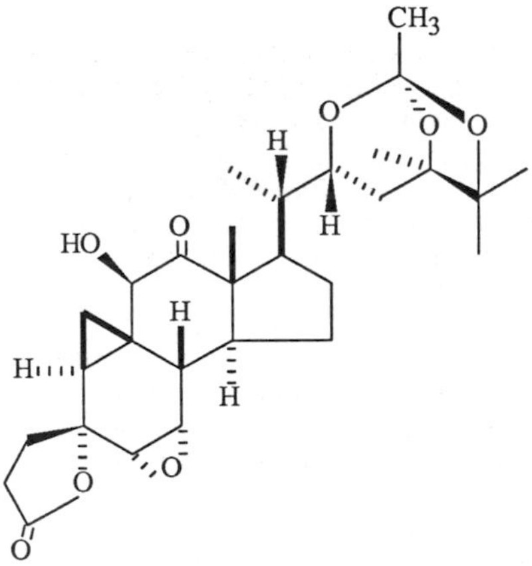

XXII

Figure 6. Structures of petuniolides from *P. integrifolia* and *P. parodii*.

Biogenesis and Interconversions of the *Petunia* Steroids

Inasmuch as no direct evidence for specific biosynthetic pathways in the formation of these steroids is available at the present time, any proposals on this subject are necessarily rather speculative. Still, speculations on certain of these pathways are nonetheless valuable in suggesting experimental approaches for their elucidation. The acid catalyzed conversion of the 22-acyloxy-24,25-epoxides into side chain orthoesters indicated previously (Figure 2) is undoubtably analogous to the process occurring within the plant. It is significant that only a few of these side chain epoxides have been found, and that they are of only two petuniasterone types, namely the 1,4-dien-3-ones and 1-acetoxy-4-en-3-ones (11). Both of these types have only the 7α-hydroxy group as an additional nuclear substituent. From this, it may be concluded that the more highly substituted petuniasterones arise after formation of the orthoester group. The relationship between the 1-acetoxy-4-ene-3-ones and the dienones of the A-ring is not completely clear since either type of compound may arise from the other. Elimination of acetate at position-1 from the former type can give the dienone, and addition of acetate, Michael-wise, at this position of the dienone yields the acetoxy compound. However, only one of these 1-acetoxy derivatives that also bears an orthoester on the side chain (PS- E) has ever been isolated (12). This suggests that the dieneones are also formed at a very early stage by the elimination of acetate. The numerous hydroxy and acetoxy derivatives as well as the ring epoxides are probably elaborated after dienone formation. As mentioned earlier, the D-ring epoxides (Figure 5) are probably not transformed biogenetically into the D-ring ketones (X & XI) and oxetanes (XII-XV); however, it is possible to perceive a role for these epoxides in the formation of the seco-steroids VIII and IX.

Petuniasterone O (VII) is particularly interesting from a biogenetic standpoint because its 5α-hydroxy and 6α,7α-epoxy groups suggest that it occupies a position midway between the usual petuniasterones and the petuniolides. If we assume that VII could eventually be functionalized to a $\Delta^{9,11}$-12-ketone such as XXIII (Figure 7) then

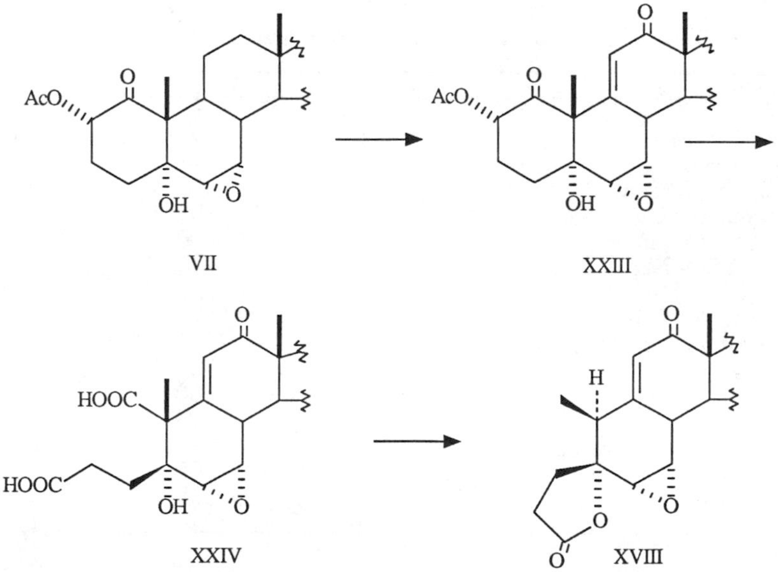

VII XXIII

XXIV XVIII

Figure 7. Proposed biosynthetic scheme for conversion of petuniasterone O into petuniolide C.

oxidative cleavage of ring A between carbons-1 and -2 could give the diacid XXIV, which in turn, would undergo decarboxylation at C-10 (facilitated by the vinylogous keto group at position-12). Cyclization of the pendant carboxy-bearing chain to form petuniolide C (XVIII) can occur easily. We have found that reclosure of a similar open chain derivative of petuniolide C having an additional 7,8-double bond (obtained by base treatment of petuniolide C) is very rapid. The implication of the pathway proposed above is that $\Delta^{9,11}$-12-ketones are precursors of the remaining petuniolide types. This is not inconsistent with the structures of the known petuniolides.

Biological Activity of the *Petunia* Steroids

Growth reduction of *Heliothis zea*. In our bioassays we have tested the *Petunia* steroids against the corn earworm (*H. zea*) as well as the fall armyworm (*Spodoptera frugiperda* Smith) and the pink bollworm (*Pectinophora gossypiella* Saunders). These bioassays were carried out using pure substances deposited on cellulose powder and incorporated into artificial diet (10) at several levels. Ten neonate larvae were used for each concentration. A dose-response curve was generated in this way for each compound, relating insect weight gain (compared to controls) to the dietary concentration of added material. The dose of additive required to reduce the growth of insect larvae to fifty percent of control values is termed ED_{50}. For the earworm and armyworm, response curves obtained under these conditions of chronic toxicity were smoothly sigmoidal, and even at doses above the ED_{50} all larvae survived. However, the pink bollworm gave essentially an all or nothing response with mortality occurring without much weight curtailment of the survivors. Most of our results are from studies with the corn earworm; however, the armyworm was qualitatively similar, but somewhat less sensitive for the compounds selected.

We have found, for the compounds tested, that only those substances having the bicyclic orthoester moiety on the side chain are significantly active in reducing larval growth of *H. zea*. Table I presents bioassay results for some of these petuniasterones.

Table I. Growth Inhibition of *Heliothis zea* by Petuniasterones

Compound	ED_{50} (PPM)
PS-A (I)	130 (150)*
PS-A 7-acetate	144
17β-Hydroxy PS-A	>400
17β-Hydroxy PS-A 7-acetate	>800
30-Hydroxy PS-A	>400
16-Keto PS-A 7-acetate	700
PS-D (II)	130 (325)*
12-acetoxy PS-D 7-acetate	115
11-hydroxy-12-acetoxy PS-D 7-acetate	93
PS-I (III)	125
PS-J (IV)	>1000
PS-K (V)	>800
PS-N (IX)	75
PS-O (VII)	165
PS-Q (VIII)	185

* Value obtained for *S. frugiperda*.

It may be seen that activity is not dependent upon whether a (methylthio)carbonyl substituent is present on the orthoacetate grouping. We have examined two examples of orthopropionates (not shown in Table I) which are analogous to PS-D (II) and 12-acetoxy PS-D and found them to be somewhat more active than are the orthoacetates. Placement of a hydroxy α- to the thiolester group in 30-hydroxy PS-A was significant in reducing activity since larvae fed on diets containing 400 ppm of that compound attained over 90 percent of normal growth. Generally, nuclear substitution in rings A, B and C does not greatly affect activity. The notable exception, PS-J (IV), is very insoluble, and its true dietary concentration may be considerably less than the amount incorporated into the diet. Substitution in ring D does decrease inhibitory activity as is shown by 17β-hydroxy PS-A, 16-keto PS-A acetate and by PS-K (V). Nevertheless, the D-ring can undergo considerable modification with retention of activity. The 17-oxa-steroid, PS-N (IX), is actually slightly more inhibitory than PS-A even though the side chain has been displaced from its usual position in respect to the steroid nucleus. Preliminary results on the entirely open analog, PS-Q (VIII) indicate that much less insect growth inhibition occurs in this case (ED_{50} 185 ppm). At the other end of the molecule, complete alteration of ring A as in PS-O (VII), which has an entirely different substitution pattern than most petuniasterones, has little effect in reducing activity.

That the A-ring in the *Petunia* steroids may be extensively modified without loss of biological activity is convincingly demonstrated by the petuniolides (Table II). All but one of these lactones is impressively active against *H. zea*, and the least effective compound is still comparable to most of the usual petuniasterones in this respect.

Table II. Growth Inhibition of *Heliothis zea* by Petuniolides

Petuniolide	ED_{50} (PPM)
A (XVI)	13
B (XVII)	10
C (XVIII)	3 (69)*
D (XIX)	2
E (XX)	21
F (XXI)	170
G (XXII)	12

* Value obtained for *S. frugiperda*.

Other insects and a crustacean. The fall armyworm (*S. frugiperda*) is much more resistant toward the effect of petuniolide C than is *H. zea* (Table II), but the pink bollworm (*P. gossypiella*) is more susceptible. We observe, however, that even though 70% kill took place with this compound at a dietary level of 0.25 ppm, surviving *P. gossypiella* larvae grew to 74% of control weights. This effect is typical for substances that are toxic to early instar larvae, but which do not adversely influence older animals. In all our studies, older lepidopteran larvae were found to be more resistant toward plant allelochemicals than were younger individuals. The pink bollworm is a highly specialized feeder, and its extreme susceptibility toward petuniolide C probably is a consequence of adaptation to a few plants at the expense of tolerance for allelochemicals produced by plants other than its usual hosts.

Petuniolide C (XVIII) was available in reasonable amounts, and we submitted samples of the compound elsewhere for tests upon other insects. The tobacco hornworm (*M. sexta*) was susceptible to this compound in artificial diets (as might have been expected from early observations of larvae on the plant), with only 50% survival of animals to the prepupal stage at a dosage of 20 ppm (J. Oliver, personal communication). Another caterpillar, the variegated cutworm (*Peridroma saucia* Hübner), was insensitive to petuniolide C. However, it was learned that this insect is a natural pest of *Petunia* and may be evolutionarily adapted to compounds of this sort (M. Isman, personal communication). The Bertha armyworm (*Mamestra configurata* Walker) is about as susceptible as is the corn earworm toward artificial diets containing XVIII (Isman). Feeding experiments were also carried out using neonate larvae of the Colorado potato beetle (*L. decimlineata*). A methanolic solution of XVIII was sprayed on to leaves of susceptible tomato plants, and larvae were applied after solvent evaporation was complete. After four days, the plants were examined for surviving larvae, and their developmental stage was determined. Mortality was 70-90%, and no larvae developed beyond first instar. Controls on untreated leaves exhibited 100% survival and had reached second instar (W. Cantelo, personal communication).

We also examined the effect of petuniolide C upon the brine shrimp, *Artemia salina*, in order to test the response of another class of arthropod toward this chemical. This crustacean has been suggested as a general experimental subject in toxicity determination of chemical substances toward invertebrates, and for the indication of cytotoxicity (21, 22). Solutions for the test were prepared by adding stock solutions of XVIII in ethanol to containers of artificial salt water (1% EtOH final concentration) with sonic agitation. Even with added ethanol, the solubility limit of the steroid was about 1.0 ppm. Newly hatched brine shrimp were added to these solutions, and controls were run using salt water and salt water containing 1% ethanol. After 24 hr., no mortality was observed for the EtOH controls, and no toxic effect was seen for petuniolide C at 1.0 ppm.

Acute effects. When the concentrations of active substances in artificial diets are increased to levels well above the ED_{50}, acute effects are observed on test subjects. *Heliothis zea* showed immediate distress after ingesting only a few mouthfuls of diet containing 1000 ppm of PS-A, and comparable results were seen at levels of 50 ppm for petuniolide C. Larvae thus affected stop feeding and exhibit violent convulsive movements. This is accompanied by regurgitation and copious diuresis leading to dehydration and loss of body turgor. Although the animals become moribund fairly soon, death is not rapid, and some larger larvae have been observed to remain alive for several days after the onset of symptoms. At lower test concentrations, third to fifth instar larvae generally survive treatment if they are transferred to control diets.

Heliothis zea larvae were subjected to topical applications of petuniolide C to determine if contact with diet treated at higher levels of this compound could produce symptoms even though no ingestion of the diet occurred. Solutions of petuniolide C in acetone were applied in 1.0 µl aliquots to larvae having an average weight of 10 mg; the amounts were 0.5 µg and 1.0 µg (50 and 100 mg/kg of body weight). Application was to the middle of the larva well back from the head to minimize accidental ingestion, and controls received an equivalent amount of acetone. No immediate effect was noted. After 48 hr., all larvae were alive, but the 100 mg/kg subjects were slightly shrunken in appearance. It was possible to observe a greater toxic effect at an application level of 330 mg/kg. In this case also, larvae remained alive 24 hr. after application; however, 80% were dead at 48 hr. Clearly, topical absorption of petuniolide C is not an particularly important mode of entry of this compound into the insect.

Occurrence of the *Petunia* Steroids in the Plant

We have observed that the concentrations of the *Petunia* steroids varies considerably between horticultural varities of *P. hybrida* and between various species of this plant (9). Table III shows the approximate content of a number of insect-inhibitory petuniasterones in leaves of certain *Petunias* that we have examined. In the examples selected, the individual compounds were isolated by HPLC from extracts of anhydrous leaves and their respective amounts represent a very conservative minimum quantity. Weights were converted to concentrations on the fresh basis assuming a water content of 80% in leaves of the plant. We had previously noted that the effect of active petuniasterones is additive in nature, and it is clear that the aggregate concentration of these compounds present in the plant is sufficient to account for substantial toxicity. Not shown in the table is the significant observation that leaves of *P. violacea* did not contain petuniasterones, and that these leaves were not toxic to *H. zea* larvae (9). A degree of uncertainty exists in the classification of this species, and it is considered by some that *P. violacea* is synonymous with *P. integrifolia* (23). However, using petuniasterone content as a taxonomic indicator, it would appear that these species are separate entities.

Table III. Petuniasterone Content (mg/kg)* of Fresh *Petunia* Leaves

Compound	Royal Cascade	*P. axillaris*	*P. parodii*	*P. integrifolia*
PS-A (I)	120	85	40	90
17β-hydroxy PS-A	10	**	**	**
PS-D (II)	30	10	40	**
12-acetoxy PS-D 7-acetate	100	50	55	**
PS-N (IX)	40	**	150	320

*80% water assumed for fresh leaf. **Value not obtained.

Although the petuniasterones are important resistance factors, we feel that the petuniolide content of the plant represents a better measure of insect resistance because of the much greater toxicity of the latter compounds (except for petuniolide F). We isolated (dry basis) about 300 mg/kg of petuniolide C (XVIII) from consolidated batches of *P. parodii* leaves and have obtained petuniolides A (XVI), B (XVII) and D (XIX) from *P. integrifolia* in quantities of 100, 430 and 200 mg/kg, respectively. Petuniolides E (XX), F (XXI) and G (XXII) were isolated from *P. parodii* in respective amounts of 45, 240 and 190 mg/kg. If it is assumed that the effect of these substances is additive, then their content in fresh leaves of *P. parodii* would be at least twenty-times ED_{50}, and is easily enough to account for the acute symptoms which are observed for larvae eating this plant.

Of the petuniolides, petuniolide C is the most important in *P. parodii* because of its high content and low ED_{50} value. We desired to survey its concentration in this species since the isolated quantity (above) varied from batch to batch, and it was suspected that this might represent seasonal variation in the plant. We examined samples of freeze-dried leaf material obtained at two week intervals from plants grown in outside beds over a five-month period. Analysis was by HPLC on a microsorb 5μ C-18 column with an acetonitrile-water gradient optimized for this compound, and detection was by UV absorption at 232 nm. The concentration of petuniolide C (dry basis) in *P. parodii* ranged from 240 to 920 mg/kg (average 415), but no seasonal trend could be found. Greenhouse grown plants of this species had an average content of 505 mg/kg, whereas plants of *P. axillaris* had a content of 175 mg/kg under this condition.

Content of petuniolide C (XVIII) in plant organs. During the course of our field study, a natural infestation of *Heliothis virescens* Fabr. (tobacco budworm) occurred on surrounding plants (mainly tomato and *Physalis* species). We found that whereas no leaf feeding took place on *Petunia*, considerable damage to the flowers did occur. Also, we noticed that seed capsules of this plant were entered by the larvae, and immature seeds were consumed. Inasmuch as it appeared likely that this selective feeding was a consequence of differing Petuniolide C content in different parts of the plant, we performed analyses for the compound in various plant organs. Neither the flowers nor the immature seed contained any detectable amount of XVIII. Additionally, it was determined that the wall of the capsule and its inner structure were also free of the compound. Older seed from dry opened capsules did not contain any XVIII, but seedling plants of 50 to 75 mm height had 330 mg/kg. The content of XVIII in mature plant stems was below 15 mg/kg, and that of the root was 510 mg/kg. The nil content of petuniolide C in the plant parts that are consumed by the larvae is consistent with the observation that aversive learning by insect larvae is possible, and that larvae migrating from adjoining plants are able to detect the compound without suffering fatal toxicosis.

Implications for Improvement of Crop Plants

Classically, plant breeders selecting for insect resistance in crop plants have utilized crosses between varieties within the same species or between plants of closely related species. Higher levels of chemical resistance factors can indeed be built up in this way. However, insects can potentially become adapted to these substances as a consequence of previous exposure at lower concentrations, thus being preadapted to detoxify or otherwise tolerate these chemicals. We desire to introduce completely exotic chemicals, *i. e.* substances to which specific pests have never been exposed. This could be done by crossing plants that are only distantly related to the crop plant and would include species from other genera, possibly from other families.

The purpose of the present work was to develop a basis for the transfer of resistance factors from *Petunia* into plants of economic significance. Toward this end it was important to establish the identity of the allelochemical agents, to determine the efficacy of these substances and to develop means for their analysis. We have shown that petuniolides are very significant agents of resistance, and that petuniolide C is the most toxic of these substances from *P. parodii.* The introduction into other plants of the biochemical processes involved in formation of the petuniolides is thus of great interest. A number of gene transfers in plants using the Ti plasmid of *Agrobacterium tumefaciens* have been carried out for one gene. However, it is obvious that a large number of enzymatic steps are necessary for the transformation of steroidal precursors into the eventual products, and that a large number of genes would therefore have to be transferred. At the present time, we are entirely ignorant of the detailed enzymology involved within the petuniasterone/petuniolide pathway, and even if the required knowledge were available, it would still be an impractical undertaking to introduce individually the genes controlling each enzymatic step. We are currently exploring methods for the insertion of rather large quantities of *Petunia* genetic material into the genome of tomato and potato. Protoplast fusion is being examined; but this method, which essentially combines the entire contents of two cells, may indeed be excessive in its mixing of genetic information. Even if viable plants were to be obtained after regeneration of hybrid protoplasts, it is unlikely that these plants would have economically satisfactory properties, or that they would be easily propagated. A procedure that offers more promise is that of microinjection of chromosomes or chromosomal fragments from *P. parodii* into individual protoplasts of a recipient species. Less disruption of cellular physiology would take place than in the case of protoplast fusion, and it should be possible to maintain the original chromosome number. In this way a large number of genes can be transferred in one step. In an interspecific hybridization, Griesbach (24) successfully transformed *P. hybrida* protoplasts by microinjection of chromosomes from *P. alpicola* Juss. This changed the level of production of various flavonoids and phenolic acids, reflecting simultaneous changes in the activity of several enzymes. It may be possible in a similar manner to introduce from *Petunia* the complement of enzymes required for alteration of the usual steroid biosynthesis in a recipient plant such as tomato and potato, and obtain production of the petuniolides. We feel that intergeneric (or interfamilial) hybridizations of this sort applied to crop plants will provide needed host plant resistance while still maintaining the desirable economic characteristics of these plants.

Acknowledgments

We thank M. Benson for the determination and interpretation of NMR spectra, R. Y. Wong for X-ray crystallographic structure analysis and W. H. Haddon for mass spectral determinations. Ing. Agr. Hugo A. Cordo kindly furnished samples and seed of *P. integrifolia*. Seeds of *Petunia* species were also furnished by the National Seed Storage Laboratory, Fort Collins Co.

Literature Cited

1. Fraenkel, G. F. Science 1959, 129, 1466-70.
2. Fraenkel, G.; Nayar, J.; Nalbandov, O.; Yamamoto, R. T. Symp. Insect Chem. 1960, 3, 122-26.
3. Fraenkel, G. Tagungsber. Deut. Akad. Landswirtshaftswiss. Berlin 1961, 27, 297-307.
4. Parr, J. C.; Thurston, R. J. Econ. Entomol. 1968, 61, 1525-31.
5. Thurston, R. J. Econ. Entomol. 1970, 63, 272-74.
6. Dethier, V. G.; Yost, M. T. Physiol. Entomol. 1979, 4, 125-30.
7. Dethier, V. G. Physiol. Entomol. 1980, 5, 321-25.
8. Shapiro, A. M. Ann. Entomol. Soc. Am. 1968, 61, 1221-24.
9. Elliger, C. A.; Waiss, A. C., Jr. In Insecticides of Plant Origin; Arnason, J. T.; Philogène, B. J. R.; Morand, P., Eds.; ACS Symposium Series No. 387; American Chemical Society; Washington, DC, 1989; pp 188-205.
10. Chan, B. G.; Waiss, A. C., Jr.; Stanley, W. L.; Goodban, A. E. J. Econ. Entomol. 1978, 71, 366-68.
11. Elliger, C. A.; Benson, M. E.; Haddon, W. F.; Lundin, R. E.; Waiss, A. C. Jr.; Wong, R. Y. J. Chem. Soc. Perkin Trans 1 1988, 711-17.
12. Elliger, C. A.; Benson, M.; Lundin, R. E.; Waiss, A. C., Jr. Phytochemistry 1988, 27, 3597-3603.
13. Elliger, C. A.; Benson, M.; Haddon, W. F.; Lundin, R. E.; Waiss, A. C., Jr.; Wong, R. Y. J. Chem. Soc. Perkin Trans. 1 1989, 143-49.
14. Elliger, C. A.; Haddon, W. F.; Waiss, A. C., Jr.; Benson, M. J. Nat. Prod. 1989, 52, 576-80.
15. Elliger, C. A.; Waiss, A. C., Jr.; Wong, R. Y.; Benson, M. Phytochemistry 1989, 28, 3443-52.
16. Elliger, C. A.; Wong, R. Y.; Benson, M.; Waiss, A. C., Jr. J. Nat. Prod. 1989, 52, 1345-49.
17. Kirson, I.; Glotter, E. J. Nat. Prod. 1981, 44, 633-47.
18. Ronchetti, F.; Russo, G.; Longhi, R.; Sportolini, G. J. Lab. Comp. & Radiopharm. 1978, 14, 687-96.
19. Elliger, C. A.; Wong, R. Y.; Waiss, A. C., Jr.; Benson, M. J. Chem. Soc. Perkin Trans. 1 (in press).
20. Glotter, E.; Kirson, I.; Lavie, D.; Abraham, A. In Bioorganic Chemistry; Van Tameln, E. E., Ed.; Academic Press: New York, 1978; Vol. II pp. 57-95.
21. Meyer, B. N.; Ferrigni, N. R.; Putnam, J. E.; Jacobsen, L. B.; Nichols, D. E.; McLaughlin, J. L. Planta Medica 1982, 45, 31-34.
22. Alkofahi, A.; Rupprecht, J. K.; Anderson, J. E.; McLaughlin, J. L.; Mikolajczak, K. L.; Scott, B. A. In Insecticides of Plant Origin; Arnason, J. T.; Philogène, B. J. R.; Morand, P., Eds.; ACS Symposium Series No. 387; American Chemical Society; Washington, DC, 1989; pp 25-43.
23. Sink, K. C. in Petunia; Sink, K. C., Ed.; Springer: Berlin 1984; pp. 3-9.
24. Griesbach, R. J. Plant Science 1987, 50, 69-77.

RECEIVED May 16, 1990

Chapter 15

Arthropod-Resistant and -Susceptible Geraniums

Comparison of Chemistry

D. Hesk[1], L. Collins[2], R. Craig[3], and R. O. Mumma[1]

[1]Department of Entomology, [2]Department of Chemistry, and [3]Department of Horticulture, Pesticide Research Laboratory and Graduate Study Center, Pennsylvania State University, University Park, PA 16802

Glandular trichome exudate from insect resistant and susceptible geraniums (*Pelargonium xhortorum*) was chemically analyzed to determine differences between the two plant lines. Hplc analysis of the exudate from resistant plants showed that it was predominantly made up of unsaturated anacardic acids, with the C_{22} ω5 and C_{24} ω5 anacardic acids contributing nearly 80% of the total. By contrast, the exudate from the susceptible plants was chiefly saturated, with the C_{22} and C_{24} saturated anacardic acids contributing nearly 50% of the total. The C_{22} ω5 and C_{24} ω5 anacardic acids were only present in trace amounts. In addition a number of other significant peaks observed in the chromatograms of extracts from susceptible plants, and also seen in small amounts in the resistant profile, were isolated and characterized by mass spectrometry and nmr spectroscopy. Two unsaturated anacardic acids, which contributed nearly 30% of the total exudate in the susceptible plant, and only seen in trace amounts in the resistant plant, were identified as C_{24} ω6,9 and C_{24} ω9 anacardic acids. A number of odd chain length and branched chain anacardic acids, were also isolated and identified.

0097–6156/91/0449–0224$07.75/0
© 1991 American Chemical Society

The common garden geranium (*Pelargonium xhortorum)* has been shown to possess a potent chemical defense consisting of glandular trichomes which secrete a viscous sticky exudate.(1-5) This strategy, which is primarily a physical entrapment mechanism, rather than a toxic one, is effective against small arthropod species, and may also provide resistance against larger herbivores. The exudate from the trichomes was found to be comprised of a mixture of anacardic acids, which in the mite resistant plants, was shown to be mainly C_{22} and C_{24} $\omega 5$ unsaturated anacardic acids. Small amounts of the saturated analogues were also found in the exudate from resistant plants(1-4). Analysis of exudate from mite susceptible geraniums, showed that it contained the same four compounds, but with the saturated materials predominating (5, 6).

Analyses were performed on a number of resistant and susceptible plants in order to determine whether the chemical composition was constant within each plant line. While it was found that the chemical profile of the resistant plants was fairly constant, the exudate from susceptible plants exhibited variability. Furthermore, on close examination of the gas chromatographic trace of the exudate from susceptible plants, it was determined that in addition to the four previously identified compounds, there were a number of other significant peaks, which were unidentified. In addition, as all the peaks eluted close together, it was possible that the apparent variability was due to unresolved peaks (5, 6).

Hence, the objectives of this study were to develop an improved separation of the anacardic acids, to characterize the unknown compounds in order to fully compare the resistant and susceptible plants and to determine whether the exudate composition was indeed constant within both the resistant and susceptible lines.

Experimental

Plant Source. Geraniums were maintained in a greenhouse environment using standard cultural practices. The plant line 71-17-7 and its corresponding self-pollinated progeny, family 87-5, had been previously determined to be mite and aphid resistant, and plant line 71-10-1 and its corresponding self-pollinated progeny, families 85-26, 87-13 and 87-14, had previously been determined to be mite and aphid susceptible (7). Newly opened inflorescences, from 12 plants, 6 resistant and 6 susceptible, were selected from random locations in the greenhouse for chemical analysis.

Isolation and purification of exudate. The exudate from both the resistant and susceptible plants was collected by immersing the inflorescences (minus the petals) in methylene chloride for one hour. The extract was then dried over anhydrous sodium sulfate, filtered and evaporated to leave the crude anacardic acid residue. This procedure had previously been shown to produce a representative extract of anacardic acids (5, 6).

The carboxyl group in the anacardic acids was methylated (diazomethane-diethyl ether, 5 minute treatment) before the acids were separated from the other extracted material by thin layer chromatography (tlc). Merck 20x20cm silica gel F254 plates with a pre-concentration zone were used. The plates were developed in a two step solvent system consisting of benzene: diethyl ether: ethanol: acetic acid (50: 40: 1: 0.5, by volume) followed by hexane: diethyl ether: acetic acid (84: 15: 1, by volume). Development was allowed to proceed to about half the length of the plate in the first step and the entire length of the plate in the second. The monomethylated anacardic acid band was located by viewing at 254nm and comparing the Rf value with that of a monomethyl anacardic acid standard, before it was scraped from the plate and eluted with methylene chloride (3x2mL). The methylene chloride extract was filtered through a pasteur pipette containing a glass wool plug, and then evaporated to leave the purified anacardic acids. Finally the phenolic group on the anacardic acids was methylated (diazomethane-diethyl ether plus a few drops of methanol, 30 minute treatment (8)) to give the dimethylated anacardic acids.

Bulk Extraction of Anacardic Acids for NMR. Separate batches (2.5Kg each) of resistant and susceptible inflorescences were soaked in methylene chloride for one hour. The methylene chloride extracts were dried, filtered and evaporated as described above to leave crude resistant and susceptible exudates. A preliminary clean up of each was performed using column chromatography. Each crude extract was dissolved in hexane (15mL) and 5mL of that was applied to a silica gel chromatography column (18cm x 2cm) and eluted successively with (i) hexane (100mL), (ii) 4% diethyl ether, 1% acetic acid in hexane (200mL), (iii) 10% diethyl ether, 1% acetic acid in hexane (200mL), (iv) 50% diethyl ether, 1% acetic acid in hexane (200mL) and (v) 25% methanol in diethyl ether (100mL). The column eluent was collected in 10mL tubes and each was analyzed by tlc in order to determine the elution point of the anacardic acids. The tubes containing the anacardic acids were combined and subsequently pooled with the acids collected from the other two column runs.

The partially purified anacardic acids, from both the resistant and susceptible plants were further cleaned up prior to hplc separation using Merck 20x20cm, 2mm thickness preparative tlc plates. The compounds were applied to several plates as the dimethylated compounds (diazomethane, 30 minute treatment (8)), and run in the same two step solvent system described above. The plates were viewed at 254nm and the anacardic acid bands were cut and eluted from the silica using methylene chloride (3x30mL). The purified anacardic acids were then ready for separation by preparative hplc.

HPLC Analysis of Anacardic Acids. The anacardic acids collected from both the resistant and susceptible plants were analyzed by high performance liquid chromatography (hplc) using a 25cm x 4.6mm Supelcosil 5µ LC8 DB column. The solvent system used consisted of isopropanol, acetonitrile, 0.01M acetic acid (49: 14: 37, by volume) and was run isocratically at 1.2mL/min. Detection was achieved using a Waters 490 uv detector at 212nm, with the trace being recorded on a Shimadzu 3A recorder-integrator. In preparative chromatography of the the bulk extract, a 25cm x 10mm Supelcosil 5µ LC8 DB semi-preparative column was used. In this case the mobile phase was modified to isopropanol, acetonitrile, 0.01M acetic acid (49: 15: 30, by volume) and run isocratically at 3mL/min. The eluting solvent from each absorption peak was collected into separate vials, and the solvent was removed by rotary evaporation. Each anacardic acid was further purified by hplc, to achieve a purity of greater than 95%.

GC-MS Analysis. The dimethylated anacardic acids were analyzed by gas chromatography-mass spectrometry (gc-ms). All analyses were performed in the EI mode, using a Kratos model MS-25 mass spectrometer, with a 30m x 0.53mm RTX-5 (Restek Corp., cross bonded 95% dimethyl-5% diphenyl polysiloxane) capillary column with a temperature program of 250-280°C at 5°C/minute.

NMR Analysis. The dimethylated anacardic acids were analyzed by proton nmr spectroscopy. All samples were dissolved in deuterochloroform and were run at 360Mz on a Bruker AM 360.

Dimethyl Disulfide Derivatization. Dimethyl disulfide (DMDS) derivatization was performed on the unsaturated anacardic acids in order to locate the position of the double bond (9). A solution of the anacardic acid (0.05-0.1mg) was dissolved in hexane (1.75mL). DMDS (2.5mL) and iodine solution (0.25mL of a 60mg I_2/mL ether) were added to the reaction vial, and the reaction was allowed to run overnight at 40°C. After the reaction was complete, 5mL of a 5% (w/v) solution of sodium thiosulfate was added to the reaction mixture to remove the excess iodine. The organic layer was removed, and evaporated to leave the crude DMDS derivative, which was purified by hplc, before being submitted for mass spectrometric analysis.

Results.

HPLC Analysis. Hplc analysis of the resistant exudate (Figure 1) showed that there are two major compounds which are known to be the C_{22} ω5 (compound B) and C_{24} ω5 (compound I) unsaturated anacardic acids from earlier work, and a number of minor compounds of which only the C_{22} saturated (compound G) and C_{24} saturated (compound M) anacardic acids have previously been identified (2-6). Of the uncharacterized compounds, E and J are of particular interest, as they are present in similar quantities to the C_{22} and C_{24} saturated anacardic acids. This analysis is in good agreement with the capillary gc method used by Walters, who identified the four major anacardic acids and found them in similar quantities to those seen in the hplc trace.

Hplc analysis of the susceptible exudate by contrast revealed some differences in the composition to that reported by Walters (5, 6). A very complicated profile was seen, with at least ten peaks each contributing over 1% to the total exudate composition (Figure 2). Of those, only the C_{22} saturated (compound G) and C_{24} saturated (compound M) anacardic acids were known and hence the majority of the anacardic acids making up the exudate were unknown structures. Thus there was clearly a need to characterize the unknown compounds to obtain a full chemical profile of the susceptible exudate. Of particular interest was the almost complete lack of the C_{22} ω5 (compound B) and C_{24} ω5

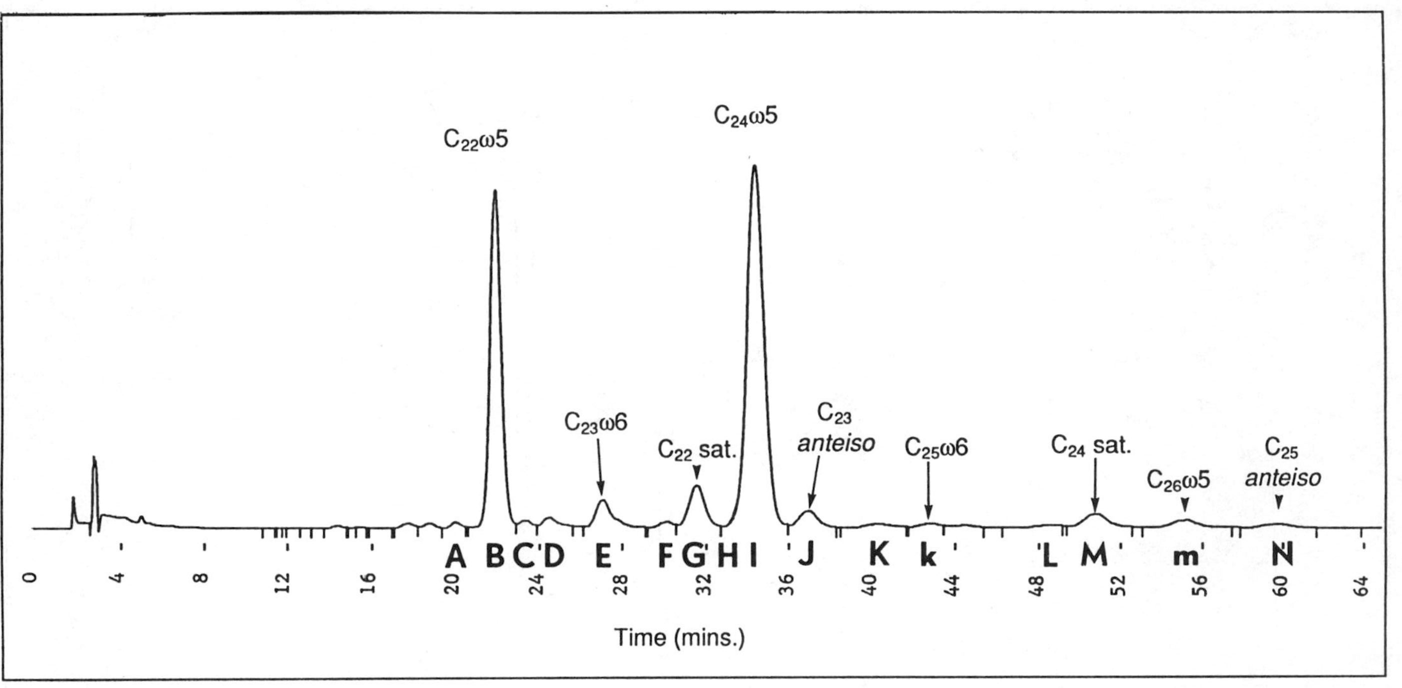

Figure 1: HPLC Profile of Glandular Trichome Exudate From Resistant Geraniums

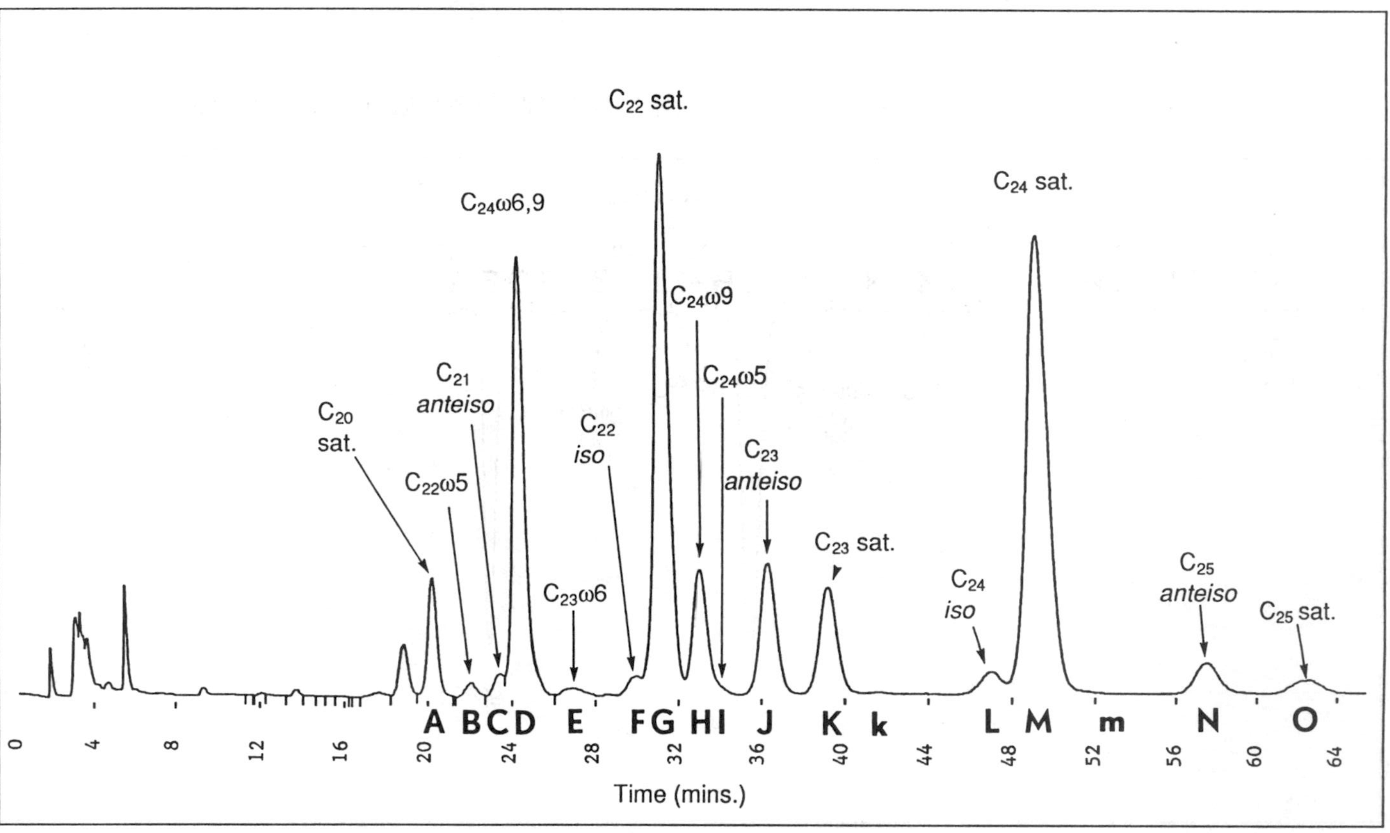

Figure 2: HPLC Profile of Glandular Trichome Exudate From Susceptible Geraniums

(compound I) anacardic acids, contrasting with the analysis by Walters, who observed that they were present in amounts varying from 30% to 60% of the total susceptible exudate composition (5, 6).

In order to resolve this ambiguity, the major purified anacardic acids from the hplc were analyzed under the same gc conditions used by Walters (Table I).

Table I: HPLC and GC Retention Times of Anacardic Acids

Code	Name	HPLC Rt(min)	GC Rt(Min)
A	C_{20} sat	19.0	nd
B	C_{22} ω5	20.5	7.74
C	C_{21} *anteiso*	21.9	nd
D	C_{24} ω6,9	22.8	10.11
E	C_{23} ω6	25.2	nd
F	C_{22} *iso*	27.8	nd
G	C_{22} sat.	29.0	7.02
H	C_{24} ω9	30.7	9.50
I	C_{24} ω5	31.5	9.82
J	C_{23} *anteiso*	33.6	7.81
K	C_{23} sat	36.1	8.11
L	C_{24} *iso*	43.2	8.73
M	C_{24} sat	45.2	9.29
N	C_{25} *anteiso*	52.5	9.94
O	C_{25} sat	56.7	nd

It was found that compound J, a major peak in the susceptible co-chromatographed with the C_{22} ω5 (compound B) peak on gc and compounds H and N closely chromatographed with the C_{24} ω5 (compound I) peak. Thus it appears that the difference in results can be accounted for by the improved separation achieved on the hplc over the capillary gc method, and hence what were believed to be ω5 anacardic acids in the susceptible exudate were really other compounds which happened to co-chromatograph with them. Such a problem is not encountered in the resistant exudate, as the ω5 anacardic acids are by far the major components.

Mass Spectral and NMR Analysis.

The mass spectral and nmr data from each of the unknown anacardic acids are given in Tables II and III respectively, and the results for each compound are discussed in detail below. The spectra were also obtained for the previously known anacardic acids (Compounds B, G, I, and M) and are listed in Tables II and III for comparative purposes. However because their structures are already known (5, 6), they are not discussed below.

Compound A. The mass spectrum of dimethylated A showed a molecular ion at *m/z* 348 and a significant peak at *m/z* 317, indicating a loss of $-OCH_3$ (Table II). The region from 100 to 200 *m/z* is practically identical in all the dimethylated anacardic acids under investigation, because as can be seen from Figure 3, none of the ions in this region contain the side chain, which differentiates each anacardic acid. The base peak ion occurs at *m/z* 161 produced by the loss of water from the ion produced by α-cleavage of the side chain from the aromatic ring (*m/z* 179). There are also substantial peaks at *m/z* 148 and *m/z* 121, indicating losses of $-OCH_3$ and $-(O{=}C){-}OCH_3$, respectively from the *m/z* 179 fragment.

The principal resonances seen in the nmr spectrum of A, together with all the other anacardic acids isolated, are shown in Table III. For the purposes of clarity, the two signals corresponding to methylated carboxyl (δ3.88ppm singlet) and phenol (δ3.79ppm singlet) groups have been omitted, as have the aromatic signals. In all cases the aromatic region consisted of two pairs of doublets at δ6.73 and δ6.79ppm, and a double doublet at δ7.23ppm, fully consistent with an ABC type system.

The nmr spectrum of A shows a triplet at δ2.51ppm corresponding to the methylene group adjacent to the ring (Table III). The methylene group beta to the ring can also be seen as a multiplet at δ1.54ppm, with remainder of the methylene groups present in an intense broad resonance at δ1.23ppm. No resonances in the vinylic region of the spectrum are seen, thus confirming the result from mass spectrometry that A is saturated. The remaining signal in the spectrum is the terminal methyl group at δ0.87ppm, which in this case is a simple triplet, integrating for three protons. Thus as there is only one methyl group present it can be concluded that the side chain in A is unbranched, and hence is identified as the C_{20} straight chain saturated anacardic acid.

Table II: Mass Spectrometric Analysis of Anacardic Acids in Glandular Trichome Exudate

Code	Name	Mwt	INTENSITY (% BASE PEAK ION) M	M-31(2)	M/Z 180	M/Z 161	M/Z 148	M/Z 121
A	C_{20} sat.	348	46	31	51	100	15	19
B	C_{22} ω5	374	20	15.	47	100	79	34
C	C_{21} *anteiso*	362	73	44	52	100	15	18
D	C_{24} ω6,9	400	29	62	64	100	35	38
E	C_{23} ω6	388	43	27	91	100	33	36
F	C_{22} *iso*	376	41	24	51	100	15	23
G	C_{22} sat.	376	34	23	55	100	14	15
H	C_{24} ω9	402	41	31	98	100	30	45
I	C_{24} ω5	402	20	16	37	100	33	32
J	C_{23} *anteiso*	390	40	22	50	100	15	23
K	C_{23} sat.	390	51	30	55	100	14	17
k	C_{25} ω6	416	31	31	41	100	20	31
L	C_{24} *iso*	404	31	16	51	100	14	22
M	C_{24} sat.	404	36	20	49	100	56	24
m	C_{26} ω5	430	19	19	32	100	21	34
N	C_{25} *anteiso*	418	39	18	52	100	15	22
O	C_{25} sat.	418	61	29	59	100	14	19

Table III: NMR Analysis of Anacardic Acids in Glandular Trichome Exudate*

Compound	CH_3	$CH_2C\mathit{H}CH_2$	$C\mathit{H}_2$-CH-Ar	$C\mathit{H}_2$-CH=	CH_2-Ar	CH=CH
A	0.87*t* (*Straight*)	1.23*m*	1.54*m*		2.51*t*	
B	0.87*t* (*Straight*)	1.24*m*	1.55*m*	1.99*m*	2.51*t*	5.32*m* *J=4.6, 4.7Hz*
C	0.85*m* (*Anteiso*)	1.24*m*	1.54*m*		2.52*t*	
D	0.89*t* (*Straight*)	1.30*m*	1.57*m*	2.04*m* 2.78*dd*	2.54*t*	5.38*m*
E	0.86*t* (*Straight*)	1.27*m*	1.53*m*	1.99*m*	2.50*t*	5.32*m* *J=5.2, 5.0Hz*
F	0.83*d* (*Iso*)	1.24*m*	1.54*m*		2.51*t*	
G	0.86*t* (*Straight*)	1.23*m*	1.55*m*		2.51*t*	
H	0.85*t* (*Straight*)	1.26*m*	1.55*m*	1.98*m*	2.51*t*	5.32*m* *J=3.6, 3.8Hz*
I	0.87*t* (*Straight*)	1.23*m*	1.55*m*	2.00*m*	2.51*t*	5.33*m* *J=4.5, 4.6Hz*
J	0.83*m* (*Anteiso*)	1.22*m*	1.54*m*		2.51*t*	
K	0.89*t* (*Straight*)	1.29*m*	1.56*m*		2.54*t*	
k	0.89*t* (*Straight*)	1.29*m*	1.57*m*	2.01*m*	2.54*t*	5.36*m* *J=4.5, 4.6Hz*
L	0.84*d* (*Iso*)	1.22*m*	1.52*m*		2.51*t*	
M	0.86*t* (*Straight*)	1.23*m*	1.55*m*		2.51*t*	
m	0.90*t* (*Straight*)	1.29*m*	1.56*m*	2.03*m*	2.54*t*	5.36*m* *J=4.5, 4.7Hz*
N	0.83*m* (*Anteiso*)	1.23*m*	1.55*m*		2.52*t*	
O	0.86*t* (*Straight*)	1.23*m*	1.53*m*		2.51*t*	

* *m*=multiplet, *t*=triplet, *d*=doublet, *dd*=double doublet

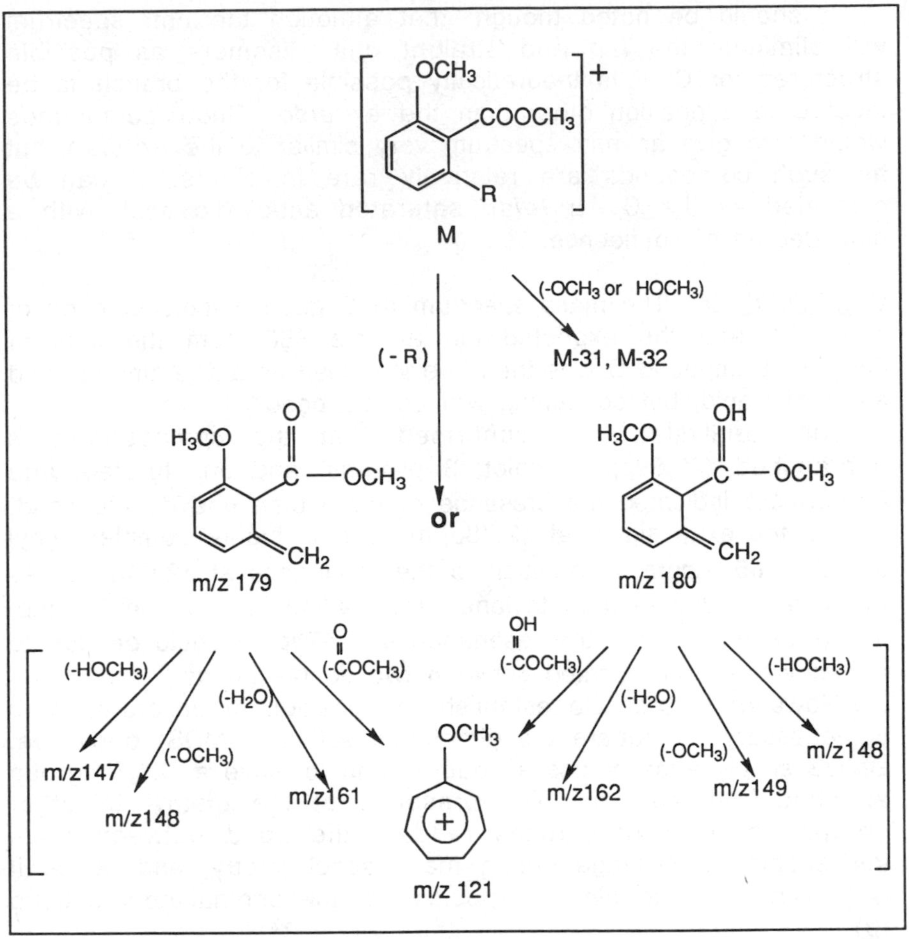

Figure 3: Mass Spectral Fragmentation of Dimethylated Anacardic Acids

<u>Compound C</u>. The mass spectrum again showed the common anacardic acid ions, and a molecular ion of *m/z* 362, which suggested that Compound C was a C_{21} saturated anacardic acid. The nmr spectrum also indicated that C was saturated, and the terminal methyl signal at δ0.85ppm was complexed and represented 6 protons. Hence C was identified as a C_{21} branched chain anacardic acid, and by comparison with the nmr spectra of *iso* and *anteiso* branched chain fatty acid standards, the terminal methyl signal was identical to the terminal methyl resonance of an *anteiso* group, thus C is in all probability the C_{21} *anteiso* saturated anacardic acid.

It should be noted though, that although the nmr spectrum will eliminate the *iso* and straight chain isomers as possible structures for C, it is theoretically possible for the branch to be located in a position other than the *anteiso*. Such compounds would also give an nmr spectrum very similar to the *anteiso*, but as such compounds are relatively rare in plants, C can be classified as the C_{21} *anteiso* saturated anacardic acid, with a high degree of confidence.

Compound D. The mass spectrum of D gave a molecular ion of m/z 400, and the expected ion at m/z 368 from the loss of CH_3OH. Compound D was therefore identified as a C_{24} unsaturated anacardic acid, but containing two double bonds.

Nmr analysis of D confirmed that the compound was unbranched (δ0.89ppm, triplet, 3 protons) and unsaturated, and furthermore indicated the presence of more than one double bond, due to the extra signal at δ2.78ppm, and a highly complex vinyl signal at δ5.38ppm, in addition to the resonance at δ2.04ppm. As the latter is due to a methylene group adjacent to a vinyl group, it was postulated that the resonance at δ2.78ppm could be due to a methylene group located between two double bonds.

However in order to establish the position of unsaturation, it is necessary to prepare the dimethyl disulfide (DMDS) derivative. DMDS simply adds across a double bond to leave a CH_3S- group attached both carbons. The presence of the methyl thioether groups on adjacent carbons renders the bond between them susceptible to cleavage during mass spectrometry, and hence it is possible to determine the position of the original double bond (9).

DMDS has previously been used to establish the location of double bonds in mono-unsaturated compounds (9); however it appeared to work very well in the case of compound D, giving one major peak on the hplc. The mass spectrum turned out to be much more complicated than expected, because of partial decomposition of the derivative under the conditions employed. From Figure 4, the ion at *m/z* 291, which is derived from the loss of CH_3OH from *m/z* 323 places one double bond 9 carbons in from the terminal methyl group. The position of the second double bond was localized from the ion *m/z* 347, which arises from the loss of a CH_3SCH_3 unit from the 9,10 carbon bond to leave a thio-epoxide group, a break across the 6,7 carbon bond and a loss of CH_3SH. The ion at *m/z* 315 arises from the loss of CH_3OH from *m/z* 347. Hence the second double bond is located 6 carbons in from the terminal methyl group and therefore compound D is

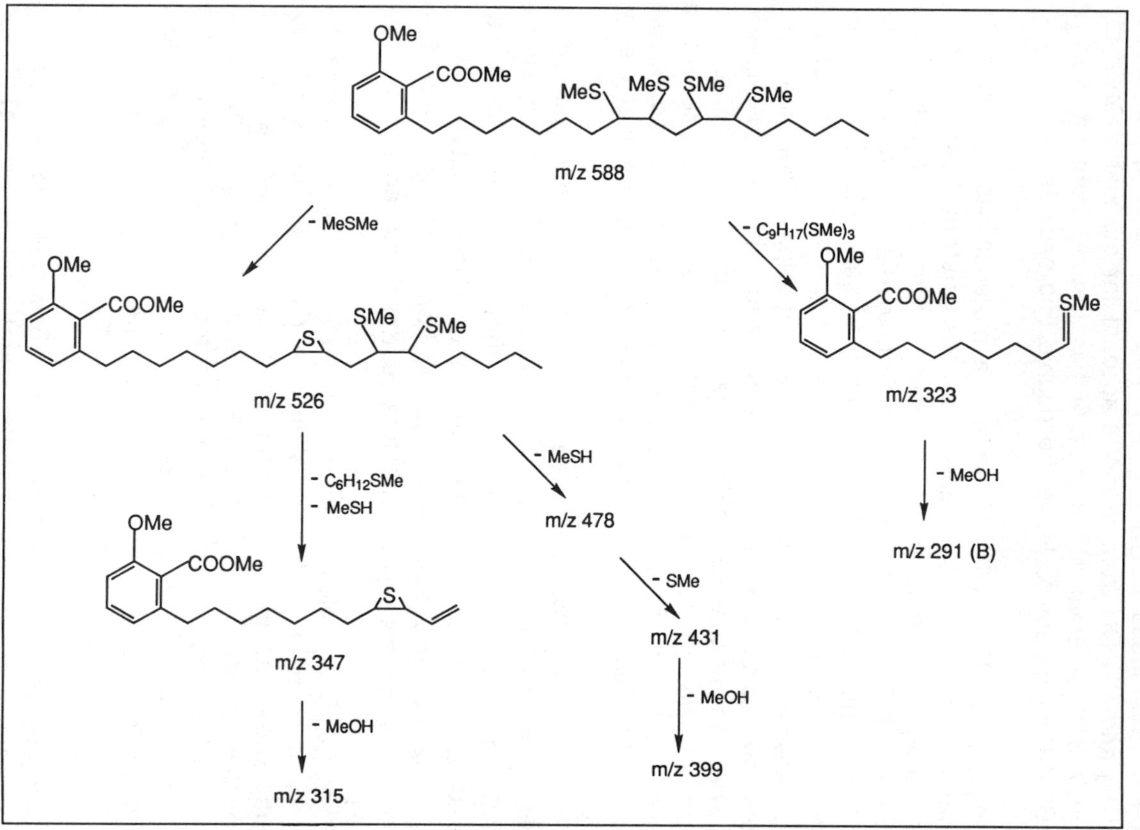

Figure 4: Mass Spectral Fragmentation of the DMDS Derivative of D

identified as being a C_{24} ω6,9 unsaturated anacardic acid. The other major ions in the spectrum can be accounted for by loss of CH_3SCH_3 from the molecular ion (not seen) to leave *m/z* 526. That ion itself loses CH_3SH to give *m/z* 478, and further losses of SCH_3 and CH_3OH from *m/z* 478 account for the ions at *m/z* 431 and *m/z* 399. A small ion at m/z 464 is formed from the loss of two CH_3SCH_3 groups to leave the double thio-epoxide.

Compound E. E was found to have a molecular ion at *m/z* 388 indicating that the structure was consistent with a C_{23} unsaturated anacardic acid. The mass spectrum of the DMDS derivative of E showed the expected molecular ion at *m/z* 482, and two significant fragments at *m/z* 319, and *m/z* 131. The ion at *m/z* 131 is the fragment expected from the cleavage of the bond between adjacent methyl thioether groups, when the bond occurs six carbons from the methyl end of the side chain. However, as anacardic acids containing branched chains had already been identified, *m/z* 131 could also represent a fragment, with an *iso* or *anteiso* structure and hence the double bond would then occur 5 carbon atoms from the terminal methyl group.

This ambiguity was easily solved by the nmr spectrum, which showed that the terminal methyl signal (δ0.86ppm) was a simple 3 proton triplet, which hence proved that the double bond was located at the ω6 position. In addition coupling constants of 5.2 and 5.0Hz indicated that the the double bond was *cis.* Hence E is identified as the *cis* C_{23}ω6 unsaturated anacardic acid.

Compound F. A molecular ion at *m/z* 376 and the knowledge that the retention time of F was slightly earlier than G, which was already known as the C_{22} straight chain saturated anacardic acid, strongly suggested that Compound F was a C_{22} branched chain saturated anacardic acid. The nmr of the terminal methyl signal (δ0.83ppm) showed it contained 6 protons, and was split into a simple doublet, consistent with an '*iso* type' end chain. Hence F is the C_{22} *iso* saturated anacardic acid.

Compound H. The mass spectrum shown in Figure 5 has a molecular ion at *m/z* 402 and the usual ions at *m/z* 180, 161, 148 and 121 indicating that H is a C_{24} unsaturated anacardic acid. The ion at *m/z* 370 is formed from the loss of CH_3OH from the molecular ion, The DMDS derivative gave the spectrum shown in Figure 6. The expected molecular ion at *m/z* 496 was seen and the fragment at *m/z* 173 located the position of unsaturation as

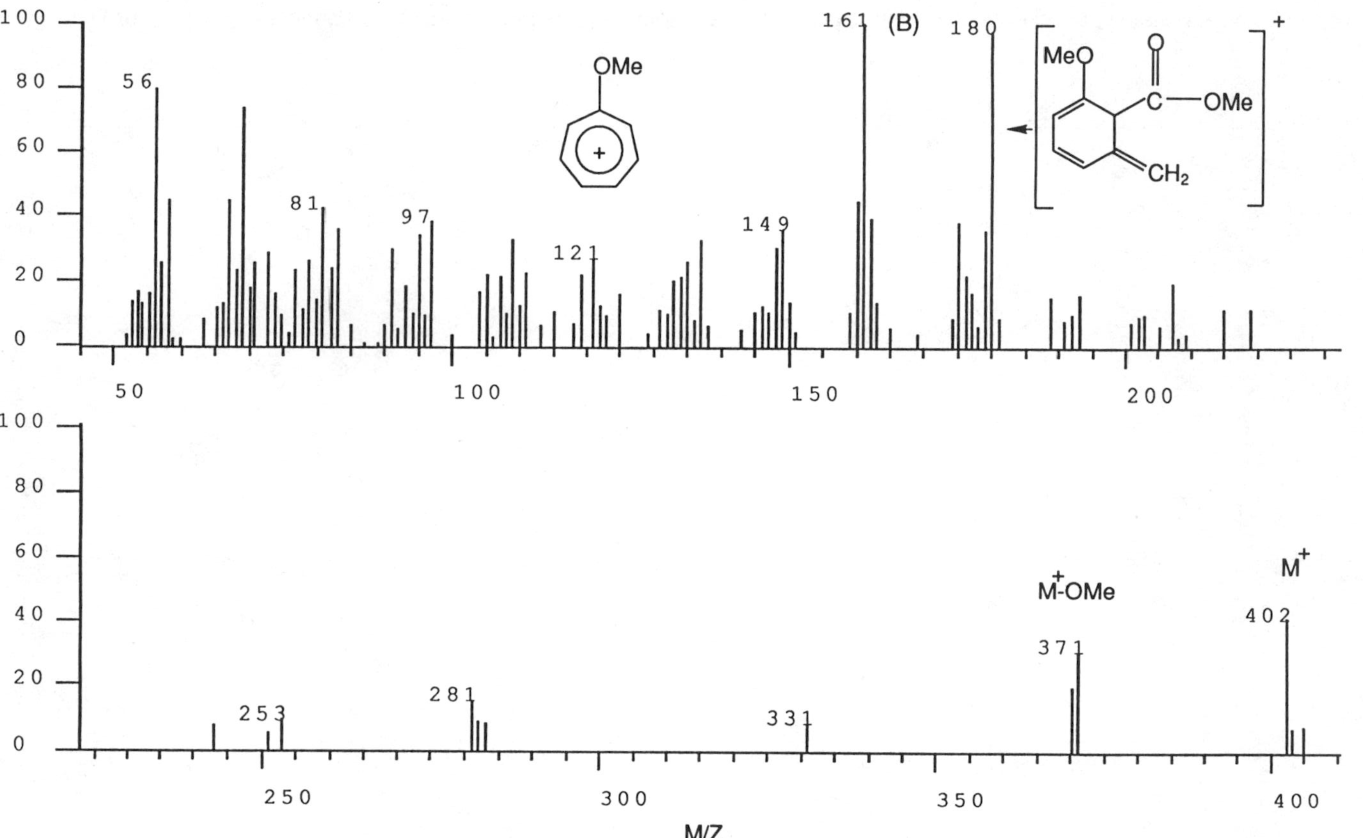

Figure 5: EI Mass Spectrum of Dimethylated Compound H from Geranium

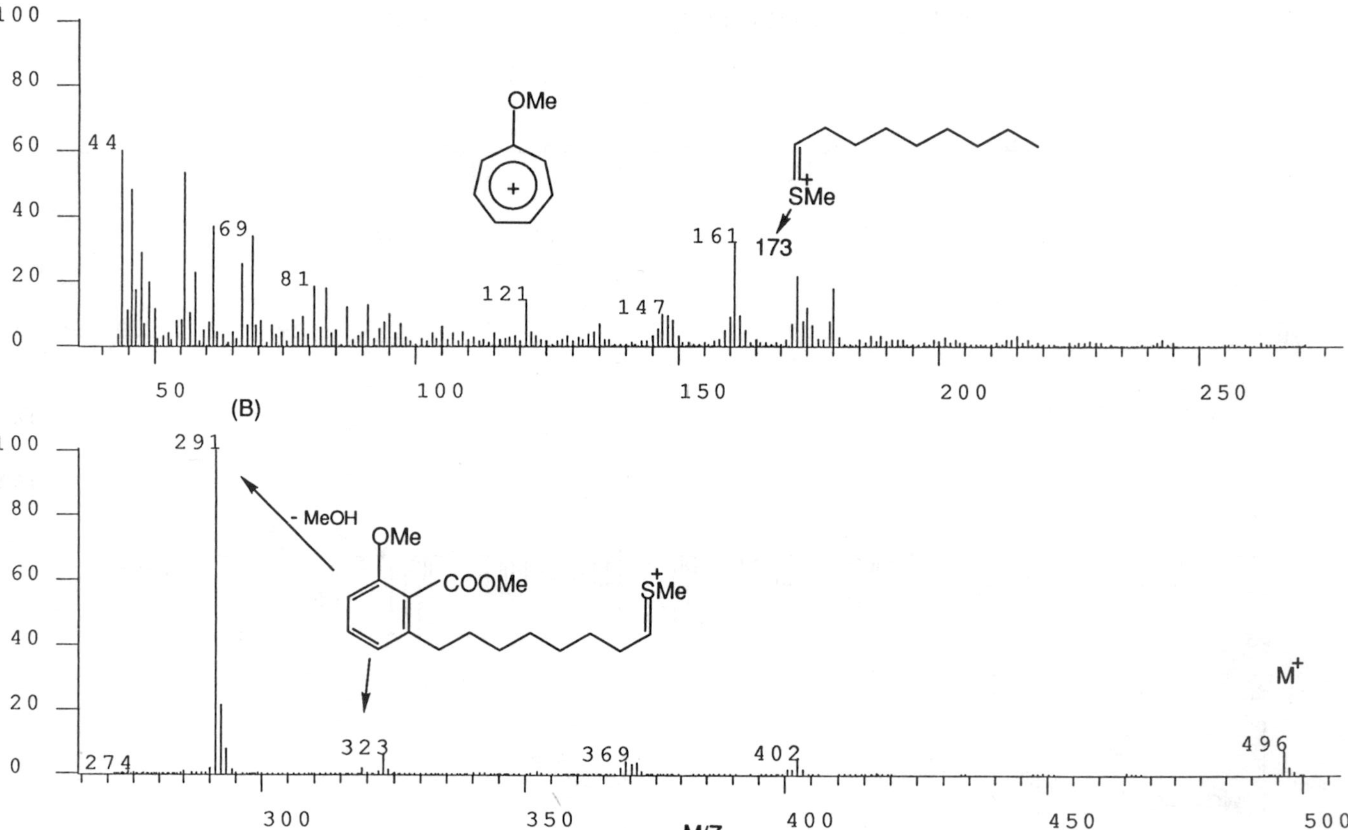

Figure 6: EI Mass Spectrum of the Dimethyl Disulphide Derivative of Dimethylated Compound H from Geranium

being 9 carbon atoms from the end of the terminal methyl group. The fragment at *m/z* 291, formed from the loss of CH_3OH from *m/z* 323, supported this conclusion. In addition the nmr spectrum showed that the side chain was unbranched (δ0.85ppm, triplet), and also that the geometry of the double bond was *cis*, due to *J* values of only 3.6 and 3.8Hz. Hence compound H is identified as the *cis* C_{24} ω9 unsaturated anacardic acid.

Compound J. J was found to have a molecular ion at *m/z* 390, together with the characteristic ions in the 100 to 200 *m/z* region, suggesting that the structure was a C_{23} saturated anacardic acid. As can be seen from Figure 7a nmr analysis confirmed that J was saturated, and examination of the terminal methyl group signal (δ0.83ppm multiplet), showed it integrated for 6 protons and contained a splitting pattern consistent with *anteiso* branching. Hence J is more than likely the C_{23} *anteiso* saturated anacardic acid.

Compound K. The mass spectrum was practically identical to that of compound J, with the same molecular ion at *m/z* 390, but with a slightly later hplc retention time, and hence it was thought likely that K was a C_{23} straight chained saturated anacardic acid. Confirmation of this was provided from the nmr spectrum which showed a simple three proton terminal methyl triplet at δ0.89ppm, as well as the absence of any signals in the vinyl region. Hence K is the C_{23} straight chain saturated anacardic acid.

Compound k. This minor peak in resistant exudate, and not seen in the susceptible exudate, gave a molecular ion at *m/z* 416 suggesting that the structure was a C_{25} unsaturated anacardic acid. The mass spectrum of the DMDS derivative gave the expected molecular ion at *m/z* 510, a fragment at *m/z* 131 and an ion at m/z 347, indicating that a 6 carbon fragment had been lost. However as in the case of Compound E, it was also possible that this 6 carbon fragment was branched, thus meaning that the double bond would be effectively located in the ω5 position. This ambiguity was easily cleared up by the nmr spectrum which showed a simple triplet which integrated for 3 protons at δ0.89ppm, and also because of the small *J* values (4.5, 4.6Hz), the geometry of the double bond was shown to be *cis*. Hence k is the *cis* C_{25} ω6 unsaturated anacardic acid.

Compound L. The mass spectrum gave a molecular ion at *m/z* 404 together with the other expected anacardic acid ions. Furthermore as compound L had an earlier retention time on the hplc than compound M, the known C_{24} saturated straight chained anacardic acid, it was thought likely that L was a C_{24} branched chain saturated anacardic acid. The exact structure was confirmed by nmr spectroscopy (Figure 7b), which showed the terminal methyl group (δ0.84ppm) to be a 6 proton doublet, a splitting pattern consistent with *iso* branching. Thus L is the C_{24} *iso* saturated anacardic acid.

Compound m. This was another compound only seen in the resistant exudate, and from a molecular ion of *m/z* 430 it was identified as a C_{26} unsaturated anacardic acid. The spectrum of the DMDS derivative gave a molecular ion of *m/z* 524 and the fragment at *m/z* 117 placed the double bond 5 carbons in from the terminal methyl group The ion at *m/z* 375, was also consistent with a 5 carbon fragment being lost. The nmr spectrum showed a simple triplet, integrating for 3 protons, at the terminal methyl position, showing that the chain was unbranched, and thus confirming the ω5 double bond position. In addition the small *J* values of 4.5 and 4.7 Hz placed the double bond geometry as *cis*, and hence m is identified as the *cis* C_{26} ω5 unsaturated anacardic acid.

Compound N. The spectrum showed the compound had a molecular weight of 418, consistent with a C_{25} saturated anacardic acid. Nmr analysis showed that the splitting of the terminal methyl group (δ0.83ppm multiplet, 6 protons) was complex and identical to the terminal methyl group splitting in the nmr spectrum of the *anteiso* standard. Hence N is in all probability the C_{25} *anteiso* saturated anacardic acid.

Compound O. This gave a practically identical spectrum to N, with the same molecular weight, and because of the slightly later retention time on the hplc it was strongly suspected that O was the C_{25} straight chain isomer. This was readily confirmed by nmr as the terminal methyl group at δ0.86ppm was a simple 3 proton triplet, meaning that O is the C_{25} straight chain saturated anacardic acid.

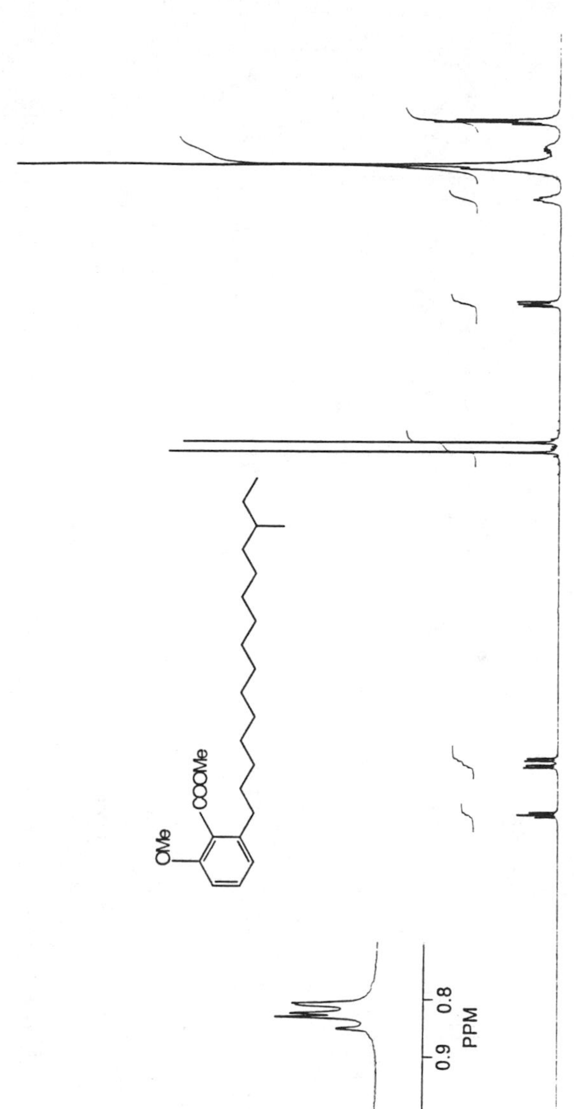

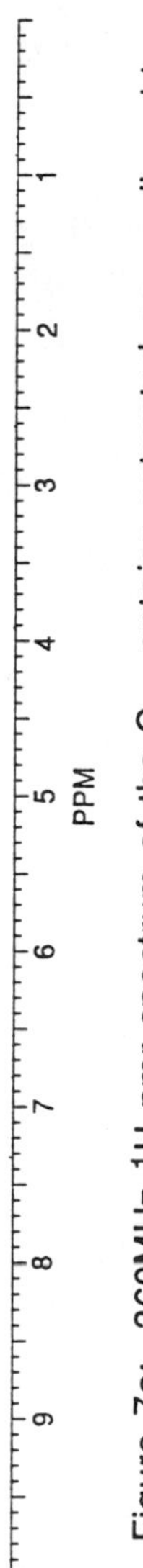

Figure 7a: 360MHz ^1H nmr spectrum of the C_{23} *anteiso* saturated anacardic acid

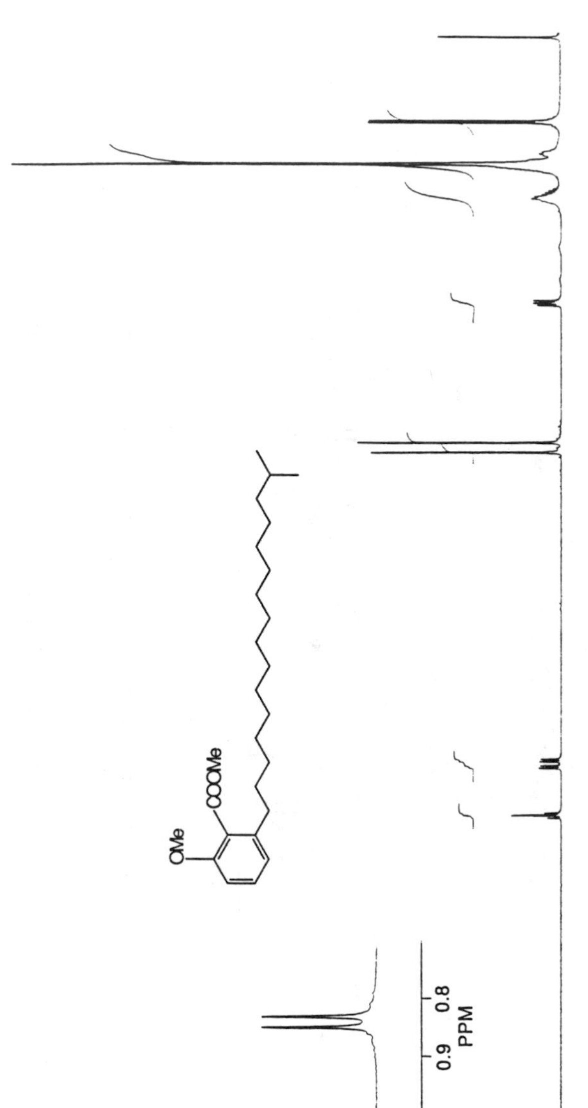

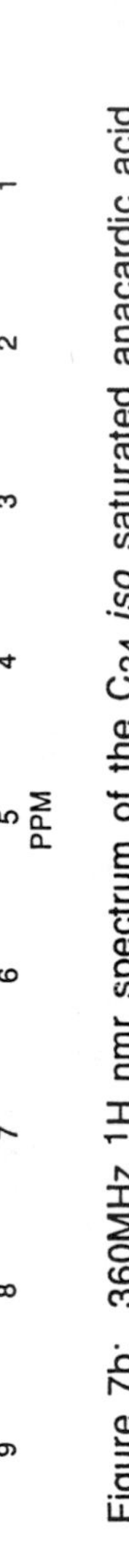

Figure 7b: 360MHz ^{1}H nmr spectrum of the C_{24} *iso* saturated anacardic acid

Discussion.

Having identified the majority of the anacardic acids in both resistant and susceptible exudate, it is possible to more fully highlight the differences between the two, than was previously the case.

Table IV: Percentage Composition of Anacardic Acids in Geranium Glandular Trichome Exudate.

Code	Name	Resistant	Susceptible
A	C_{20} sat	0.38±0.06	3.28±0.31
B	C_{22} ω5	30.55±0.10	0.46±0.08
C	C_{21} *anteiso*	0.55±0.03	0.78±0.07
D	C_{24} ω6,9	0.95±0.10	16.62±0.64
E	C_{23} ω6	3.43±0.05	0.42±0.04
F	C_{22} *iso*	0.53±0.05	0.96±0.10
G	C_{22} sat.	5.20±0.04	23.86±0.81
H	C_{24} ω9	0	7.88±1.26
I	C_{24} ω5	49.83±0.27	0.14±0.14
J	C_{23} *anteiso*	2.03±0.05	7.54±0.88
K	C_{23} sat.	0.65±0.07	5.62±0.28
k	C_{25} ω6	0.45±0.05	0
L	C_{24} *iso*	0.35±0.05	1.84±0.20
M	C_{24} sat.	2.63±0.09	27.26±1.54
m	C_{26} ω5	1.53±0.03	0
N	C_{25} *anteiso*	0.90±0.04	2.40±0.22
O	C_{25} sat.	0	1.00±0.13

From Table IV, resistant exudate is predominantly unsaturated, with the C_{22} ω5 and C_{24} ω5 anacardic acids making up 80% of the total exudate. In addition there is a small amount of C_{26} ω5 anacardic acid, which brings the total amount of ω5 unsaturated anacardic acids to 82%. There interestingly is also a significant quantity of the C_{23} ω6 anacardic acid and a small amount of the C_{25} ω6 anacardic acid, which together contribute nearly 4% to the exudate composition. The remainder of the exudate is composed of saturated anacardic acids, the major compounds being the C_{22} and C_{24} saturated and C_{23} *anteiso* saturated anacardic acids.

The identification of ω6 unsaturated compounds is interesting because all previous studies on the biosynthesis of anacardic acids pointed to an ω5 desaturase, forming C_{16} and C_{18} ω5 unsaturated fatty acids which would then be processed into the C_{22} and C_{24} ω5 unsaturated anacardic acids (5, 10, 11,). The studies also suggest there is a mechanism whereby the ω5 fatty acids, which are only present in small amounts (<1%) in comparison with the total fatty acid pool, are specifically selected in preference to all the other fatty acids, and synthesized into ω5 anacardic acids. Hence the finding of ω6 unsaturated anacardic acids in significant amounts may suggest that there is some flexibility in the selection mechanism.

The susceptible exudate, by contrast is predominantly saturated, but unsaturated anacardic acids, different from those observed in the resistant plants, still compose about 30% of the exudate. Of the saturated compounds, the C_{22} and C_{24} saturated anacardic acids are the major components, making up about 50% of the exudate. The remaining 20% is a complex mixture of odd chain length and branched chain anacardic acids, of which the C_{23} *anteiso* and the C_{23} straight chain saturated anacardic acids make up the majority, although the C_{24} *iso* and C_{25} *anteiso* compounds are also present in significant amounts.

Of the remaining 30% of the exudate which is unsaturated, the most striking thing to note is the almost complete lack of any ω5 unsaturated anacardic acids. The C_{22} ω5 is present, but only contributes around 0.5% to the total exudate, and the C_{24} ω5 is not seen at all. The unsaturated exudate is instead composed of two compounds, the C_{24} ω6,9 and the C_{24} ω9 anacardic acids, the former contributing nearly 17% to the total exudate composition, and the latter nearly 8%. From the biosynthetic scheme proposed in earlier work, it is more than likely that these two compounds are derived from the C_{18} ω6,9 and ω9 unsaturated fatty acids, linoleic and oleic acids (5, 10, 11). The C_{24} ω6,9 anacardic acid is also present in the resistant exudate, but only contributes around 1% of the total exudate, and thus is not a major component in the resistant plant.

The analyses have been performed on a number of resistant and susceptible plants, which included the parents and first and second generation self-pollinated progeny, in order to assess variability within each plant line. Previous work by Walters showed that the chemical exudate composition was held relatively constant within the resistant line, and the results presented in Table IV would support that conclusion (5, 6).

Clearly from the standard deviations, there is very little variation in the exudate chemistry within the resistant line, implying that the chemistry is under tight genetic and biochemical control, and the plant line is homozygous for the pertinent gene or genes.

The results from the susceptible line, indicate that the chemistry of the exudate is also relatively constant, although from the standard deviations, it is apparent that there is more variability than is observed in the resistant line. However, the results suggest that the plant line is homozygous, which is essential for any inheritance studies (Walters, D. S. et al. In Pesticides and Alternatives; Elsevier: Amsterdam, in press).

These findings with the susceptible plant line contrast with those found by Walters, who as previously mentioned found significant amounts of the ω5 anacardic acids, and also observed great variability within the susceptible line (5, 6). The ω5 compounds in the susceptible line which were reported by Walters (5, 6), are most likely compounds J and H, which co-chromatograph under the gc conditions used. In no instance using the hplc method, were any significant amounts of the ω5 anacardic acids detected in the susceptible plants. The variability seen by Walters could have been influenced by the small amount of exudate obtained from the susceptible line by the bicarbonate wash technique and the lack of tlc purification of the extract.

Some further comments on the chemical composition of the exudate are warranted, particularly concerning the branched chain and the odd chain length anacardic acids.

Before doing this, however, it is useful to briefly consider the biosynthesis of anacardic acids. It is believed that anacardic acids are synthesized from long chain fatty acids via a modification of the methyl salicylate synthetase scheme (11, 12). In effect a fatty acid, such as palmitate (C_{16}) is elongated by six carbons by the addition of three acetate units in a series of condensation and dehydration steps, followed by ring closure and aromatization to form a C_{22} saturated anacardic acid (5, 10, 11). Similarly, margaric acid (C_{17}) is synthesized to a C_{23} saturated anacardic acid. A comparative study on the biosynthesis of anacardic acids in resistant and susceptible plants, using ^{14}C fatty acid precursors will be the subject of a separate publication.

Hence it is now possible to define odd chain length and also branched chain anacardic acids as ones which are derived from odd chain length and branched chain fatty acids. The presence of significant amounts of C_{23} and C_{25} anacardic acids in the susceptible plant, suggests that odd chain length fatty acids are firstly present in the plant, which is interesting since they are not commonly found in biological systems, and that they can be processed into anacardic acids with relative ease. However it can be seen that there is still a preference for even chain length fatty acids, as indicated by the greater quantities of even chain length anacardic acids.

One striking thing to note from the structures of the branched chain compounds is the fact that the C_{21}, C_{23} and C_{25} branched chain anacardic acids are all *anteiso*, and the C_{22} and C_{24} branched compounds are *iso*. At this point in the work it is not clear why this should be the case, but the findings would suggest that odd chain length and even chain length branched chain fatty acids are synthesized from different precursors.

It is relatively rare to find significant quantities of both odd chain length and branched chain fatty acids in biological systems. However such compounds have been isolated and identified from the epicuticular wax from Brussels sprout leaves, with *anteiso*-C_{17} and *anteiso*-C_{19} making up 36% of the total saturated fatty acid fraction (13). Similarly Radunz has isolated and identified *iso* and *anteiso* mono methyl branched fatty acids in phospholipids, contained in yellow-white leaves and petals of the plastome mutants "*Prasinizans*" of *Antirrhinum majus* and "*Xanthi*" of *Nicotiana tabacum* (14). In addition small amounts of odd chain length fatty acids and alcohols have been found in the surface waxes of *Zea mays* husks (15).

Summary.

The analysis of the resistant line confirms the findings of Walters by showing that the exudate was predominantly unsaturated, with the C_{22} and C_{24} ω5 anacardic acids being the major compounds. A number of minor components in the resistant exudate have been identified, most of which are other ω5 and ω6 unsaturated anacardic acids. The findings with the susceptible plant exudate agreed with Walters on the aspect that the exudate was predominantly saturated, but differed in the respect that the ω5 compounds were shown to be virtually absent. The major components were the C_{22} and C_{24} saturated anacardic acids, but also there were large amounts of a C_{24} ω6,9 unsaturated and an *anteiso* C_{23} compound, along with significant quantities of other anacardic acids. It should be noted that many of these peaks were seen by Walters on gc, but for the purposes of the study at that time, it was considered necessary to focus on the four anacardic acids identified from the resistant plant (5, 6).

Finally, the chemical composition of the exudate was found to be virtually constant within both the resistant and susceptible plant lines, suggesting that the exudate composition is genetically controlled. A preliminary inheritance study has been performed by Walters and will be the subject of a future publication.

Acknowledgments

This work was supported by Northeastern Regional Project NE-115 and the United States Department of Agriculture Grant No. 87-CRCR-1-2423. The Authors would also like to thank Dr. R. Minard, Dr. A. Freyer, Mr. E. Bogus, Miss M. Hardy and Miss J. Harman for assistance.

Literature Cited.

1. Stark, R. S. Ph. D. Thesis, Pennsylvania State University, University Park, 1975.
2. Gerhold, D. L.; Craig, R.; Mumma, R. O. J. Chem. Ecol. 1984, 10, 713.
3. Craig, R.; Mumma, R. O.; Gerhold, D. L.; Winner, B.; Snetsinger, R. In Natural Resistance of Plants to Pests: Role of Allelochemicals; Green, M. B.; Heden, P. A., Eds.; ACS Symposium Series No. 296; American Chemical Society: Washington, DC, 1986; p168.
4. Walters, D. S.; Minard, R.; Craig, R.; Mumma, R. O. J. Chem. Ecol. 1988, 14, 743
5. Walters, D. S. Ph.D. Thesis, Pennyslvania State University, University Park, 1988.
6. Walters, D. S.; Craig, R.; Mumma, R. O. J. Chem. Ecol. 1990, 16, 877.
7. Walters, D. S.; Grossman, H.; Craig, R,; Mumma, R. O. J. Chem. Ecol. 1989, 15, 357.
8. Gellerman, J. L.; Schlenk, H. Anal. Chem. 1968, 40, 739.
9. Buser, H.; Am, H.; Guerin, P.; Rauscher, S. Anal. Chem. 1983, 55, 818.
10. Geissman, T. A. In Biogenesis of Natural Compounds, Bernfield, P., Ed.; Macmillan: New York, 1963; p563.

11. Walters, D. S.; Craig, R.; Mumma, R. O. Phytochemistry 1990, 29, 1815.
12. Vogel, G.; Lynen, G. Meth. Enzymol. 1975, 38, 520.
13. Baker, E. A.; Holloway, P. J. Phytochemistry 1975, 14, 2463.
14. Radunz, A. In The Metabolism, Structure and Function of Plant Lipids, Stumpf, P. K.; Mudd, J. B.; Nes, W. D., Eds.; Plenum Press: New York, 1987; p197.
15. Bianchi, G.; Avato, P.; Salamini, F. Cereal Chem. 1984, 61, 45.

RECEIVED July 2, 1990

Chapter 16

Maysin in Corn, Teosinte, and Centipede Grass

R. C. Gueldner[1], Maurice E. Snook[1] B. R. Wiseman[3], N. W. Widstrom[3], D. S. Himmelsbach[2], and C. E. Costello[4]

[1]Tobacco Quality and Safety Research Unit, and [2]Plant Structure and Composition Research Unit, Agricultural Research Service, U.S. Department of Agriculture, P.O. Box 5677, Athens, GA 30613

[3]Plant Resistance–Germplasm Enhancement Research Unit, Agricultural Research Service, U.S. Department of Agriculture, P.O. Box 748, Tifton, GA 31793

[4]Department of Chemistry, Massachusetts Institute of Technology, Cambridge, MA 02139

Maysin, a flavonoid *C*-glycoside originally found in the silks of the Mexican corn, Zapalote Chico, is also found in the leaves of corn, teosinte and centipedegrass, which are resistant to the corn earworm and the fall armyworm, the major insect pests of corn in the southeastern United States. The two insect pests feed in the whorl which has a lower maysin content than earlier leaves. The maysin content of corn leaves is highest in the tips of the leaves connected just below the ear-bearing nodes. In some varieties of corn luteolin *C*-glycosides related to maysin are shown to be present by HPLC-UV analysis in silks and/or leaves. These varieties are potential sources for the variant luteolin *C*-glycosides that are likely resistant factors for the corn earworm and the fall armyworm.

The two most important pests of corn in the southeastern United States are the corn earworm (*Heliothis zea* [Boddie]) and the fall armyworm (*Spodoptera frugiperda* [J. E. Smith]). The corn earworm damages the leaves in the early stages of corn growth, and later, as the corn develops, infests the ears. The fall armyworm damages all parts of the plant and in late season causes severe damage in the form of broken leaves as well as reduced leaf photosynthetic activity. Grasses are also targets of the fall armyworm which is a leaf feeder on relatively susceptible plants such as bermudagrass (*Cynodon dactylon* [L.] Pers.). Variations in resistance have been studied in different clones of bermudagrass (1). Centipede grass (*Eremochloa ophiuroides* (Munro [Hack.]) (2,3) and teosinte (*Zea mays* L.ssp. *mexicana* [Schrad.]) are examples of Gramineae which are very resistant to the fall armyworm. Species of Gramineae that are resistant to the fall armyworm sustain less damage than susceptible plants and are readily identified by lack of feeding damage. Factors that cause resistance may be complex and often require prolonged scrutiny by teams of entomologists and chemists to define these factors. These factors may be physical or chemical. For instance, Walter's White, a sweet corn, is an example of mechanical resistance to the corn earworm which rarely reaches the kernels of the ear because of the large quantity and excellent quality of silks that provide the larvae with adequate nourishment.

In the realm of chemical resistance just three types of compounds occurring

**This chapter not subject to U.S. copyright
Published 1991 American Chemical Society**

in corn have been shown to be inhibitory to insects in laboratory bioassays: the hydroxamic acid family, Hx, typified by DIMBOA (2,4-dihydroxy-7-methoxy-2H-1,4-benzoxazin-3[4H]-one) (Fig. 1a) which occurs in corn in the form of a glucoside, caffeic acid derivatives typified by chlorogenic acid (Fig. 1b) and flavonoid glycosides typified by maysin, (Fig. 1c). The inhibition of growth of the larvae in laboratory bioassays is taken as evidence of a resistance factor in corn when it can be shown that the factor occurs at a significant level in the area of the plant that the larvae feed upon. The criterion of larval growth inhibition as the measure of the contribution of a chemical factor to resistance in a plant remains at least partially subjective because it is impossible to exactly match the nutritive value of the intact plant in the artificial medium used in the laboratory bioassay.

By the above criteria DIMBOA is not very inhibitory to the corn earworm and the fall armyworm, but is effective against the first brood European corn borer (Ostrinia nubilalis [Hubner]) when DIMBOA is at its highest level at the seedling stage. Since the sizeable literature concerning the Hx family has been recently reviewed (5) no further consideration of the Hx family will be presented in this discussion.

Chlorogenic acid (Fig. 1b), the main representative of the second class of inhibitor, is a derivative of caffeic acid, and shows growth-inhibiting activity against the corn earworm (6,7,8). The effect of growth inhibition seems to be a combination of inhibition of feeding and post-ingestive phenomena (9). Chlorogenic acid is a growth inhibitor of the fall armyworm also (3).

Maysin (Fig. 1c), the main representative of the third class of inhibitory compounds, the flavonoid glycosides, was shown to be active against the corn earworm and was identified as the antibiotic factor in the exotic strain of corn, Zapalote Chico (10). The mode of action of maysin in inhibiting the growth of the corn earworm is similar to that of chlorogenic acid (9).

Maysin is a luteolin *C*-glycoside identified by Elliger, et al (11) as 2"-O-α-rhamnosyl-6-*C*-(6-deoxy-xylo-hexos-4-ulosyl)-luteolin (Fig. 1c). Elliger, et al (12), also showed that the presence of [3', 4']-ortho dihydroxy groups on the B ring of the aglycone enhanced the inhibitory activity of flavonoids in contrast to the monohydroxy (apigenin) compounds. This structural feature contributes to activity against both the corn earworm and the fall armyworm and is also present in chlorogenic acid which has the ortho-dihydroxy grouping on the caffeic acid moiety. When the 3' hydroxy group of maysin is methylated, activity is cut in half; when the luteolin moiety is methylated at both the 3' and 4' positions the resulting compound is reduced in activity against the corn earworm (12).

In searching for resistance factors in the resistant plants, teosinte and centipede grass, and, in examining the leaves of corn varieties, we conclude that the *C*-glycosyl flavonoids, typified by maysin, are a major biosynthetic pathway, one which we believe can possibly be genetically amplified to develop corn with resistance to both the corn earworm and the fall armyworm. We wish to report on the occurrence of maysin and related compounds in the leaves and silks of corn and discuss possible relationships to corn resistance for the corn earworm and the fall armyworm.

MATERIALS AND METHODS

Collection, Storage and Treatment of Plant Material

Individual corn silks, approximately 5 g per ear, were analyzed by reversed

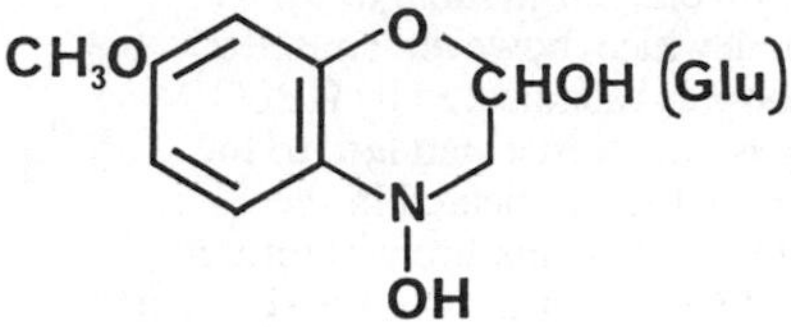

a

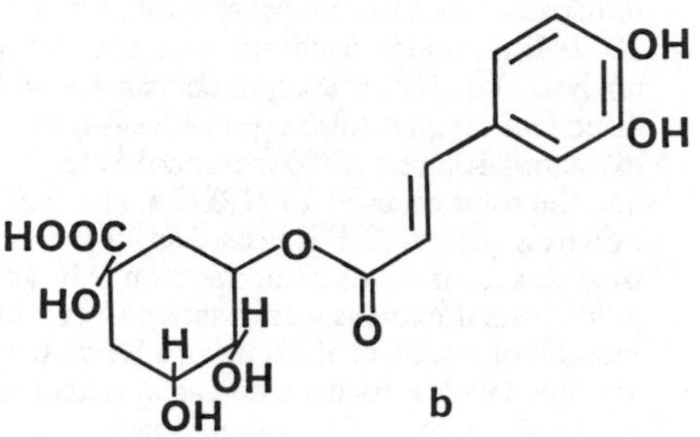

b

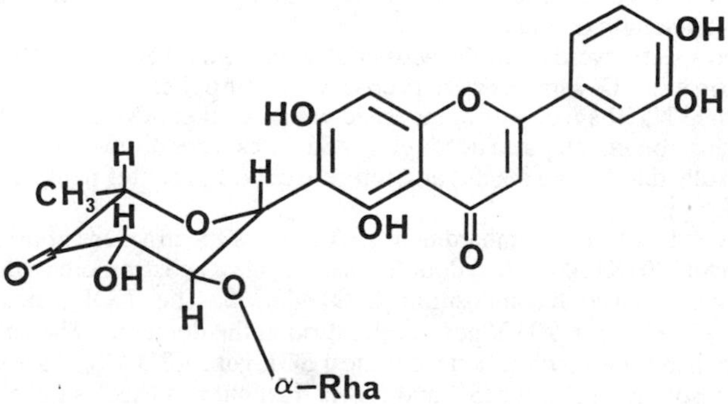

c

Figure 1. Structures of three insect inhibitory compounds from corn.

phase HPLC by the method of Snook et al (13). The silks, in methanol, were stored at O°C, allowed to warm to room temperature, and ultrasonicated for 20 min just prior to sampling for HPLC analysis. Corn leaves were collected at various stages except for the large samples intended for bulk extraction which were extracted at the 10 to 14 leaf stage. Various drying methods were used for the leaf materials. For obtaining leaf phenolic profiles freeze drying was used. For small analytical samples (1-10 g) microwave drying (3-5 min at high power with a standard kitchen microwave oven) was found to be convenient and preserved the profile of fresh plant material very well. Large samples for bulk extractions were most conveniently air-dried, which, however, caused a 10-20 % reduction in chlorogenic acid and flavonoids as indicated by HPLC analysis. For HPLC analysis chrysin was added as an internal standard to the dried leaf samples (also used with silks) which were then sonicated in the extraction solvent, 50:50 methanol:water for 30 min. The mixture was filtered and the solution used for HPLC analysis. Fresh leaf material was ground with a Polytron grinder and sonicated as for the dried plant material. The methanol used was from Burdick and Jackson (Muskegon, MI, USA)* "distilled-in-glass grade". Leaf extracts were analyzed by reversed phase HPLC similar to the method of Snook et al (9) using a Hewlett-Packard 1090M liquid chromatograph equipped with a 1040A diode array detector and Chem Station computerized data collection system. The column was a Nucleosil 5μ (4.6 X 25 cm) ODS. A solvent gradient was used starting at 20:80 methanol: water and ending at 100% methanol in 40 min. Both solvents contained 0.1% H_3PO_4. The flow rate was optimized at 0.8 mL/min for best resolution according to Meyer (14).

Spectra

UV spectra were taken on-the-fly with the 1040A diode array detector at a monitoring wavelength of 340 nm.

Infrared spectra were recorded with an Analect Model FX-60 FTIR spectrophotometer. IR samples were prepared as KBr pellets.

Carbon-13 NMR spectra were recorded with a Bruker AM-250 spectrometer at 62.9 mHz using composite pulse decoupling. Samples were disolved in d6-dimethylsulfoxide (ca. 40 mg/ml) and referenced to the center peak of the solvent at 39.5 ppm.

Positive ion fast atom bombardment (FAB) mass spectra were obtained with the JEOL HX110/HX110 tandem double-focusing mass spectrometer with JEOL gun and collisionally induced decomposition (CID) MS/MS. The JEOL instrument was operated at +10 kV with -20 kV postacceleration at the detector. The xenon neutral beam had 6 kV acceleration from the JEOL gun. CID MS/MS was performed with 1:1000 resolution in both MS-1 and MS-2. Helium was used as the collision gas at a pressure sufficient to reduce precursor ion abundance by 75%. Samples were dissolved in methanol-glycerol (1:1, v/v) for all FABMS analyses.

Bioassays

Insect bioassays were conducted according to the method of Wiseman, et al (15).

Large-Scale Extraction and Purification

Teosinte was grown to about the 12 leaf stage in Tift Co. Georgia. The leaves were stripped from the stalks and allowed to air-dry. To extract, 4.3 kg of the dry teosinte leaves were ground in a Wiley mill and soaked with intermittent stirring in 8 L 50:50 methanol: water. After 24 hours most of the solvent was siphoned off and a fresh batch of solvent was used for further extraction. After the third soaking for a total of 72 hrs the liquid extract was concentrated on a rotary evaporator to 8 L of aqueous extract. The aqueous extract was applied to a column of bonded-phase octadecylsilane on silica gel (ODS or C-18) from Waters Associates as a Prep-Pak 500 cartridge. After the adsorption of the sample the column was eluted with 1.8 L water, 2 L of 20% methanol in water, and 1 L each of 30,35,40,45,50,55,60,65 and 70% methanol. The compounds of interest eluted mainly in the 40-50 % methanol-water fractions. About 9 g of residue from the 40-50% fractions was deposited on 15 g of silica gel (J. T. Baker) and this sample was placed at the top of 100 g silica gel in a chromatographic column. Stepwise gradient elution with hexane, ethyl acetate, and ethyl acetate-acetone yielded two compounds which were further separated by droplet counter-current chromatography (DCCC) using $CHCl_3$:MeOH:n-PrOH:H2O (5:6:1:4). The upper aqueous layer was the stationary phase and the lower, mainly $CHCl_3$ layer was the mobile phase. Fractions (10 mL) were collected every hour. Fractions 71-79 contained 35.2 mg of a maysin isomer and fractions 89-96 contained 38.3 mg of maysin.

Walter's White corn plants were grown until the silks had dried. The leaves were processed and extracted in a fashion similar to teosinte. The crude extract, however, was deposited directly onto silica gel, eluted from a silica gel column as with the teosinte extract and then further purified on an open reversed phase ODS column. Luteolin glycosides were obtained in 30-50% methanol:water fractions. Further purification of small amounts for FAB-MS was achieved by thin-layer chromatography using aluminum backed silica gel plates (Whatman cat. # 4420 222) with EtOAc:MEK:formic acid: water, 5:3:1:1 as developing solvent. The compounds in the scraped bands were eluted with methanol.

RESULTS AND DISCUSSION

Maysin was found in the leaves of teosinte, corn, and centipede grass as shown by the HPLC chromatograms (Fig. 2) with UV spectra (see Fig. 3) acquired on-the-fly. In these HPLC separations obtained with a water:methanol gradient with 0.01% H_3PO_4 in each solvent, the early peaks were identified as caffeic acid derivatives, mainly chlorogenic acid and its isomers, and the later peaks were identified as mainly flavonoids with the detecting wavelength at 340 nm. In centipede grass the major phenolic compound was chlorogenic acid but maysin and other luteolin derivatives (Fig. 2c) were present as indicated by peak retention times and UV spectra. Teosinte (Fig. 2a), on the other hand, contained a high level of maysin in the leaves and silks, and two other luteolin derivatives in the leaves which eluted just before and just after the maysin peak in the HPLC chromatograms. Very little chlorogenic acid or its isomers was found. Maysin was isolated from teosinte leaves and its ^{13}C NMR spectrum (Fig. 4a) compared and found to be identical to the original ^{13}C NMR spectrum (unpublished) obtained by Elliger (11). In addition another isomer of maysin was separated from maysin by DCCC and its ^{13}C NMR spectrum recorded (Fig. 4b). Maysin has been detected at various levels by HPLC in the lower, early (pre-whorl) leaves of all but 24 out of more than 300 varieties, inbreds and populations. However, in the silks, maysin was usually much more concentrated in the 297 samples of inbreds and populations analyzed. The fall armyworm and corn earworm both prefer to feed on whorl leaf tissue. However, late in the season when population densities are high, the fall armyworm will also feed more towards the tip of the leaf which contains higher levels of maysin.

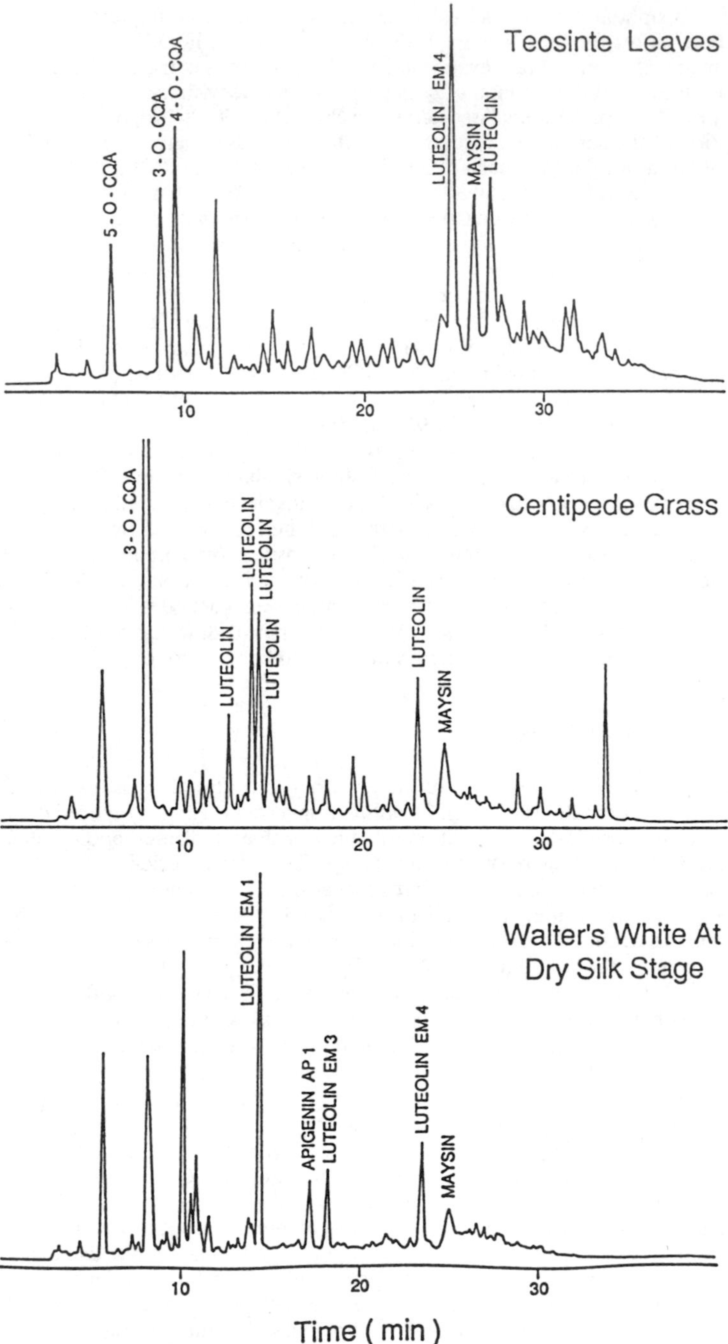

Figure 2. HPLC profiles of leaf extracts (340 nm).

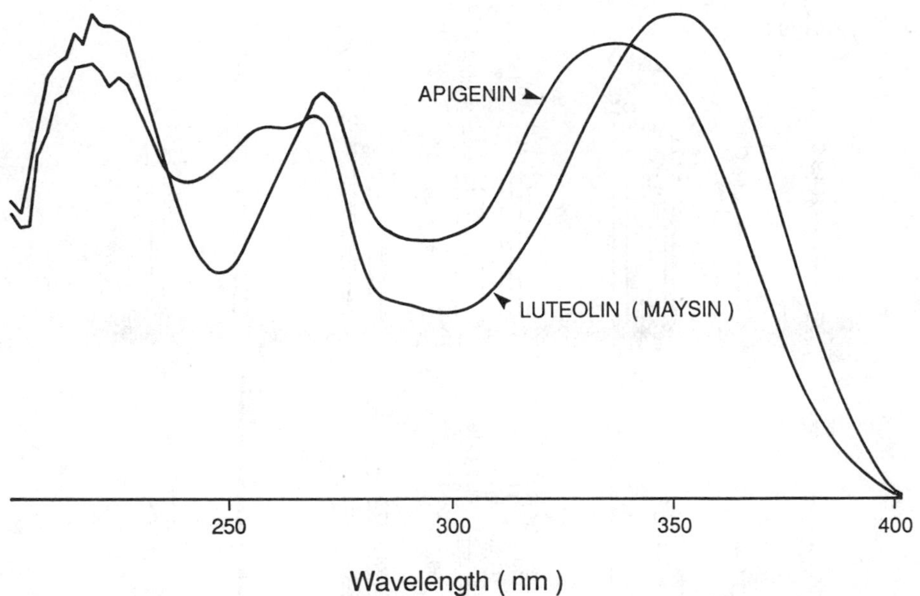

Figure 3. UV spectra of luteolin (maysin) and apigenin compounds.

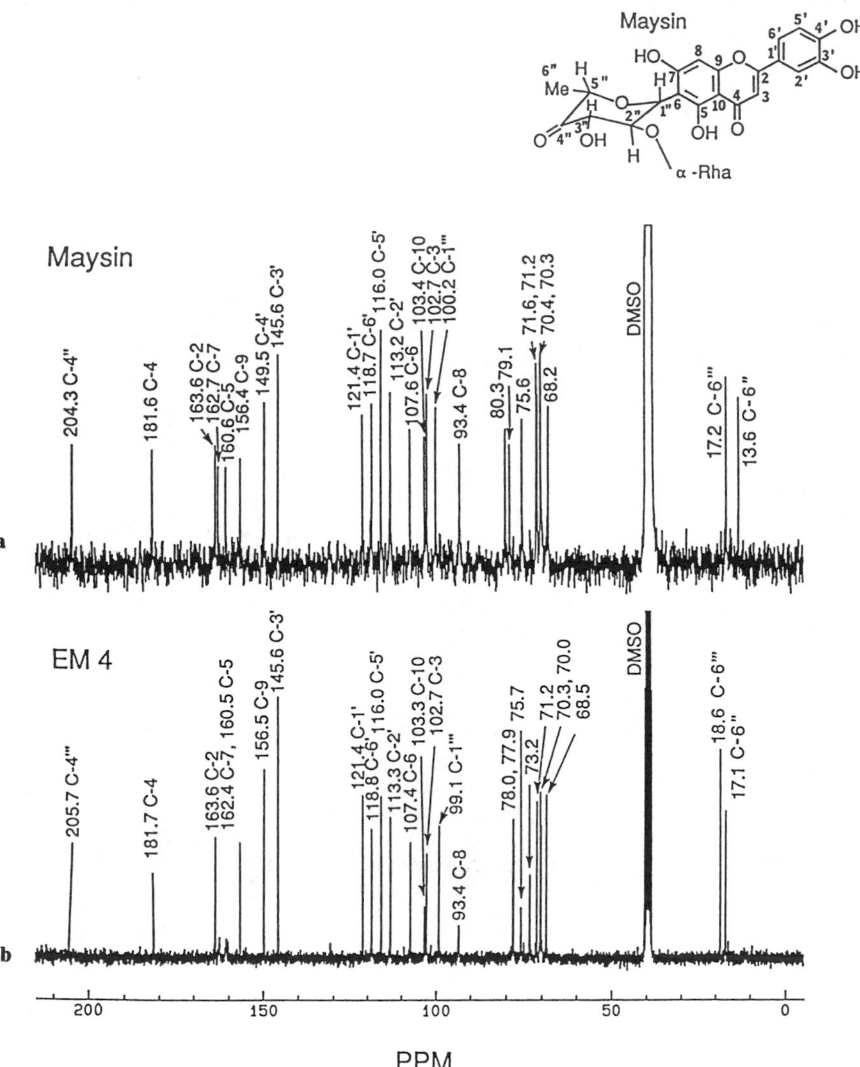

Figure 4. ^{13}C NMR spectra of a) maysin and b) EM4.

Corn Mutants with Non-Browning Silks

Prior to the elucidation of the structure of maysin the lack of luteolin derivatives in some silks was noted by Levings and Stuber (16) in mutants with nonbrowning silks. Levings and Stuber noted further that the "browning response is controlled by a single dominant locus which permits, primarily, the synthesis of [3',4'] dihydroxyl flavones by the silks and that when the silks are injured these flavones are enzymatically oxidized resulting in brown pigment formation. The nonbrowning type [was] shown to be due to a recessive allele which when homozygous blocks synthesis of the [3',4'] dihydroxyl flavones. Absence of the flavone substrates precludes brown pigment formation in response to wounding." It was further concluded by Levings and Stuber that the nonbrowning mutant silks contained the necessary oxidative enzymes since added [3',4'] dihydroxyl substrates were converted to the brown pigment by a crude enzyme preparation of the nonbrowning mutant silks. The homozygous recessive (nonbrowning) condition, fvfv, does not block all flavonoid synthesis since some inbred lines contained an anthocyanin in their silks. The fvfv genotype does not block flavone synthesis in the leaves and tassel as shown by comparison of two dimensional paper chromatograms of leaf and tassel extracts to silk chromatograms of the browning (FvFv, dominant) genotype.

Our HPLC analyses of silks and leaves are consistent with the above observations of Levings and Stuber. Recently we have confirmed by HPLC that silks of WF9, a browning type used in the study by Levings and Stuber, contain maysin, at 2.8% dry weight, as well as 3.2% dry weight of chlorogenic acid. Another variety, Ab18, produces silks which have almost no flavonoids but which contains maysin at low levels in the leaves. The backcross (Ab18 X Z. Chico) X Ab18 produces some progeny which have a high level of an apigenin derivative in the silks similar to the levels of maysin ordinarily found in Z. Chico silks. In the Levings and Stuber scenario Ab18 could be a mutant with little or no ability to synthesize flavones or anthocyanins, indicating a mutant with a metabolic block only in the silk at the chalcone synthase step (gene c2) or steps immediately following. In the (Ab18 X Z. Chico) X Ab18 cross apigenin synthesis occurs and the corresponding anthocyanin, pelargonidin, would be possible but the silk would be of the nonbrowning type because luteolins are not synthesized. Apparently Z. Chico provides the gene or genes which increase flavonoid synthesis in the Ab18 backcross. Also desirable would be the addition of the gene or genes coding the synthesis of the 3'-hydroxylase enzyme to convert apigenin derivatives to luteolin derivatives to increase the inhibitory potential of the silks (and also the leaves). (See Figure 5.)

Other Luteolin Derivatives of Corn

In contrast to the silk profile the corn leaf HPLC profile (Fig. 2b) is generally more complex. In many silks maysin comprises 70-80% of the absorbance at 340 nm of the total flavonoids (chlorogenic acid and its isomers are usually barely detectable) as determined by HPLC. In leaves, luteolin derivatives other than maysin are present, especially when senescence is approaching. A list of flavonoids, most not completely characterized, but having the UV spectrum of luteolin or apigenin, is presented in Table 1 which contains properties of compounds found in mature leaves (after silks have dried) of Walter's White corn or teosinte leaves. Accordingly maysin, EM 4, and EM 1 all have a hydrolyzable rhamnose moiety and a non-pyrone carbonyl group (Table 1). (In our parlance the unknown compounds are called EM which means "early maysin" in reference to the order of elution relative to maysin

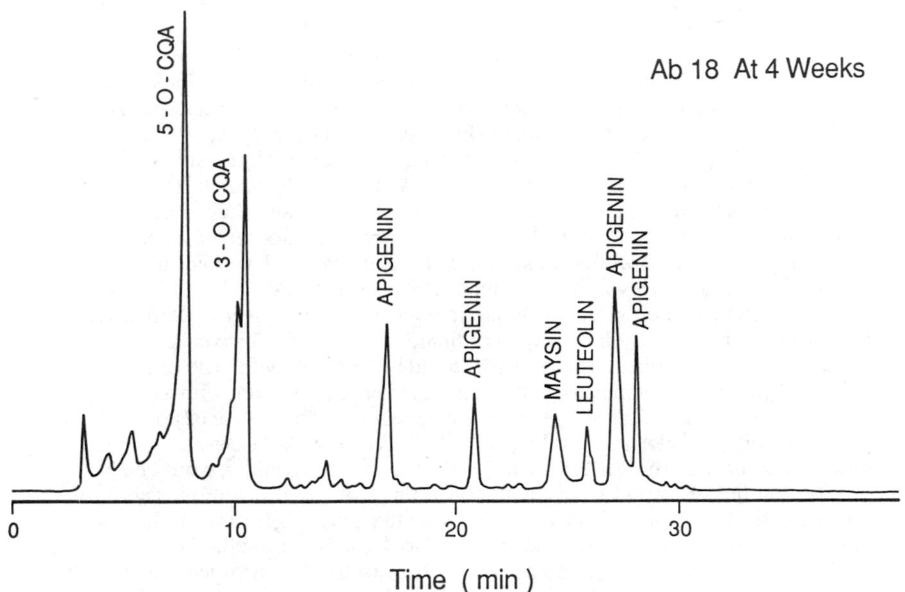

Figure 5. HPLC profile of leaf extracts of Ab18 at 340 nm.

Table 1. Properties of Luteolin in C-Glycoside Analogs of Maysin

Compound Name or Code	Plant Leaf Origin	Molecular Formula	Molecular Weight	HPLC Retention Time	TLC Retention Time	UV Type Maxima, nm	IR C=O on Sugar	Hydrolyzable Sugars*	FAB-MS
AP-1	Teosinte	--		30.0	.66	Apigenin 269,339	-	-	
Maysin	Teosinte	$C_{27}H_{28}O_{14}$	576	25.0	.55	Luteolin 269,351	+ $1737 cm^{-1}$	Rha	Loss of 146
EM4	Teosinte Walters White (W.W.)	$C_{27}H_{28}O_{14}$	576	23.5	.65	Luteolin 269,351	+ $1732 cm^{-1}$	Rha	Loss of 146
EM3	W.W.†	$C_{27}H_{30}O_{14}$	578	18.1		Luteolin 269,348	-	-	
AP-2	W.W.†	--		17.1		Apigenin 270,336			
EM1	W.W.†	$C_{27}H_{30}O_{15}$	594	14.2	.33	Luteolin 268,350	+ $1718 cm^{-1}$	Rha	Loss of 146

†Walter's White Sweet Corn.
*GC analysis as TMS derivatives.

upon HPLC analysis on a reversed phase ODS column.) In contrast EM 3 has no hydrolyzable rhamnose, no non-pyrone carbonyl. When purified EM 3 readily crystallized to a solid melting at about 280°C., and had a molecular weight consistent with a di-*C*-rhamnosyl luteolin.

SUMMARY

In the absence of other known inhibitory compounds we postulate that maysin and its derivatives and chlorogenic acid and its isomers are important factors contributing to the resistance to the fall armyworm in leaves of centipede grass, teosinte and corn. Potential variations of luteolin compounds similar to maysin are more numerous than chlorogenic acid and its two isomers. Therefore, the future emphasis is likely to be more concerned with the genetics of the luteolin compounds.

Silks of corn containing maysin at levels of 2% dry weight or higher depress the weight of corn earworm larvae. In teosinte leaves maysin and related compound levels are high and probably account for the small amount of leaf damage by the fall armyworm, although whorl feeding does occur. In corn, maysin content varies over a wide range in a single leaf and by stalk position of the leaf, see Fig. 6. The leaf feeding sites of the corn earworm and the fall armyworm coincide with the lowest levels in the corn leaf ie, the base of the whorl leaves, and these maysin levels may be too low to significantly inhibit the larvae. Analyses of several varieties over a two year period also indicate variability in maysin levels from year to year: for example in 1988 levels were higher than in 1989. The key question concerning resistance to both the corn earworm and the fall armyworm then, is whether the maysin, chlorogenic acid or other inhibitor content can be increased in the whorl of agronomically acceptable varieties. Recently two inbreds have been identified which have much higher levels of maysin in the whorls than the five varieties shown in Fig. 6.

When we found other luteolin derivatives in the leaves that were harvested at the stage of silk drying, we initially questioned whether these compounds had any relevance to resistance to the fall armyworm or the corn earworm. It was subsequently observed that the fresh silks of a small number of varieties

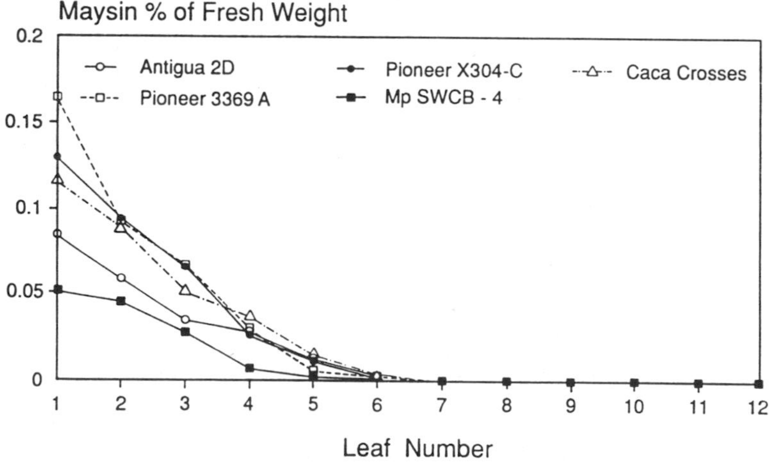

Figure 6. Maysin levels by leaf number (average of four replicates).

also contained luteolin derivatives with retention times similar (but not necessarily identical) to the leaves mentioned above. One example is the silk of T218 which showed six luteolin compounds in its HPLC profile. This example and others indicate a wide variation in biosynthetic pathways producing luteolins and provide a probable basis for breeding corn with enhanced resistance to possibly both the corn earworm and the fall armyworm in a single variety. (See Figure 7.)

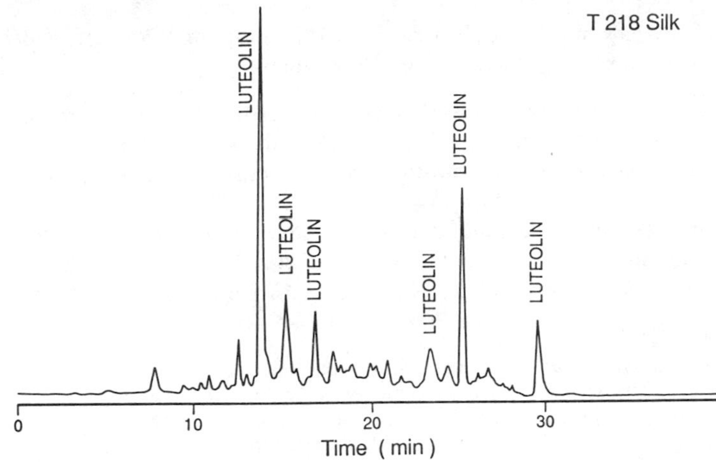

Figure 7. HPLC profile of extracts of T218 silks (340 nm).

LITERATURE CITED

1. Lynch, R.E.; Monson, W.G.; Wiseman, B.R.; Burton, G.W. Environ. Entomol., 1983, 12, 1837-1840.
2. Wiseman, B.R.; Gueldner, R.C.; Lynch, R.E. J. Econ. Entomol., 1982, 75, 245-247.
3. Wiseman, B.R.; Gueldner, R.C.; Lynch, R.E.; Severson, R. F. J. Chem. Ecol., in press.
4. Niemeyer, H.M. Phytochem. 1988, 27, 3349-3358.
5. Elliger, C.A.; Wong, Y.; Chan, B.G.; Waiss, A.C. Jr. J. Chem. Ecol. 1981, 7, 753-758.
6. Isman, M.B.; Duffey, S.S. Entomologia Exp. Appl. 1982, 31, 370-376.
7. Isman, M.B.; Duffey, S.S. J. Insect Physiol. 1983, 29, 295-300.
8. Reese, J. C.; Chan, B. G.; Waiss, A. C. Jr. J. Chem. Ecol. 1982, 12, 1429-1436.
9. Waiss, A.C. Jr.; Chan, B.G.; Elliger, C.A.; Wiseman, B.R.; McMillian, W.W.; Widstrom, N.W.; Zuber, M.S.; Keaster, A. J. J. Econ. Entomol. 1979, 72, 256-258.
10. Elliger, C.A.; Chan, B.G.; Waiss, A.C. Jr.; Lundin, R.E.; Haddon, W.F. Phytochem. 1980, 19, 293-297.
11. Elliger, C.A.; Chan, B.G.; Waiss, A.C. Jr. Naturwiss., 1980, 67, 358-360.
12. Snook, M.E.; Widstrom, N.W.; Gueldner, R.C. J. Chromatog. 1988, 477, 439-447.
13. Meyer, V.R. J. Chromatog. 1985, 334, 197-209.
14. Wiseman, B.R.; Gueldner, R.C.; Lynch, R.E. J. Econ. Entomol. 1984, 78, 328-332.
15. Levings, C. S.; Stuber, C. W. Genetics, 1971, 69, 491-498.

RECEIVED July 18, 1990

Chapter 17

Ovipositional Behavior of Tobacco Budworm and Tobacco Hornworm

Effects of Cuticular Components from *Nicotiana* Species

R. F. Severson[1], D. M. Jackson[2], A. W. Johnson[3], V. A. Sisson[2], and M. G. Stephenson[4]

[1]R. B. Russell Agricultural Research Center, U.S. Department of Agriculture, P.O. Box 5677, Athens, GA 30613
[2]Crops Research Laboratory, U.S. Department of Agriculture, P.O. Box 1555, Oxford, NC 27565
[3]Pee Dee Research and Education Center, Clemson University, U.S. Department of Agriculture, Route 1, Box 531, Florence, SC 29501
[4]Nematodes, Weeds and Crops Research, U.S. Department of Agriculture, P.O. Box 748, Tifton, GA 31793

The USDA - *Nicotiana* species germplasm collection was evaluated for tobacco budworm [*Heliothis virescens (F.)*] and tobacco hornworm [*Manduca sexta (L.)*] field infestation levels and for ovipositional responses of moths within screened field cages. Species which did not produce observable trichome exudates did not receive as many budworm or hornworm eggs as a flue-cured tobacco in paired choice or no-choice experiments. Qualitative and quantitative data on the cuticular chemistries were obtained. The cuticular extracts of all the *Nicotiana* species contained a series of aliphatic hydrocarbons. Major components in the trichome exudates were found to be diterpenes (duvane and/or labdane types) and/or sugar esters (sucrose and/or glucose). Eight different general types of sucrose esters and two types of glucose esters were identified. All sugar ester types contained a complex mixture of C3-C8 fatty acids attached to the 2,3, and 4 positions of the glucose moiety. Cuticular components were isolated and tested for tobacco budworm ovipositional response. Several cuticular diterpenes (α- and β-4,8,13-duvatriene-1,3-diols, α- and β-4,8,13-duvatrien-1-ols, manool and labda-13-ene-8α-15-diol) and two sucrose ester types (6-0-acetyl-2,3,4-tri-0-acyl- sucrose and 2,3,4-tri-0-acyl-4'-0-acetyl-sucrose) were found to increase oviposition by tobacco budworm moths when these materials were sprayed onto a leaf devoid of them. We believe that these components are contact ovipositional stimulants.

During the 1970's the USDA *Nicotiana tabacum* germplasm collection was evaluated in field plots at the Clemson University Pee Dee Research and Education Center, Florence, SC for their resistance to the tobacco hornworm, *Manduca sexta (L.)* (1), and the tobacco budworm, *Heliothis virescens* (F.) (2). During this investigation a large variation in leaf trichome types and density was observed. Johnson et al. (3, 4) classified the major trichome types from the various tobacco types as simple trichomes without exudates, glandular trichomes without exudates, glandular trichomes with exudates, and

This chapter not subject to U.S. copyright
Published 1991 American Chemical Society

small trichome hydathodes [See Johnson et al. (4) or Severson et al. (5), for trichome photographs]. In general, *N. tabacum* plants which lack glandular trichomes with exudates are resistant to insects (5, 6). However, some tobacco types with observable trichome exudates are also resistant to insect damage. Thus, the surface chemistry of insect-resistant and susceptible tobacco types with the different types of trichomes was investigated (7).

As shown in Table I (5,6,9), tobacco cultivars and introductions with glandular secreting trichomes, NC 2326, Golden Burley, NFT, TI 1223, TI 1341, TI 165 and TI 1396 produce duvane diterpenes and/or labdane diterpenes and/or sucrose esters (See Figures 1 and 2 for structures and References 5, 7 and 8 for capillary gas chromatograms). The tobacco introductions with simple trichomes, TI 1112, I-35, and those with nonsecreting glandular trichomes, TI 1024 and TI 1406, produce low levels of diterpenes and sucrose esters. The cuticular extracts of all *N. tabacum* plants studied, independent of trichome type, contained a series of C_{25}-C_{36} aliphatic hydrocarbons consisting of a series of straight-chain and iso- and anteiso-methyl-branched-hydrocarbons (5, 7). These components apparently are not associated with insect resistance.

Controlled larval feeding and oviposition tests (Table I) showed that a major mode of resistance to tobacco budworms in TI 1112, I-35, TI 1024, and NFT is ovipositional non-preference (antixenosis). The budworm resistance observed with TI 165 and TI 1396 appears to result from larval antibiosis (5,6,9). Field studies with naturally occurring populations of tobacco hornworms conducted at Oxford, NC; Florence, SC; and Tifton, GA in 1985 showed that TI 1112 and I-35 received only 11% and 16%, respectively, of the hornworm eggs relative to those deposited on NC 2326, a commercial flue-cured cultivar (9). Thus the high level of hornworm resistance observed with TI 1112 and I-35 also appears to be due to ovipositional non-preference.

In this report, we will discuss investigations of the cuticular components from *Nicotiana* species and their effects on tobacco hornworm and tobacco budworm moth oviposition. The response of tobacco budworm moths to specific cuticular isolates from a *Nicotiana* species also will be discussed.

Experimental

All *N. tabacum* plants evaluated for insect resistance and cuticular chemistries were grown under field conditions normally used for the production of flue-cured tobacco at the Clemson University Pee Dee Research and Education Center, Florence, SC; the Crops Research Laboratory, Oxford, NC and the University of Georgia Coastal Plain Experiment Station, Tifton, GA. Other *Nicotiana* species were evaluated at Oxford, NC or Tifton, GA. From 1984-1987 70 accessions of 64 *Nicotiana* species were planted to evaluate their effects on tobacco budworm and hornworm oviposition in field plots and in choice tests versus NC 2326 in cages (20). In 1985 and 1986 the different *Nicotiana* species were grown in the field and cuticular chemical extracts were obtained and analyzed in 1985. Field plots were also screened for natural infestations of insect pests.

Quantitation, Isolation and Characterization of Cuticular Components

About six weeks after transplantation, cuticular components from field-grown plants of each *Nicotiana* species were extracted by dipping young leaves in 8 oz. wide mouth bottles containing methylene chloride. After removal of the methylene chloride, the extract residue was treated with 1:1

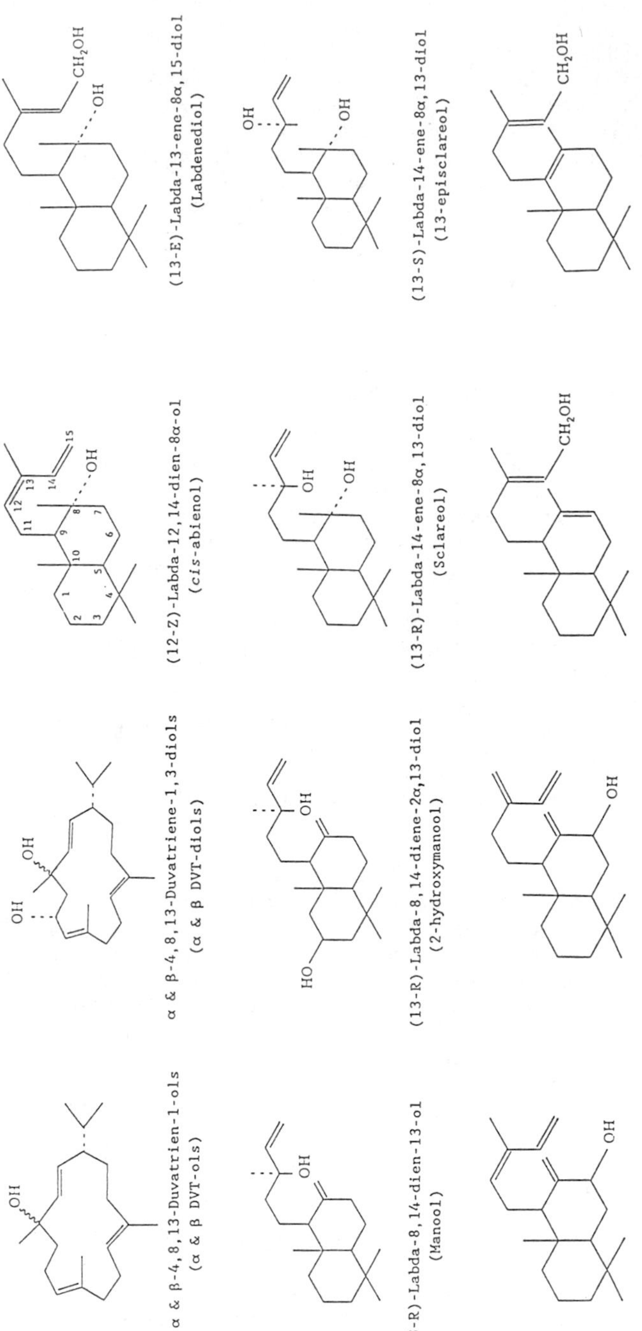

α & β-4,8,13-Duvatrien-1-ols
(α & β DVT-ols)

α & β-4,8,13-Duvatriene-1,3-diols
(α & β DVT-diols)

(12-Z)-Labda-12,14-dien-8α-ol
(*cis*-abienol)

(13-E)-Labda-13-ene-8α,15-diol
(Labdenediol)

(13-R)-Labda-8,14-dien-13-ol
(Manool)

(13-R)-Labda-8,14-diene-2α,13-diol
(2-hydroxymanool)

(13-R)-Labda-14-ene-8α,13-diol
(Sclareol)

(13-S)-Labda-14-ene-8α,13-diol
(13-episclareol)

(12-E)-Labda-8(20),12,14-trien-7β-ol
(Raimonol)

Labda-8(20),13(16),14-trien-7β-ol
(Isoraimonol)

(13-E)-Labda-7,13-dien-15-ol
(Setchelol)

(13-E)-Labda-8,13-dien-15-ol
(Isosetchelol)

Figure 1 Cuticular Diterpenes of *Nicotiana* Species.

TABLE I Comparison of the Cuticular Chemistries of Various *N. tabacum* Types to Resistance Ratings of Tobacco Budworm and Hornworm (5,6,9)

	Resistance Ratings[a]							
	Tobacco Budworm			Tobacco Hornworm	Cuticular Chemistry ($\mu g/cm^2$)[b]			
N. tabacum	Field Plant Damage	Laboratory Larval Feeding	Cage Oviposition	Field Plant Damage	α- & β- DVT-ols	α- & β- DVT-diols	Cis-abienol	Sucrose[c] Esters
NC 2326	S	S	S	S	0.8	46.0	-	2.0
Golden Burley	S	S	S	S	0.4	21.0	-	1.0
TI 1112	R	MR	R	R	0.1	0.6	-	-
I-35	R	MR	R	R	-	0.2	-	-
TI 1024	MR	MR	R	MR	-	0.8	-	0.9
TI 1406	MR	MR	R	MR	0.4	0.8	0.3	1.0
NFT	MR	S	R	MR	0.1	0.6	21.0	9.0
TI 1223	S	S	S	S	20.0	3.0	23.0	7.0
TI 1341	S	S	S	S	59.0	34.0	-	21.0
TI 165	R	R	S	MR	2.0	97.0	-	31.0
TI 1396	MR	R	S	MR	3.0	61.0	23.0	35.0

[a]Relative to NC 2326: R=resistant; MR=moderately resistant; S=susceptible.
[b]Six weeks after transplantation, Oxford, NC and Tifton, GA, 1982 and 1983.
[c]6-O-acetyl-2,3,4-tri-O-acyl-sucrose.

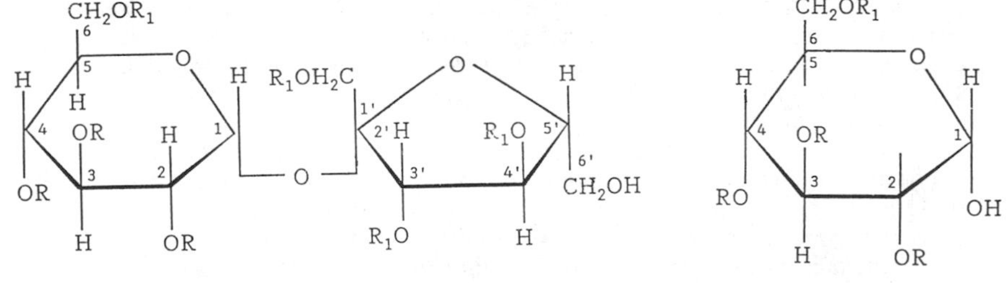

$R = C_3 - C_9$ Acyl groups
$R_1 = H$ or Acetyl

Sucrose Ester Types

A 6-O-acetyl-2,3,4-tri-O-acyl-sucrose
B 6-O-acetyl-2,3,4-tri-O-acyl-3'-O-acetyl-sucrose
C 6-O-acetyl-2,3,4-tri-O-acyl-4'-O-acetyl-sucrose
D 2,3,4-tri-O-acyl-sucrose
E 2,3,4-tri-O-acyl-3'-O-acetyl-sucrose
F 2,3,4-tri-O-acyl-4'-O-acetyl-sucrose
G 2,3,4-tri-O-acyl-3',4'-di-O-acetyl-sucrose
H 2,3,4-tri-O-acyl-1',3',4'-tri-O-acetyl-sucrose

Glucose Esters

A' 6-O-acetyl-2,3,4-tri-O-acyl-glucose
D' 2,3,4-tri-O-acyl-glucose

Figure 2 Cuticular Glucose and Sucrose Esters of *Nicotiana* Species.

N,O-bis(trimethylsily) trifluoroacetamide:dimethylformamide to convert hydroxylated components to trimethylsilylethers. Samples were analyzed by capillary gas chromatography as described by Severson et al. (7). Larger quantities of cuticular components for characterization and bioassay studies were obtained by dipping whole plant tops (upper 1/3) into methylene chloride. Components were isolated from the cuticular extracts using solvent partitioning between hexane and 80% MeOH-H_2O and/or a combination of alumina, silicic acid and Sephadex LH-20 column chromatography. Specific methodology for the isolation of components from *N. tabacum* (10, 11), and *N. glutinosa* (12, 13) have been described. The cuticular extracts of the other *Nicotiana* species were solvent partitioned between hexane and 80% MeOH-H_2O, and resulting fractions were characterized by GC retention and GC/MS data. The sucrose esters (and glucose esters) of *N. kawakamii (14), N. otophora (15), N. setchellii (15), N. tomentosa (15), N. tomentosiformis (14),* and *N. clevelandii (16)* were isolated from MeOH-H_2O soluble fractions using Sephadex LH-20-$CHCl_3$ column chromatography and confirmed by GC/MS analyses. The MeOH-H_2O soluble fractions containing the cuticular sugar esters were hydrolyzed and the resulting fatty acids were analyzed as butyl esters as described by Severson et al. (17)

Ovipositional Bioassays

Ovipositional bioassays were conducted using potted plants treated with cuticular isolates from *N. tabacum* accessions in 2.4 X 2.4 X 2 m screened cages at Oxford, NC as previously described (18, 19). Treatment plants, TI 1112 sprayed with a cuticular component, and control plants, TI 1112 sprayed with solvent blank, were placed at opposite corners of a cage. Cuticular components in 0.5 ml hexane-methylene chloride (3:1) were diluted with 9.5 or 14.5 ml of carrier solution [water:acetone (1:3)] and sprayed onto test plants with an air-brush (Badger Air-Brush Co., Model 250). Before dark, 10 mated females were released into each cage and the following morning plants were examined for eggs. Insects were from a laboratory colony started from larva collected from tobacco near Oxford, NC, and were reared for 7-10 generations on artificial diet before use.

Twelve smaller (0.46 x 1.31 m), hemicylindrical cages were used to bioassay cuticular components from *N. glutinosa* for tobacco budworm ovipositional preference. These experiments were run in a glass greenhouse in which the walls were covered with black plastic film to block extraneous lights. In the center of each end of a cage was a 10.2-cm diam. hole through which was placed a plastic frustum (11.7-cm OD bottom; 9.5-cm OD top; 8.9-cm high) holding a tobacco leaf so that it's abaxial (lower) surface was exposed. Thus, each leaf disk exposed 71 square centimeters of leaf area to the ovipositing moths. The leaf disks at opposite ends of a cage (1.3 m apart) were bioassayed for ovipositional preference in a choice-test situation. The cuticular isolates in 1 ml of acetone were mixed with 1 ml of acetone:water(1:1) and sprayed onto the leaf disks as described. Five female moths were introduced into each cage in the late afternoon, and the number of eggs on each treatment were counted the next morning. The insects were reared and prepared for bioassay as described for the outdoor oviposition cages.

RESULTS

Cuticular leaf chemistries for the different *Nicotiana* species are given in Table II. Also included in the table are the percent of tobacco budworm and hornworm eggs deposited on each species relative to the tobacco cultivar, NC 2326, in cages in choice tests (20). In the choice tests none of the *Nicotiana* species were significantly more attractive than NC 2326 to budworm or hornworm oviposition. However, in field evaluations *N. kawakamii* was more susceptible to budworm damage and several species were more susceptible to hornworm damage than NC 2326 (Table III). Excluding *N. tabacum*, 20 and 21 of the *Nicotiana* species were as attractive as NC 2326 to budworm and hornworm oviposition, respectively.

All other species which did not produce observable trichome exudates (*Nicotiana* spp. Nos. 19, 23, 25, 27, 31, 44, 47, 52, 53, 63, 64, and 71) were not attractive to budworm oviposition. Excluding *tabacum* types, only *N. sylvestris* produced significant levels of α- and β-DVT-diols.

Nine of the *Nicotiana* species produced labdane diterpenes, *N. tabacum* cv *Samsun* produced *cis*-abienol and labdenediol. Major cuticular labdanes on *N. glutinosa* 24 are manool, 15-hydroxy manool, sclareol, 13-episclareol and labdenediol. *N. glutinosa* 24A produces only the sclareol and labdenediol The labdane diterpenes of two other *Nicotiana* species, *N. raimondii*, raimonol and iso-raimonol (22) and *N. setchellii*, setchelol and iso-setchelol (23), have been characterized.

Most of the *Nicotiana* species with observable trichome exudates produced sugar esters. However, as shown in Figure 2 and Table IV, large variations in sugar ester types and distribution of ester moieties were found. We identified eight different general types of sucrose esters and two types of glucose esters (Table IV). The glucose esters are further complicated by the presence of α- and β forms. All types characterized to date have a complex mixture of C_3 to C_8 fatty acids attached to the 2,3 and 4 positions of the glucose moiety. These acids consist of normal chains, and iso- and anteiso methyl-branched isomers. For most species the major acyl group on the glucose moiety were methyl-branched C_4 to C_8 isomers. *N. hesperis* was the only *Nicotiana* species where the major sugar ester acyl groups were normal chain acids. Low levels of unsaturated acyl groups were detected in the sugar ester hydrolysates of several of the species. The unsaturated acyl group, 2-methyl-2-butenoyl, was a major component in the sugar ester isolates from *N. hesperis*.

Results of the ovipositional response of tobacco budworm moths to various cuticular isolates from *Nicotiana* spp. when applied to the leaves of the non-preferred TI 1112 are shown in Table V. The hexane soluble fraction from the cuticular extract from NC 2326 did not stimulate budworm oviposition (18). Previously, we reported that the α- and β-DVT-diol mixture, α-DVT-diol and a mixture of α- and β-DVT-ols produced a significant ovipositional response ($P<0.01$) (18). Similar results with the same compounds were obtained in this study. *Cis*-abienol isolated from *N. tabacum* cv NFT did not increase the number of budworm eggs deposited on TI 1112. The major labdane diterpenes in the cuticular extract of the *N. glutinosa* 24 and 24A were isolated and tested. At a 50 $\mu g/cm^2$ application rate, significantly ($P=0.05$) more eggs were observed on TI 1112 leaves sprayed with manool and labdenediol than were on TI 1112 leaves sprayed only with solvent blank. However, the labdenediol was not active at a 12.5 $\mu g/cm^2$ application rate, and 15-hydroxymanool and the sclareol mixture were inactive at the 50 $\mu g/cm^2$ rate. Sucrose ester isolates from *N. tabacum* TI 165 (6-O-acetyl-2,3,4-tri-O-acly-sucrose) and two sucrose ester isolates from *N. glutinosa* (2,3,4-tri-O-acly-sucrose and 2,3,4-tri-O-acyl-3'-O-acetyl-sucrose) also increased budworm oviposition when sprayed onto TI 1112 leaves.

TABLE II Comparison of Percentages of Eggs Deposited by Tobacco Budworm and Tobacco Hornworm on *Nicotiana* Species in Choice Tests with *N. tabacum* cv NC 2326 to Cuticular Chemistries of *Nicotiana* Species

Subgenus Section Species (Number)[a]	Percent of Eggs on N. Species[b]		Cuticular Chemistry of N. Species[c]			
					Sugar Esters	
	Budworms	Hornworms	Total Duvanes	Total Labdanes	Glucose	Sucrose
			------------Levels ($\mu g/cm^2$)--------			
Tabacum						
Genuinae						
tabacum (NC 2326)	50	50	105.0	-	-	5.0
tabacum (TI 1112)	24++[d]	-	Trace	-	-	Trace
tabacum (Samsun)	57	54	93.0	24.0	-	57.0
tabacum (I-35)	16++	-	Trace	-	-	-
Tomentosae						
glutinosa (24)	51	47	-	79.0	-	85.0
glutinosa (24A)	46	39	-	144.0	-	20.0
glutinosa (24B)	52	45	-	-	-	38.0
kawakamii (72)	52	51	-	38.0	88.0	24.0
otophora (38)	44	48	1.0	-	-	32.0
setchellii (51)	49	36	-	56.0	10.0	67.0
tomentosa (58)	34	48	-	6.2	1.0	21.0
tomentosiformis (59)	39+	46	-	20.0	-	40.0
Rustica						
Paniculatae						
benavidesii (8)	35++	26++	-	18.0	-	-
cordifolia (15)	7+	27++	Trace	-	-	-
glauca (23)	1++	7++	-	-	-	-
knightiana (27)	11++	43	-	-	-	-
paniculata (40)	44	40	-	-	Trace	5.0
raimondii (45)	29++	11++	-	20.0	-	-
solanifolia (52)	15++	18++	-	-	-	-
Rusticae						
v. brasilia (48)	39	32+	1.0	-	24.0	25.0
v. pavonii (44)	26++	46	-	-	-	-
v. pumila (49)	49	30++	-	-	2.0	20.0
Petunioides						
Alatae						
alata (3)	50	39	-	-	-	56.0
bonariensis (11)	41	44	Trace	-	2.0	40.0
forgetiana (21A)	52	27++	-	-	-	15.0
langsdorfii (28A)	42	44	-	-	-	8.0
longiflora (30)	31+	34+	-	-	-	2.0
plumbaginifolia (43A)	31++	27++	-	-	-	30.0
sylvestris (55)	48	52	57.0	-	-	-

Table II. Continued

Subgenus Section Species (Number)[a]	Percent of Eggs on N. Species[b]		Cuticular Chemistry of N. Species[c]			
	Budworms	Hornworms	Total Duvanes	Total Labdanes	Sugar Esters Glucose	Sugar Esters Sucrose
			------------Levels (µg/cm²)--------			
Petunioides (Continued)						
Trigonophyllae						
palmerii (39)	48	26++	-	-	6.0	40.0
trigonophylla (60)	54	42	-	-	4.0	154.0
Undulatae						
arentsii (6)	14++	45	Trace	-	-	-
undulata (61A)	8++	30++	2.0	-	-	1.0
wigandioides (63)	10++	35	-	-	-	-
Acuminatae						
acuminata (2)	40	39	-	-	-	179.0
attenuata (7)	45	46	-	-	8.0	44.0
miersii (33)	22++	11++	-	-	292.0	-
pauciflora (41)	37+	39	-	-	5.0	113.0
Bigelovianae						
bigelovii (10A)	26++	29++	-	-	15.0	75.0
clevelandii (14)	7++	27++	-	-	7.0	66.0
Nudicaules						
nudicaulis (36)	20++	25++	-	-	-	90.0
Suaveolentes						
africana (71)	29+	24++	-	-	-	-
amplexicaulis (65)	35+	13++	-	-	1.0	3.0
benthamiana (9)	45	41	-	-	1.0	2.0
cavicola (68)	35+	32+	-	-	18.0	162.0
debneyi (17)	12++	29++	-	-	2.0	15.0
excelsior (19)	2++	3++	-	-	-	-
exigua (20)	24++	19++	-	-	1.0	10.0
fragrans (22)	24++	7++	-	-	-	526.0
goodspeedii (25)	9++	27++	-	-	-	-
gossei (26)	38++	29++	-	-	3.0	8.0
hesperis (67)	18++	14++	-	-	-	8.0
ingulba (64)	13++	13++	-	-	-	-
maritima (31)	18++	29+	-	-	-	Trace
megalosiphon (32)	25++	28++	-	-	-	Trace
occidentalis (37)	46	32+	-	-	6.0	45.0
rosulata (53)	11++	27+	-	-	-	-
rotundifolia (47)	27++	11++	-	-	-	Trace
simulans (66)	41	25++	-	-	-	10.0
suaveolens (55)	10++	12++	-	-	-	Trace
umbratica (69)	46	42	-	-	16.0	126.0
velutina (62)	23++	38++	-	-	-	5.0

Continued on next page

Table II. Continued

Subgenus Section Species (Number)[a]	Percent of Eggs on *N*. Species[b]		Cuticular Chemistry of *N*. Species[c]			
					Sugar Esters	
	Budworms	Hornworms	Total Duvanes	Total Labdanes	Glucose	Sucrose
			Levels ($\mu g/cm^2$)			
Petunioides (Continued)						
Noctiflorae						
acaulis (1)	6++	12++	-	-	-	32.0
noctiflora (35)	22++	19++	-	-	10.0	110.0
petunioides (42)	6++	6++	-	-	Trace	56.0
Repandae[e]						
nesophila (34A)	45	32++	Trace	-	-	-
repanda (46)	39+	29++	Trace	-	-	-
stocktonii (54)	43	42	2.0	-	-	-

[a]USDA - National Plant Germplasm System collection site number.
[b]Paired choice tests with 4 plants of *Nicotiana* species versus 4 plants of NC 2326 flue-cured tobacco in same cage, Oxford, NC 1984-87. Average of 10 to 30 replications.
[c]Field plants, Oxford, NC, 1986. Average of two replications.
[d]Paired *t* test; ++=Significantly different ($P<0.01$); +=Significantly different ($P<0.05$).
[e]Major cuticular components of species in section Repandae are C_{12}-C_{15} hydroxacylnornicotines (24).

TABLE III Percentages of *Nicotiana* spp. Plants Infested with Tobacco Budworm or Hornworm Larvae in Field Plots at Oxford, NC and Tifton, Ga., 1985-86.

Nicotiana Species	Percent Tobacco Budworm Infested Plants[a]	*Nicotiana* Species	Percent Tobacco Hornworm Infested Plants[b]
kawakamii	46.3	*kawakamii*	72.3
tabacum cv. samsun	33.9	*glutinosa* (24B)	48.6
tabacum cv. NC 2326	23.9	*tomentosiformis*	48.0
alata	21.3	*glutinosa* (24)	40.1
debneyi	18.0	*setchellii*	36.8
glutinosa (24)	17.6	*tabacum cv. samsun*	33.7
glutinosa (24A)	13.7	*tabacum cv. NC 2326*	32.0
bigelovii	13.6	*glutinosa* (24A)	28.8
clevelandii	9.6	*amplexicaulis*	26.6
otophora	9.1	*bigelovii*	25.8
10 species	5.1-9.0	*sylvestris*	21.1
19 species	0.1-5.0	4 species	10.1-20.0
31 species	0.0	28 species	0.1-10.0
		27 species	0.0

[a]Averaged over 3 data sets: Oxford, NC, 1985; Tifton, GA, 1985; and Tifton, GA, 1986. 3 replications per location; 12-plant plots.
[b]Averaged over 2 data sets: Oxford, NC, 1985 and Tifton, GA, 1986. 3 replications per location; 12-plant plots.

TABLE IV Comparison of Sugar Ester Types and the Major Ester Acyl Groups Found in the Cuticular Sugar Fractions of the *N.* Species

Section Species (Number)	Sugar Ester Type[a] Sucrose	 Glucose	Major Sugar Esters Acyl[b] Groups
Genuinae			
tabacum (NC 2326)	A[c]	-	3-MeC$_4$, 2-MeC$_4$, isoC$_4$
tabacum (TI 165)	A[c]	-	3-MeC$_5$, 3-MeC$_4$, 2-MeC$_4$
Tomentosae			
glutinosa (24)	D,E[c]	-	4-MeC$_6$, 5-MeC$_6$, 4-MeC$_5$
glutinosa (24A)	D,E[c]	-	4-MeC$_6$, 5-MeC$_6$, 4-MeC$_5$
glutinosa (24B)	D,E[c]	-	4-MeC$_6$, 5-MeC$_6$, 4-MeC$_5$
kawakamii (72)	A,D[c]	A′,D′[c]	3-MeC$_5$, 2-MeC$_4$, 4-MeC$_5$
otophora (38)	A,B,D,E[c]	-	4-MeC$_6$, 5-MeC$_6$, 3-MeC$_5$
setchellii (51)	D,E,F,G,H[c]	D′[e]	3-MeC$_5$, 3-MeC$_4$, 2-MeC$_4$
tomentosa (58)	A,B,D,E[c]	-	3-MeC$_5$, 4-MeC$_6$, 5-MeC$_7$
tomentosiformis (59)	A,B,D[c]	-	3-MeC$_5$, 2-MeC$_4$, 3-MeC$_4$
Paniculatae			
paniculata (40)	nc[d]	-	3-MeC$_5$, 2-MeC$_4$, 3-MeC$_4$
Rusticae			
v. brasilia (48)	nc	nc	3-MeC$_5$, 3-MeC$_4$, 4-MeC$_5$
v. pumila (49)	nc	nc	3-MeC$_5$, 3-MeC$_4$, 4-MeC$_5$
Alatae			
alata (3)	A,C,D,F[e]	nc	2-MeC$_4$, 3-MeC$_4$, isoC$_4$
bonariensis (11)	nc	nc	3-MeC$_5$, 2-MeC$_4$, 3-MeC$_4$
forgetiana (21A)	nc	nc	2-MeC$_4$, 4-MeC$_6$, 5-MeC$_6$
plumbaginifolia (43A)	nc	nc	4-MeC$_6$, 2-MeC$_4$, 5-MeC$_7$
Trigonophyllae			
palmerii (39)	D,E,F,G[e]	D′[e]	5-MeC$_7$, 3-MeC$_5$, 4-MeC$_6$
trigonophylla (60)	D,E[e]	D′[e]	5-MeC$_7$, 4-MeC$_6$, 2-MeC$_4$
Acuminatae			
acuminata (2)	B[e]	-	3-MeC$_5$, 3-MeC$_4$, isoC$_4$
attenuata (7)	A,B,D,E[e]	A′,D′[e]	3-MeC$_5$, isoC$_4$, 4-MeC$_5$
miersii (33)	-	A′,D′[e]	2-MeC$_4$, 2-MeC$_4$$^{1=}$, 3-MeC$_4$
pauciflora (41)	nc	nc	
Bigelovianae			
bigelovii (10A)	D,E[e]	D′[e]	3-MeC$_5$, 2-MeC$_4$, isoC$_4$
clevelandii (14)	E,G[c]	D′[c]	3-MeC$_5$, 2-MeC$_4$, isoC$_4$
Nudicaules			
nudicaulis (36)	nc	-	4-MeC$_5$, 3-MeC$_4$, isoC$_4$

Continued on next page

Table IV. Continued

Section Species (Number)	Sugar Ester Type[a] Sucrose	 Glucose	Major Sugar Esters Acyl[b] Groups
Suaveolentes			
benthamiana (9)	nc	-	6-MeC_7, 5-MeC_7, 5-MeC_6
cavicola (68)	nc	nc	5-MeC_7, 5-MeC_6, 6-MeC_7
debneyi (17)	nc	nc	3-MeC_5, 2-MeC_4, 3-MeC_4
exigua (20)	nc	nc	5-MeC_6, 4-MeC_6, 4-MeC_5
fragrans (22)	nc	-	4-MeC_5, isoC_4, 3-MeC_4
gossei (26)	nc	nc	5-MeC_7, 5-MeC_6, 4-MeC_5
hesperis (67)	nc	-	C_7, C_8, 5-MeC_7
occidentalis (37)	nc	nc	2-MeC_4, 3-MeC_5, 3-MeC_4
simulans (66)	nc	-	5-MeC_6, 4-MeC_6, 4-MeC_5
umbratica (69)	nc	-	3-MeC_5, 4-MeC_5, 2-MeC_4
Noctiflorae			
noctiflora (35)	nc	nc	3-MeC_4, 5-MeC_6, 3-MeC_5
petunioides (42)	nc	nc	3-MeC_5, 2-MeC_4, 4-MeC_6

[a]See Figure II for type designations.
[b]Listed in order of relative abundance. Analyzed by capillary GC as butyl esters after the hydrolysis of the sugar ester fraction. Characterized by GC retention and GC/MS data.
C_4 = butanoyl, C_5 = pentanoyl, C_6 = hexanoyl, C_7 = heptanoyl, C_8 = octanoyl, and 2-$MeC_4^{1=}$ = 2-methyl-2-butenoyl.
[c]Isolated and characterized by GC/MS as trimethylsilyl ethers.
[d]nc = not characterized.
[e]Characterized from GC/MS data obtained from trimethylsilylated MeOH-H_2O fraction of the cuticular extract.

TABLE V. Ovipositional Response of Tobacco Budworm Moths to Cuticular Components from *Nicotiana* spp. Applied to Nonpreferred TI 1112 (Entry A) in Paired Choice Tests with TI 1112 Sprayed with Solvent Blank (Entry B).

Cuticular Components Applied to Entry A	Amount Applied to Entry A ($\mu g/cm^2$)	Ovipositional Response When Compared With Solvent Blank
Non Polar Lipids		
Hexane Solubles[a]	25.0	NS[b]
Diterpenes		
α & β-DVT-diols[c]	50.0	++
α-DVT-diol[d]	37.5	++
β-DVT-diol[d]	12.5	++
Oxidized DVT-diols[e]	50.0	NS
α & β-DVT-ols[f]	37.5	++
Cis-abienol[g]	37.5	NS
Labdene diol[h]	12.5	NS
Labdene diol[h]	50.0	+
Manool[i]	50.0	+
15-OH Manool[j]	50.0	NS
Sclareols[j,k]	50.0	NS
Sucrose Esters		
6-O-acetyl-2,3,4-tri-O-acyl-sucrose[l]	12.5	+
6-O-acetyl-2,3,4-tri-O-acyl-sucrose[l]	50.0	+
2,3,4-tri-O-acyl-4′-O-acetyl-sucrose[j]	50.0	++
2,3,4-tri-O-acyl-sucrose[j]	50.0	+

[a]Isolated from NC 2326 consisting of hydrocarbons, fatty alcohols, and wax esters (21).
[b]T-test significance; NS=numbers of eggs on Entry A and Entry B not significantly different at the 5% level; +=significantly more eggs on Entry A at the 5% level; ++=significantly more eggs on Entry A at the 1% level.
[c]Isolated from the cuticular extract of NC 2326 (72% α, 23% β, and 5% oxidized diols); sprayed onto TI 1112 at an application rate of 50$\mu g/cm^2$ (19).
[d]Isolated from NC 2326 (11).
[e]A complex mixture of oxidative degradation products of the α- & β-DVT-diols, including hydroxyepoxy, hydroxyoxy and trihydroxy degradation products, isolated from NC 2326.
[f]Isolated from TI 1341 (98%; α & β ratio 9:1) (11).
[g]Isolated from NFT (98%) (11).
[h]Isolated from *N. glutinosa* #24A (98%) (13).
[i]Obtained from Aldrich Chemical Co.
[j]Isolated from *N. glutinosa* #24 (13).
[k]A mixture of sclareol and 13-episclareol (13).
[l]Isolated from TI 165 (10).

DISCUSSION

The data presented show that certain cuticular components of *Nicotiana* spp. increase the tobacco budworm ovipositional frequency. It also indirectly indicates that these compounds may affect tobacco hornworm oviposition. The duvane diterpenes, α- & β-DVT-diols and α- & β-DVT-ols are very active ovipositional stimulants for the tobacco budworm. Analysis of the cuticular chemistry of numerous *N. tabacum* cultivars, breeding lines and TI's showed a large variation in cuticular duvane production (5,7). Sixty-eight of these tobacco types versus NC 2326 were tested in cage choice tests for ovipositional activity by tobacco budworm . A significant correlation (r=0.74) between the ovipositional response relative to NC 2326 and the log of the total cuticular duvane levels was observed (8). A similar highly significant positive correlation (r=0.94) was reported between the log of eight levels of α- & β-DVT-diols sprayed on TI 1112 plants and tobacco budworm ovipositional response (8). Little tobacco budworm ovipositional activity occurs on tobacco types which produce only duvanes when cuticular levels below 5 $\mu g/cm^2$ are found.

In contrast to the duvanes, a similar dose response ovipositional activity relationship was not observed for sucrose ester isolates from TI 165. As shown in Table V, a four-fold increase in application rate of 6-O-acetyl-2,3,4-tri-O-acyl-sucrose did not significantly affect ovipositional activity. Also, the type of sucrose ester and/or the composition of its 2,3,4-tri-O-acyl moieties appears to affect tobacco budworm ovipositional activity. These factors are possible explanations for the large variations in ovipositional frequency observed with the *Nicotiana* spp. (Table II). However, other differences in physical characteristics among the *Nicotiana* spp., such as leaf size, hairiness and plant growth characteristics, could further complicate the correlation of surface chemistries with ovipositional frequency. The activities of different types of compounds can only be determined when the compounds are evaluated in the same matrix.

Several other labdane diterpenes which we have not tested, may also affect ovipositional activity. The lack of damage in the field on and ovipositional response to *N. raimondii* and *N. benavidesii* indicate that the cuticular labdanes produced by these plants do not positively affect budworm or hornworm oviposition. We are currently isolating the labdanes from *N. kawakamii* and *N. setchellii* for ovipositional bioassay.

Knowledge of plant cuticular components which modify insect behavior will be useful in the control of a given pest. When the component is not a valuable quality factor for consumer acceptance, breeding of plants that lack ovipositional stimulants will reduce pest damage. The use of plant breeding to increase levels of insect ovipositional stimuli could produce plants which will be useful as trap crops. This could lead to the reduction in the use of pesticides which increases production costs and environmental contamination. The information presented here should benefit research efforts on other crops, such as corn, cotton, soybeans, peanuts, and vegetables that are attacked by the same insect pests.

In this paper we have only discussed chemical contact cues used by insects to identify ovipositional sites. In nature, olfactory, mechanical, and visual stimuli may also be important in the location, recognition, and acceptance of host plants. Much work remains to be done in this area. We believe that multidisciplinary teams are needed to work toward a more complete understanding of insect-plant interactions. With the knowledge obtained, plants can be more readily designed to naturally resist insect damage.

LITERATURE CITED

1. Johnson, A. W. J. Econ. Entomol. 1979, 72:914-915.
2. Johnson, A. W.; Chaplin, J. F. Tob. Sci. 1982, 26:157-158.
3. Johnson, A. W.; Severson, R. F. Tob. Sci. 1982, 26:98-102.
4. Johnson, A. W.; Severson, R. F.; Hudson, J.; Garner, G. R.; Arrendale, R. F. Tob. Sci. 1985, 29:67-72.
5. Severson, R. F.; Johnson, A. W.; Jackson, D. M. Recent Adv. in Tob. Sci. 1985, 11, 105-74.
6. Johnson, A. W.; Severson, R. F. J. Agric. Entomol. 1984, 1:23-32.
7. Severson, R. F.; Arrendale, R. F.; Chortyk, O. T.; Johnson, A. W.; Jackson, D. M.; Gwynn, G. R.; Chaplin, J. F.; Stephenson, M. G. J. Agric. Food Chem. 1984, 32:566-570.
8. Cutler, H. G.; Severson, R. F.; Cole, P. D.; Jackson, D. M.; Johnson, A. W. In Natural Resistance of Plants to Pests; Green, M. B.; Hedin, P. A., Eds.; ACS Symposium Series No. 296; American Chemical Society: Washington, DC, 1986; pp 178-196.
9. Jackson, D. M.; Severson, R. F.; Johnson, A. W. Recent Adv. in Tob. Sci. 1989, 15:26-116.
10. Severson, R. F.; Arrendale, R. F.; Chortyk, O. T.; Green, C. R.; Thome, F. A.; Steward, J. L.; Johnson, A. W. J. Agric. Food Chem. 1985, 33:870-875.
11. Severson, R. F.; Stephenson, M. G.; Johnson, A. W.; Jackson, D. M.; Chortyk, O. T. Tob. Sci. 1988, 32:99-103.
12. Arrendale, R. F.; Severson, R. F.; Sisson, V. A., Costello, C. E.; Leary, J. A.; Himmelsbach, D. S.; van Halbeek, H. J. Agric. Food Chem. 1990, 38:75-85.
13. Severson, R. F.; Mueller, S. J.; Sisson, V. A.; Jackson, D. M. Abstract of Papers, Tob. Chem. Res. Conf. 1988, 42:29.
14. Severson, R. F.; Arrendale, R. F.; Sisson, V. A.; Stephenson, M. G. Abstract of Papers, Tob. Chem. Res. Conf. 1987, 41:19.
15. Severson, R. F.; Arrendale, R. F.; Chortyk, O. T.; Sisson, V. A.; Stephenson, M. G. Abstract of Papers, Tob. Chem. Res. Conf. 1989, 43:19.
16. Arrendale, R. F.; Severson, R. F.; Sisson, V. A.; Stephenson, M. G. Abstract of Papers, Tob. Chem. Res. Conf. 1985, 41:44.
17. Severson, R. F.; Sisson, V. A.; Smith, L. B.; King, B. G; Chortyk, O. T. Abstract of Papers, Tob. Chem. Res. Conf. 1988, 42:39.
18. Jackson, D. M.; Severson, R. F.; Johnson, A. W.; Chaplin, J. F.; Stephenson, M. G. Environ. Entomol. 1984, 13:1023-1030.
19. Jackson, D. M.; Severson, R. F.; Johnson, A. W.; Herzog, G. A. J. Chem. Ecol. 1986, 12:1349-59.
20. Jackson, D. M.; Severson, R. F. In IBPGR Training Courses: Lecture Series. 2. Scientific Management of Germplasm: Characterization, Evaluation and Enhancement. Stalker, T.; Chapman, C., Eds.; International Board Plant Genetic Resources: Rome, Italy, 1989; pp 101-124.
21. Arrendale, R. F.; Severson, R. F.; Chortyk, O. T.; Stephenson, M. G. Beitr. Tabakforsch. Int. 1988, 14:67-84.
22. Noma, M.; Suzuki, F.; Gamou, K.; Kawashima, N. Phytochemistry 1982, 21:395-97.
23. Noma, M.; Suzuki, F.; Gamou, K.; Kawashima, N. Phytochemistry 1983, 22:133-35.
24. Severson, R. F.; Arrendale, R. F.; Cutler, H. G.; Jones, D.; Sisson, V. A.; Stephenson, M. G. In Biologically Active Natural Products Potential Use in Agriculture; Cutler, H. G., Ed.; ACS Symposium Series No. 380; American Chemical Society: Washington, DC, 1988; p 335.

RECEIVED July 31, 1990

Chapter 18

Corn Rootworm Feeding on Sunflower and Other Compositae

Influence of Floral Terpenoid and Phenolic Factors

Christopher A. Mullin, Ali A. Alfatafta, Jody L. Harman, Anthony A. Serino, and Susan L. Everett

Pesticide Research Laboratory and Graduate Study Center, Department of Entomology, Pennsylvania State University, University Park, PA 16802

Antifeedants for adult western corn rootworm, *Diabrotica virgifera virgifera* LeConte (Coleoptera: Chrysomelidae), were isolated and identified from inflorescences of cultivated sunflower, *Helianthus annuus* L., and Canadian goldenrod, *Solidago canadensis* L. Fractionation of floral principles was guided by a bioassay using treated disks from squash flowers containing cucurbitacins, potent rootworm feeding stimulants. Sequential surface extraction of sunflower inflorescences followed by solvent partitioning of residues yielded ethyl acetate solubles rich in antifeedant activity. Further chromatography on Toyopearl TSK HW-40F and/or silica gel gave over 65 compounds, from which 15 active structures were identified. Feeding deterrency decreased in order of sesquiterpenes >> diterpenes > flavonoids > dicaffeoylquinic acids, of which the most potent were sesquiterpene lactone angelates including argophyllin A and 3-methoxyniveusin A, the diterpenoic acids grandifloric acid and its 15-angelate, and the flavonoids nevadensin and quercetin ß-7-O-glucoside. Similarly, kaempferol was identified as a weak antifeedant from Canadian goldenrod. Two of the electrophilic germacranolide angelates with 4,5-unsaturation, when injected into rootworm adults, gave neurotoxic symptoms (hyperexcitability, enhanced egg expulsion, tarsal tetany) similar to picrotoxinin, a sesquiterpene lactone epoxide known to act on the γ-aminobutyric acid-gated chloride channel. These neurotoxic antifeedants may explain both the seven-fold decreased tolerance of western corn rootworm to aldrin and its decreased longevity when fed on floral tissues of sunflower in comparison to corn. Relevance of these results to other herbivore-phytochemical associations, particularly those with chrysomelids, will be discussed.

Phytochemicals produced from secondary metabolic pathways are major mechanisms by which plants are protected from excessive herbivory. The role of foliar chemicals in retarding or preventing consumption of leaves, the primary photosynthetic organs of plants, has been clearly established. However, few studies have addressed the negative effects of floral chemistry on insect herbivory. Reproductive structures should, expectedly, be well-defended to assure adequate propagation of plant genes (1, 2). Yet the attributes of flowers that attract pollinators (i.e. visual or volatile cues, nectar and pollen quality) have dominated study in the chemical basis for insect-floral relationships. The considerable amounts of flavonoids, carotenoids and steroids in pollen (3, 4), alkaloids and phenolics in nectars (5), UV-quenching flavonoid nectar guides (6, 7), and floral fragrances (e.g. 8) are most often associated with attraction and rewarding of essential pollinators and not with defense against floral

0097–6156/91/0449–0278$06.00/0
© 1991 American Chemical Society

consumption. This general phytochemical enhancement of pollination is consistent with animals functioning as pollinators for the majority of angiosperm species, but tends to neglect the defense of plant propagules from florivores particularly where self-pollination is evident. In the ensuing discussion, we will emphasize terpenoid and phenolic factors that protect Compositae (Asteraceae) flowers from excessive consumption by insects.

Terpenoids as Regulators of Herbivory - Associations with Chrysomelidae

Plant sesquiterpenes and other terpenoids are major determinants of insect-plant interactions (9-16). Many insecticidal and antifeedant terpenoids are epoxides including monoterpene (17, 18), sesquiterpene (10, 19-23), diterpene (11, 24) and triterpene derivatives (25-27) typified by the potent antifeedant azadirachtin (28-30). Most biological effects have been determined with Lepidoptera and non-chrysomelid Coleoptera. Occasionally, the same compound, while normally inhibitory to herbivores, may for adapted insect species or at low concentrations have a stimulatory effect (13). Insects, in turn, synthesize their own defensive (31, 32) and pheromonal (33) terpenoids. Plants may utilize insect pheromones such as the sesquiterpene alarm pheromone, *trans*-ß-farnesene, in their own defense (34, 35). Inhibitory cyclic sesquiterpenes (Table I) and diterpenes (Table II) for insect herbivores have been identified from at least 28 genera of the terpenoid-rich Compositae. These studies were largely confined to extrafloral tissues.

Floral chemistry is increasingly being associated with antiherbivore actions, particularly among terpenoids. In *Gossypium*, phenolic sesquiterpenoid-derived heliocides and the dimeric gossypol are anti-lepidopteran factors found in high concentration in flower parts (9, 36). The monoterpene-derived iridoid glycosides protect nectar of *Catalpa* from consumption by non-pollinators (37). Floral concentrations of terpenoids higher than that of leaves and externally compartmentalized into trichomes have been noted (36, 38). In the Compositae, a plant family characterized by many self-pollinated species, both monoterpene derivatives such as the insecticidal pyrethrins from *Chrysanthemum* spp. (39, 40) and toxic sesquiterpene lactones and diterpenoic acids concentrated in the floret achenes of wild *Helianthus* spp. (41-44) are clearly protecting flowers from excessive herbivory. Both niveusin A from *H. niveus* and 8ß-sarracinoyloxycumambranolide from *H. maximiliani* deter feeding of the sunflower moth, *Homoeosoma electellum* (43). It is thought that sesquiterpene lactones in glandular trichomes of the anther prevent pollen-feeding by this sunflower pest; foliar sesquiterpenes including the epoxide, argophyllin A, **I**, from *H. argophyllus* (44), and diterpenoic acids (45) may explain antibiosis in sunflower for this pest and others (Tables I and II) including the chrysomelid *Zygogramma exclamationis* (F.) (41).

Among the most noted of chrysomelid-terpenoid investigations have been *Diabrotica* spp. feeding associations with squash cucurbitacins, triterpenoidal-derived electrophiles that serve as potent feeding stimulants for corn rootworms (46, 47). Cucurbitacin contents are particularly high within the anther and filament of *Cucurbita maxima*, a much preferred squash species for *Diabrotica* spp. as a pollen-source of food (48). Interestingly, for other chrysomelids such as the Cruciferae leaf beetles (49) and Colorado potato beetle, *Leptinotarsa decemlineata* (50), these compounds are strong feeding deterrents. Work with other squash-feeding Diabroticine chrysomelids has identified a number of potent antifeedants including the neem tetranortriterpenoids (Meliaceae) for striped cucumber beetle, *Acalymma vittatum* (F.) and southern corn rootworm, *D. undecimpunctata howardi* Barber (51), and the sesquiterpenoid celangulin from Chinese bittersweet (Celastraceae) for *Aulacophora femoralis chinensis* (52). Neem (29, 30, 53) as well as citrus (26) limonoids generally deter feeding of chrysomelid species. While large amounts of dietary sesquiterpene lactones from *Encelia farinosa* (Table I) deter the growth of the specialist herbivore, *Trirhabda geminata*, natural resistance by this composite species to this chysomelid appears more associated with elevated chromene levels (54). Antifeedant and toxic sesquiterpenes for the Colorado potato beetle have been identified from the wild tomato *Lycopersicum hirsutum* (55), the sagebrush *Artemisia tridentata* (50), and from other Compositae and some Apiaceae species (56 and refs therein). Antifeedant diterpenoids for this chrysomelid are also known (57). The goldenrod diterpenoids, in turn, are antifeedant to the *Solidago* specialist, *Trirhabda canadensis* (24, 58). Various monoterpenes and cardenolides are also important as stimulators or inhibitors of chrysomelid herbivory (59), and some compounds from these terpenoid classes as well as the cucurbitacins are utilized in beetle defense against natural enemies (47, 60).

Table I. Cyclic Sesquiterpenes from Compositae that Deter Insect Herbivores

Plant Genera Sesquiterpene	Insect species	Inhibitor of	References
Achillea			
Caryophyllene	*Locusta migratoria*	Feeding	61
Artemisia			
Absinthin, Achillin Caryophyllene ar-Curcumene Desacetoxymatricarin Dehydroleucodin α–Santonin	*Heliothis zea, Hypochlora alba, Leptinotarsa decemlineata, Melanoplus sanguinipes, Pieris rapae, Spodoptera littoralis, Sitophilus granarius, Tribolium confusum, Trogoderma granarium*	Feeding Growth	50, 62-66
Centaurea			
Cnicin Salonitenolide	*Sitophilus granarius, Tribolium confusum, Trogoderma granarium*	Feeding Longevity	63, 67
Chrysanthemum			
Artecanin Canin	*Sitophilus granarius, Tribolium confusum, Trogoderma granarium*	Feeding	63
Cichorium			
8-Deoxylactucin Lactucopicrin	*Schistocerca gregaria*	Feeding	68
Encelia			
Farinosin	*Trirhabda geminata*	Growth ?	54
Eupatorium			
1-Desoxy-8-epi-ivangustin seco-Eudesmanolide Eupatoriopicrin Euponin Cadinene type	*Atta cephalotes, Drosophila melanogaster, Philasomia ricini, Sitophilus granarius, Tribolium confusum, Trogoderma granarium*	Feeding Growth Longevity Oviposition	56, 67 69-73
Grossheimia			
Grossheimin	*Sitophilus granarius, Tribolium confusum, Trogoderma granarium*	Feeding Longevity	63, 67
Helenium			
Helenalin Linifolin A Tenulin	*Epilachna varivestis, L. decemlineata, Melanoplus sanguinipes, Ostrinia nubilalis, Peridroma saucia, Sitophilus granarius, T. confusum, Trogoderma granarium*	Feeding Growth Longevity Oviposition	63, 74-77
Helianthus			
Argophyllins A & B Budlein A, Eupatolide Cumambranolide ester Desacetyleupasserin Niveusin A	*Homoeosoma electellum, Melanoplus sanguinipes, Spodoptera eridania, S. litura*	Feeding Growth Longevity	42-44, 78
Homogyne			
Bakkenolide A	*Leptinotarsa decemlineata, Peridroma saucia, Sitophilus granarius, Tribolium confusum, Trogoderma granarium*	Feeding Growth Longevity	56, 71, 77

Table I (cont'd). Cyclic Sesquiterpenes from Compositae that Deter Insect Herbivores

Plant Genera Sesquiterpene	Insect species	Inhibitor of	References
Inula			
Alantolactone Isoalantolactone	*Sitophilus granarius, Tribolium confusum, Trogoderma granarium*	Feeding Longevity	56, 79, 80
Iva			
Coronopilin	*Sitophilus granarius, Tribolium confusum, Trogoderma granarium*	Feeding Longevity	63, 75
Jurinea			
Alatolide	*Sitophilus granarius, Tribolium confusum, Trogoderma granarium*	Feeding Longevity	63, 67
Lasianthaea			
Lasidiol angelate	*Atta cephalotes*	Feeding	81
Melampodium			
Caryophyllene oxide Guaianol, Melampodin A Melampodinin A, Spathulenol	*Atta cephalotes,* *Spodoptera frugiperda*	Feeding Growth Longevity	82, 83
Onopordon			
Onopordopicrin	*Sitophilus granarius, Tribolium confusum, Trogoderma granarium*	Feeding	63
Parthenium			
Conchosins A & B Confertin, Coronopilin Isochiapin B, Ligulatin C Parthenin and derivatives Tetraneurins A, B & E	*Heliothis zea* *Melanoplus sanguinipes* *Spodoptera exigua* *Tribolium confusum* Other species	Feeding Growth Heartbeat Longevity	75, 84-88
Petasites			
Petasitolide A	*Sitophilus granarius*	Feeding	56
Schkuhria			
Schkurins I & II	*E. varivestis, Spodoptera exempta*	Feeding	89
Tithonia			
Tagitinin C	*Philasomia ricini*	Feeding	90
Venidium			
Hirsutolide	*Sitophilus granarius, Tribolium confusum, Trogoderma granarium*	Feeding	63
Vernonia			
Glaucolide A 11,13-Dihydro-vernodalin	*Diacrisia virginica, Sibine stimulea, Spodoptera eridania, exempta, frugiperda S. ornithogalli, Trichoplusia ni*	Feeding Growth Oviposition	20, 91, 92
Xanthium			
8-Epi-xanthatin, Xanthumin Xantholide A (= Ziniolide)	*Drosophila melanogaster*	Growth	93, 94
Xeranthemum			
Xerantholide	*Sitophilus granarius, Tribolium confusum, Trogoderma granarium*	Feeding	63

Table II. Diterpenes from Compositae that Deter Insect Herbivores

Plant Genera Diterpene	Insect species	Inhibitor of	References
Chrysothamnus			
18-Hydroxygrindelic acid	*Leptinotarsa decemlineata*	Feeding	57
18-Succinyloxygrindelic acid			
Grindelia			
6α & ß-Hydroxygrindelic acids	*Schizaphis graminum*	Feeding	95
Helianthus			
Angeloylgrandifloric acid	*Heliothis virescens, H. zea,*	Growth	44, 45, 96
Ciliaric acid, *cis*-Ozic acid	*Homoeosoma electellum,*	Longevity	
Kaur-16-en-19-oic acid	*Pectinophora gossypiella*		
16-Hydroxykaurane			
16-Hydroxykaur-11-en-19-oic acid			
Trachylobanoic acid			
Lasianthaea			
Kaur-16-en-19-oic acid	*Atta cephalotes*	Feeding	97
Melampodium			
Kolavenol	*Atta cephalotes*	Feeding	97
Solidago			
16-Hydroxykaurane	*Trirhabda canadensis*	Feeding	24, 58
15-Hydroxykaur-16-en-19-oic acid			
17-Hydroxykaur-15-en-19-oic acid			

Phenolic - Insect Associations: Relevance to Chrysomelidae

Many polyhydroxylated flavonoids and related phenolics have been shown to limit insect herbivory (98-100, Hesk et al., this volume). Inhibitory actions by phenolics often require both the high concentrations naturally present in plants and chemical structures bearing adjacent (ortho) hydroxyl groups (cf 101), although exceptions to both these trends occur with aphids (102,103). At lower dosages or with phenolic specialists, stimulatory rather than inhibitory effects on feeding may result (cf 98, 104). While this tendency is also found among chrysomelids, other structural features such as the type of sugar and its position of attachment may be more important in influencing activity. Flavonoid glycosides are known that both stimulate and inhibit chrysomelid feeding (105, 106). Simple phenolics such as chlorogenic acid have been shown to deter a Salicaceae-feeding leaf beetle (107). Strongly UV-absorbing flavonoids (108) and other phenolic derivatives (109) with pro- or anti-insect activities are increasingly being found within floral tissues, suggesting that their adaptive roles extend beyond the visual orientation of pollinators (6, 110, 111).

Compositae-Corn Rootworm Interactions

Adult northern corn rootworm, *D. barberi* Smith & Lawrence (NCR), readily feed on flowers of the Compositae (Asteraceae) that are barely acceptable to western corn rootworm, *Diabrotica virgifera virgifera* LeConte (WCR). Rearing adult WCR continuously on inflorescences of cultivated sunflower *Helianthus annuus* L. var. Giant Gray Stripe, or Canadian goldenrod *Solidago canadensis* L. var *canadensis*, reduces its longevity by 40% and 70%, respectively, to that on corn ears. By contrast, NCR's longevity is not significantly affected by host shifts from corn to Compositae (112). Also, an antifeedant action of this food was observed for WCR in the short term (< 24 hr). A two-dimensional thin-layer chromatographic (tlc) method developed by us to analyze flavonoid and phenolic acid aglycones within small amounts (< 30 mg) of plant or insect tissue (103) gave a lavender fluorescing compound (366 nm) at R_f coordinates (0.21, 0.25) that accumulates in WCR after long-term feeding on sunflower petals (Figure 1). The free acid has been identified (confirmed by UV, EIMS and ^{1}H-NMR) as *trans*-caffeic acid both within the plant and insect after preparative isolation on silica gel and co-chromatography in four solvent systems (P. R. Urzua, W. R. Wenerick and C. A. Mullin, unpubl.). A gold fluorescing compound at R_f coordinates (0.23, 0.29), that is sequestered from sunflower petals by rootworm, is the aglycone form of a quercetin flavonoid (see below).

Isolation and characterization of feeding deterrents. A more systematic fractionation of floral principles responsible for the feeding deterrent and toxic effects of sunflower was then conducted, guided by a squash disk bioassay where relative consumption after 5, 24 and 48 hr by adult WCR of solvent- or compound-treated flower disks was measured. This bioassay was designed to detect only highly active antifeedants that counteract the potent feeding stimulatory effect of cucurbitacins. In 1988 studies, residues from one-week extracts of petals and florets using ice-cold 95% ethanol were dissolved in water and partitioned in order by chloroform, ethyl acetate and *n*-butanol. Most of the original antifeedant and longevity-reducing activities concentrated into the ethyl acetate fraction, and were isolated by Toyopearl TSK HW-40F using methanol-water (75:25) to give three major phenolic components and a number of unbound polar terpenoids (dashed sunflower profile, Figure 2). Two of the phenolics had UV spectra resembling caffeoyl esters, and the other exhibited UV characteristics resembling a glycoside of quercetin with a free 3-hydroxy group. Through use of ^{1}H- and ^{13}C-NMR in d_6-DMSO, the flavonol was identified as quercetin ß-7-O-glucoside, **II**, and the phenolic acids as 3,5-dicaffeoyl-, **III**, and 3,4-dicaffeoylquinic acids, **IV**. Only the former was antifeedant for WCR

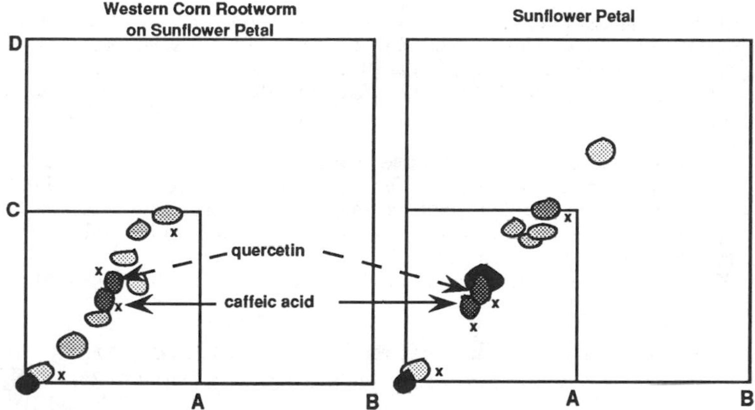

Figure 1. Two-dimensional tlc of phenolic acids and flavonoids in WCR that had fed for one month on sunflower petals. Degree of shading indicates quench at 254 nm; x = apparent cochromatography between insect and plant. See ref. 103 for details on development solvents.

Quercetin ß-7-O-glucoside
II

3,5-Dicaffeoylquinic acid **III**
3,4-Isomer **IV**

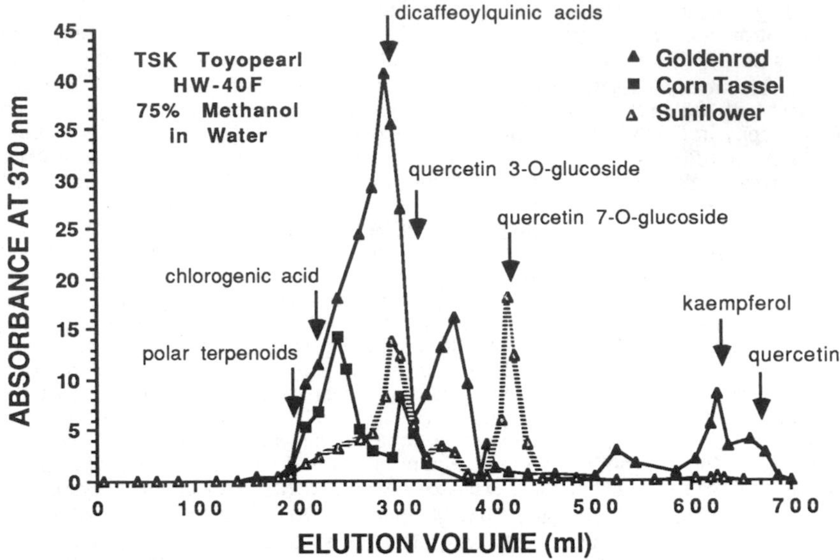

Figure 2. Composite chromatogram of floral ethyl acetate residues on Toyopearl HW-40F.

in a squash disk bioassay (Table III). Interestingly, a mixture of 3,5- and 3,4-dicaffeoylquinic acids, the major phenolics within the floral tissues of sunflower, was stimulatory to rootworm feeding at a low dose (32 µg/disk), but not at a higher dose (129 µg/disk) more representative of the intact flower. Quercetin 7-O-glucoside has previously been isolated from the flowers (7, 113) and pollen (114) of *H. annuus*, and 3,5-dicaffeoylquinic acid has been identified in sunflower seeds (115).

By similar procedures we isolated and identified free kaempferol from the ethyl acetate residue of an ethanolic extract of goldenrod inflorescences (solid triangle profile, Figure 2) as a weak rootworm deterrent . This flavonol was previously reported from *S. canadensis* (116). In order of richness (amounts per weight tissue) in floral phenolics, goldenrod was greater than sunflower which was richer than corn tassel (Figure 2).

The unbound terpenoids from sunflower (Figure 2) were further purified by silica gel

Table III. Effect of Sunflower Floral Chemicals on Rootworm Consumption of Blue Hubbard Squash Disks[a]

Compound	Dose (µg/disk)	Relative Consumption (treated/control)[b] 5 hr	24 hr	48hr
Argophyllin A (+ some **VI**)	40	0.32*	0.23*	0.24*
3-Methoxyniveusin A (+ some **VI**)	40	0.39*	0.26*	0.26*
Quercetin ß-7-O-glucoside	114	0.81	0.86*	0.74*
3,5+3,4-Dicaffeoylquinic acids	32	1.25	1.65*	2.83*

[a]Dual choice tests with 8 µl of solvent or compound per 1.5 cm flower disk
[b] * = significantly different from methanol control at $p < 0.05$ based on area consumed.

chromatography and identified as the sesquiterpene lactones argophyllin A, **I**, and 3-methoxyniveusin A, **V**, each contaminated with the lesser antifeedant 1-methoxy-4,5-dihydroniveusin A, **VI**. These sesquiterpene lactone angelates are greater than five times more potent than quercetin 7-ß-O-glucoside as antifeedants for WCR (Table III).

In 1989, whole sunflower heads were extracted by solvent immersion for only two minutes to optimize the isolation of labile cuticular terpenoids probably occurring as trichome exudates. Sequential two-minute surface extraction of 84 inflorescences by petroleum ether, methylene chloride-methanol (3:1 v/v) and methanol followed by partitioning of residues (22 g) from the combined polar extracts between water and ethyl acetate, and then the ethyl acetate residues (16 g) between 90% aqueous methanol and petroleum ether yielded methanolic solubles (8 g) rich in antifeedant activity. Column, flash and thin-layer silica gel chromatography of these solubles gave more than 65 compounds, which in decreasing order of abundance were primarily diterpenoic acids, sesquiterpene lactone angelates, and methoxylated flavonoids. Compounds were identified through use of 360 MHz ^{1}H-NMR (including NOE, spin-decoupling), 126 MHz ^{13}C-NMR (including GASPE), MS and UV/Vis spectroscopy as necessary. Purity was assessed by tlc and by reversed-phase high performance liquid chromatography (hplc) on C8 and C18 columns using acetonitrile-water gradients. Indeed, the flower of this annual species of *Helianthus* was quite chemically complex.

Thirty-four of these compounds were bioassayed for deterrency to WCR as above, with the order of potency; sesquiterpenes (7 compounds) >> diterpenes (6) > methoxylated flavonoids (4). Bioassay data (Table IV) and structures are included here for the more potent

Table IV. Floral Feeding Deterrents for Adult Western Corn Rootworm Among Ethyl Acetate Soluble Chemicals from the Sunflower[a]

CHEMICAL CLASS **Structure**		Relative Consumption after 24 hr (treated/control)[b]	
	compound name	40 µg/disk	80 µg/disk
SESQUITERPENES			
I	argophyllin A	0.23 ± 0.04	----
V	3-methoxyniveusin A	0.75 ± 0.08	0.30 ± 0.09
VI	1-methoxy-4,5-dihydroniveusin A	0.97 ± 0.09[d]	0.71 ± 0.02
VII	15-hydroxy-3-dehydrodesoxytifruticin	0.97 ± 0.16[d]	0.75 ± 0.08
VIII	3-oxo- derivative of **VI**	----	0.72 ± 0.08
IX	10-methoxy-3-oxo-derivative of **VI**	0.80 ± 0.05	0.64 ± 0.02
DITERPENES			
X	kaur-16-en-19-oic acid	1.05 ± 0.21	0.85 ± 0.13[d]
XI	grandifloric acid	0.91 ± 0.05	0.61 ± 0.14
XII	grandifloric acid angelate[c]	0.92 ± 0.04	0.70 ± 0.13
XIII	trachylobane	----	1.00 ± 0.01[d]
XIV	15-hydroxytrachyloban-19-oic acid	----	0.96 ± 0.03[d]
XV	7-oxo-trachyloban-15,19-diol	1.00 ± 0.01	0.98 ± 0.01[d]
FLAVONOIDS			
XVI	nevadensin	----	0.93 ± 0.06[d]
XVII	5-hydroxy-4,6,4'-trimethoxyaurone	----	1.01 ± 0.01[d]

[a]Dual choice tests with the Blue Hubbard squash disk bioassay using 8 µl of solvent or compound solution per 1.5 cm flower disk.

[b]Mean ± SEM for 4 replicates per dose; consumption based on area.

[c]Dosages were 50 and 100 µg/disk, respectively.

[d]Although inactive at 24 hr, substantial feeding deterrency observed at 5 hr.

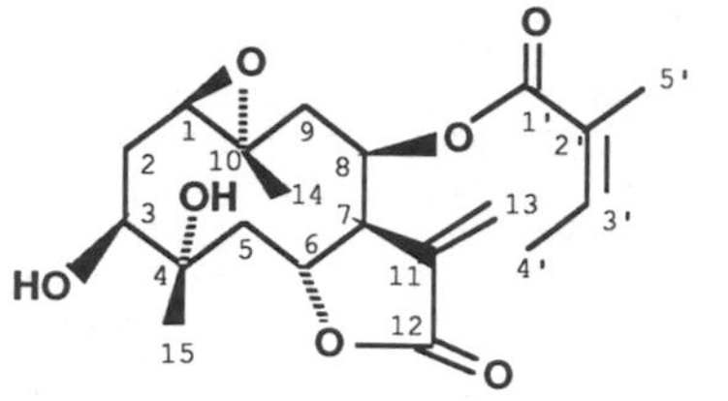

I Argophyllin A

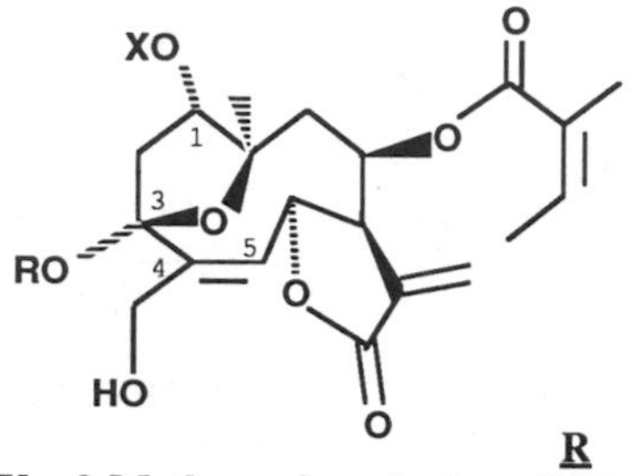

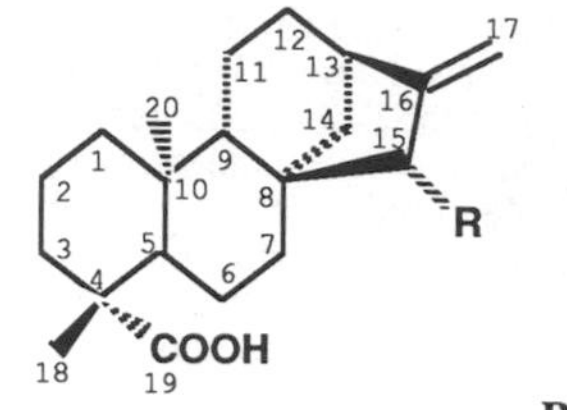

		R
X	Kaur-16-en-19-oic	H
XI	Grandifloric acid	OH
XII	Grandifloric angelate	OAng

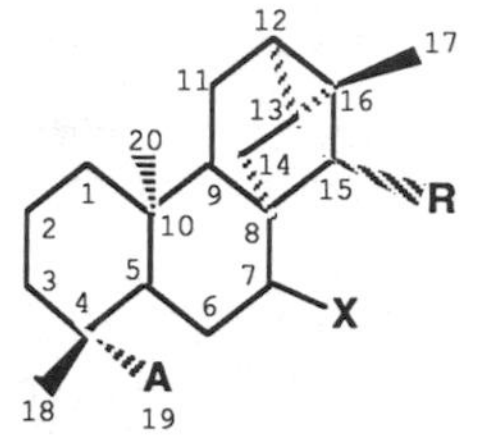

		R	X
V	3-Methoxyniveusin A	CH3	H
VI	1-Methoxy-4,5-dihydro-niveusin A	H	CH3

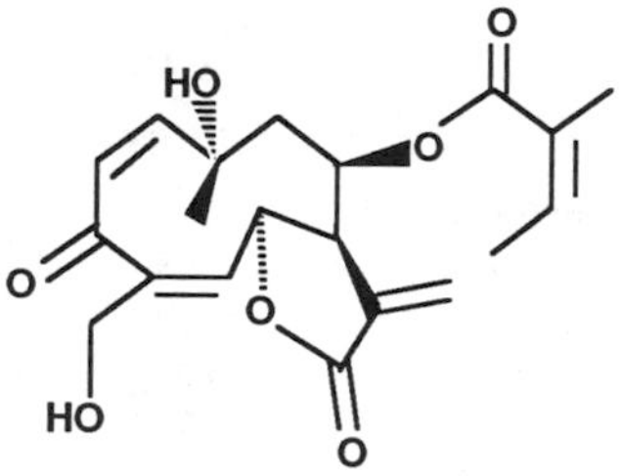

		A	X	R
XIII	Trachylobane	CH3	H	H
XIV	15-Hydroxytrachylo-ban-19-oic	COOH	H	OH
XV	7-Oxo-trachyloban-15,19-diol	CH2OH	=O	OH

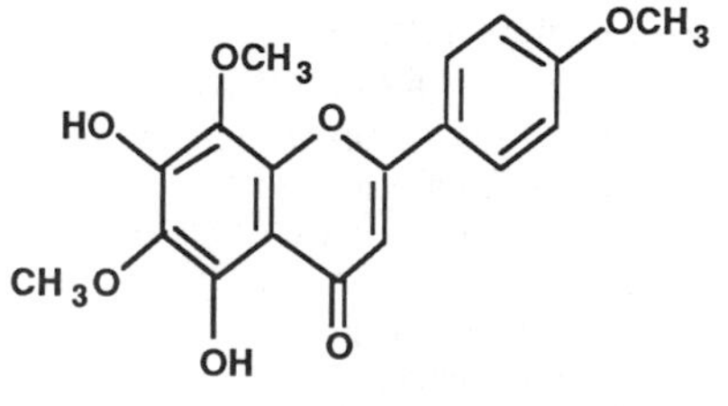

VII 15-Hydroxy-3-dehydro-desoxytifruticin

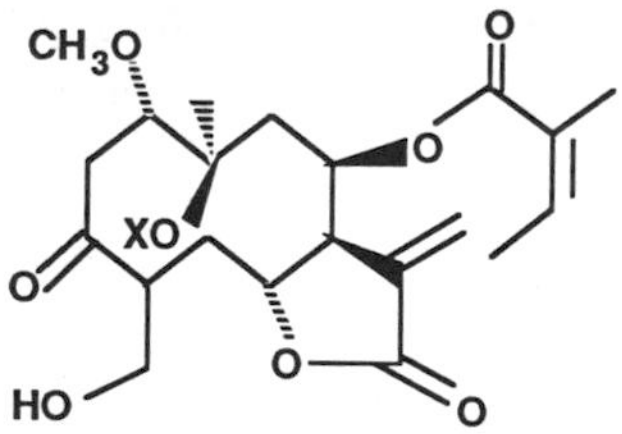

XVI Nevadensin

		X
VIII	3-Oxo- derivative of VI	H
IX	10-Methoxy-3-oxo-VI	CH3

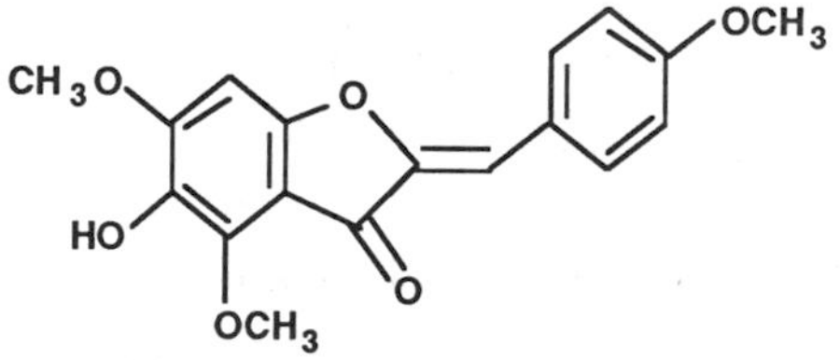

XVII 5-Hydroxy-4,6,4'-trimethoxyaurone

antifeedants. Included among these are a sesquiterpene, **IX**, a diterpene, **XV**, and an aurone, **XVII**, new to science. The most potent feeding deterrents included sesquiterpene lactone angelates of the germacranolide type with epoxidation or 4,5-unsaturation. For the diterpenoic acids, the 16-kaurene system appeared more active than the trachylobane type, and 15-hydroxylation in both types improved feeding deterrence. In general, the diterpenoic acids and the methoxylated flavonoids had less residual activity (i.e. good bioactivity up to 48 hr) than the sesquiterpenes. Also, the polar phenolics isolated in 1988 from the more extensive extraction of floral tissues were much reduced in the 1989 surface extracts, indicating the expected intracellular localization of the flavonol glucosides and caffeoylquinic acids.

Natural versus Artifactual Sesquiterpenoids from *Helianthus*. Many of the sesquiterpenes were isolated as methoxylated derivatives that may have resulted from the interaction of methanol with precursor epoxides under acid conditions. Since the more abundant diterpenoic acids, probably present with the sesquiterpenes in the same trichomes (44, 117, 118), could provide the requisite low pH for these additions/rearrangements, we hypothesized that isolating the terpenoids under buffered conditions in the absence of alcoholic solvents may result in chemistry more representative of the intact plant. Thus, 139 sunflower inflorescences (with many opened disk florets) were extracted for 30 seconds with 4 by 4 L of a heterogeneous solvent (75% ethyl acetate: 25% 50 mM potassium phosphate pH 8 buffer) at the Rock Springs Field Lab immediately after cutting. Extracts were transported on ice to the lab, and the ethyl acetate residue fractionated by silica gel chromatography as before. This residue proved to be enriched with at least ten different sesquiterpene lactones. While characterization remains incomplete, it is clear that less methoxy-substituted sesquiterpene angelates occur if methanol is absent indicating that these compounds, some of which are reported by others from sunflower (118), may be artefacts of the isolation method. Nevertheless, 3-methoxyniveusin A, **V**, is present, and appears to be synthesized *de novo* in the plant.

Recent work in two other laboratories has led to the identification of eight germacranolides in extrafloral aerial tissue (primarily leaves) of *H. annuus* . Melek *et al.* (117) have identified argophyllin A, **I**, as the major and another epoxide, argophyllin B, and niveusin B as minor components, whereas Spring *et al.* (118) have identified 15-hydroxy-3-dehydrodesoxytifruticin, **VII**, and its hemiketal as major and argophyllin B, niveusin C, 1-methoxy-4,5-dihydroniveusin A, **VI**, and its anhydrido analog as the minor components. Part of the discrepancy between these labs might be due to cultivar differences since a wild variety was used in the former and var *giganteus* was used in the latter study. However, cyclic sesquiterpene epoxides similar to the argophyllins (119, 120) are sensitive to both acid and base catalyzed rearrangements that form tetrahydrofurans and ultimately conjugated systems such as **VII**. Sufficient acidity for these reactions may result from co-occurring diterpenoic acids on the plant surface. Thus, the argophyllins or more labile epoxides may be the actual or, at least, predominant forms in which these C-6 lactonized germacranolide angelates are present in sunflower.

Interactions between antifeedant sesquiterpenes and other plant allelochemicals. Binary combinations of one of the more potent WCR feeding deterrents with another at a dose that gives weak feeding deterrence were explored with eight combinations of chemicals in the squash disk bioassay. No synergistic or antagonistic interactions for combinations of deterrents within or between the sesquiterpene (**V-VII, IX**), diterpene (**XI, XII**) and flavonoid (**XVII**) classes were noted. This indicates that the suite of antifeedants present in sunflower inflorescences act jointly in an additive fashion.

Neurotoxicity of antifeedant sesquiterpenes. The sesquiterpene, **V**, and the conjugated lactone, **VII**, when injected as DMSO solutions (200 nl) into WCR adults at dosages of 2.5 μg or greater per insect (avg. live wt of 17 mg), gave neurotoxic symptoms at 24 hr (excitability, hyperextension of ovipositor, egg expulsion, tarsal tetany) similar to that of picrotoxinin, a known γ-aminobutyric acid (GABA) antagonist, but not like that of avermectin (sluggish movements, paralysis), a GABA agonist (121, 122). The acute toxicity of these sesquiterpenes ($LD_{50}s$ > 500 μg/g insect), although low compared to avermectin and picrotoxinin (Table V), was substantial considering that the most active compound, **I**, and coadministration of synergists was not tested. Picrotoxinin, a sesquiterpene epoxide lactone from *Anamirta cocculus* L., has some structural similarities to that of *Helianthus* germacranolides. The latter sesquiterpene lactones, as is the case for other antifeedants (123, 124), have electrophilic centers including allylic hemiketal, conjugated ketone, and epoxide sites in addition to the conjugate lactone which may interact with critical nucleophiles such as thiol (125) and amino groups (126) of sensory receptors. Based on the structure-activity data presented above, the lactone site is not solely responsible for feeding deterrency or probably neurotoxicity. It remains to be determined if electrophilicity is associated with the GABA-like effects on the central nervous system. Other terpenoids are known to inhibit acetylcholinesterase (127). This is the first report of a putative GABA antagonist (i.e. convulsant) for an insect within its native food plant.

Deductions: Dietary Phytochemicals, Insecticide Resistance and Corn Rootworm Control

Pioneer populations of WCR in central Pennsylvania, depending on their food, are 90 to 1200 times more resistant to aldrin than an endemic population of NCR. Susceptibility of WCR to aldrin increased at least seven times when the adults consumed inflorescences of sunflower or other Compositae species rather than corn, whereas the northern species was equally susceptible to aldrin on sunflower or goldenrod (129). The more frequent consumption of Compositae pollen and floral tissues by the northern over the more corn-specializing western species could, over many generations, have led to the loss of aldrin resistance in NCR, which had similarly high resistance as WCR prior to cancellation of the cyclodienes for rootworm control (130). The terpenoid-rich flowers of the Compositae may provide the responsible chemistry that results in increased susceptibility to the chlorinated cyclodienes. These insecticides are believed to act via their epoxides at the same GABA-regulated chloride ionophore site as picrotoxinin (121, 122). Our studies with WCR indicate a low cross-resistance between this plant neuroexcitant and the cyclodienes, but the resistance ratio between species for picrotoxinin (4 times) is two orders of magnitude less than that observed for aldrin (Table V) and does not argue solely for an insensitive GABA site in mediating cyclodiene resistance. These rootworm populations are equally susceptible to the acetylcholinesterase inhibitors (Table V).

Neurotoxic antifeedants from Compositae should provide important leads into strategies that ameliorate the control of the *Diabrotica* complex. Phytochemicals with combined effects that result in loss of insecticide resistance, reduced feeding, decreased life span, and neurotoxicity in rootworms may be a practical avenue to low chemical input strategies for corn production. Also, phytochemical antagonism of cyclodiene resistance may have important consequences to future control of corn rootworm by insecticides such as avermectins and pyrethroids (e.g. tefluthrin) which, certainly in the former case (121) and at least secondarily in the latter case (122), act on the GABA gated-chloride ionophore complex.

Table V. Susceptibility of Adult Corn Rootworms in Central PA to Neurotoxicants

Rootworm species	Topical LD_{50} (μg/g insect)[a]					
	GABA-A chloride channel ligands			Acetylcholinesterase inhibitors		
	Aldrin	Picrotoxinin[b]	Avermectin[b]	Carbofuran	Terbufos	Isofenphos
Western	1980	111	58	1.16	2.91	3.39
Northern	6.0	26.2	24	1.05	2.78	4.58

[a]50% mortality determinations at 24 hr by probit analysis.
[b]Estimated by injection; for picrotoxinin (128), a 2 hr prior topical treatment with 5 μg/insect of the cytochrome P450 inhibitor piperonyl butoxide was used.

Acknowledgments

This work was made possible through the support of the U.S.D.A.(CRGO 89-37263-4567 and NEPIAP 88-34050-3361) and the Pennsylvania Agricultural Experiment Station.

Literature Cited

1. McKey, D. In Herbivores, Their Interaction with Secondary Plant Metabolites; Rosenthal, G. A.; Janzen, D. H., Eds.; Academic: New York, 1979; pp 55-133.
2. Bazzaz, F. A.; Chiariello, N. R.; Coley, P. D.; Pitelka, L. F. BioScience 1987, 37, 58-67.
3. Barbier, M. Prog. Phytochem. 1971, 2, 1-34.
4. Stanley, R. G.; Linskens, H. F. Pollen: Biology, Biochemistry, Management; Springer: New York, 1974.
5. Baker, H.G.; Baker, I. In The Biology of Nectaries; Bentley, B.; Elias, T., Eds.; Columbia University Press: New York, 1983; pp 126-152.
6. Thompson, W. R.; Meinwald, J.; Aneshansley, D.; Eisner, T. Science 1972, 177, 528-530.
7. Harborne, J. B.; Smith, D. M. Biochem. Syst. Ecol. 1978, 6, 287-291.
8. Whitten, W. M.; Williams, N. H.; Armbruster, W. S.; Battiste, M. A.; Strekowski, L; Lindquist, N. Syst. Bot. 1986, 11, 222-228.
9. Stipanovic, R. D.; Bell, A. A.; Lukefahr, M. In Host Plant Resistance to Pests; Hedin, P.A., Ed.; ACS Symposium Series No. 62; American Chemical Society: Washington, DC, 1977; pp 197-214.
10. Mabry, T. J.; Gill, J. E. In Herbivores, Their Interaction with Secondary Plant Metabolites; Rosenthal, G. A.; Janzen, D. H., Eds.; Academic: New York, 1979; pp 501-537.
11. Kubo, I.; Nakanishi, K. In Advances in Pesticide Science, Part 2; Geissbuhler, H.; Brooks, G. T.; Kearney, P. C., Eds.; Pergamon: New York, 1979, pp 284-294.
12. Koul, O. Indian Rev. Life Sci. 1982, 2, 97-125.
13. Brattsten, L. B. In Plant Resistance to Insects; Hedin, P.A., Ed.; ACS Symposium Series No. 208; American Chemical Society: Washington, DC, 1983; pp 173-195.
14. Rodriguez, E. In Bioregulators for Pest Control; Hedin, P.A., Ed.; ACS Symposium Series No. 276; American Chemical Society: Washington, DC, 1985; pp 447-453.
15. van Beek, T. A.; de Groot, A. Recl. Trav. Chim. Pays-Bas 1986, 105, 513-527.
16. Reese, J. C.; Holyoke, C. W., Jr. In Handbook of Natural Pesticides, Vol. 3B; Morgan, E. D.; Mandava, N. B., Eds.; CRC Press: Boca Raton, 1987; pp 21-66.
17. Grundy, D. L; Still, C.C. Pestic. Biochem. Physiol. 1985, 23, 383-388.
18. Bowers, M. D. In Chemical Mediation of Coevolution; Spencer, K. C., Ed.; Academic: New York, 1988; pp 133-165.
19. Rodriguez, E.; Towers, G. H. N.; Mitchell, J.C. Phytochemistry 1976, 15, 1573-1580.
20. Burnett, W. C., Jr.; Jones, S. B., Jr.; Mabry, T. J. In Biochemical Aspects of Plant and Animal Coevolution; Harborne, J.B., Ed.; Academic: London, 1978, pp 233-257.
21. Doskotch, R. W.; Odell, T. M.; Girard, L. In The Gypsy Moth: Research Toward Integrated Pest Management; Doane, C. C.; McManus, M. L., Eds.; U.S. Dept. Agric. Tech. Bull. No. 1584; Washington, DC, 1981; pp 657-666.
22. Ivie, G. W.; Witzel, D. A. In Handbook of Natural Toxins, Vol. 1; Keeler, R. F.; Tu, A. T., Eds.; Marcel: New York, 1982; pp 543-584.
23. Picman, A. K. Biochem. Syst. Ecol. 1986, 14, 255-281.
24. Cooper-Driver, G. A.; Le Quesne, P. W. In Allelochemicals: Role in Agriculture and Forestry; Waller, G. R., Ed.; ACS Symposium Series No. 330; American Chemical Society: Washington, DC, 1987; pp 534-550.
25. Dreyer, D. L. In Chemistry and Chemical Taxonomy of the Rutales; Waterman, P. G.; Grundon, M. F., Eds.; Academic: New York, 1983; pp 215-245.
26. Bentley, M. D.; Rajab, M. S.; Alford, A. R.; Mendel, M. J.; Hassanali, A. Entomol. Exp. Appl. 1988, 49, 189-193.

27. Elliger, C. A.; Waiss, A. C., Jr. In Insecticides of Plant Origin; Arnason, J. T.; Philogene, B. J. R.; Morand, P., Eds.; ACS Symposium Series No. 387; American Chemical Society: Washington, DC, 1989; pp 188-205.
28. Yamasaki, R. B.; Klocke, J. A. J. Agric. Food Chem. 1987, 35, 467-471.
29. Jacobson, M., Ed. Focus on Phytochemical Pesticides, Vol. 1. The Neen Tree; CRC: Boca Raton, 1989; 178pp.
30. Saxena, R. C. In Insecticides of Plant Origin; Arnason, J. T.; Philogene, B. J. R.; Morand, P., Eds.; ACS Symposium Series No. 387; American Chemical Society: Washington, DC, 1989; pp 110-135.
31. Eisner, T.; Kluge, A. F.; Ikeda, M. I.; Meinwald, Y. C.; Meinwald, J. J. Insect Physiol. 1971, 17, 245-250.
32. Prestwich, G. D. Annu. Rev. Entomol. 1984, 29, 201-232.
33. Bowers, W. S. In Biotechnology, Biological Pesticides and Novel Plant-Pest Resistance for Insect Pest Management; Roberts, D. W.; Granados, R. R., Eds.; Cornell University: Ithaca, 1988; pp 123-131.
34. Gibson, R. W.; Pickett, J. A. Nature 1983, 302, 608-609.
35. Ave, D. A.; Gregory, P., Tingey, W. M. Entomol. Exp. Appl. 1987, 44, 131-138.
36. Stipanovic, R. D. In Plant Resistance to Insects; Hedin, P.A., Ed.; ACS Symposium Series No. 208; American Chemical Society: Washington, DC, 1983; pp 69-100.
37. Stephenson, A. G. J. Chem. Ecol. 1982, 8, 1025-1034.
38. Kelsey, R. G. In Biology and Chemistry of Plant Trichomes; Rodriguez, E.; Healey, P. L.; Mehta, I., Eds.; Plenum: New York, 1984; pp 187-241.
39. Head, S. W. In Pyrethrum, The Natural Insecticide; Casida, J. E., Ed.; Academic: New York, 1973;. pp 25-53.
40. Casida, J.E. Environ. Health Persp. 1980, 34, 189-202.
41. Rogers, C. E. In Biology and Breeding for Resistance to Arthropods and Pathogens in Agricultural Plants; Harris, M. K., Ed.; Texas Agric. Expt. Stn. Misc. Publ. 1451, 1980; pp 359-389.
42. Gershenzon, J.; Rossiter, M.; Mabry, T. J.; Rogers, C. E.; Blust, M. H.; Hopkins, T. L. In Bioregulators for Pest Control; Hedin, P.A., Ed.; ACS Symposium Series No. 276; American Chemical Society: Washington, DC, 1985; pp 433-446.
43. Rossiter, M.; Gershenzon, J.; Mabry, T. J. J. Chem. Ecol. 1986, 12, 1505-1521.
44. Rogers, C. E.; Gershenzon, J.; Ohno, N.; Mabry, T. J.; Stipanovic, R. D.; Kreitner, G. L. Environ. Entomol. 1987, 16, 586-592.
45. Elliger, C. A.; Zinkel, D. F.; Chan, B. G.; Waiss, A. C., Jr. Experientia 1976, 32; 1364-1366.
46. Metcalf, R. L.; Metcalf, R. A.; Rhodes, A. M. Proc. Natl. Acad. Sci., U.S.A. 1980, 77, 3769-3772.
47. Metcalf, R. L. J. Chem. Ecol. 1986, 12, 1109-1124.
48 Andersen, J. F.; Metcalf, R. L. J. Chem. Ecol. 1987, 13, 681-699.
49. Nielsen, J. K. Entomol. Exp. Appl. 1978, 24, 41-54.
50. Jermy, T.; Butt, B. A.; McDonough, L.; Dreyer, D. L.; Rose, A. F. Insect Sci. Appl. 1981, 1, 237-242.
51. Reed, D. K.; Warthen, J. D., Jr.; Uebel, E. C.; Reed, G. L. J. Econ. Entomol. 1982, 75, 1109-1113.
52. Wakabayashi, N.; Wu, W. J.; Waters, R. M.; Redfern, R. E.; Mills, G. D., Jr.; DeMilo, A. B.; Lusby, W. R.; Andrzejewski, D. J. Nat. Prod. 1988, 51, 537-542.
53. Yamasaki, R. B.; Klocke, J. A. J. Agric. Food Chem. 1989, 37, 1118-1124.
54. Wisdom, C. S.; Rodriguez, E. Biochem. Syst. Ecol. 1983, 11, 345-352.
55. Carter, C. D.; Gianfagna, T. J.; Sacalis, J. N. J. Agric. Food Chem. 1989, 37, 1425-1428.
56. Nawrot, J.; Harmatha, J.; Novotny, L. Biochem. Syst. Ecol. 1984, 12, 99-101.
57. Rose, A. F. Phytochemistry 1980, 19, 2689-2693.
58. Le Quesne, P. W.; Cooper-Driver, G. A.; Villani, M.; Do, M. N. In New Trends in Natural Products Chemistry 1986; Rahman, A-U.; Le Quesne, P. W., Eds.; Elsevier: Amsterdam, 1986; pp 271-282.
59. Mitchell, B. K. J. Insect Physiol. 1988, 34, 213-225.
60. Pasteels, J. M.; Rowell-Rahier, M.; Braekman, J.-C.; Daloze, D. Biochem. Syst. Ecol. 1984, 12, 395-406.

61. Bernays, E. A.; Chapman, R. F. Ecol. Entomol. 1977, 2, 1-18.
62. Wada, K.; Munakata, K. Agric. Biol. Chem. 1971, 35, 115-118.
63. Nawrot, J.; Bloszyk, E.; Grabarczyk, H.; Drozdz, B. Prace Nauk. Inst. Ochr. Ros. 1982, 24, 27-36.
64. Wisdom, C. S.; Smiley, J. T.; Rodriguez, E. J. Econ. Entomol. 1983, 76, 993-998.
65. Yano, K. J. Agric. Food Chem. 1987, 35, 889-891.
66. Blust, M. H.; Hopkins, T. L. Entomol. Exp. Appl. 1987, 45, 37-46.
67. Nawrot, J.; Bloszyk, E.; Harmatha, J.; Novotny, L. Z. Ang. Ent. 1984, 98, 394-398.
68. Rees, S. B.; Harborne, J. B. Phytochemistry 1985, 24, 2225-2231.
69. Nakajima, S.; Kawazu, K. Heterocycles 1978, 10, 117-121.
70. Nakajima, S.; Kawazu, K. Agric. Biol. Chem. 1980, 44, 2893-2899.
71. Harmatha, J.; Nawrot, J. Biochem. Syst. Ecol. 1984, 12, 95-98.
72. Okunade, A. L.; Wiemer, D. F. Phytochemistry 1985, 24, 1199-1201.
73. Bordoloi, M. J.; Shukla, V. S.; Sharma, R. P. Tetrahedron Lett. 1985, 26, 509-510.
74. McGovran, E. R.; Mayer, E. L. U. S. Dept. Agric. Bur. Entomol. Plant Quarantine, Entomol. Tech. 1942, E-572.
75. Picman, A. K.; Picman, J. Biochem. Syst. Ecol. 1984, 12, 89-93.
76. Arnason, J. T.; Isman, M. B.; Philogene, B. J. R.; Waddell, T. G. J. Nat. Prod. 1987, 50, 690-695.
77. Isman, M. B.; Brard, N. L.; Nawrot, J.; Harmatha, J. J. Appl. Entomol. 1989, 107, 524-529.
78. Watanabe, K.; Ohno, N.; Yoshioka, H.; Gershenzon, J.; Mabry, T. J. Phytochemistry 1982, 21, 709-713.
79. Picman, A. K.; Elliott, R. H.; Towers, G. H. N. Biochem. Syst. Ecol. 1978, 6, 333-335.
80. Streibl, M.; Nawrot, J.; Herout, V. Biochem. Syst. Ecol. 1983, 11, 381-382.
81. Wiemer, D. F.; Ales, D. C. J. Org. Chem. 1981, 46, 5449-5450.
82. Smith, C. M.; Kester, K. M.; Fischer, N. H. Biochem. Syst. Ecol. 1983, 11, 377-380.
83. Hubert, T. D.; Wiemer, D. F. Phytochemistry 1985, 24, 1197-1198.
84. Sharma, R. N.; Joshi, V. N. Biovigyanam 1977, 3, 225.
85. Picman, A. K.; Elliott, R. H.; Towers, G. H. N. Can. J. Zool. 1981, 59, 285-292.
86. Isman, M. B.; Rodriguez, E. Phytochemistry 1983, 22, 2709-2713.
87. Isman, M. B.; Rodriguez, E. Environ. Entomol. 1984, 13, 539-542.
88. Isman, M. B. Pestic. Biochem. Physiol. 1985, 24, 348-354.
89. Pettei, M. J.; Miura, I.; Kubo, I.; Nakanishi, K. Heterocycles 1978, 11, 471-480.
90. Sarma, D. N.; Barua, N. C.; Sharma, R. P. Chem. Ind. 1985, 167-168.
91. Burnett, W. C., Jr.; Jones, S. B., Jr.; Mabry, T. J.; Padolina, W. G. Biochem. Syst. Ecol. 1974, 2, 25-29.
92. Ganjian, I.; Kubo, I.; Fludzinski, P. Phytochemistry 1983, 22, 2525-2526.
93. Kawazu, K.; Nakajima, S.; Ariwa, M. Experientia 1979, 35, 1294-1295.
94. Tahara, T.; Sakuda, Y.; Kodama, M.; Fukazawa, Y.; Ito, S.; Kawazu, K.; Nakajima, S. Tetrahedron Lett. 1980, 21, 1861-1862.
95. Rose, A. F.; Jones, K. C.; Haddon, W. F.; Dreyer, D. L. Phytochemistry 1981, 20, 2249-2253.
96. Herz, W.; Kulanthaivel, P.; Watanabe, K. Phytochemistry 1983, 22, 2021-2025.
97. Wiemer, D. F. Rev. Latinoamer. Quim. 1985, 16, 98-102.
98. Hedin, P. A.; Waage, S. K In Plant Flavonoids in Biology and Medicine: Biochemical, Pharmacological, and Structure-Activity Relationships; Cody, V.; Middleton, E., Jr.; Harborne, J. B., Eds.; Alan Liss: New York, 1986; pp 87-100.
99. Duffey, S. S. In Insects and the Plant Surface; Juniper, B.; Southwood, Sir R., Eds.; Edward Arnold: London, 1986; pp 151-172.
100. Harborne, J. B. In Plant Flavonoids in Biology and Medicine II: Biochemical, Cellular, and Medicinal Properties; Cody, V.; Middleton, E., Jr.; Harborne, J. B.; Beretz, A., Eds.; Alan R. Liss: New York, 1988; pp 17-27.
101. Elliger, C. A.; Chan, B. C.; Waiss, A. C., Jr. Naturwiss. 1980, 67, 358-360.
102. Dreyer, D. L.; Jones, K. C. Phytochemistry 1981, 20, 2489-2493.
103. Mullin, C. A. In Molecular Aspects of Insect-Plant Associations; Brattsten, L. B.; Ahmad, S., Eds.; Plenum: New York, 1986; pp 175-209.

104. Schultz, J. C. In Chemistry and Significance of Condensed Tannins; Hemingway, R. W.; Karchesy, J. J., Eds.; Plenum: New York, 1989; pp 417-433.
105. Nielson, J. K. In Biology of Chrysomelidae; Jolivet, P.; Petitpierre, E.; Hsiao, T. H., Eds.; Kluwer Academic: Dordrecht, 1988; pp 25-40.
106. Matsuda, K. In Biology of Chrysomelidae; Jolivet, P.; Petitpierre, E.; Hsiao, T. H., Eds.; Kluwer Academic: Dordrecht, 1988; pp 41-56.
107. Matsuda, K.; Senbo, S. Appl. Ent. Zool. 1986, 21, 411-416.
108. Hedin, P. A.; Miles, L. R.; Thompson, A. C.; Minyard, J. P. J. Agric. Food Chem. 1968, 16, 505-513.
109. Nitao, J. K.; Zangerl, A. R. Ecology 1987, 68, 521-529.
110. Rieseberg, L. H.; Schilling, E. E. Amer. J. Bot. 1985, 72, 999-1004.
111. Harborne, J. B. In Introduction to Ecological Biochemistry; 3rd Ed. Academic: London, 1988; pp 42-81.
112. Siegfried, B. D.; Mullin, C. A. Environ. Entomol. 1990, 19, 474-480.
113. Sando, C. E. J. Biol. Chem. 1925, 64, 71-74.
114. Ohmoto, T.; Udagawa, M.; Yamaguchi, K. Shoyakugaku Zasshi 1986, 40, 172-176.
115. Mikolajczak, K. L.; Smith, C. R., Jr.; Wolff, I. A. J. Agric. Food Chem. 1970, 18, 27-32.
116. Batyuk, V. S.; Kovaleva, S. N. Khim. Prir. Soedin. 1985, 566-567.
117. Melek, F. R.; Gage, D. A.; Gershenzon, J.; Mabry, T. J. Phytochemistry 1985, 24, 1537-1539.
118. Spring, O.; Benz, T.; Ilg, M. Phytochemistry 1989, 28, 745-749.
119. Geissman, T. A. Rec. Adv. Phytochem. 1973, 6, 65-95.
120. Herz, W. Israel J. Chem. 1977, 16, 32-44.
121. Matsumura, F.; Tanaka, K.; Ozoe, Y. In Sites of Action for Neurotoxic Pesticides ; Hollingworth, R. M.; Green, M. B., Eds.; ACS Symposium Series No. 356; American Chemical Society: Washington, DC, 1987; pp 44-70.
122. Eldefrawi, M. E.; Abalis, I. M.; Sherby, S. M.; Eldefrawi, A. T. In Neuropharmacology and Pesticide Action; Ford, M. G.; Lunt, G. G.; Reay, R. C.; Usherwood, P. N. R., Eds.; Ellis Horwood: Chichester, 1986; pp 154-173.
123. Ma, W.-C. Physiol. Entomol. 1977, 2, 199-207.
124. Schoonhoven, L. M.; Fu-Shun, Y. J. Insect Physiol. 1989, 35, 725-728.
125. Norris, D. M. In Sterol Biosynthesis Inhibitors and Anti-Feeding Compounds; Haug, G.; Hoffmann, H., Eds.; Springer: Berlin, 1986; pp 97-146.
126. Lam, P. Y.-S.; Frazier, J. L. Tetrahedron Lett. 1987, 28, 5477-5480.
127. Ryan, M. F.; Byrne, O. J. Chem. Ecol. 1988, 14, 1965-1975.
128. Siegfried, B. D.; Mullin, C. A. Pestic. Biochem. Physiol. 1990, 36, 135-146.
129. Siegfried, B. D.; Mullin, C. A. Pestic. Biochem. Physiol. 1989, 35, 155-164.
130. Metcalf, R. L. In Pest Resistance to Pesticides; Georghiou, G. P.; Saito, T., Eds; Plenum: New York, 1983; pp 703-733.

RECEIVED July 18, 1990

Chapter 19

Insecticidal Constituents of *Azadirachta indica* and *Melia azedarach* (Meliaceae)

S. Mark Lee[1], James A. Klocke[2], Mark A. Barnby[3], R. Bryan Yamasaki[4], and Manuel F. Balandrin[5]

National Product Sciences, Inc., University of Utah Research Park, 417 Wakara Way, Salt Lake City, UT 84108

The insecticidal constituents of the Indian neem tree (Azadirachta indica) and the related chinaberry tree (Melia azedarach) were investigated using bioassay-guided fractionation and isolation techniques. Azadirachtin was isolated as the principal insecticidal and antifeedant constituent of A. indica seeds. In addition, 25 volatile compounds were identified as constituents of crushed neem seeds. The major volatile constituent identified, di-n-propyl disulfide, was shown to be larvicidal to three species of insects. Furthermore, two new insecticidal compounds, 1-cinnamoylmelianolone and 1-cinnamoyl-3,11-dihydroxymeliacarpin, were isolated from the fruit of M. azedarach. The insecticidal activities of these new compounds, compared with those of azadirachtin and several of its derivatives, suggest structure-activity relationships and a mode of action that may be useful in the design of synthetic analogs.

The neem tree, Azadirachta indica A. Juss., is a tropical and subtropical species indigenous to India and Southeast Asia (1) which is now widely distributed in many tropical and subtropical regions of both the Old and New Worlds (2-4). The chinaberry tree, Melia azedarach L., is a native of tropical Asia (5), but is now also widely distributed in drier regions of the southern and western United States (6) (e.g., Texas, Arizona, southeastern

[1]Current address: Chemistry Laboratory Services, California Department of Food and Agriculture, 3292 Meadowview Road, Sacramento, CA 95832

[2]Current address: Division of Entomology and Parasitology, College of Natural Resources, University of California, Berkeley, CA 94720

[3]Current address: Department of Entomology, University of Georgia, Athens, GA 30602

[4]Current address: SERES Laboratories, Inc., 3331 Industrial Drive, Santa Rosa, CA 95403

[5]Current address: Natural Product Sciences, Inc., University of Utah Research Park, 420 Chipeta Way, Salt Lake City, UT 84108

0097–6156/91/0449–0293$06.00/0
© 1991 American Chemical Society

Nevada, and southwestern Utah). The two species are closely related taxonomically, with both belonging to the same subfamily (Melioideae) and tribe (Melieae) in the family Meliaceae (mahogany family) (7-9). Recent phytochemical studies have also illustrated the close similarities which exist between their secondary metabolites, especially with regard to their insecticidal tetranortriterpenoid limonoid constituents (meliacins) (7-11).

Both plant species have long been recognized for their medicinal (12-16) and insecticidal properties (11,15,17,18). Indeed, these two species have long been recognized as possessing among the most outstanding antifeedant and insecticidal activities among members of the Meliaceae (17-20).

In this chapter, our recent studies on the insecticidal constituents from these two plant species will be presented, concentrating first on the volatile organosulfur compounds recently isolated from neem seeds (21), and then on the azadirachtin-type tetranortriterpenoid limonoids present in both species (22-26). Our recent studies on the insecticidal meliacins of the azadirachtin type suggest structure-activity relationships and a mode of action which may be useful in the design of synthetic analogs.

Volatile Organosulfur Compounds of Crushed A. indica Seeds

Crushed and/or ripening neem tree fruits and expressed neem seed oil give off a strong alliaceous (garlic-like) odor. This characteristic alliaceous odor has also been reported to be present in the leaves, inner bark, timber, and heartwood of this tree species (1,14,21). Some of the reputed medicinal properties of neem seed oil have been attributed to the sulfurous compounds that it contains. Although the presence of sulfur-containing compounds in neem oil had previously been reported (12,14), no detailed chemical studies of neem volatile organosulfur compounds had been reported prior to our recent study (21).

The volatiles from crushed neem seeds were purged with nitrogen, trapped onto Amberlite XAD-4 resin traps at room temperature, recovered and concentrated into diethyl ether using a Kuderna-Danish evaporative concentrator at 30°C, and analyzed by means of capillary gas chromatography/mass spectrometry (GC/MS). For comparative purposes, similar volatile concentrates were prepared from freshly chopped onion bulbs and garlic cloves, and blank controls were simultaneously prepared for each of the three test samples.

A total of 25 compounds were identified by GC/MS analysis as constituents of the crushed neem seed volatile concentrate, 22 of which were organosulfur compounds. Most of the sulfur-containing compounds were either aliphatic disulfides (eight in all) or higher polysulfides (nine trisulfides and three tetrasulfides). A number of the di- and polysulfides present in the neem seed volatile concentrate were mixed unsymmetrical aliphatic alkyl (methyl, propyl, and butyl) and alkenyl (propenyl) containing derivatives. However, no monosulfides or dimethyl polysulfides (often observed in onion and garlic extracts) were found among the neem seed volatiles, and onion-type lachrymatory factors were also absent. Nevertheless, apart from these exceptions, the neem seed volatile constituents were found to generally closely resemble those of onions (21).

Most of the neem seed volatiles identified (19 out of 25) were present in concentrations of less than 1% of the total area detected in the GC/MS total ion chromatogram. Many of these minor components (e.g., isomeric dimethylthiophenes and various polysulfides (i.e., tri- and tetrasulfides)) are probably heat-generated artifacts produced upon GC/MS analysis of thermolabile precursors.

The major volatile constituents of crushed neem seeds were identified by means of GC/MS as di-n-propyl disulfide (75.74% of the total area detected), n-propyl-trans-1-propenyl disulfide (9.67%), and n-propyl-cis-1-propenyl disulfide (2.76%) (Figure 1). Together, these three closely eluting compounds accounted for 88.17% of the total area detected (all of the 25 compounds identified accounted together for 98.74% of the total area detected). The corresponding trisulfides (2.19, 4.22, and 2.20%, respectively) were the next most abundant group of compounds detected. However, as previously alluded to above, they probably

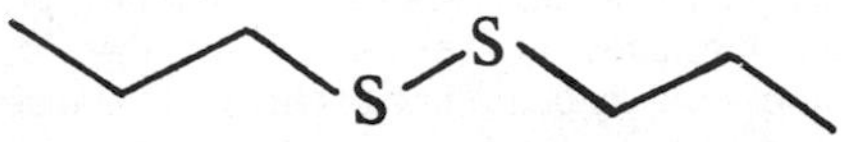

Di-*n*-propyl disulfide

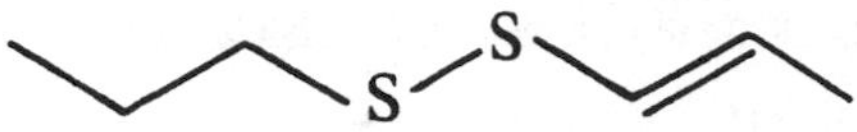

n-Propyl-*trans*-1-propenyl disulfide

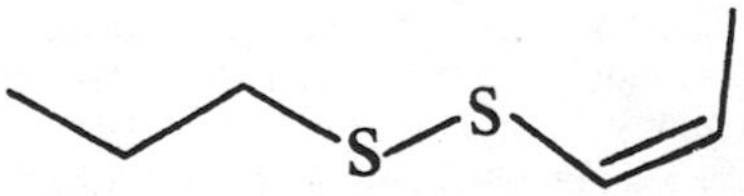

n-Propyl-*cis*-1-propenyl disulfide

Figure 1. Chemical structures of the principal volatile organosulfur compounds isolated from crushed neem (*Azadirachta indica*) seeds.

represent heat-generated artifacts. Furthermore, the principal neem seed volatiles (i.e., the disulfides) identified by GC/MS are probably artifacts generated from nonvolatile precursor substrates by enzymatic activity when the seeds are crushed, as is the case with onion and garlic tissues (21,27).

The neem seed volatiles may be responsible for the reputed insect repellent properties attributed to neem seeds. For example, neem seeds have been reported to repel (generally at a distance, without contact) various types of insects, including book mites, locusts, planthoppers, white ants, cockroaches, moths, mosquitoes, cattle flies, post-harvest insects, and stored-products pests (1,12,14,21). We found that the major volatile component of crushed neem seeds, i.e., di-n-propyl disulfide, exhibited larvicidal properties in bioassays against the larvae of three species of insects, i.e., the yellow fever mosquito (Aedes aegypti; LC_{50} (lethal concentration necessary to kill 50% of the treated insects), 66 ppm), the tobacco budworm (Heliothis virescens; LC_{50}, 1000 ppm), and the corn earworm (H. zea; LC_{50}, 980 ppm) (21).

The most probable chemical-ecological explanation for the presence of biologically active volatile organosulfur compounds in neem seeds is that these compounds, like many other plant secondary metabolites, may play a role in chemical defense mechanisms against herbivorous insects, higher herbivorous predators, and invading microorganisms. Because of their extreme volatility and pungency, these compounds may serve as repellents to attacking insects (27) before they can cause significant injury to neem trees, their leaves, or their seeds. In addition, the neem seed volatiles (and/or their nonvolatile precursors) may be responsible, at least in part, for some of the reputed medicinal properties attributed to neem seeds (21).

Insecticidal Tetranortriterpenoids of A. indica and M. azedarach

The potent tetranortriterpenoid limonoids of A. indica (neem) and M. azedarach (chinaberry) (i.e., meliacins) are considered to be among the most promising and interesting of the plant-derived insecticidal compounds yet discovered (17). Azadirachtin (Figure 2) is considered to be the prototype compound for this class of feeding deterrent (antifeedant) and insecticidal substances (17,18,28-31).

Azadirachtin and Derivatives. The potent and specific effects of azadirachtin against a wide variety of insects, including a number of economically important species of insect pests, have warranted its evaluation as both a source of and a model for new commercial insect control agents. The potential of azadirachtin lies in its several advantages, which include its potency, which is comparable to that of the most potent conventional synthetic insecticides. Furthermore, azadirachtin has a high degree of specificity (affecting behavioral, developmental, and biochemical processes peculiar to insects), it is non-mutagenic (in the Ames test) (32,33), it is biodegradable, and it has systemic activity in plants, being translocated throughout the plant following absorption through the leaves (34) and/or root system (20). Azadirachtin has several effects on susceptible insects, including potent feeding deterrency, growth inhibition, and ecdysis inhibition (disruption of the molting process) (17,34).

Several analogs of azadirachtin (both natural and semi-synthetic) have been shown to exhibit activities comparable to those of the parent compound. For example, two semi-synthetic derivatives of azadirachtin, i.e., 22,23-dihydroazadirachtin and 2',3',22,23-tetrahydroazadirachtin, were prepared and tested as growth inhibitors and toxicants against H. virescens larvae (22,26), and as antifeedants against Spodoptera frugiperda (fall armyworm) (35). The two hydrogenated derivatives proved to be as active as azadirachtin itself in all of these bioassays (22,26,35). Furthermore, 22,23-dihydroazadirachtin also had activity comparable to that of azadirachtin against Epilachna varivestis (Mexican bean beetle) (36). In addition, other related derivatives, such as the semi-synthetic 3-deacetylazadirachtin,

	R_1	R_2	R_3
Azadirachtin	Ac	CO_2CH_3	Tig
1-Cinnamoyl-3,11-dihydroxymeliacarpin	H	CH_3	Cinn

	R_1	R_2	R_3
1-Cinnamoylmelianolone	H	H	Cinn
3-Acetyl-1-cinnamoylmelianolone	Ac	H	Cinn
3,19-Diacetyl-1-cinnamoylmelianolone	Ac	Ac	Cinn
1-Decinnamoylmelianolone	H	H	H

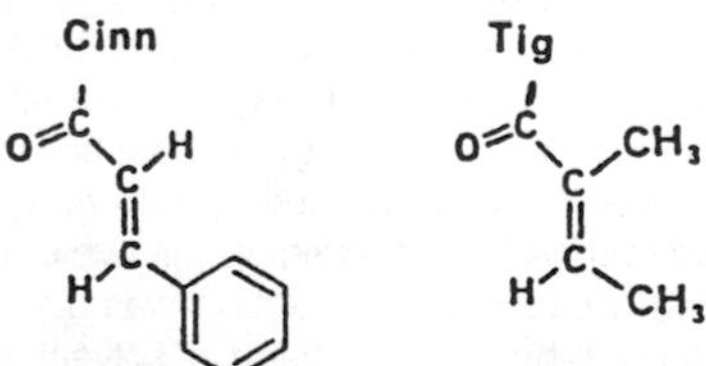

Figure 2. Stereostructures of some natural and derivatized insecticidal tetranortriterpenoids isolated from Azadirachta indica and Melia azedarach.

also have activity comparable to that of azadirachtin as growth inhibitors and toxicants against the larvae of H. virescens (22) and E. varivestis (36).

In a recent study (25), it was found that radioactively labelled (tritiated) 22,23-dihydroazadirachtin administered orally to fifth (last) instar larvae of H. virescens was rapidly absorbed through the midgut and into the hemocoel, and that excretion of the compound was slow. Data obtained by thin-layer chromatographic (TLC) analysis indicated that metabolism of the dihydroazadirachtin to a more polar form(s) probably occurred following absorption through the midgut (25).

Azadirachtin disrupts the normal growth and development of a variety of insect pests, possibly by functioning as a molting hormone (ecdysteroid) analog (17,25,34,36-38). In several species of insects treated with azadirachtin, ecdysis was prevented possibly by the disruption of the molting hormone titre (17,20,36). Azadirachtin was found to be tightly bound to binding proteins in the insect body, and its action is irreversible, one dose (1 μg) being sufficient to disrupt subsequent developmental and/or reproductive stages (25,36). For example, development to the pupal stage was disrupted by such doses (25).

Insecticidal Constituents of M. azedarach. Azadirachtin has previously been reported to be a constituent of both A. indica and M. azedarach (9,17,18,39,40). However, recent, more detailed investigations suggest that azadirachtin has never been isolated from or identified in authentic M. azedarach, and that it has so far been shown to occur only in A. indica (7,8,10,11,28,31).

We found that methanolic extracts of chinaberry (M. azedarach) fruits from southwestern Utah possessed insecticidal potency comparable or equivalent to that of A. indica seed extracts and/or to that of azadirachtin (19,23). However, after a thorough search by means of TLC and high-pressure liquid chromatography (HPLC), we were unable to detect the presence of azadirachtin in these chinaberry extracts. Instead, in an effort to isolate and identify the principal active constituents of M. azedarach, we isolated and characterized two new potent insecticidal tetranortriterpenoid limonoids. The first of these was a novel meliacin termed 1-cinnamoylmelianolone. It possesses a novel structure (skeletal arrangement) not previously observed among the azadirachtin-like meliacins (23). The second compound was a new derivative of meliacarpin (11,31), i.e., 1-cinnamoyl-3,11-dihydroxymeliacarpin (Figure 2). The insecticidal activities of these two new compounds, plus two acetylated derivatives of 1-cinnamoylmelianolone (Figure 2) were compared to those of azadirachtin and several of its derivatives using two species of lepidopterous insects, H. virescens (Table I) and S. frugiperda (Table II).

Green chinaberry (M. azedarach) fruits were collected during late September of 1986 from trees growing in Hurricane, Utah. (We have observed that chinaberries allowed to ripen and become yellow tend to ferment, giving off malodorous products) (6).

Methanolic extracts of the fresh chinaberry fruits (5.7 kg fresh weight) were prepared, concentrated, and filtered. The aqueous methanolic filtrate was partitioned successively, first with hexane and then with methylene chloride. The methylene chloride partition fraction was concentrated in vacuo at 40°C yielding an extract weighing 6.6 g which was then subjected to flash column chromatography on silica gel using a step-gradient elution involving n-hexane, ethyl acetate, isopropanol, and methanol. Effluents from this column were monitored by means of thin-layer chromatography (TLC). The fraction containing the compounds of interest (2.5 g) was rechromatographed on a low-pressure column using 60% aqueous methanol as eluting solvent system. Monitoring of collected fractions by TLC revealed the fractions containing the two compounds (meliacins) of interest (yield, 1.38 g). Further purification of this fraction was accomplished by means of droplet counter-current chromatography (DCCC) using a solvent system (mixture) incorporating chloroform-toluene-methanol-water (41), finally affording ca. 200 mg of 1-cinnamoylmelianolone and ca. 300 mg of 1-cinnamoyl-3,11-dihydroxymeliacarpin. Final purification of the compounds was accomplished by means of normal- and reversed-phase

Table I. Growth-Inhibitory and Larvicidal Effects of Natural and Derivatized Tetranortriterpenoids from Azadirachta indica and Melia azedarach Fed in Artificial Diet to First-Instar Larvae of Heliothis virescens.

Test Compound	EC_{50}[a] (ppm)	LC_{50}[b] (ppm)
Azadirachtin[c,d]	0.07	0.80
22,23-Dihydro-azadirachtin[d]	0.08	0.47
2',3',22,23-Tetra-hydroazadirachtin[d]	0.08	0.30
3-Deacetylazadirachtin[d]	0.09	0.37
1-Detigloyl-22,23-dihydro-azadirachtin[d]	0.59	2.40
Deacetylazadirachtinol[c]	0.17	0.80
1-Cinnamoylmelianolone[e]	0.12	1.50
3-Acetyl-1-cinnamoyl-melianolone	0.15	1.50
3,19-Diacetyl-1-cinnamoyl-melianolone	0.12	1.50
1-Cinnamoyl-3,11-dihydroxy-meliacarpin[e]	0.18	3.50

[a]EC_{50} is the effective concentration in ppm of additive necessary to reduce larval growth to 50% of the control values.

[b]LC_{50} is the lethal concentration in ppm of additive necessary to kill (usually by inhibiting ecdysis, the final stage of the molting process) 50% of the treated insects.

[c]Naturally occurring compound isolated from Azadirachta indica.

[d]Similar preliminary results were obtained against the corn earworm (H. zea) and the fall armyworm (S. frugiperda) (see Table II).

[e]Naturally occurring compound isolated from Melia azedarach.

Table II. Growth-Inhibitory and Larvicidal Effects of Natural and Derivatized Tetranortriterpenoids from Azadirachta indica and Melia azedarach Fed in Artificial Diet to First-Instar Larvae of Spodoptera frugiperda.

Test Compound	EC_{50}[a] (ppm)	LC_{50}[b] (ppm)
Azadirachtin[c]	0.08	1.0
1-Cinnamoylmelianolone[d]	0.04	1.3
3-Acetyl-1-cinnamoyl-melianolone	0.06	2.4
3,19-Diacetyl-1-cinnamoyl-melianolone	0.04	2.5
1-Cinnamoyl-3,11-dihydroxy-meliacarpin[d]	0.04	1.6

[a] EC_{50} is the effective concentration in ppm of additive necessary to reduce larval growth to 50% of the control values.

[b] LC_{50} is the lethal concentration in ppm of additive necessary to kill (usually by inhibiting ecdysis, the final stage of the molting process) 50% of the treated insects.

[c] Naturally occurring compound isolated from Azadirachta indica.

[d] Naturally occurring compound isolated from Melia azedarach.

HPLC methods previously described (22,24,26,42). The cinnamoyl ester-bearing compounds were detected by ultraviolet (UV) monitoring at 280 nm. Structure elucidation of the purified compounds was carried out by means of infrared (IR) and UV spectrophotometry, proton and carbon-13 nuclear magnetic resonance (NMR) spectroscopy, and electron impact (EI-) and fast-atom bombardment mass spectrometry (FAB-MS). The structures of the two new isolates and three of their derivatives were established on the basis of spectroscopic data and spectral evidence obtained on comparison with azadirachtin (23).

The two new meliacins showed general TLC chromatographic properties similar to those of azadirachtin, including giving characteristic colors typical of azadirachtin after spraying with a vanillin-H_2SO_4-ethanol solution followed by heating. However, unlike azadirachtin, the two new compounds showed the strong UV absorptions (in methanol) at 280 nm typical of an α,β-unsaturated, unsubstituted aromatic group, i.e., a *trans*-cinnamoyl group. This identification was corroborated by IR, EI- and FAB-MS, and proton and carbon-13 NMR spectra (11,23,31).

Structure-Activity Relationships of Azadirachtin and Its Analogs and Derivatives

The insecticidal activities of the two new meliacins, two acetylated derivatives of 1-cinnamoyl-melianolone, and azadirachtin and several of its analogs and derivatives tested against *H. virescens* and *S. frugiperda* are shown in Tables I and II, respectively. These data, together with other findings on the insecticidal activities of closely related compounds (20,22,25,26,35,36,43), suggest that specific ester groups are not required at positions 1 and 3 in the azadirachtin nucleus in order to maintain a high level of insecticidal activity. However, the presence of certain ester groups (e.g., tigloyl, cinnamoyl) at these positions may provide a favorable hydrophilic/lipophilic balance necessary for optimum transport across various membranes and physiological compartments as these molecules make their way to their target sites or receptors. It seems probable that the ester groups at positions 1 and 3 in ring A are hydrolyzed off enzymatically *in vivo* (e.g., by esterases) to afford the more polar deesterified metabolites which are the true ultimate molecular species involved in interaction with the appropriate receptor(s) at the molecular level.

A number of metabolic and physiological studies (17,25,34,36) strongly suggest that azadirachtin and related compounds act at hormonal (physiological) concentrations rather than at the more usual pharmacological or toxicological concentration levels (36). Furthermore, it has previously been observed that the effects of azadirachtin on susceptible insects (such as the disturbance of hormonal systems giving rise to ecdysis (molting) inhibition) are similar to those produced by the administration of certain plant-derived ecdysteroid analogs (phytoecdysones) such as ponasterone A (Figure 3) (34,44,45). In addition, recent studies with synthetic ecdysone agonists have produced hormonal disturbances and molting cycle (ecdysis) failures (46,47) similar to those observed upon administration of azadirachtin and/or the phytoecdysones (such as ponasterone A). These observations suggest that azadirachtin and related compounds may be acting as ecdysteroid (molting hormone) analogs, thus interfering with insect development by causing profound hormonal disturbances. In this regard, azadirachtin and related compounds may be acting either as ecdysteroid agonists or antagonists (or as a combination of both at various different receptors), either directly or indirectly (e.g., by negative feedback inhibition of hormone biosynthesis or metabolism, or related mechanisms).

A comparison of the chemical structure of the deesterified derivative of azadirachtin with that of the insect molting hormone, ecdysterone (Figure 3), shows several structural similarities, especially with regard to the spatial disposition of the hydroxyl groups at C-3 in the A rings and in the oxygenation pattern in the B and D rings and adjacent moieties in both molecules. It is noteworthy that although the A/B ring junction is *trans* in azadirachtin and related compounds (with the 3-OH in the axial position) and *cis* in the ecdysteroids (with the 3-OH in the equatorial position) (36), the hydroxyl groups of both molecules at position

Azadirachtin skeleton,
deesterified
(1,3-dideacylated)

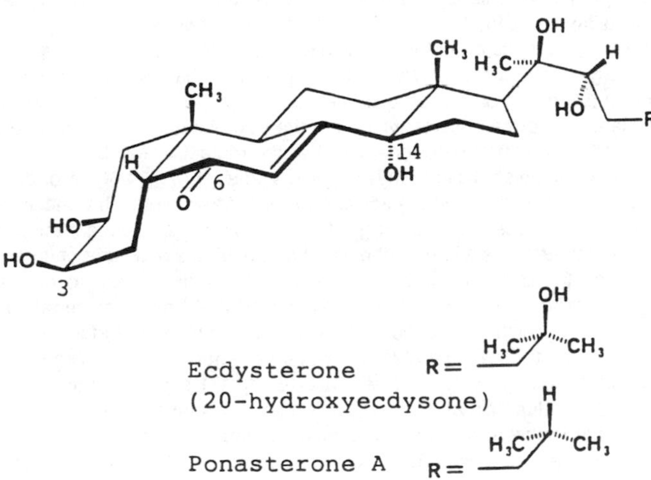

Figure 3. Stereostructural relationships between azadirachtin, ecdysterone (the insect molting hormone), and the phytoecdysone, ponasterone A.

C-3 occupy the same relative spatial relationship with regard to the rest of the structure (skeletal carbon framework) in both types of molecules (Figure 3). Thus, there may be two alternative carbon frameworks which can hold the C-3 hydroxyl oxygens in space in such a way that appropriate receptor interactions can occur. Also noteworthy is the equatorial (or pseudoequatorial) disposition of the oxygen functionalities at C-6 In the B rings and the axial (or pseudoaxial) disposition of the oxygen functionalities at C-14 in the D rings of both molecules.

These observations, both with regard to biological (insecticidal) activities and chemical structures, suggest that considerable portions of the azadirachtin structure may be required in order to elicit the hormonal disturbances previously observed. Rembold has suggested a minimum structure required to elicit the desired biological activities in this class of compounds (insecticidal tetranortriterpenoids) (36). The proposed ecdysteroid-analog mode of action and structure-activity relationships may be useful and important considerations in the design of synthetic analogs of these natural and biorational insecticides.

Acknowledgments

This work was supported in part by a grant awarded to J.A.K. by the U.S. National Science Foundation (PCM-8314500). We thank Ms. Deborah Camomile for typing this manuscript.

Literature Cited

1. Kunkel, G. Flowering Trees in Subtropical Gardens; Dr. W. Junk: The Hague-Boston-London, 1978; p 258.
2. Ahmed, S.; Grainge, M. Econ. Bot. 1986, 40, 201-209.
3. Lewis, W.H.; Elvin-Lewis, M.P.F. Econ. Bot. 1983, 37, 69-70.
4. Ahmed, S.; Bamofleh, S.; Munshi, M. Econ. Bot. 1989, 43, 35-38.
5. Munz, P.A.; Keck, D.D. A California Flora and Supplement; University of California: Berkeley, 1973; p 999.
6. Duffield, M.R.; Jones, W.D. Plants for Dry Climates; H.P. Books: Tucson, 1981; p 118.
7. Taylor, D.A.H. In Chemistry and Chemical Taxonomy of the Rutales; Waterman, P.G.; Grundon, M.F., Eds.; Academic: New York, 1983; pp 353-375.
8. Silva, M.F. das G.F. da; Gottlieb, O.R.; Dreyer, D.L. Biochem. Syst. Ecol. 1984, 12, 299-310.
9. Silva, M.F. das G.F. da; Gottlieb, O.R. Biochem. Syst. Ecol. 1987, 15, 85-103.
10. Taylor, D.A.H. Prog. Chem. Org. Nat. Prod. 1984, 45, 1-102.
11. Kraus, W. In New Trends in Natural Products Chemistry 1986; Atta-ur-Rahman; Le Quesne, P.W., Eds.; Studies in Organic Chemistry, Vol. 26; Elsevier Science: Amsterdam, 1986; pp 237-256.
12. Nadkarni, K.M.; Nadkarni, A.K. Indian Materia Medica, 3rd ed.; Popular Book Depot: Bombay, India, 1954; Vol. 1, pp 776-784.
13. Chopra, R.N.; Nayar, S.L.; Chopra, I.C. Glossary of Indian Medicinal Plants; Council of Scientific and Industrial Research: New Delhi, 1956; pp 31-32, 163-164.
14. Dey, K.L.; Mair, W. The Indigenous Drugs of India, 2nd ed.; Pama Primlane, Chronica Botanica: New Delhi, 1973; pp 186-188.
15. Perry, L.M.; Metzger, J. Medicinal Plants of East and Southeast Asia: Attributed Properties and Uses; Massachusetts Institute of Technology: Cambridge, Massachusetts, 1980; pp 260-262.
16. Kapoor, L.D. Handbook of Ayurvedic Medicinal Plants; CRC: Boca Raton, Florida, 1990; pp 59-60, 225-226.
17. Schmutterer, H. In CRC Handbook of Natural Pesticides, Vol. III, Insect Growth Regulators, Part B; Morgan, E.D.; Mandava, N.B., Eds.; CRC: Boca Raton, Florida, 1987; pp 119-170.
18. Warthen, J.D., Jr.; Morgan, E.D. In CRC Handbook of Natural Pesticides, Vol. VI, Insect Attractants and Repellents; Morgan, E.D.; Mandava, N.B., Eds.; CRC: Boca Raton, Florida, 1990; pp 23-134.

19. Lee, S.M.; Stone, G.A.; Klocke, J.A. 190th American Chemical Society National Meeting, Chicago, Illinois, September 8-13, 1985; American Chemical Society: Washington, DC, 1985; Agrochemicals Division Abstract No. 76.
20. Klocke, J.A. In Allelochemicals: Role in Agriculture and Forestry; Waller, G.R., Ed.; ACS Symposium Series No. 330; American Chemical Society: Washington, DC, 1987; pp 396-415.
21. Balandrin, M.F.; Lee, S.M.; Klocke, J.A. J. Agric. Food Chem. 1988, 36, 1048-1054.
22. Yamasaki, R.B.; Klocke, J.A. J. Agric. Food Chem. 1987, 35, 467-471.
23. Lee, S.M.; Klocke, J.A.; Balandrin, M.F. Tetrahedron Lett. 1987, 28, 3543-3546.
24. Klocke, J.A.; Barnby, M.A.; Yamasaki, R.B.; Lee, S.M.; Balandrin, M.F. Proc. 31st Internat. Congr. Pure Appl. Chem., Section 4, Chemistry and Biotechnology of Biologically Active Substances of Medical and Veterinary Importance, 1987, pp 150-169.
25. Barnby, M.A.; Klocke, J.A.; Darlington, M.V.; Yamasaki, R.B. Entomol. Exp. Appl. 1989, 52, 1-6.
26. Barnby, M.A.; Yamasaki, R.B.; Klocke, J.A. J. Econ. Entomol. 1989, 82, 58-63.
27. Block, E. Scient. Am. 1985, 252(3), 114-119.
28. Taylor, D.A.H. Tetrahedron 1987, 43, 2779-2787.
29. Turner, C.J.; Tempesta, M.S.; Taylor, R.B.; Zagorski, M.G.; Termini, J.S.; Schroeder, D.R.; Nakanishi, K. Tetrahedron 1987, 43, 2789-2803.
30. Bilton, J.N.; Broughton, H.B.; Jones, P.S.; Ley, S.V.; Lidert, Z.; Morgan, E.D.; Rzepa, H.S.; Sheppard, R.N.; Slawin, A.M.Z.; Williams, D.J. Tetrahedron 1987, 43, 2805-2815.
31. Kraus, W.; Bokel, M.; Bruhn, A.; Cramer, R.; Klaiber, I.; Klenk, A.; Nagl, G.; Pöhnl, H.; Sadlo, H.; Vogler, B. Tetrahedron 1987, 43 2817-2830.
32. Jacobson, M. In Natural Pesticides from the Neem Tree (Azadirachta indica A. Juss.); Schmutterer, H.; Ascher, K.R.S.; Rembold, H., Eds.; German Agency for Technical Cooperation: Eschborn, Germany, 1980; pp 33-42.
33. Jongen, W.M.F.; Koeman, J.H. Environ. Mutagen. 1983, 5, 687-694.
34. Kubo, I.; Klocke, J.A. Agric. Biol. Chem. 1982, 46, 1951-1953.
35. Klocke, J.A.; Balandrin, M.F.; Barnby, M.A.; Yamasaki, R.B. In Insecticides of Plant Origin; Arnason, J.T.; Philogène, B.J.R.; Morand, P., Eds.; ACS Symposium Series No. 387; American Chemical Society: Washington, DC, 1989; pp 136-149.
36. Rembold, H. In Insecticides of Plant Origin; Arnason, J.T.; Philogène, B.J.R.; Morand, P., Eds.; ACS Symposium Series No. 387; American Chemical Society: Washington, DC, 1989; pp 150-163.
37. Ruscoe, C.N.E. Nature New Biol. 1972, 236, 159-160.
38. Leuschner, K. Naturwissenschaften 1972, 59, 217-218.
39. Morgan, E.D.; Thornton, M.D. Phytochemistry 1973, 12, 391-392.
40. Morgan, E.D.; Wilson, I.D. In CRC Handbook of Natural Pesticides: Methods, Vol. II, Isolation and Identification; Mandava, N.B., Ed.; CRC: Boca Raton, Florida, 1985; pp 3-81.
41. Lee, S.M.; Klocke, J.A. J. Liq. Chromatogr. 1987, 10, 1151-1163.
42. Yamasaki, R.B.; Klocke, J.A.; Lee, S.M.; Stone, G.A.; Darlington, M.V. J. Chromatogr. 1986, 356, 220-226.
43. Kubo, I.; Matsumoto, A.; Matsumoto, T.; Klocke, J.A. Tetrahedron 1986, 42 489-496.
44. Kubo, I.; Klocke, J.A. In Plant Resistance to Insects; Hedin, P.A., Ed.; ACS Symposium Series No 208; American Chemical Society: Washington, DC, 1983; pp 329-346.
45. Kubo, I.; Klocke, J.A. In Natural Resistance of Plants to Pests: Roles of Allelochemicals; Green, M.B.; Hedin, P.A., Eds.; ACS Symposium Series No. 296; American Chemical Society: Washington, DC, 1986; pp 206-218.
46. Wing, K.D. Science 1988, 241, 467-469.
47. Wing, K.D.; Slawecki, R.A.; Carlson, G.R. Science 1988, 241, 470-472.

RECEIVED September 10, 1990

Chapter 20

Toxicity and Neurotoxic Effects of Monoterpenoids

In Insects and Earthworms

Joel R. Coats, Laura L. Karr[1], and Charles D. Drewes

Department of Entomology and Department of Zoology, Iowa State University, Ames, IA 50011

The insecticidal activity of several monoterpenoids from essential oils was evaluated against insect pests. Toxicity tests illustrated the bioactivity of *d*-limonene, α-terpineol, ß-myrcene, linalool, and pulegone against insects, including the house fly, the German cockroach, the rice weevil, and the western corn rootworm. Bioassays were conducted to assess their toxicity via topical application, fumigation, ingestion, and ovicidal exposures. Growth, reproduction and repellency were also evaluated in the German cockroach. Non-invasive electrophysiological recordings were used with an earthworm to investigate neurotoxic effects of the monoterpenoids. Relevant monoterpenoid bioassay results in the literature are also discussed.

Many essential oils from plants possess biological activity against pests that could be harmful to the plant. Some exhibit acute toxicity, while others demonstrate repellent, antifeedant, or antioviposition effects or inhibition of growth, development or reproduction (1). Many of the fragrant volatile oils contain ten-carbon hydrocarbons, or their related alcohols, ketones, aldehydes, carboxylic acids, and oxides, and are termed monoterpenoids. Most are considered secondary plant chemicals, with little direct metabolic importance, but with considerable coevolutionary significance (2). Plant-insect interactions have been studied for many years, but a better understanding of these complex coadaptive relationships could provide a basis for using plant-derived chemicals in biorational approaches for better management of pest organisms (3). Botanical insecticides such as pyrethrins and rotenone have proven to be both safe and effective in controlling insect pests.

An improved knowledge of the monoterpenoids (as well as sesquiterpenes, diterpenes, triterpenes, tetraterpenes) and their effects on insects contributes to unraveling the intricate interactions that have shaped the coevolution of insects and plants. It also provides leads for possible utility of these safe, degradable compounds in modern pest control, and, as more advanced genetic engineering capabilities develop, the potential for

[1]Current address: DowElanco, Walnut Creek, CA 94598

0097–6156/91/0449–0305$06.00/0
© 1991 American Chemical Society

exploiting their efficacy by transferring genes to different crops or by selecting for the protective chemicals in breeding programs. Figure 1 shows the structures of 2 cyclic and 2 acyclic monoterpenoids.

Knowledge of the spectrum of insecticidal activity is limited for most of the terpenoids. The results of *d*-limonene trials against a wide range of insect groups indicate that this important constituent of citrus oil is toxic to some life stages of some species via some routes of exposure (4,5,6,7). Its utility as a broad-spectrum insecticide, however, does not seem feasible. Spectrum of repellent activity has been evaluated for several types of terpenes, and reproductive effects have been described for some chemicals (1,8).

Mechanisms of acute toxicity have not been elucidated for the monoterpenoids, but the onset of symptoms is usually rapid, manifested as agitation, hyperactivity, and quick knockdown (4,5). Our investigations have also included electrophysiological studies of the toxic effects of monoterpenoids on nerves.

Methods

Topical applications to house flies (*Musca domestica*) and German cockroaches (*Blattella germanica*), fumigation of German cockroaches and rice weevils (*Sitophilus oryzae*) and repellency to German cockroaches were conducted by methods described previously (4). The toxicity by ingestion and effects on growth and reproduction were evaluated by incorporation of the chemicals into the ground cat chow diet of the German cockroaches (9). Lavicidal and ovicidal activity against the western corn rootworm (*Diabrotica vergifera vergifera*) were tested in petri dishes of soil and on moist blotter paper, respectively (4).

Repellency of terpenoids was evaluated with German cockroaches in choice tests, using pairs of plastic boxes (9 X 8.5 X 2cm) connected by plastic tubing. Treated filter paper was placed in one box, and a control (acetone treated) filter paper was put in the other box (4). Hedgeapple, bay leaves, and spearmint chewing gum were tested using the weight of products (in μg) per unit of volume of the box (in cm^3) for determination of exposure concentrations ($1\ \mu g/cm^3 = 1$ ppm).

Neurotoxicity was assessed in the earthworm *Eisenia fetida* using non-invasive electrophysiological techniques described previously (10). The earthworms were exposed in a vapor/contact method in which a small volume of terpenoid was delivered onto a moist filter paper in a vial which was then closed tightly (11). Periodically, worms were placed on an etched circuitboard recording grid for electrophysiological testing, and giant fiber activity associated with the escape response was monitored (12). The method has been used previously to examine sublethal effects of pesticides on earthworm neural activity (13).

Bioassay responses were calculated as LD_{50}'s, ED_{50}'s or ET_{50}'s using the trimmed Spearman-Karber method (14). Duncan's multiple range test and analysis of variance were used for the repellency trials, and the paired comparison t-test was used to analyze the food preference experiments.

Results and Discussion

I. Acute Toxicity. The utility of an insecticide has often been judged by its immediate and acute actions on pest species of insects. Terpenoids can effect toxicity symptoms very rapidly via contact or vapor exposures, including hyperactivity and tremors. Their degree of potency is significantly less than conventional synthetic organic insecticides, often by orders of magnitude. However, their actions can be very effective under circumstances that allow brief high-concentration uses of generally safe

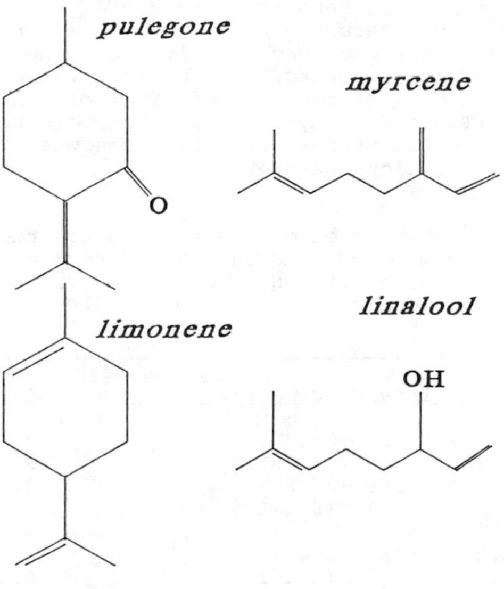

pulegone

myrcene

O

limonene

linalool

OH

Fig. 1 - Structures of 4 monoterpenoids

chemicals (e.g., greenhouses, animal shampoos and dips, fumigations).

Research has been limited to a few insect species and a few terpenoids, but the results of the topical treatments, fumigations, and repellency studies indicate that the monoterpenoids can exert substantial toxicity alone, or with a synergist, and demonstrate considerable repellent activity as well. Very little is known about their mode of action.

Topical Exposure. Dosing of female house fly females with five monoterpenoids yielded toxic effects when applied alone at high doses. *d*-Limonene was the most active of the five (Table I). Use of the synergist piperonyl butoxide enhanced the activity of *d*-limonene, pulegone, and linalool considerably, by 17, 21, and >14 fold, respectively. These results indicate that those three terpenoids' insecticidal activity is expressed more fully when the oxidative detoxification process is inhibited. It is not surprising that flies can detoxify them rapidly, considering the relatively simple hydrocarbon structures of the monoterpenoids.

In the male German cockroach trials, pulegone was shown to be the most potent compound tested, but it required 260 μg/insect at the median lethal dose. *d*-Limonene and linalool also demonstrated slight toxicity (Table I). Myrcene, α-terpineol and *l*-limonene (not shown) were the least toxic. *d*-Limonene was slightly synergized, while linalool was not.

Table I. Acute toxicity of monoterpenoids to the house fly, *Musca domestica*, and the German cockroach, *Blattella germanica*, by topical application (24-h mortality data), alone or with piperonyl butoxide (p.b.)

Treatment	LD_{50}(μg/insect)[a] female house fly	male German cockroach
d-Limonene	90(70-130)	700(610-810)
1 *d*-Limonene:5 p.b.	5.2(3.7-7.5)	300(180-560)
Pulegone	166(131-201)	260(230-300)
1 Pulegone: 5 p.b.	7.7(5.4-11)	
Linalool	>100	550(410-730)
1 Linalool: 1 p.b.	32(23-44)	
1 Linalool: 5 p.b.	7.2(4.1-12)	610(530-700)
Myrcene	360(300-430)	>1,580
α-Terpineol	310(260-380)	1,070(680-1,690)

[a]95% confidence intervals

Contact toxicity trials with *d*-limonene against adult cat fleas illustrated extremely fast knockdown time and mortality. A synergistic ratio of 3.2 was observed when piperonyl butoxide was also used on treated filter papers (5).

Southern pine beetles, *Dendroctonus frontalis*, were assayed for susceptibility to 16 terpenoids from pine oleoresin (15). The LD_{50} of *d*-limonene was 0.47 μg/insect, while *l*-limonene was slightly less toxic (0.55 μg/insect), as was myrcene (0.62 μg/insect). The most toxic chemical tested was limonene dioxide (0.24 μg/insect), indicating that oxidation of limonene may be an activation process (15).

Dose-mortality studies with constituents of lyophilized lemon oil demonstrated notable toxicity to the adult cowpea weevil, *Callosobruchus maculatus*. The LD_{50} of the oil was 16.4 μg/insect, while three potent compounds isolated from the oil by thin-layer chromatography had LD_{50}'s from 4.73 to 2.66 μg/insect (16).

Nematicidal activity of some essential oils and their major constituents has been reported (17). Eugenol showed the most efficacy against 3 species of nematodes, while linalool, geraniol, and menthol were more effective against the root-knot nematode.

Fumigation. The monoterpenoids, as a family, are volatile which makes them potentially quite useful as fumigants. Most are also pleasantly odiferous and of low toxicity to mammals, properties which also are consistent with fumigation usages on produce, grain, clothing, buildings, ships and soil. Laboratory trials against adult rice weevils illustrate the relative potencies of five monoterpenoids in Table II.

Table II. Acute toxicity of monoterpenoids to the rice weevil (*Sitophilus oryzae*) and the German cockroach (*Blattella germanica*) by fumigation (24-h mortality data)

Treatment	LC_{50} (ppm)[a,b] rice weevil	German cockroach[c]
d-Limonene	19(13-27)	23(17-31)
Pulegone	3.1(2.7-3.5)	9.6(1-113)
		♀ 17(10-30)
		♂ 4.5(0.6-36)
Linalool	14	12
Myrcene	>100	>100
α-Terpineol	>100	>100

[a]95% confidence intervals
[b]mg/liter
[c]both sexes included (50:50) except where noted

The most toxic of the five in a vapor form was pulegone with an LC_{50} of 3.1 ppm (mg/liter of air). Linalool and *d*-limonene were also effective, while myrcene and α-terpineol were ineffective.

Against German cockroaches, pulegone was, again, the most efficacious, followed by linalool and limonene (Table II). Male cockroaches were four times as susceptible as the females to pulegone.

Vapor-exposure assays for adult cat fleas also showed that *d*-limonene was effective at inducing rapid knockdown and mortality as a fumigant (5). The larvae were also relatively susceptible to *d*-limonene vapors, while the eggs were less susceptible and pupae were relatively tolerant of this chemical.

Vapors of monoterpenes from pine were evaluated for bioactivity against the Western pine beetle, *Dendroctonus brevicomis*. The four-day bioassay determined that limonene was the most toxic among the five compounds tested (7).

Larvicidal and Ovicidal Activity. Acute toxicity testing of terpenoids on immature stages of insects has been limited to a few studies. Larvae (third instar) of the western corn rootworm, *Diabrotica vergifera vergifera*, were assayed in treated soil. The 48-h LC_{50} for *d*-limonene was 12.2 μg/g soil (4), which is approximately 10-fold less potent than the standard organophosphorus and carbamate compounds currently used for rootworm control (19). The eggs of that species were exposed to treated blotter paper to assess contact ovicidal activity. The LC_{50}'s at 28 days for *d*-limonene and linalool were 1.8% active ingredient (a.i.) and 0.26% a.i., respectively. These 2 monoterpenoids were about 3 orders of magnitude less potent than chlordimeform (LC_{50} = 0.0006% a.i.).

Citrus oils and several individual components of them were tested against larvae of the Caribbean fruit fly, *Anastrepha suspensa* (6). Citral was the most efficaceous monoterpenoid, followed by limonene, then α-pinene and α-terpineol. One-hour immersion in a 40% (a.i.) solution of the most potent compound resulted in 50% mortality during the larval development period. The eggs of the Caribbean fruit fly have also been evaluated for ovicidal effects of terpenoids(6). α-Terpineol was the most effective of those tested (1% a.i. caused 100% mortality), followed by citral and limonene. α-Pinene had no effect.

Studies on the house fly, *Musca domestica*, revealed the considerable acute larvicidal activity of several terpenoids, especially carvacrol and *d*-limonene (19). Several metamorphosis inhibition effects were also observed, including inhibition of pupal ecdysis, uneclosed pupae, and deformed adults. Numerous compounds exhibited potency in these developmental mortalities: camphene, carvacrol, carvone, cineole, citral, citronellal, citronellol, eugenol, farnesol, geraniol, limonene, linalool, ß-phellandrene, and α-pinene. Egg hatch was also inhibited by exposure to the terpenoids; the most effective ovicidal compounds were carvacrol, citronellal and ß-phellandrene, with no acutely toxic effects noted to the embryos, but rather show inhibition of development of the embryo and inability to eclose from the egg (19).

The larvae of the cat flea, *Ctenocephalides felis*, were susceptible to *d*-limonene (LD_{50} of 226 $\mu g/cm^2$ of filter paper), and some modest synergism was observed when piperonyl butoxide was employed (LC_{50} of 157 $\mu g/cm^2$). The vapors were also toxic to the flea larvae. Pupae were less susceptible than eggs, larvae, or adults. Eggs of the cat flea were exposed to *d*-limonene at 65 $\mu g/cm^2$ and mortality was 100%. A test of its toxicity to them in a vapor chamber showed 60% mortality at 130 $\mu g/cm^2$ (5). It is evident that some terpenoids possess efficacy against insect eggs, but the degree of potency is modest at best.

It was reported that larvae of 3 species of mosquitoes were susceptible to *d*-limonene but no data were provided (5).

Repellency. A number of other plant-derived terpenoids have been demonstrated to be repellent to various insects. Many compounds in this class have been proven to be attractants to certain insects (1). Limonene at 1 or 10 mg/box repelled the cockroaches, to the untreated box, in significantly ($p<0.05$) greater numbers than the 0.001 mg rate or the controls (4). Pulegone and linalool at the 10 mg/box rate repelled significantly more individuals than the 0.0001 mg rate or the controls. Myrcene and α-terpineol did not demonstrate any repellency. Natural pyrethrins were significantly different from the controls at the 0.1 mg/box rate.

Three other substances were bioassayed for repellency based on their purported household usefulness: hedgeapple, bay leaves, and spearmint chewing gum. Figure 2 shows that at the highest exposure rate, the hedgeapple and bay leaves effectively repelled a high percentage of the cockroaches, but were considerably less potent than pyrethrins.

A recent study of repellency of some terpenoids and phenolics showed thymol and carvacrol were more effective for deterring oviposition by female *Aedes aegypti* mosquitoes than N,N-diethyl-m-toluamide (DEET). Eucalyptol (1,8-cineole) and p-cymene were considerably less repellent than DEET (20).

House fly attractant and deterrent properties have been exhibited by numerous terpenoids. The *d* isomer of limonene was repellent while the *l*-limonene was an attractant. Carvone was attractive at low concentrations and repellent at high levels (19). A similar pattern has also been observed in German cockroaches for low and high concentrations of *d*-limonene (4).

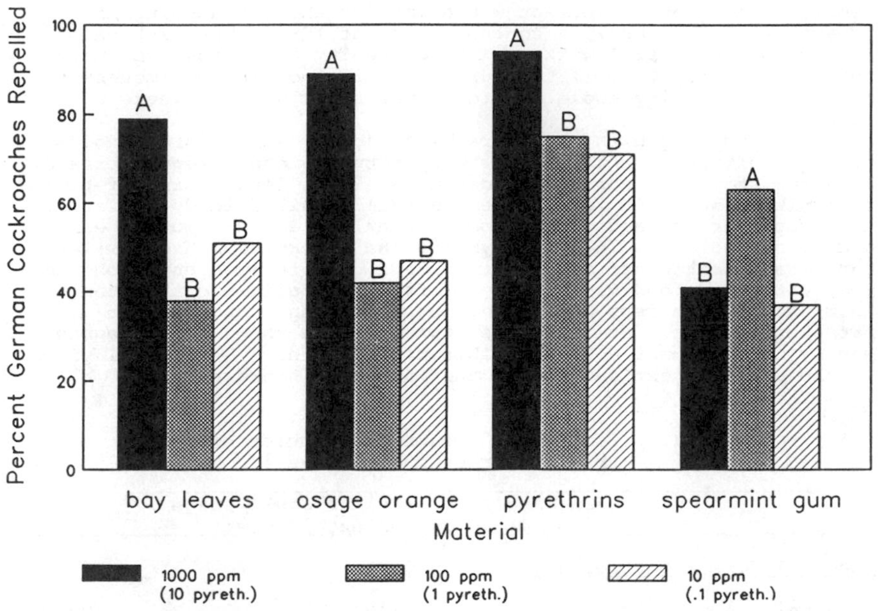

Fig. 2 - Repellency of hedgeapple, bay leaves, spearmint gum, and pyrethrins to German cockroaches.

Citrus oil components were evaluated for attractive or repellent activity toward the red scale, *Aonidiella aurantii*. Most hydrocarbon terpenes were attractive, while the alcohols, including linalool, were repellent. One aldehyde, citral, was attractive, while the other, citronellal, was repellent (21).

Antifeedant properties of the herb tansy have been observed in the Colorado potato beetle, *Leptinotarsa decemlineata*, along with some larval mortality (22). The major active constituents of oil of tansy include the monoterpenoids borneol, camphor, and thujone (23).

Effects on Insect Growth and Development. Chronic feeding studies were conducted with German cockroaches to determine if monoterpenoids have any influence on their survival or growth rate. Cohorts of newly eclosed nymphs were used in experiments that assessed the time of development to adult. Mortality was recorded weekly. The greatest effect was noted for those consuming diet treated with pulegone. A concentration of 10% in the diet caused 23% mortality after 1 week and 30% mortality at 2 weeks, compared to 50% mortality from 1% triflumuron. Linalool and *d*-limonene caused no mortality above control rates after up to 4 weeks of treatment.

Effects on growth rate were recorded for all monoterpenoids tested. Although mortality was only noted in the pulegone trials, linalool at a high dose (25%) caused a slight lengthening of time to adult. The other four terpenoids all resulted in dramatically more rapid growth and development. Linalool and *d*-limonene effects are presented in Table III. Myrcene and α-terpineol had similar but less dramatic effects (9). A feeding preference study showed that untreated food was greatly preferred, confirming that the increased growth rate was not due to a feeding attractant or arrestant effect. At 1% active ingredient in the diet, pulegone was by far the most active inhibitor of feeding. A hormonal effect is one possible explanation of the increase in growth rate.

Table III. Effects of two monoterpenoids on growth rate of German cockroaches

	Mean days from hatch to adult stage (% active ingredient in diet)			
Chemical/dose	Control(0%)	0.1%	1%	10%
Linalool	131	116	120	108
d-Limonene	144	131	123	113

Earlier work investigated growth effects from monoterpenoids dosed topically on last-instar larvae of the house fly (19). Development and metamorphosis failures occurred with several compounds. Cineole was the most active inhibitor of pupal eclosion (40% inhibition at a dose of 10 μg/insect) and of imaginal differentiation, within those uneclosed pupae. Camphene, eugenol, and α-pinene were also active. Inhibition of pupal ecdysis was highest in insects treated with camphene, although carvone and several others were also effective.

Effects of pulegone on the survival, growth and development of armyworm larvae have been investigated by Brattsten (1). Growth of the fall armyworm was markedly inhibited, and considerable mortality was noted at 0.1% in the diet.

Effects on Insect Reproduction. The reproductive cycle in insects is a complex process, often sensitive to perturbation by low doses of toxicants or hormones. In light of distinct differences between insect and mammalian reproductive systems and endocrine controls, it should be possible to exploit those differences for biorational control of insect pests. Few investigations have been conducted on the effects of monoterpenoids on insect reproductive function or success.

The German cockroach was not very susceptible to disruption of reproduction by d-limonene, linalool, myrcene, and α-terpineol (9). Exposure of males or females to lifetime feeding, to topical application, or to vapor exposure caused no effect to minimal effects on reproduction. However, topical treatment of their oöthecae resulted in considerable embryotoxicity. d-Limonene was the most toxic of the group, with linalool also affecting the development and survival of the cockroach embryos (Table IV).

Table IV. Embryotoxic effect of monoterpenoids in German cockroach following topical treatment of the oötheca

Chemical	Dose(μg)	Mean % of oöthecae producing offspring
d-Limonene	0	90.0
	210	73.3
	420	70.0
	840	16.7
Linalool	0	90.0
	215	86.7
	430	90.0
	860	70.0
Myrcene	0	90.0
	198	80.0
	395	63.3
	790	46.7
α-Terpineol	0	90.0
	235	70.0
	470	53.3
	940	80.0

Studies on the southern armyworm, *Spodoptera eridania*, proved that pulegone fed to larvae was effective in reducing the number of eggs laid by the adults (1). Housefly reproduction was affected deleteriously by several monoterpenoids. Those most potent at inhibiting egg hatch following adults' exposure to treated surfaces included linalool, carvacrol, ß-phellandrene, and citronellal. *d*-Limonene was moderately active and *l*-limonene was only slightly effective (19). A study of 31 essential plant oils on reproduction inhibition was conducted in the rice weevil. Many of the oils were active in reducing population reproduction, although 2 oils acted to stimulate growth and increase populations (24). The involvement of monoterpenoids in the juvenile hormone synthesis pathway and the role of juvenile hormone in mating and vitellogenesis support the reproductive impact of these terpenoids on insects.

Neurotoxic Effects. Symptoms of acute poisoning of insects by monoterpenoids are similar to those effected by some neurotoxic compounds. Cockroaches and house flies both exhibited overt hyperactivity, loss of coordination, and tremors. In cat fleas, trembling and paralysis of legs were also noted, followed by convulsions and death (5). Our attempts to study the neurotoxic actions of *d*-limonene in American cockroach nerve preparations yielded somewhat erratic results, apparently due to difficulties of delivering and maintaining a specific concentration of this highly volatile and saline-insoluble chemical to the nerve. The non-invasive neuroassay for earthworms had been utilized successfully to characterize and quantify neurotoxic effects of insecticides (10), and it offered advantages that could provide a superior technique for studying the neurotoxicity of monoterpenoids.

Results from short-term contact exposure of earthworms, *Eisenia fetida*, to various terpenoids are presented in Table V. The methods of exposure and recording have been presented elsewhere (11). Neurophysiological symptoms that were detected after 30 min. include: (a) decreased velocity of impulse conduction in medial (MGF) and lateral (LGF) giant nerve fibers along the worm (Figure 3), (b) decreased sensitivity to a touch stimulus, as indicated by difficulty or failure in evoking giant nerve fiber spikes in response to a light tactile stimulus, and (c) decreased amplitude or absence of the muscle electrical response that normally accompanies each MGF spike. A rebounding of action potentials was also observed in the MGF for some of the chemicals.

These neurophysiologcal symptoms were also accompanied by clear-cut behavioral and morphological symptoms (Figure 4) which included various combinations of clitellar swelling, ataxia and general limpness of the body. The neurotoxic, behavioral, and morphological effects were fully reversible at sublethal concentrations. These results suggest that the terpenoids, as a group, are neuroactive as indicated by adverse electrophysiological effects on earthworm escape reflex pathways, as well as impaired postural or locomotory function. However, the data are as yet insufficient to determine either the exact site(s) of terpenoid action or whether these compounds share a common mode of action.

Table V. Sublethal effect of short-term contact exposure to 5 μl of monoterpenoids in the earthworm *Eisenia fetida*. (c = control; *d*-lim = *d*-limonene; pul = pulegone; lin = linalool; myr = myrcene)

SYMPTOM 0 = absent + = present	TREATMENT (30 min exposure) c	 *d*-lim	 pul	 lin	 myr
1. Decreased MGF velocity[a]	0	+	+	+	0
2. Decreased LGF velocity[a]	0	0	+	+	+
3. Decreased sensitivity to touch stimulus	0	0	0	+	0
4. Loss of MGF-mediated muscle potential	0	+	+	+	+
5. Clitellar swelling	0	+	0	0	+
6. Ataxia	0	+	+	+	0
7. Limpness of body	0	0	+	+	0

[a]Velocity is < 90% of pretreatment velocity

Conclusions

It is clear that a wide range of monoterpenoids possess some degree of insecticidal acitivity. The insecticidal role of most will be limited to specialty uses. Currently, limonene, linalool, and citronellal occupy places in the competitive and dynamic pesticide market. The most promising potential for exploitation of these molecules may lie in the synthesis of derivatives and analogs through directed synthesis (25). The biological effects summarized here, i.e., the repellency, acute toxicity, fumigant activity, reproductive toxicity, and neurotoxicity reflect the wide spectrum of activities possible, each caused by an interaction of monoterpenoid at an active site in the insect. Mode-of-action studies on neurotoxicity and growth and reproductive effects are necessary to elucidate the specific toxicological bases for the

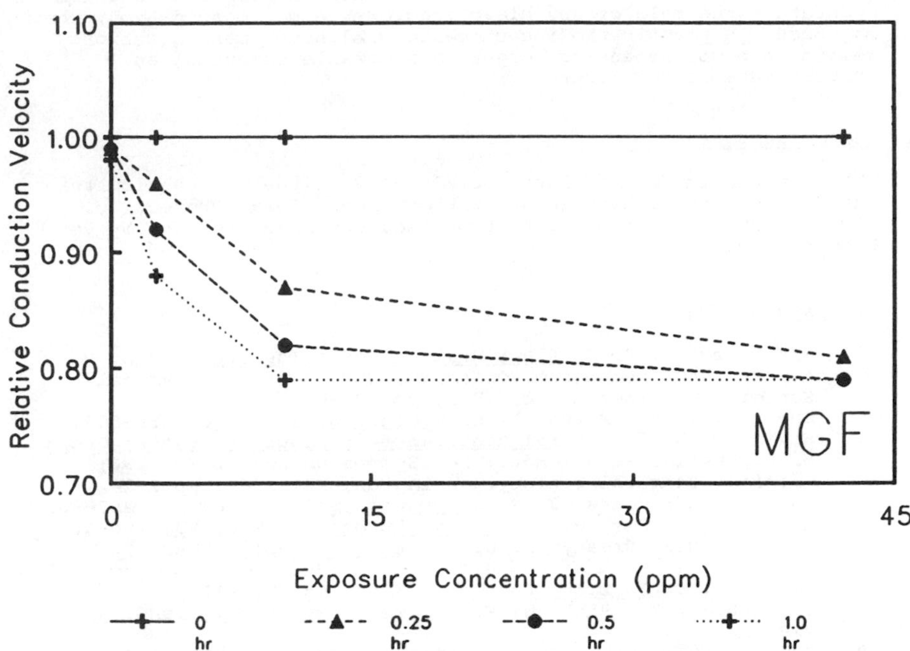

Fig. 3 - Effects of *d*-limonene concentration and exposure duration on relative conduction velocity in MGF (reproduced from 11, by permission, copyright Academic Press, 1990).

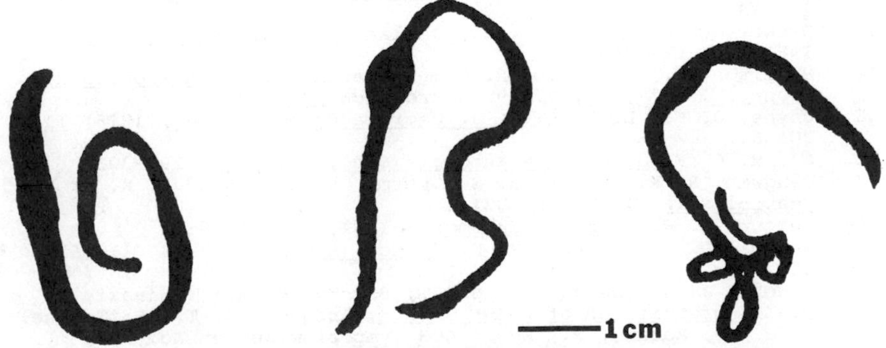

Fig. 4 - Silhouettes, from video recordings of normal untreated earthworm (left), worm at 30 min postexposure to 6.3 ppm *d*-limonene (center), and worm at 5 min postexposure to 100 ppm *d*-limonene (right).

insecticidal activity. Quantitative structure-activity relationships must be derived to optimize effectively those toxicant-receptor interactions. Metabolism studies are required to understand the role of any bioactivations, e.g., epoxidation, that may occur *in vivo* in the insect pests. Although many questions remain, the monterpenoids posses considerable potential as insecticides of the future.

Acknowledgment

This chapter is Journal Paper No. J-14132 of the Iowa Agriculture and Home Economics Experiment Station, Ames, Iowa; Project No. 2306. Partial funding for this project was also provided by Pet Chemical Inc.

Literature Cited

1. Brattsten, L. B. In *Plant Resistance to Insects*; Hedin, P. A., Ed.; ACS Symposium Series No. 208; American Chemical Society: Washington, DC, 1983; pp 173-95.
2. Ehrlich, P. R.; Raven, P. H. *Evolution*, 1964, *18*, 586-608.
3. Hedin, P. A. In *Natural Resistance of Plants to Pests*; Green, M. B.; Hedin, P. A., Eds.; ACS Symposium Series No. 296; American Chemical Society:Washington, DC, 1986; pp 2-14.
4. Karr, L. L.; Coats, J. R. *J. Pestic. Sci.*, 1988, *13*, 287-90.
5. Hink, W. F.; Fee, B. J. *J. Med. Entomol.*, 1986, *23*, 400-4.
6. Styer, S. C.; Greany, P. D. *Environ. Entomol.*, 1983, *12*, 1606-8.
7. Smith, H. S. *J. Econ. Entomol.*, 1965, *58*, 509-10.
8. Collart, M. G.; Hink, W. F. *Entomol. Exp. Appl.*, 1986, *42*, 225-30.
9. Karr, L. L.; Coats, J. R. *J. Econ. Entomol.*, 1990, (in press).
10. Drewes, C. D.; Vining, E. P.; Callahan, C. A. *Environ. Toxicol. Chem.*, 1984, *3*, 599-607.
11. Karr, L. L.; Drewes, C. D.; Coats, J. R. *Pestic. Biochem. Physiol.*, 1990, *36*, 175-86.
12. Drewes, C. D.; Vining, E. P. *Pestic. Biochem. Physiol.*, 1984, *22*, 93-103.
13. Drewes, C. D.; Zoran, M. J.; Callahan, C. A. *Pestic. Sci.*, 1987, *19*, 197-208.
14. Hamilton, M. A.; Russo, R. C.; Thurston, R. V. *Environ. Sci. Technol.*, 1977, *11*, 714-9. Correction *12*, 417.
15. Coyne, J. F.; Lott, L. H. *J. Georgia Entomol. Soc.*, 1976, *11*, 301-5.
16. Su, H. C. F. *J. Georgia Entomol. Soc.*, 1976, *11*, 297-301.
17. Sangwan, N. K.; Verma, B. S.; Verma, K. K.; Dhindsa, K. S. *Pestic. Sci.*, 1989, *28*, 331-5.
18. Sutter, G. R. *J. Econ. Entomol.*, 1982, *75*, 489-91.
19. Sharma, R. N.; Saxena, K. N. *J. Med. Entomol.*, 1974, *11*, 617-621.
20. Klocke, J. A.; Balandrin, M. F.; Barnby, M. A.; Yamasaki, R. B. In *Insecticides of Plant Origin*; Arnason, J. T.; Philogene, B. J. R.; Morand, P., Eds.: ACS Symposium Series No. 387; American Chemical Society: Washington, DC, 1989; pp 136-49.
21. Salama, H. S.; Saleh, M. *Z. Ang. Ent.*, 1984, *97*, 393-8.
22. Hough-Goldstein, J. A. *Environ. Entomol.*, 1990, *19*, 234-8.
23. Shearer, W. R. *J. Nat. Prod.*, 1984, *47*, 964-9.
24. Singh, D.; Siddiqui, M. S.; Sharma, S. *J. Econ. Entomol.*, 1989, *82*, 727-33.
25. McLaren, J. S. *Pestic. Sci.*, 1986, *17*, 559-78.

RECEIVED September 19, 1990

Chapter 21

Novel Fish Toxins of the Cadinane Sesquiterpene Class

From the Philippine Mangrove Plant *Heritiera littoralis*

D. Howard Miles[1], Vallapa Chittawong[1], Armando A. de la Cruz[2], Allen Matthew Payne[1], and Edgardo Gomez[3]

[1]Department of Chemistry, University of Central Florida, Orlando, FL 32816
[2]Department of Biological Science, Mississippi State University, Mississippi State, MS 39762
[3]Marine Science Center, University of the Philippines, Quezon City, Philippines

The Philippine mangrove plant *Heritiera littoralis* (Steraliaceae) has been used as a fish and spearhead poison. Crude extracts of this plant were shown to be toxic to the fish *Tilapia nilotica*. Four novel sesquiterpenes of the cadinane type with an unusual oxygenation pattern and an aromatic ring have been previously reported. A fifth novel sesquiterpene of this type which was assigned the name Vallapianin has been identified. The structure elucidation of Vallapianin is reported.

A search for biodegradable compounds which protect crops is of high priority to the world wide agricultural community. Naturally occurring substances are known to be biode-gradable. Therefore there is a renewed emphasis upon obtainment of biologically active substances from natural sources such as plants.

Plants are known to produce compounds such as rotenone (I) which have been useful as pesticides. However rotenone and the plants from which it is derived were first known for their toxicity to fish. Thus fish toxicity is a useful measure of possible pesticidal activity and the potential of the plant to provide crop protection.

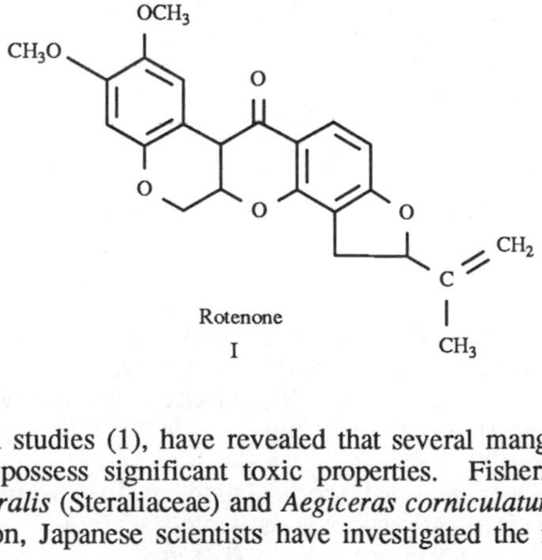

Rotenone
I

Ethnobotanical studies (1), have revealed that several mangrove species which grow in Southeast Asia possess significant toxic properties. Fisherman have used leachates from *Heritiera littoralis* (Steraliaceae) and *Aegiceras corniculatum* (Aegicerataceae) as fish poisons. In addition, Japanese scientists have investigated the ichthyotoxic properties of

0097–6156/91/0449–0317$06.00/0
© 1991 American Chemical Society

Excoecaria agallocha (Euphorbiaceae) (2). In Australia, *Derris trifoliata* stupified fish in shallow water (3). Investigators have also examined the ethanolic extracts of *D. urucu* and *D. nicou*, which contained rotenone as the main constituent. The hexane extracts of *D. sericea* and *D. araripensis* (4) were also examined. While the hexane extracts contained no rotenone, they were as toxic to fish as those extracts containing rotenone. Several scientists have recently reiterated the presence of such toxic substances in five mangrove plants, namely *H. littoralis*, *D. trifoliata*, *E. agallocha*, *A. corniculatus*, and *A. floridium* (5). The chemistry of the mangrove plant *H. littoralis* has been recently investigated (6-10). *H. littoralis* (11) is a moderate-sized, evergreen seashore tree. The flowers are axillary and hang in yellowish tassels. The leaves are brown, curvy, and 6-25 mm long. The bark is pinkish grey, smooth, and becomes fissured and flaky on older trees. The apex of the leaf is rounded or slightly heart shaped. The fruit is purplish brown, measuring 4-7 cm long. The young twigs are brown and curvy.

Previous studies of this plant considered that the salinity is one of the major factors influencing the vegetation of the mangrove swamps. Accordingly observations were undertaken of distributions of a number of mangrove species in the tidal rivers in Northern Queensland (12). Free amino acids, total methylated onium compound (TMOC) and total nitrogen were investigated in young and old leaves of this plant from Northern Queensland, Australia, by Popp (13-15). Inorganic ions and organic acids were also found in *H. littoralis*, which is regarded as a brackish water species. Leaf age did not appear to effect Na^+ and Cl^- storage. Low molecular weight carbohydrates occurring in both young and old leaves of this species were identified using gas chromatography.

In this chemical investigation[1] extracts of *H. littoralis* obtained by the preliminary fractionation shown in Figure 1 were bioassayed with the quick screening test (16). A 100% mortality during the first 24 hours was used as the criterion for activity. Using this criterion activity was detected in the hexane, chloroform, and aqueous extracts of the roots. This toxicity prompted an investigation of the chemical constituents of *H. littoralis*.

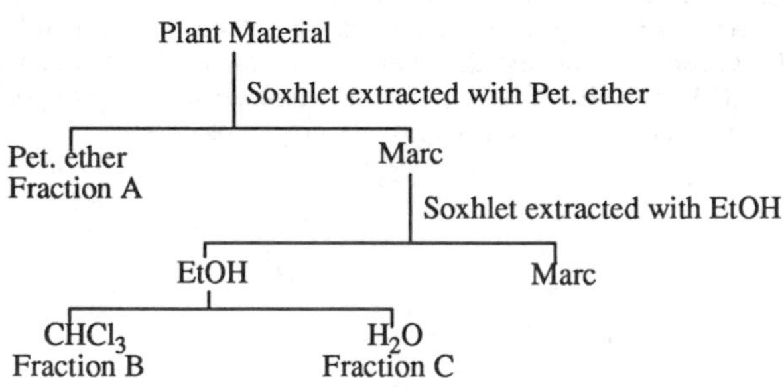

Figure 1. Fractionation Scheme for *Heritiera littoralis*

In 1985, Lho (6,7) reported the isolation from the chloroform extract of a novel cadinane sesquiterpene with an unusual oxygenation pattern and an aromatic ring. This compound was assigned the name heritol (II). Heritol demonstrated ichthyotoxicity (7) to *Tilapia nilotica* fingerlings (25-35 mm length, 0.05-0.25 g dry weight) at a concentration of 20 ppm (90 min).

Miles (8) et al. later reported the isolation of two additional novel cadinane sesquiterpenes with the same basic skeleton as heritol (I) from the hexane extract. These compounds were assigned the names heritonin (III) and heritianin (IV).

Heritianin (IV) showed toxicity (9) to fish (*T. nilotica*) with a total mortality of 2 hr at a dose of 100 ppm, while heritonin required 12 hr for total mortality at the same dose level.

In 1987, Chittawong (10) reported, from the aqueous extract, the isolation and identification of a fourth novel cadinane sesquiterpene with the same basic skeleton as

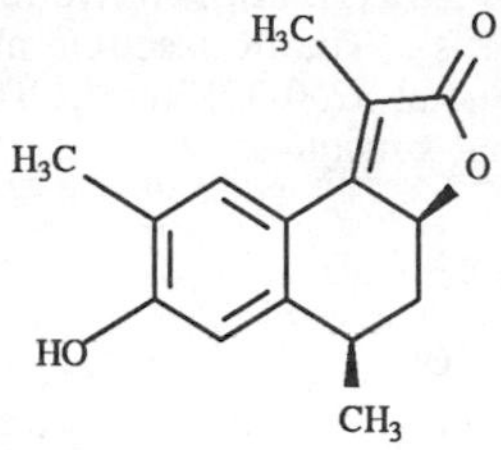

Heritol
II

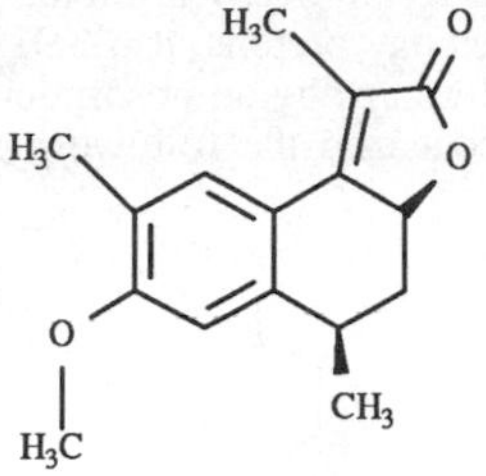

Heritonin
III

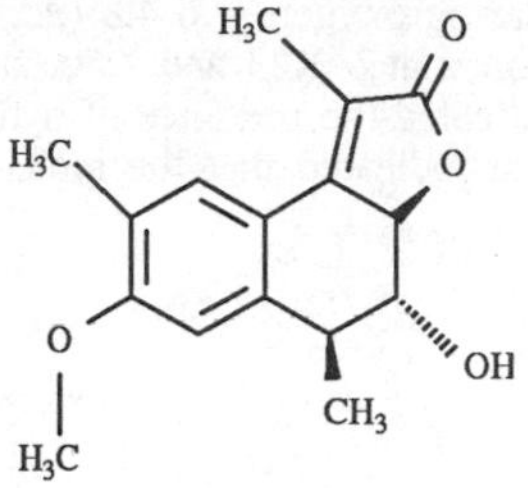

Heritianin
IV

heritol (I). This compound was assigned the name vallapin (V). Vallapin showed 80% inhibition of feeding against the cotton boll weevil (*Anthonomus grandis*) by the Hedin method (17).

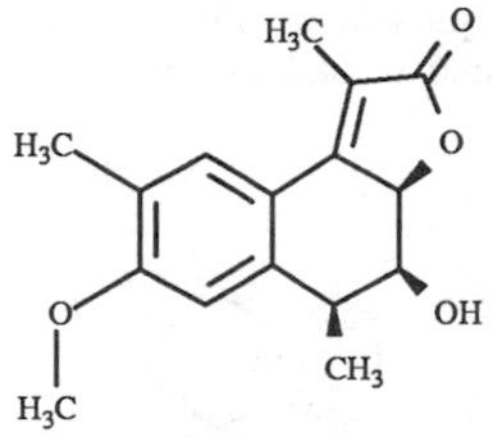

Vallapin
V

Materials and Methods

The roots of *H. littoralis* were collected from the Mangrove Forest Reserve in Pagbilao, Quezon, in the Philippines, voucher number 80987. Twenty-one kilograms of chopped air dried roots were extracted with hexane to yield 98.2 g of crude hexane extract. The marc was then extracted with 95% EtOH to yield 524 g of crude extract. The ethanol extract was partitioned between $CHCl_3$:H_2O to yield 38 g of $CHCl_3$ fraction, 70 g of water extract, and 400 g of insoluble. The insoluble fraction was washed with $CHCl_3$:EtOH (3:1), $CHCl_3$:EtOH (1:1), and $CHCl_3$:EtOH (1:3).

Twenty grams of the $CHCl_3$:EtOH (3:1) fraction of the roots of *H. littoralis* was chromatographed on an open column with 400 g of silica gel. The column was eluted with a hexane-chloroform-methanol solvent system.

Results and Discussion

Compound VI was isolated as a white powder from the 20% methanol-chloroform fraction, mp 182° C. A molecular formula of $C_{16}H_{18}O_5$ was determined by high resolution mass spectrometry (M^+ m/e 290.115). This formula indicated eight degrees of unsaturation. The presence of aromaticity was indicated by IR bands at 1600 cm^{-1} and 1490 cm^{-1}. IR bands at 1750 cm^{-1} and 1640 cm^{-1} indicated the presence of an α,β-unsaturated-γ-lactone. The aromatic nature of compound VI was confirmed by the ^1H NMR spectrum, which showed resonances at δ 6.89 (1H, s) and 7.58 (1H, s) for two isolated protons on an aromatic ring.

A further study of the ^1H NMR spectrum provided evidence of two non-equivalent methyl resonances at δ 1.55 (3H, d, J=7 Hz) and 2.15 (3H, s). A singlet at δ 2.15 indicated that this methyl group was attached to a quatenary carbon. The doublet at δ 1.55 indicated that this methyl group was attached to a methine carbon. The ^1H NMR also showed resonances for methylene protons at δ 4.71 (2H, s), a proton on a carbon bearing oxygen at δ 4.8 (dd, J=1.8 Hz), a benzylic proton at δ 3.0 (1H, m), two hydroxyl groups at δ 1.23 and 1.60, and methoxy protons at δ 3.91 (3H, s). The IR spectrum also indicated the presence of hydroxyl groups by an absorption band at 3250-3350 cm^{-1}. This data indicated that the molecule contained the following partial structures:

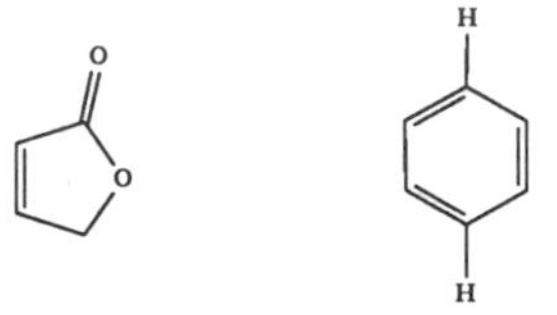

2 X $-CH_3$

2 X $-OH$

$-CH_2$

$-OCH_3$

a b c

Partial structures a and b account for seven degrees of unsaturation, which suggests that an additional ring might be present in order to obtain an unsaturation number of 8. Combination of the fragment a and b leads to the basic skeleton shown below.

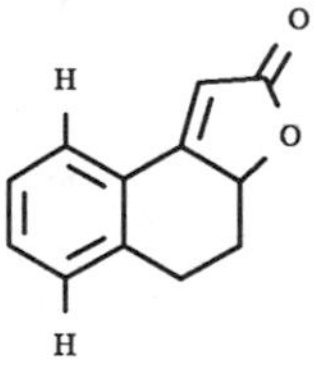

Following the isoprene rule, the structure is composed of the three isoprene units which are combined head-to-tail as a cadinane skeleton. Accordingly, compound VI should have the basic skeleton shown below.

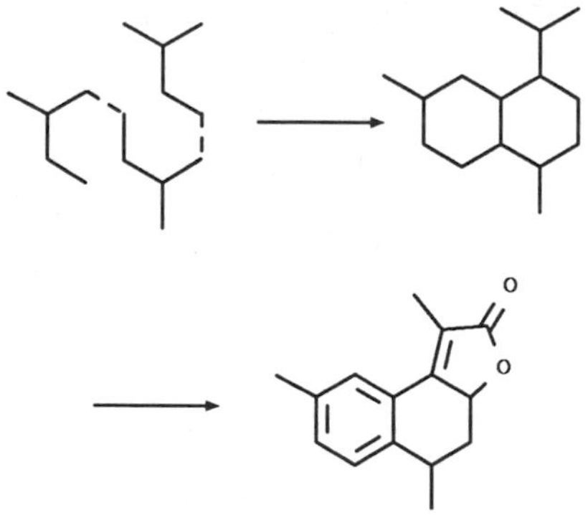

Additional justification of this assignment was that the IR spectrum was identical with heritianin except for the presence of a much larger band for a hydroxy group. The mass spectrum fragmentation pattern of compound VI also resembled that of heritianin. The presence of peaks at m/e 259, 141, 128, 115, 91, and 77 indicated that the strucutres of compound VI and heritianin were similar. The ^{1}H NMR spectrum of compound VI was identical with that of heritianin (Table I) except for the absence of a methyl group at δ 2.25 and the addition of a methylene group at δ 4.71. The -OH group was placed on C-14 since the methyl group at δ in heritianin (IV) was not present. The structure of compound VI was therefore assigned as the novel structure shown below.

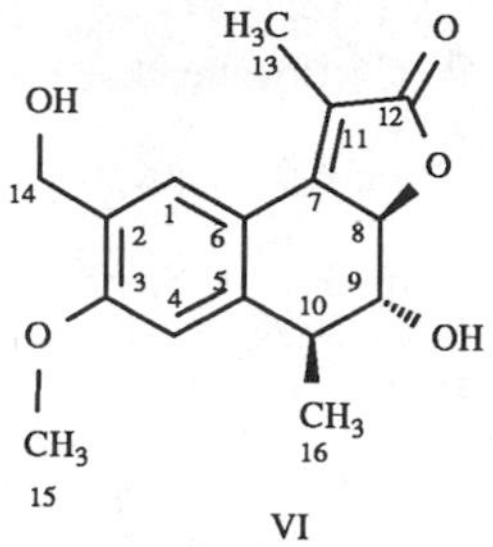

VI

Table I. Comparison of ^{1}H NMR of Heritianin (IV) and Compound VI

Proton	Chemical Shift	
	Heritianin (IV)	Compound VI
1	6.85 (s)	6.89 (s)
4	7.40 (s)	7.58 (s)
8	4.80 (dd, J = 1,8)	4.80 (dd, J = 1,8)
9	3.49 (d, J = 7)	3.49 (t, J = 7)
10	3.01 (m)	3.01 (m)
13	2.15 (s)	2.15 (s)
15	3.90 (s)	3.91 (s)
16	1.55 (d, J = 7)	1.55 (d, J = 7)
14 (OH)		1.60 (s)

Conclusions

There is a great need for new biodegradable agrochemicals which could be compatible with the environment. Toxic compounds such as vallapianin (VI), heritol (II), heritonin (III), heritianin (IV), and vallapin (V) have potential as natural pesticides. Plants from tropical regions of the world offer particularly intriguing possibilities in this regard since they are subjected to severe disease and insect pressure. This is especially true for mangrove plants because of their proximity to water. Vallapianin (VI), like the others isolated from this plant is of special interest since it occurs in a mangrove plant and possesses a novel sesquiterpene structure containing an α,β-unsaturated-γ-lactone.

Acknowledgments

This study was supported by the Division for International Programs, National Science Foundation (INT - 8202732). We are indebted to Drs. Catherine Costello and Thomas Dorsey of the MIT Mass Spectrometry Facility for their invaluable assistance in obtaining the mass spectrum. We thank Dr. Jim Dechter and Ms. Vicki Farr of the University of Alabama nmr facility for their help in obtaining ^{1}H nmr data.

Literature Cited

1. de la Cruz, A. A.; Chapatwalla, K. D.; and Miles, D. H. *Life Sciences*, **1981**, *29*, 1997.
2. Kawasima, T.; Tahahaski, T.; Insue, Y.; Koclama, M.; and Ito, S. *Phytochemistry*, **1971**, *18*, 3308.
3. de la Cruz, A. A. *The Function of Mangroves in Mangrove and Estaurine Vegetation in South Asia*, Biotrop. Spec. Publ. No. 10, Bogor, Indonesia, 1979.
4. Mars, W. B.; Donascimento, M. C.; Dovalle, J. R.; Arogao J. A. *Crime Cult*, **1973**, *25*, 647.
5. de la Cruz, A. A.; Miles, D. H.; and Chapatwala, K. D. *Life Sciences*, **1985**, *30*, 1805.
6. Lho, D. S. Ph. D. Dissertation, Mississippi State University, 1985.
7. Miles, D. H.; Lho, P. S.; de la Cruz, A. A.; Gomes, E. D.; Weeks, J. A.; and Atwood J. L. *J. Org. Chem.* **1987**, *52*, 2930.
8. Miles, D. H.; de la Cruz, A. A.; Ly, A. M.; Lho, D. S.; Gomez, E. D.; Weeks, J. A.; Atwood, J. *Toxicants from Mangrove Plants Ichthyotoxins from the Philippine Plant* Heritiera Littoralis, *Allelochemicals: Role in Agriculture and Forestry*; Walker, G. R., Ed.; ACS Symposium Series 300 1985.
9. Ly, A. M. Ph.D. Dissertation, Mississippi State University, 1985.
10. Chittawong, V. Ph.D. Dissertation, Mississippi State University, 1987.
11. Whitmore, T. C. Ed. *Tree Flora of Malaysia*, Forest Department, Ministry of Primary Industries: Malaysic, 1973, Vol II, pp. 361-363.
12. Bunt, J. S.; Williams W. T.; and Clay H. T. *Aust. J. Bot.*, **1984**, *30*, 401.
13. Popp, M. *Z. Pflonzenphysiol. Bd.*, **1984**, *113*, 395.
14. Popp, M. *Z. Pflanzenphysiol. Bd.*, **1984**, *113*, 411.
15. Popp, M.; Larker, F.; and Wergel P. Z. *Pflanzenphysiol. Bd.*, **1984**, *114*, 15.
16. de la Cruz, A. A.; Gomez, E. D.; Miles, D. H.; Casipi, G. J. B.; Chavez, V. B. *Int. J. Ecol. Environ. Sci.*, **1985**.
17. Hedin, P. A.; Thompson, A. C.; Minyard, J. P. *J. Econ. Etomol.* **1966**, 59, 181.

RECEIVED May 16, 1990

PHYTOALEXINS AND PHOTOTOXINS IN PLANT PEST CONTROL

Chapter 22

Binding a Phytoalexin Elicitor to DNA

A Model Approach

Francis H. Witham[1] and David L. Gustine[2]

[1]Department of Horticulture, Pennsylvania State University, University Park, PA 16802

[2]U.S. Regional Pasture Research Laboratory, Agricultural Research Service, U.S. Department of Agriculture, University Park, PA 16802

Computer models of an active elicitor of phytoalexin production, hexa (ß-D-glucopyranosyl)-D-glucitol, and six inactive oligosaccharide analogs were superimposed separately on a model of base paired nucleotides constructed according to parameters consistent with the classical Watson-Crick B-DNA. The elicitor-DNA complex showed the alignment of rotated residues of the elicitor parallel to the base pairs with specific hydroxyl groups in position for hydrogen bonding at opposite phosphate oxygens on complementary DNA strands. Elevation of the active oligosaccharide above base pairs, due to the B-1,6- and B-1,3-linkages, and partial intercalation of the glucitol residue was especially evident with space-filling models. A comparison of the "fit" of the model elicitor to nucleotides with the "nonfits" of the inactive oligosaccharides may have relevance to an understanding of elicitor-induced phytoalexin production *in vivo* and to the design of new and nonobvious molecules which exhibit elicitor activity.

Disease resistance in plants may be due in part to the action of chemicals (elicitors) that are released from a pathogen and induce the host to produce a substance(s) that operates as an antibiotic(s) on the invading organism. Since specific disease resistance chemicals are induced (*de novo* production) in the host cells, it is possible that elicitors bind to DNA and through binding induce template modifications which influence gene product formation for the development of disease resistance in plants.

Sharp et al. (1) reported that the partial hydrolysis of mycelial walls of *Phytophthora megasperma* f. sp. *glycinea* produced a mixture of soluble substances containing elicitors of phytoalexin production in soybean cotyledons. Reduction of a hepta glucoside fraction with $NaBH_4$, followed by purification and structural characterization, showed the presence of one active and seven

0097–6156/91/0449–0324$06.00/0
© 1991 American Chemical Society

chemically similar though inactive oligosaccharides which were all hexa (ß-D-glucopyranosyl)-D-glucitols (1-2). The active compound and six of the inactive hexa (ß-D-glucopyranosyl)-D-glucitols consist of ß-1,6- and ß-1,3-linked glucopyranosyl residues with a glucitol terminus. The other inactive oligosaccharide consists of ß-1,4-linked glucopyranosyl residues and glucitol terminus (2). In addition, Ossowski and colleagues (3) synthesized a hexa (ß-D-glucopyranosyl)-D-glucitol which was identical in structure and biological activity to the mycelial wall-derived elicitor (4).

Another synthesized and chemically similar compound, octa-ß-D-glucopyranoside, was only about one third as active as the cell wall-derived elicitor and the active synthetic hexa (ß-D-glucopyranosyl)-D-glucitol (4). The structure-activity relationship of the elicitor in contrast to that of the inactive oligosaccharide analogs (1-2-3-4) as well as reports of elicitor-regulated induction of protein (5) or other products necessary for the expression of disease resistance, prompted us to construct and evaluate computer and Corey, Pauling, Koltun (CPK) space-filling models of the hexa (ß-D-glucopyranosyl)-D-glucitols. The glucitol derivatives, although produced from the reducing conditions during extraction of the cell wall components, were studied because previous work relating to elicitor activity involved the use of the reduction products (1-2). Also, the spatial features of glucitol would be very similar to those of a ß-D-glucopyranoside reducing terminal.

According to genetic theory, selective gene expression among plants and animals and indeed among different cells within an organism is a controlling feature of differentiation and development. Different cell types, with their array of structural and functional proteins and widely different specific enzyme and receptor sites, require differential DNA transcription as an important mechanism of gene expression. Hadwiger (6) has proposed that transcriptionally poised genes within loop structures of nuclear DNA can be induced directly by elicitors released from the plant pathogen.

Thus, models of the active hexa (ß-D-glucopyranosyl)-D-glucitol and inactive analogs were bound through simulated interactions to model DNA nucleotides in a fashion similar to that previously reported in studies of small molecule interactions with DNA (7-8-9).

Methods

Computer-simulated chemical models of compounds characterized by Sharp et al. (1-2) were constructed according to the parameters governing bond lengths and angles of the software program, Chem 3D Plus (Version 2.0, Cambridge Scientific Computing, Inc., Cambridge, MA). The Chem 3D software program was used in a Macintosh II computer (Apple Computer, Inc.) with hard disk and five megabytes of memory. The ChemDraw program (version 2.01, Cambridge Scientific Computing, Inc.) was used to illustrate the general arrangement of the inactive compounds and for labelling purposes. In all computer models, carbon atoms are shaded and oxygen atoms are stippled, unless otherwise indicated. Also, for purposes of clarity, rectification (hydrogen atom placement) of the computer structures is not shown. Reproduction of the screen image for this and all other computer models was accomplished with a Laser Writer II NTX (Apple Computer, Inc.).

CPK space-filling atoms were obtained from Schwarz/Mann (Cambridge, MA). The scale of each constructed model is 1.25 cm per A. Accuracy for each of the following is: bond angles ± 0.30′; covalent radii ± 0.01 A; van der Wall's radii ± 0.03 A, with respect to nominal values chosen for each species. Models were constructed according to known information and conventions and consistent with the parameters of the CPK atom design.

Structural Features of the Active and Inactive Oligosaccharides

As reported by Sharp and colleagues (1-2), the general structure of the active elicitor (V) and arrangement of six of the seven elicitor-inactive hexa (ß-D-glucopyranosyl)-D-glucitols (II through IV and VI through VIII), show significant structural similarities (Figure 1). The elicitor activity of oligosaccharide V is likely related to the relative placement of the ß-1,6-linked termini, (residue A to the second residue B and the glucitol (E) to the fourth residue D) and branched residues, F and G, which are ß-1,3-linked to the glucosyl residues, B and D, respectively. Conversely, the inactive oligosaccharides exhibit residue branches (equivalent to F and G of the elicitor molecule) located at different sites of the molecule.

Configuration of the Model Elicitor

A Chem 3D computer molecular model illustrates the conformation of the active elicitor, oligosaccharide V, after rotation of each terminal and branched residue approximately 90 degrees about its respective y-axes (Figure 2). In this conformation, the rotated residues (A and F linked to residue B; glucitol (E) and residue G linked to residue D) are aligned perpendicularly to the ß-D-glucopyranosides B, C and D. Further, the intermolecular distance between the oxygen of C6 of residue A and the oxygen of C6 of residue F, place the hydroxyls in position for hydrogen bonding to double-stranded DNA at each of the phosphate oxygens on opposite strands. Similarly, the distance of the C1 oxygen of the rotated glucitol from the oxygen of the C6 of residue G is approximately equal to the distance between phosphate oxygens of double-stranded DNA. For purposes of discussion, the arrangement of two oligosaccharide residues (for example, residues A and F) that are rotated in this manner and which exhibit a collective internuclear distance approximating the distance from one phosphate to another of double-stranded DNA is referred to as an "anchor."

Interaction of Computer and CPK Models of the Elicitor and DNA Nucleotides

A computer model of the active hexa (ß-D-glucopyranosyl)-D-glucitol with rotated residues was superimposed on a model of three base paired nucleotides constructed according to parameters consistent

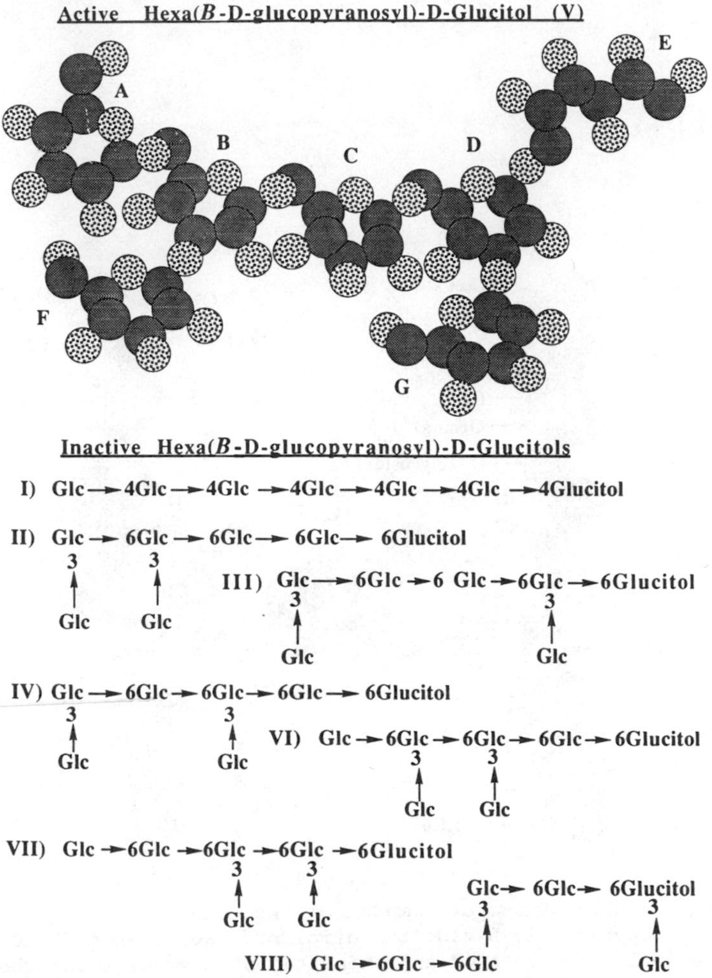

Active Hexa(*B*-D-glucopyranosyl)-D-Glucitol (V)

Inactive Hexa(*B*-D-glucopyranosyl)-D-Glucitols

I) Glc→4Glc→4Glc→4Glc→4Glc→4Glc→4Glucitol

II) Glc→6Glc→6Glc→6Glc→6Glucitol
3↑Glc (on first Glc→6Glc), 3↑Glc (on second)

III) Glc→6Glc→6 Glc→6Glc→6Glucitol
3↑Glc (on first Glc), 3↑Glc (on fourth)

IV) Glc→6Glc→6Glc→6Glc→6Glucitol
3↑Glc (on first Glc), 3↑Glc (on third)

VI) Glc→6Glc→6Glc→6Glc→6Glucitol
3↑Glc (on second), 3↑Glc (on third)

VII) Glc→6Glc→6Glc→6Glc→6Glucitol
3↑Glc (on third), 3↑Glc (on fourth)

VIII) Glc→6Glc→6Glc
3↑
Glc→6Glc→6Glucitol
3↑Glc

Figure 1. Chem 3D computer model of active hexa (ß-D-glucopyranosyl)-D-glucitol (flat structure). Letters indicate individual glucopyranosyl residues (A through G and glucitol, E). With the exception of the ring oxygens and glycoside linkages all other oxygens are hydroxyl oxygens. Residues A through E show ß-1,6-linkages. Branched residues F and G are linked ß-1,3- to residues B and D, respectively.

General arrangement of inactive hexa (ß-D-glucopyranosyl)-D-glucitols. Arrows indicate ß-1,4-; ß-1,6-; and ß-1,3-linkages between D-glucopyranosyl and glucitol residues. The primary differences among oligosaccharides, II through VIII are the positions of the ß-1,3-linked glucopyranosyl (branched) residues in each molecule. Glucopyranosyl residues of the inactive oligosaccharides are referred to in the text according to the letter system indicated for the elicitor (A through E of the chain and branched residues from left to right, G and F).

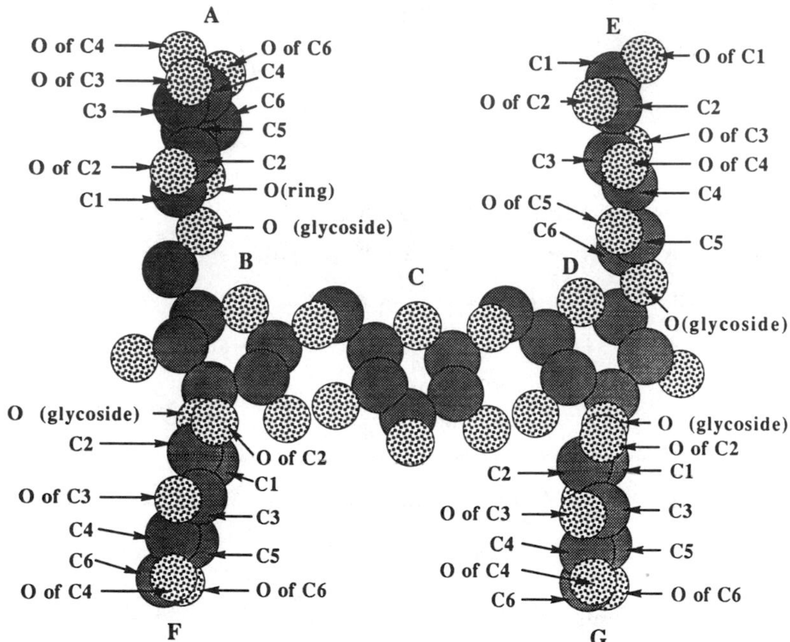

Figure 2. Chem 3D computer model of the hexa (ß-D-glucopyranosyl)-D-glucitol elicitor. Residues A, F, G and glucitol (E) were rotated approximately 90 degrees about their Y axes. The internuclear distance from the C6 oxygen of A to the C6 oxygen of F is referred to as an anchor (see text). Similarly, E and G constitute an anchor. Although not shown in the model, the C2, C3 and C4 atoms with their respective hydroxyl oxygens of residue C would be oriented up and above the plane of the C2 and C4 oxygens of residues B and D, respectively in order to avoid steric hindrances and a less stable molecular conformation.

with the classical Watson-Crick B-DNA (Figure 3). The nucleotide sequence (5'-dTdTdT-3') was selected because the nonrotated portions of the backbone of the elicitor molecule could be positioned over the adenine bases of the antiparallel strand without interference encountered by the bulky guanine or methyls of thymine. Other nucleotide combinations, however, might be appropriate. The elicitor model with shaded atoms bound to model DNA nucleotides (Figure 3, bottom) illustrates more clearly the general characteristics of the elicitor-DNA complex:

1. The first anchor (residues A and F, respectively) was positioned with possible hydrogen bonding, via the C6 Hydroxyls, to the phosphate oxygens of each strand (top left and bottom left). The internuclear distance between the C6 hydroxyls approximated the distance between the phosphate oxygens of each strand.

2. The second anchor (consisting of glucitol, top right and the glucopyranosyl branch (G), bottom right) is two base pairs removed (left to right) from the first anchor. It also showed an approximate internuclear distance equivalent to the distance across DNA strands with alignment that would be along the cavity between two base pairings. In this alignment, hydrogen bonding to the phosphate oxygens is via the C1 hydroxyl of the glucitol and the C6 hydroxyl of the branched residue (G). The flat portion of the elicitor molecule consisting of three glucosyl residues (B, C, and D) connecting one anchor to the other is perpendicular to the base pair alignment of the DNA nucleotides.

As in the computer molecules, a CPK space-filling model of the active elicitor, with rotated terminals and branched residues, is positioned for binding to DNA (Figure 4). Each anchor, oriented parallel to the base pairs, occupies a distance that is approximately the same as the distance across DNA strands from one phosphate to the other. Further, the anchors are bound to the phosphate oxygens of each strand via hydrogen bonding of the hydroxyls of C6 of each rotated residue (A and F) and the hydroxyls of the other anchor (located at C1 of the glucitol (E) and C6 of the residue (G). In contrast to the computer models, elevation of portions of the oligosaccharide above the base pairs due to the ß-1,6- and ß-1,3-linkages and partial intercalation of the rotated glucitol residue (bottom right) is only evident with CPK space-filling models.

Positioning of Active and Inactive Oligosaccharides to Phosphates of DNA

Hydrogen bonding, according to the pattern observed, appears to be important for the binding of the active molecule to DNA. It should be expected, therefore, that inactive molecules would lack the necessary chemistry and/or geometry for optimum hydrogen bonding. In this regard, a comparison of the "fit" of the model elicitor (V) with the model inactive molecules (ß-1,4 linked compound not included) with rotated residues positioned in relation

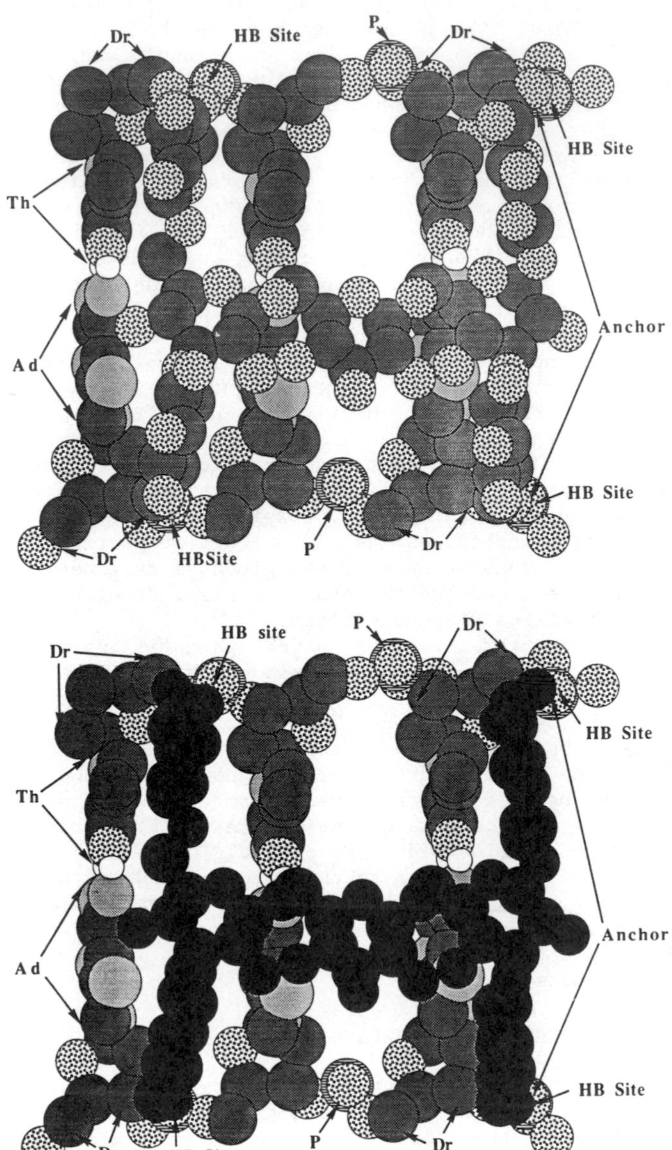

Figure 3. Computer model of the elicitor with rotated residues forming anchors bound to a model of DNA nucleotides (top). The abbreviations are: Ad, adenine; Dr, deoxyribose; HB, hydrogen bonding site; P, phosphorus; and Th, thymine. The top strand of the segment of model DNA is oriented 3'dTdTdT-5', left to right and the bottom strand orientation is 5'-dAdAdA-3', left to right.

Computer model (bottom) of elicitor (all atoms darkened) bound to a model of DNA nucleotides.

Figure 4. CPK space-filling model of the elicitor bound to model DNA. The DNA is oriented 5′dGdAdAdA-3′, left and top to bottom and 3′-dCdTdTdT-5′, right and top to bottom. The arrows left to right (top) indicate hydrogen bonding of the C6 hydroxyls of residues F and A, respectively to phosphate oxygens of DNA. The bottom arrows, from left to right, indicate the hydrogen bonding of the C6 and C1 hydroxyls of residue G and glucitol, respectively to the phosphate oxygens of DNA.

The glucitol (bottom right) is partially intercalated while the other portions of the elicitor are elevated above the nucleotides.

to the phosphates found at fixed distances in double-stranded DNA reveals striking differences in hydrogen bonding characteristics (Figure 5).

Each oligosaccharide molecule was constructed separately with the Chem 3D program according to parameters governing bond lengths and angles (see methods). Rotations of terminals and branched residues were 90 degrees about their Y axes similar to the elicitor molecule (see Figure 2). The phosphates were copied, with distances maintained, from previously constructed computer DNA models and transferred to ChemDraw. The oligosaccharides were transferred from Chem 3D to ChemDraw and superimposed on a set of phosphates so that initially the C6 oxygen of residue A would be in position to facilitate hydrogen bonding to the first phosphate of a given set. The positioning of the remaining portion of each molecule with respect to the other phosphates then depended upon its individual geometry.

The inactive though chemically similar compounds do not conform to the "fit" described for the elicitor (Figure 5). The binding of oligosaccharide II to model DNA, for example, shows that the upper portion of the first anchor (rotated residues A and F) extends considerably beyond the phosphate, although hydrogen bonding to the phosphate oxygen through the hydroxyl of C6 of residue A is possible. However, the lower portion of the same anchor, consisting of the rotated second branch residue which is ß-1,3-linked to residue B, is in alignment with the phosphate oxygen of the bottom strand (Figure 5). Although not shown, CPK models of the same complex show that residue F (ß-1,3-linked to A) extends above as well as beyond the DNA strand. In addition, the rotated glucitol residue (E) binds to one phosphate oxygen, but unlike the active elicitor molecule, there is no second complete anchor due to the absence of an appropriate residue.

The other models of the inactive hexa (ß-D-glucopyranosyl)-D-glucitols show similar "nonfits" by the absence of a complete anchor and/or extension of an oligosaccharide above and beyond the phosphates of the model DNA structure (II, III, IV, VIII). The extension of an anchor in itself, however, may result in partial but not complete loss of elicitor activity as inferred by Sharp et al. (9) from the structure-activity relationship of the octa (ß-D-glucopyranoside). In addition, nonalignment to a phospate might result in a steric hindrance between a branched residue and a nucleotide base (oligosaccharides IV, VI, and VII). Thus, the residue positioning of the different inactive oligosaccharides with respect to that of the phosphates of DNA may result in partial or inappropriate binding for activity.

Discussion

The construction of computer and CPK space-filling molecular models is based on established physical and chemical parameters. While models cannot substitute for direct experimental evidence, they can be used to describe rather successfully structures of biochemicals, molecular interactions, reaction products and many aspects of molecular dynamics. Models may be particularly helpful in depicting the chemistry of active sites of enzymes, antibodies and receptors. Accordingly, the interaction of models described

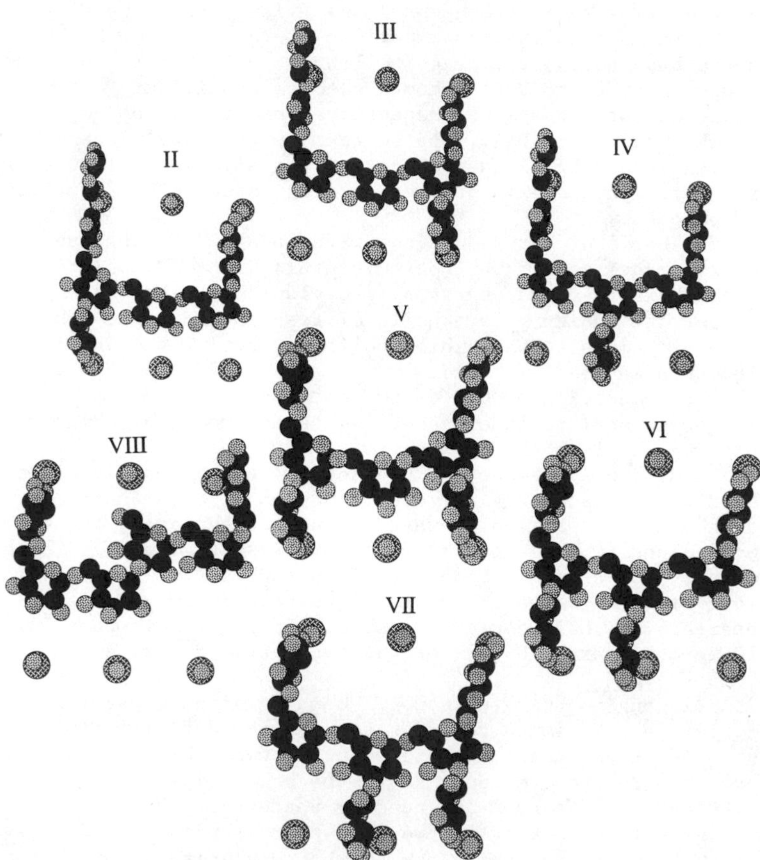

Figure 5. The active elicitor (V) and six other but inactive hexa (ß-D-glucopyranosyl)-D-glucitols (II, III, IV, VI, VII, and VIII). Oligosaccharides are each shown in relation to six phosphates (three representing a DNA strand) as they would be present at prescribed distances (to scale) in a model of DNA. All models show rotated terminals and branched residues similar to the molecular model of the elicitor.

herein strongly suggests that the binding of an elicitor (at least one chemical type) of disease resistance to DNA is chemically feasible and highly probable. In fact, the proposed binding of the elicitor, hexa (ß-D-glucopyranosyl)-D-glucitol to DNA, based on the rotation of four residues to form two anchors with internuclear distances equal to that from one DNA strand to the other, the establishment of four attachment sites (via hydrogen bonding) along two DNA strands, and the accommodation of the elicitor backbone above the base pairs is very compelling. In contrast, the inactive compounds (nonfits to DNA), though chemically similar to the active molecule, do not possess the necessary geometry for optimum binding (hydrogen bonding) to DNA. The suggestion that the placement of the terminals and branched residues of the elicitor molecule is important for phytoalexin production (10) is supported by this model approach.

On the basis of our deductions with models, the observed structural requirements for optimum interaction of the active elicitor, hexa (ß-D-glucopyranosyl)-D-glucitol, with DNA may be useful for the design of synthetic oligosaccharides and other nonobvious chemicals that exhibit elicitor activity. These requirements may be:

1. Accommodation by DNA of two anchors (rotated ß-D-glucopyranosyl residues according to the model of compound V), oriented parallel to the base pairs.
2. A distance between the two anchors approximately equivalent to the distance necessary for accommodation of each anchor between and/or above base pairs plus that distance occupied by the elicitor backbone along two base pairs of the DNA (see Figure 3). Larger compounds, however, which consist of multiples (1x, 2x, etc.) of the same distance between anchors of Compound V on a longer oligosaccharide chain could be active. In fact, this possibility should be considered in work on the synthesis of active analogs of compound V.
3. Binding to DNA via hydrogen bonding of at least four appropriately positioned residues for optimum activity.
4. Elevation of the nonrotated portion of the oligosaccharide backbone and rotated residues above the base pairs (in this case a result of the ß-1,6-and ß-1,3-bonding pattern).
5. Effective masking of two base pair sequences and three cavities (between base pairs) by the bound elicitor.

ß-glucan elicitors are thought to be released from fungal cell walls via the action of endo ß-glucanases (10). Initially, elicitors may bind to plasma membrane receptor proteins (11-12) and may be carried or diffuse into the cytoplasm of the host cell. By unknown mediators, elicitors may move to the nuclear membrane and then into the nucleoplasm where they could be positioned along nucleotides by a specific protein(s) or bind directly to DNA. Another possibility is that elicitor molecules enter the nucleus directly in the absence of any mediators. Passage of the elicitor-active hexa (ß-D-glucopyranosyl)-D-glucitol through the nuclear membrane is possible since pores in the nuclear envelope are greater than 100 angstroms (13), a size that is sufficiently larger than the elicitor molecule.

The specificity of binding of an elicitor as well as proteins and many other biologically active molecules to DNA (at specific nucleotide sequences, major groove or other location) may depend upon the nature of the receptor and carrier, pH, extent of coiling and hydration of the DNA and numerous other factors, which must be studied experimentally. Nevertheless, in view of the apriori genetic rationale concerning the selective expression of genes and the flow of biological information in a cell (i.e., DNA to RNA to protein to other regulatory chemicals), it is possible that the oligosaccharide elicitor(s) of disease resistance and other biologically active molecules that induce the formation of gene products cause template modifications (frame shifts, etc.) after binding to DNA.

Acknowledgments

We thank John H. Pazur and Charles D. Boyer for their helpful suggestions in preparation of the manuscript and Charles W. Heuser for his assistance with the illustrations. This research was supported by the Department of Horticulture and the Pennsylvania Experiment Station of The Pennsylvania State University.

Literature Cited

1. Sharp, J. K.; Valent, B.; Albersheim, P. J. Biol. Chem. 1984, 259, 11312.
2. Sharp, J. K.; McNeil, M.; Albersheim, P. J. Biol. Chem. 1984, 259, 11321.
3. Ossowski, P.; Pilotti, A.; Garegg, P. J.; Lindberg, B. J. Biol. Chem. 1984, 259, 11337.
4. Sharp, J. K.; Albersheim, P.; Ossowski, P.; Pilotti, A.; Garegg, P.; Lindberg, B. J. Biol. Chem. 1984, 259, 11341.
5. Lamb, C. J.; Lawton, M. A.; Dron, M.; Dixon, R. A. Cell 1989, 56, 215.
6. Hadwiger, L. A. Phytopathology 1988, 78, 1009.
7. Hendry, L. B.; Witham, F. H.; Chapman, O. L. Perspect. Biol. Med. 1977, 21, 120.
8. Hendry, L. B.; Witham, F. H. Perspect. Biol. Med. 1979, 22, 333.
9. Witham, F. H.; Hendry, L. B.; Chapman, O. L. Origins Life 1978, 9, 7.
10. Mauch, F.; Staehelin, L. A. The Plant Cell 1989, 1, 447.
11. Schmidt, W. E.; Ebel, J. Proc. Nat. Acad. Sci. U.S.A. 1987, 84, 4117.
12. Yoshikawa, M.; Keen, N. T.; Wang, M. Plant Physiol. 1983, 73, 497.
13. Newport, J. W.; Forbes, D. J. Annu. Rev. Biochem. 1987, 56, 535.

RECEIVED May 2, 1990

Chapter 23

Desoxyhemigossypol, A Cotton Phytoalexin

Structure–Activity Relationship

R. D. Stipanovic, M. E. Mace, M. H. Elissalde, and A. A. Bell

Southern Crops Research Laboratory and Veterinary and Toxicology Research Laboratory, Agricultural Research Service, U.S. Department of Agriculture, Route 5, Box 805, College Station, TX 77840

Desoxyhemigossypol (dHG), a phytoalexin produced by cotton in response to infection, has been shown to be: highly toxic to the pathogen, *Verticillium dahliae*; soluble in water at the pH of infected xylem vessels at concentrations higher than those required to kill conidia and mycelia; and present *in vivo* at the site of infection and in contact with the mycelia. However, the structure-activity relationship of this compound is unknown. Recent evidence suggests dHG readily forms a free radical by autoxidation, which can be initiated by trace quantities of transition metals such as iron. Reducing agents that stabilize dHG have been shown to reduce significantly its toxicity to *V. dahliae*. Thus, the dHG free radical is implicated as an essential element in the mechanism of cytotoxicity. Phytoalexins, such as the pterocarpans, are postulated to operate by a similar mechanism.

Verticillium dahliae is a serious and frequently devastating plant pathogenic fungus that attacks many important agricultural crops (1). In the U.S. it is the major pathogen affecting cotton, being most troublesome in areas under irrigated cultivation with cool temperatures (mean < 28°C) during the growing season (2).

The fungus attacks the plant through the root system, where it penetrates 1-2 mm from the root tip and grows toward the vascular system (3). Hyphae in the xylem vessels produce spores that are transported upward in the xylem fluids and eventually infect the foliar plant parts (4). To prevent this systemic infection, the plant must seal the infected vessels and kill the fungus.

Research reported herein was initiated to determine the relative toxicity of the cotton phytoalexins and address specific questions which have plagued the study of phytoalexins: 1) solubility in an aqueous medium, 2) location *in vivo* at the site of infection, 3)

This chapter not subject to U.S. copyright
Published 1991 American Chemical Society

physical contact between the pathogen and phytoalexin *in vivo*, and 4) the structure-activity relationship of the phytoalexins.

Our research has led us to propose an anatomical-biochemical model to explain the resistance mechanism in *Gossypium barbadense* cotton to infection by *V. dahliae*. We have proposed that the resistance response is effected by the sequential, integrated action of tylose occlusion of infected stem xylem vessels and accumulation of toxic levels of phytoalexins at sites of infection (5). Phytoalexins appear initially in solitary paratracheal parenchyma cells and subsequently diffuse into the vessel lumen. Staining reagents that react with these compounds show the phytoalexins coating the pathogen mycelium (6,7). The phytoalexins in cotton vascular tissue have been identified as hemigossypol (HG), desoxyhemigossypol (dHG), hemigossypol-6-methyl ether (MHG), and desoxyhemigossypol-6-methyl ether (dMHG) (8,9). Their structures are shown in Figure 1. Ten days after inoculation of the resistant cotton, *G. barbadense*, concentrations (μg/ml of water in stele) of the phytoalexins were: dHG, 25; HG, 26; dMHG, 57; MHG, 79 (10).

On examination of the structures of these phytoalexins, one might intuitively expect that HG and MHG with their reactive aldehyde groups would be more toxic than dHG and dMHG, respectively. However, experiments on the toxicity of these compounds to mycelia growth and inhibition of conidial germination of *V. dahliae* have shown the opposite to be true (10). Of the four compounds, dHG is the most toxic. Thus, dHG at a concentration of 4 μg/ml inhibited 95% of conidial germination; 45 μg/ml of MHG and 10 μg/ml of HG or dMHG was required to achieve this level of inhibition. Similarly, at 15 μg/ml dHG killed all mycelia, but 25 μg/ml of dMHG, 35 μg/ml of HG and 45 μg/ml of MHG were required to kill all mycelia.

The limited solubility of most phytoalexins in an aqueous medium has raised concerns as to their ability to act as effective fungicides (11). With the exception of dHG, the cotton phytoalexins are relatively insoluble in water. At pH 6.3 (the ~pH of infected xylem vessels) 50 μg/ml of dHG will dissolve in phosphate buffer; the solubility of the other three phytoalexins range from 2 to 4 μg/ml at pH 6.3 (10).

Another concern of plant pathologists is the dearth of evidence showing contact between the phytoalexin and the pathogen *in vivo* (12). We developed an antimony reagent that gives a green chelate specific for dHG and a red chelate for HG and MHG. This reagent shows a green dHG chelate deposited on *V. dahliae* mycelium in infected stele tissue. A red chelate is also evident; this is believed to be the HG chelate produced by oxidation of dHG (13).

Thus, dHG appears to meet all the criteria necessary to act as an effective phytoalexin:

1) Highly toxic to both *V. dahliae* mycelia and conidia at concentrations found in infected tissue.
2) Dissolves in water buffered to the pH of infected xylem vessels at concentrations significantly higher than those required to kill *V. dahliae* conidia and mycelia.

3) Present *in vivo* at the site of infection and in contact with *V. dahliae* mycelia.

In consideration of these findings we have concentrated our efforts on dHG, its chemistry, and structure-activity relationship.

Chemistry of dHG

In the crystalline state, dHG can be stored under an inert atmosphere at -70°C for an extended period of time (>1 year) with no significant decomposition. However, when solubilized in the aqueous medium (D-glucose, $(NH_4)_3PO_4$, $MgSO_4$, and 1% DMSO) used to grow and bioassay *V. dahliae*, it decomposes in 48 hours at room temperature in air (Figure 2a). The only recognizable product from this decomposition is HG which is obtained in 77% yield (Figure 2b). The rate of decomposition can be significantly reduced if dHG is solubilized under an inert atmosphere (Figure 2a). Decomposition is significantly faster in phosphate buffered water without glucose than in the nutrient buffer (Figure 2a).

The stabilizing effect of two substrate specific enzymes, superoxide dismutase (SOD) and catalase, on dHG have also been studied. Superoxide ($O_2^{-\cdot}$) is the specific substrate for SOD. Hydrogen peroxide (H_2O_2) or short chain (i.e. CH_3- or C_2H_5-) hydroperoxides are the specific substrates for catalase. SOD dismutates $O_2^{-\cdot}$ to give oxygen and hydrogen peroxide (Equation 1); catalase destroys hydrogen peroxide to give oxygen and water (Equation 2). Catalase effectively stabilized dHG in solution

$$2\ O_2^{-\cdot} + 2\ H^+ \xrightarrow{SOD} O_2 + H_2O_2 \qquad [1]$$

$$2\ H_2O_2 \xrightarrow{Catalase} 2\ H_2O + O_2 \qquad [2]$$

(Figure 3a). In two days, ~35% of the dHG decomposed in the growth medium that contained catalase, as compared to 94% in the medium alone. Under the same conditions SOD had essentially no stabilizing effect on dHG (Figure 3a); ~88% decomposed during two days. Strong reducing agents, such as 1.0 mM solutions of ascorbic acid, glutathione, phenylthiocarbamide, or 2,4-dithiopyrimidine significantly reduced decomposition of dHG (Figure 3b,3c). Ascorbic acid and glutathione have been used as free radical scavengers to stabilize pyrogallol (14), which readily oxidizes in air. However, 0.1 mM solutions of the weak reducing agent, thiourea, and the specific hydroxyl free radical (HO·) scavenger, sodium benzoate, failed to stabilize dHG solutions (Figure 3b).

A plausible mechanism to explain the decomposition of dHG is shown in Figure 4. The reaction is initiated by a free radical such as HO·. The hydroxyl free radical is extremely reactive, and even in the presence of scavengers, only a few radicals need escape to start the reaction by abstraction of a hydrogen atom from dHG to give the free radical [dHG]·. Propagation steps involve the reaction of [dHG]· with O_2 to give the alkyl peroxide radical [dHG]OO·. This radical continues the reaction by hydrogen atom abstraction from dHG to give the alkyl hydroperoxide [dHG]OOH and a

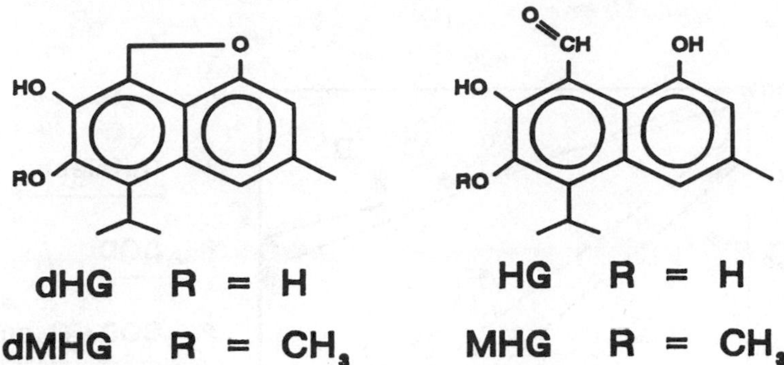

Figure 1. Structures of cotton phytoalexins, dHG = desoxyhemigossypol; dMHG = desoxyhemigossypol-6-methyl ether; HG = hemigossypol; MHG = hemigossypol-6-methyl ether.

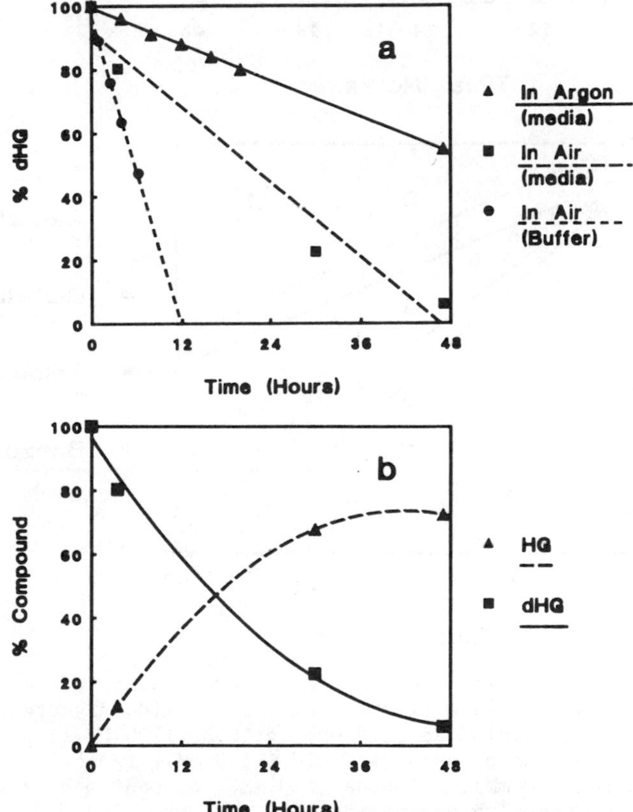

Figure 2. a) Decomposition of dHG in buffered (pH 6.3) media containing glucose under argon and air, and in buffer only in air; b) decomposition of dHG and concomitant formation of HG in buffered media in air.

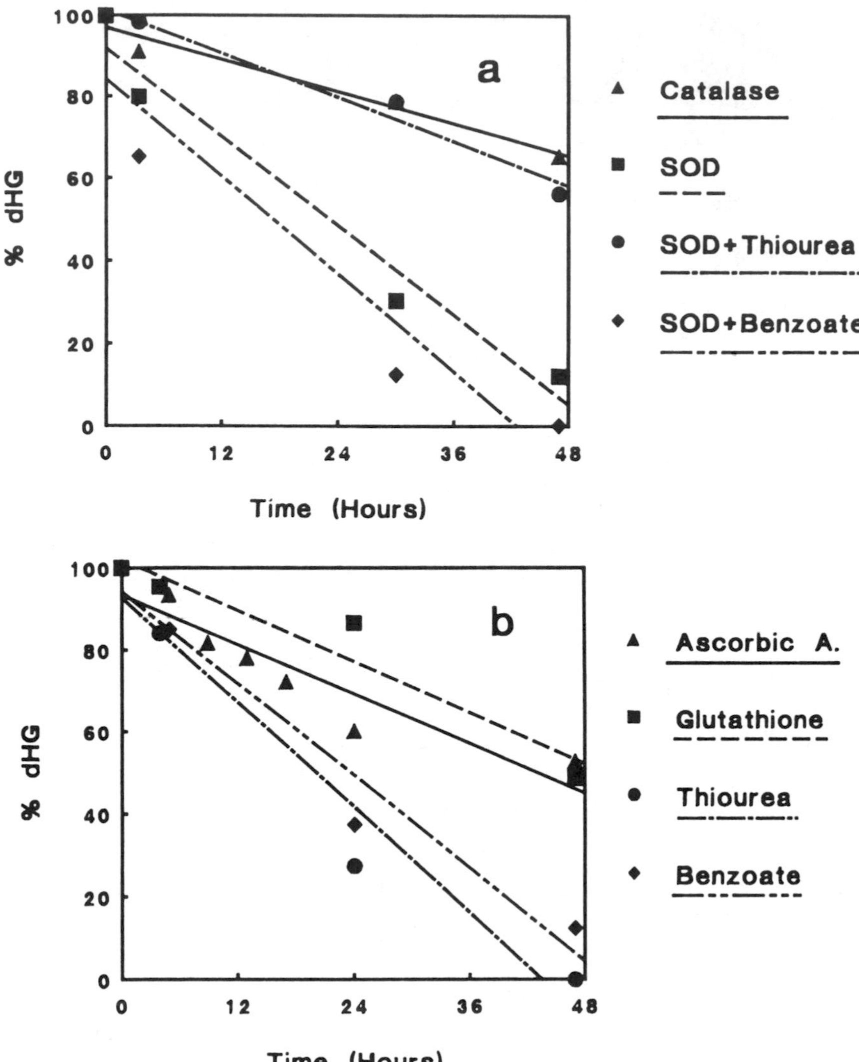

Figure 3. Decomposition of dHG in buffered media in the presence of a) catalase (1,700 U/ml), superoxide dismutase (SOD) (1700 U/ml) plus thiourea (1.0 mM), or SOD (1700 U/ml) plus benzoate (1.0 mM); b) ascorbic acid, (1.0 mM), reduced gluthathione (1.0 mM), thiourea (1.0 mM), or sodium benzoate (1.0 mM); c) phenylthiocarbamide (1.0 mM), or 2,4-Dithiopyrimidine; d) diethylenetriaminepentaacetic acid (DTPA) (1.0 mM), ethylenediaminetetraacetic acid (EDTA) (1.0 mM), or no chelator.

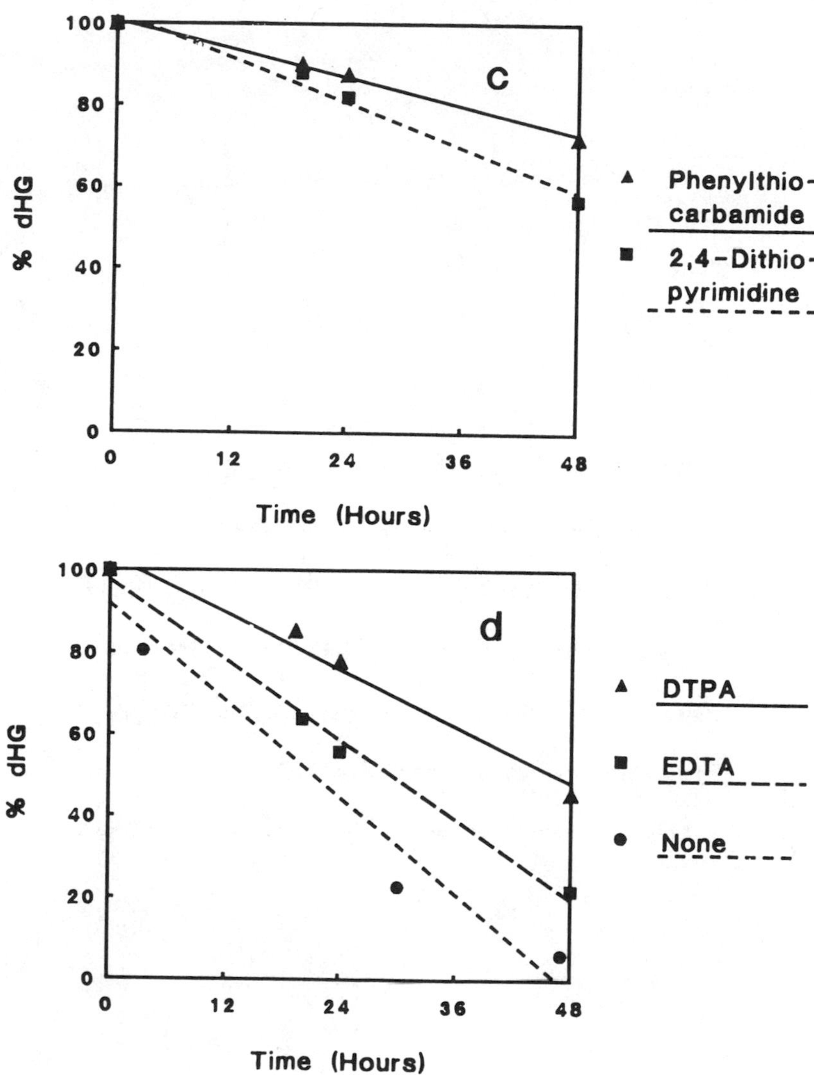

Figure 3. *Continued.*

$$HO\bullet + H\text{-}[dHG] \longrightarrow H_2O + [dHG]\bullet$$

$$[dHG]\bullet + O_2 \longrightarrow [dHG]OO\bullet$$

$$[dHG]OO\bullet + H\text{-}[dHG] \longrightarrow [dHG]OOH + [dHG]\bullet$$

$$[dHG]OOH + Fe^{2+} \longrightarrow [dHG]O\bullet + HO^- + Fe^{3+}$$

$$[dHG]O\bullet + H\text{-}[dHG] \longrightarrow [dHG]OH + [dHG]\bullet$$

$$[dHG]OOH + O_2^{\bar{\bullet}} \longrightarrow [dHG]O\bullet + O_2 + HO^-$$

Figure 4. Proposed steps in the autoxidation of dHG.

second [dHG]·. The alkyl hydroperoxide can react with trace quantities of transition metals such as ferrous ion to give [dHG]O·, HO^- and Fe^{3+}; [dHG]O· can also act to extract a hydrogen atom from dHG. The resulting alcohol [dHG]OH is the hemiacetal of HG. The chemical structures for some of the proposed intermediates are shown in Figure 5.

The classical Fenton reaction (Equations 4-6) can account for the formation of hydrogen peroxide and HO· required in the mechanism shown in Figure 4. The dismutation of $O_2^{-\cdot}$ (Equation 4) is the rate controlling step in the formation of H_2O_2 (15), but hydrogen peroxide is ultimately derived from $O_2^{-\cdot}$, and indeed the Fenton reaction depends on a source of $O_2^{-\cdot}$.

$$O_2^{-\cdot} + H^+ \rightleftharpoons HO_2^{\cdot} \quad [3]$$

$$HO_2^{\cdot} + O_2^{-\cdot} + H^+ \longrightarrow H_2O_2 + O_2 \quad [4]$$

$$O_2^{-\cdot} + Fe^{+3} \longrightarrow O_2 + Fe^{+2} \quad [5]$$

$$H_2O_2 + Fe^{+2} \longrightarrow HO^{\cdot} + {}^{-}OH + Fe^{+3} \quad [6]$$

Note, in Equations 4-6, three $O_2^{-\cdot}$ are required for each HO· produced. Superoxide is derived by a one electron reduction of O_2 (16). Trace quantities of transition metals are potential candidates to act as reductants in this reaction (i.e. Equation 7).

$$O_2 + Fe^{2+} \longrightarrow O_2^{-\cdot} + Fe^{3+} \quad [7]$$

Thus, SOD is not expected to prevent the decomposition of dHG; the product of its reaction with $O_2^{-\cdot}$ is H_2O_2 which can be a source of HO· (e.g. Equation 6). However, catalase would prevent the decomposition of dHG since it destroys H_2O_2. Baker and Gebicki (15) observed a similar phenomena when they studied the Fenton reaction by gamma irradiated Fe^{3+}-EDTA solutions. Catalase completely inhibited synthesis of HO·, but only 43% was inhibited with SOD. The formation of HO· was at a maximum at pH 4.8; this decreased to 42% at pH 7.4.

Strong reducing agents are expected to intercept the decomposition of dHG at three points: 1) by decomposing H_2O_2, 2) by reducing [dHG]·, and 3) by keeping transition metals in their reduced state and thus slowing the formation of $O_2^{-\cdot}$ (Equation 7). One might expect the HO· scavenger, benzoate, also to slow the decomposition because HO· is crucial to initiation of the chain reaction shown in Figure 4. However, once HO· abstracts a hydrogen atom from dHG, the initiated chain reaction will proceed unimpeded by simple HO· scavengers. Benzoate is specific, acting only to scavenge HO· (15), and not other free radicals. Thiourea is a weak reducing agent as compared to glutathione (17). Thus, thiourea is able to scavenge HO·, but apparently is a poor scavenger of [dHG]·.

The weak reducing agent, thiourea, when combined with SOD significantly reduces the decomposition of dHG, but the benzoate-SOD combination does not. Two routes exist for the formation of HO·:

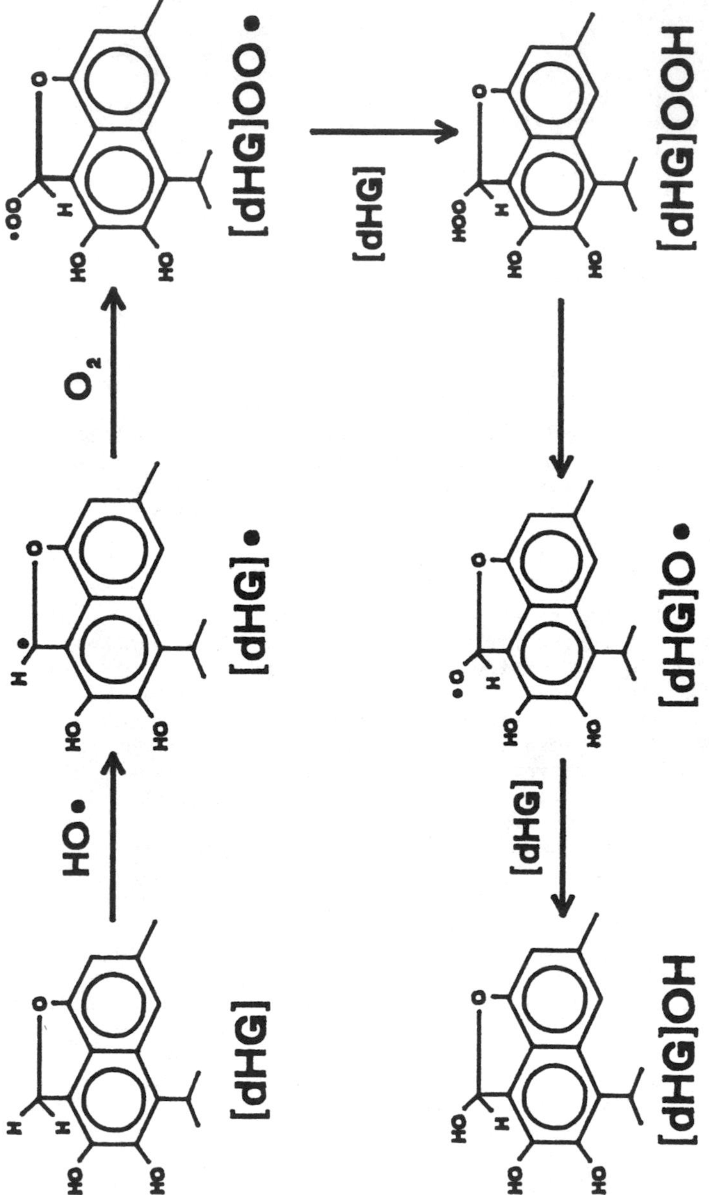

Figure 5. Structure for proposed intermediates formed during the autoxidation of dHG.

the last step in the Fenton reaction (i.e. Equation 6) and the Haber-Weiss reaction, Equation 8 (18).

$$O_2^{-\cdot} + H^+ + H_2O_2 \longrightarrow H_2O + O_2 + HO^\cdot \qquad [8]$$

Thiourea can act as a weak reducing agent to destroy H_2O_2 slowly and to keep trace quantities of transition metals such as iron in their reduced (e.g. ferrous) state and thus reduce the rate of $O_2^{-\cdot}$ formation (Equation 7). As observed by Baker and Gebicki (15), the optimization of Equation 6 is dependent on the rate of formation of $O_2^{-\cdot}$. Thus any reactant that slows Equation 7 will ultimately slow Equation 6. SOD intercedes in Equation 8, by destroying $O_2^{-\cdot}$ but produces H_2O_2 which is required in Equation 6. When acting in concert, thiourea slows the formation of $O_2^{-\cdot}$ and is able to react with the limited amount of H_2O_2 produced by SOD. This sequence of events again implicates H_2O_2 as a critical agent of decomposition.

The Fenton reactions (Equations 4-6) and the formation of $O_2^{-\cdot}$ by reduction of O_2 (i.e. Equation 7), require the intervention of a transition metal such as ferrous ion. DTPA is a strong enough chelator of ferrous ion to significantly deter the formation of $O_2^{-\cdot}$ (i.e. Equation 7), while EDTA is not (19,20). We found DTPA significantly reduced the rate of decomposition of dHG and was more effective than EDTA (Figure 3d).

Structure-Activity Relationship. The site and mode of action of phytoalexins have been investigated. Electron and light microscopy strongly implicate their involvement in disruption of the plasmalemma (21-23). Other symptoms of membrane dysfunction in pathogens treated with phytoalexins are: leakage of electrolytes and metabolites (12,24,25), loss in mycelia dry weight (24,26,27) and inhibition of oxygen uptake (24). Free radicals are cytotoxic entities that react with and destroy cell membranes. Apostol et al. (28) proposed that H_2O_2 may be involved in resistance of soybeans to *V. dahliae*, and Sun et al. (29) also implicated H_2O_2 in the photoactivation of the foliar cotton phytoalexin, 2,7-dihydroxycadalene. These observations suggest a role for free radicals in disease resistance.

Our experiments on the stability of dHG strongly suggest that dHG decomposes by a free radical mechanism. We hypothesize that dHG derives its toxicity from its ready ability to form free radicals. Extrapolations from our observations have allowed us to probe the nature of this toxicity.

We have developed a colorimetric method to assay viable conidia, which employs the tetrazolium salt, MTT (30). Using this method we measured the toxicity of dHG to *V. dahliae*, and determined an LD_{50} value of 3.5 μg/ml. The effects on the toxicity of dHG of the reducing agents, $HO^\cdot$ scavengers, and enzymes examined in the decomposition studies were assayed by this method. As predicted from the decomposition studies, neither sodium benzoate (0.1 mM), thiourea (1.0 mM), nor SOD (900 U/ml) appreciably affects the toxicity of dHG (Figure 6). Sodium benzoate was assayed at 0.1 mM because it is toxic to *V. dahliae* at higher concentration. The effects of the reducing agents glutathione and ascorbic acid, and

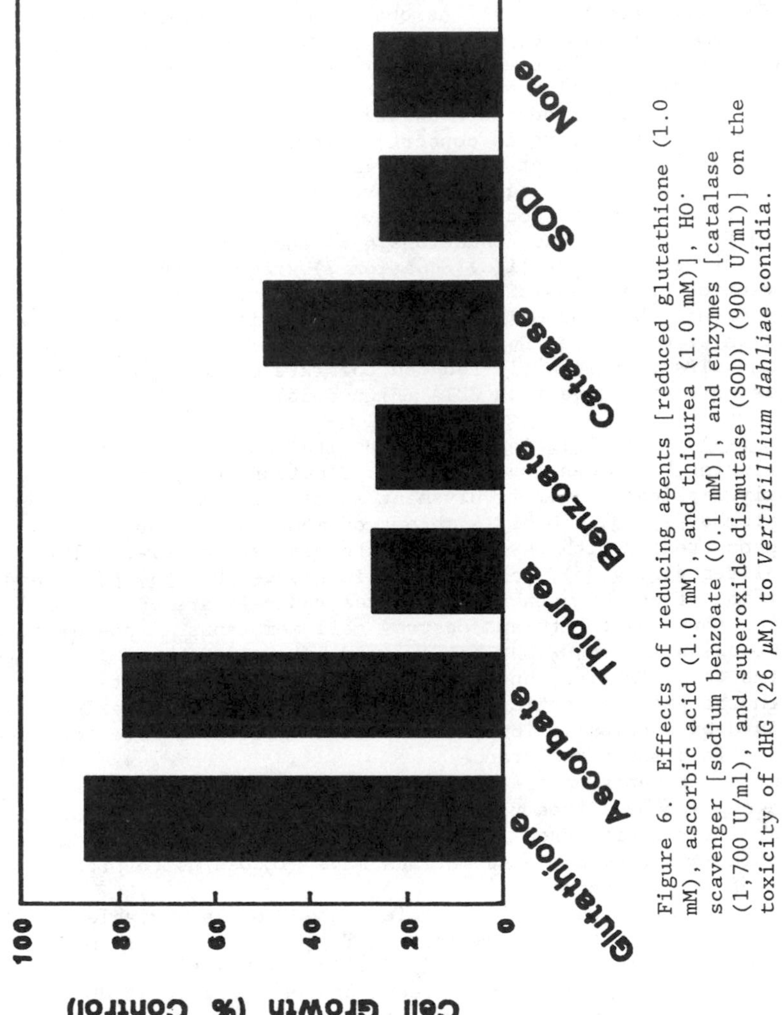

Figure 6. Effects of reducing agents [reduced glutathione (1.0 mM), ascorbic acid (1.0 mM), and thiourea (1.0 mM)], HO· scavenger [sodium benzoate (0.1 mM)], and enzymes [catalase (1,700 U/ml), and superoxide dismutase (SOD) (900 U/ml)] on the toxicity of dHG (26 μM) to *Verticillium dahliae* conidia.

the enzyme catalase also are shown in Figure 6. When 1.0 mM ascorbic acid or glutathione was added to the growth media, the toxicity of dHG was significantly reduced; catalase (1700 U/ml) also reduced its toxicity. The action of ascorbic acid, glutathione and catalase is consistent with our hypothesis.

The phytoalexin dHG is rapidly absorbed by *V. dahliae* conidia, 75% of the dHG in a 6.5 μg/ml solution being absorbed from the bioassay media by *V. dahliae* conidia (5 X 10^5 cells/ml) in one minute. Each *V. dahliae* cell would therefore contain 3.2 X 10^{10} molecules of dHG. When these cells were washed repeatedly with media, ~70% of the dHG was removed in six washes; subsequent washes removed smaller and smaller amounts.

Rapid uptake from solution also has been implicated for the phytoalexin, kievetone. Damage to the plasmalemma of *Rhizoctonia solani* hypha, as shown by electron micrographs and increased leakage of ^{14}C-labeled metabolites, occurs within 90 minutes after exposure to kievitone (31).

We hypothesize dHG is rapidly absorbed by *V. dahliae* conidia and mycelia. An initiator (e.g. HO·) begins a chain reaction in which a series of highly reactive, cytotoxic free radicals are generated. One or more of these radicals reacts with vital processes in cell membranes (probably the plasmalemma) leading ultimately to death of conidia and mycelia.

The HO· does not appear to be the toxic entity since HO· scavengers do not decrease toxicity. The diminution of toxicity by reducing agents and catalase apparently results from their ability to prevent initiation of the free radical reaction, and/or interruption of the chain reaction process.

Sterols do not appear to be involved in the interaction of phytoalexins because sterol incorporation in the media does not protect *Rhizoctonia solani* against the phytoalexins kievitone (31) or phaseolin (24). A likely site of interaction is polyunsaturated fatty acid esters in membranes which readily undergo free radical autoxidations, and the three dimensional conformation of this autoxidation product is significantly different from the starting material. Such conformational changes are expected to disrupt membranes. Oxidation of inert cellular components such as proteins by the cooxidation mechanism proposed by Pryor (32) is also possible. Peroxy radicals (ROO·), and especially the alkoxy radical (RO·) which result from lipid oxidation are proposed as the mediators that react with these rather inert cellular components (32).

The role of active oxygen and the resulting free radicals from phytoalexins may be important aspects in the defense response. For example, the production of H_2O_2 (28) and $O_2^{-\cdot}$ (33) in response to a pathogen derived elicitor has been confirmed in soybean cell suspension cultures. Tobacco leaf disc infected with tobacco mosaic virus showed a marked increase in $O_2^{-\cdot}$ generating activity (34).

The pterocarpans and the extensive research on their toxicity to plant pathogens provide an excellent example of compounds which have the potential to readily form free radicals. VanEtten carefully studied the toxicity of six pterocarpans, three 6a, 11a-dehydropterocarpans and two coumestans to *Aphanomyces euteiches* and *Fusarium solani* (35). The general structures for these isoflavonoids are shown in Figure 7. All of the pterocarpans have an ethereal benzylidene proton at C_{11a} which is part of a dihydrofuran ring. This proton should readily dissociate to form a free radical. The free radical, which is an sp^3 carbon, is planar with and stabilized by the benzene ring, a situation analogous to that found in dHG·. The dehydropterocarpans lack a proton at C_{11a}, but protons are available at C_6. These protons, which are not part of a furan ring, are capable of forming a styryl free radical. Examination of Drieding models shows the pyran ring is slightly puckered when C_6 is an sp^3 carbon. Thus these protons are expected to be less reactive than the proton at C_{11a} in the pterocarpans. The coumestans lack protons at both C_{11a} and C_6, and contain no carbons which would readily form free radicals.

In general the pterocarpans are more toxic to *F. solani* than the dehydropterocarpans, and the same is true for most of the pterocarpans to *A. euteiches*. The coumestrols are not toxic to either of these organisms. Thus the propensity to form free radicals could be used to provide some guidance as to the predicted toxicity of these compounds. There are glaring exceptions to this observation such as the loss of toxicity when the pterocarpan, pisatin ($R_1 = OCH_3$, $R_2 = O-$, $R_3 = OCH_2-$, $R_4 = OH$), is demethylated; the demethylated product is significantly less toxic than pisatin.

The reasons for the large differences in toxicity among related phytoalexins may depend on solubility considerations. As suggested by VanEtten (35), solubility in the bioassay medium is important in determination of toxicity. Indeed some toxicity results *in vitro* could be affected by microscopic precipitation of the phytoalexin. However, we suspect lipid solubility in the plasmalemma or other membranes is also important. Thus a careful balance between solubility in the aqueous xylem fluid and lipid solubility in the plasmalemma must be maintained for optimum phytoalexin toxicity. This proper balance in solubilities and a propensity to form free radicals are each expected to play an important role in phytoalexin toxicity.

The structure of dHG seems to be ideally suited to act as an effective phytoalexin in cotton stems. The water solubility of dHG allows rapid diffusion from paravascular cells into xylem vessels. Light is not necessary for the toxicity of dHG, which is consistent with its site of action in the xylem vessels. Apparently, the only constituents necessary to start the decomposition of dHG, and activate it as a toxin, are oxygen and a transition metal which are both in adequate supply in the xylem fluids. Invasion of the xylem by the pathogen triggers a series of events, one of which is the biosynthesis of dHG. The dHG reacts with oxygen to form a free radical that theoretically should react indiscriminately with both pathogen and plant tissue. To overcome dHG the pathogen could diminish its effectiveness by producing a strong reducing environment or oxidizing it at a distance before contact is made. Factors influencing this aspect of virulence require additional investigations.

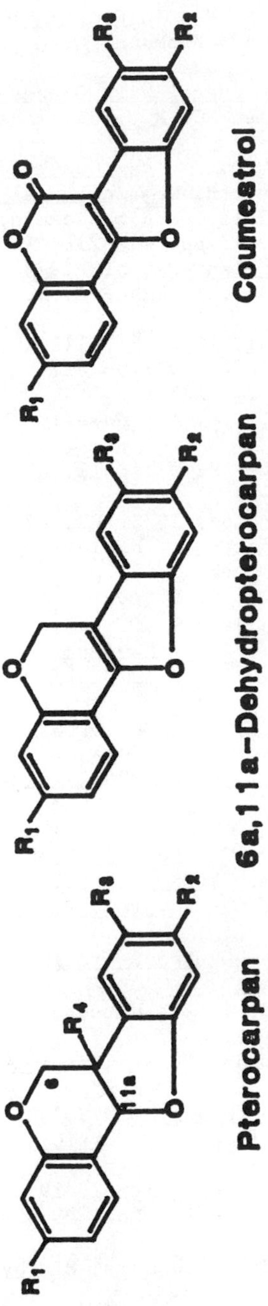

Figure 7. Structure of some isoflavonoid phytoalexins and analogues.

Literature Cited:

1. Snyder, W. C.; Smith, S. H. In Fungal Wilt Diseases of Plants; Mace, M. E., Bell, A. A.; Beckman, C. H., Eds.; Academic: New York; 1981, pp. 25-50.
2. Schnathorst, W. C. In Fungal Wilt Disease of Plants; Mace, M. E.; Bell, A. A.; Beckman, C. H., Eds.; Academic: New York, 1981; pp. 81-111.
3. Gerik, J. S.; Huisman, O. C. Phytopathology 1988, 78, 1174-78.
4. Beckman, C. H.; Talboys, P. W. In Fungal Wilt Diseases of Plants; Mace, M. E.; Bell, A. A.; Beckman, C. H., Eds.; Academic: New York, 1981; pp. 487-21.
5. Mace, M. E. Physiol. Plant Pathol. 1978, 12, 1-11.
6. Mace, M. E.; Bell, A. A.; Beckman, C. H. Can. J. Bot. 1976, 54, 2095-99.
7. Mace, M. E. New Phytol. 1983, 95, 115-19.
8. Bell, A. A.; Stipanovic, R. D.; Howell, C. R.; Fryxell, P. A. Phytochemistry. 1975, 14, 225-31.
9. Stipanovic, R. D.; Bell, A. A.; Howell, C. R. Phytochemistry. 1975, 14, 1809-11.
10. Mace, M. E.; Stipanovic, R. D.; Bell, A. A. Physiol. Plant Pathol. 1985, 26, 209-18.
11. Daly, J. M. Phytopathology 1972, 62, 392-400.
12. Smith, D. A. In Phytoalexins; Bailey, J. A.; Mansfield, J. W., Eds.; John Wiley: New York, 1982; p. 247.
13. Mace, M. E.; Stipanovic, R. D.; Bell, A. A. New Phytol. 1989, 111, 229-32.
14. Ghosh, J. C.; Rakshit, P. C. Biochem. Z. 1937, 294, 330-35.
15. Baker, M. S.; Gebicki, J. M. Arch. Biochem. Biophy. 1984, 234, 258-64.
16. Taube, H. In Oxygen, Proc. Symp. Spons. New York Heart Association; Little Brown: New York, 1965; pp. 29-52.
17. Randall, L. O. J. Biol. Chem. 1946, 164, 521-27.
18. Haber, F.; Weiss, J. J. Proc. Roy. Soc. London, Ser. A. 1934, 147, 332-51.
19. Marklund, S.; Marklund, G. Eur. J. Biochem. 1974, 47, 469-74.
20. Cohen, G.; Sinet, P. M. FEBS Letters. 1982, 138, 258-60.
21. Harris, J. E.; Dennis, C. Physiol. Plant Pathol. 1976, 9, 155-65.
22. Smith, D. A.; Bull, C. H. Proc. 3rd Int. Cong. Plant Pathol. 1978, p. 245.
23. Rossall, S.; Mansfield, J. W.; Hutson, R. A. Physiol. Plant Pathol. 1980, 16, 135-46.
24. Van Etten, H. D.; Bateman, D. F. Phytopathology. 1971, 61, 1363-1372.
25. Higgins, V. J. Phytopathology. 1978, 68, 339-45.
26. Bailey, J. A.; Deverall, B. J. Physiol. Plant Pathol. 1971, 1, 435-49.
27. Smith, D. A. Physiol. Plant Pathol. 1976, 9, 45-55.
28. Apostol, I.; Heinstein, P. F.; Low, P. S. Plant Physiol. 1989, 90, 109-16.
29. Sun, T. J.; Melcher, U.; Essenberg, M. Physiol. Mol. Plant Pathol. 1988, 33, 115-26.

30. Stipanovic, R. D.; Mace, M. E.; Elissalde, M. H. In Beltwide Cotton Production Research Conference; J. M. Brown, Ed.; National Cotton Council, Memphis: 1989; p. 39 (Abstract).
31. Bull, C. A. Ph.D. Thesis, University of Hull, U.K., 1981, and micrographs shown in Ref. 12. pp. 223-28.
32. Pryor, W. A. Fed. Proceed. 1973, 32, 1862-68.
33. Lindner, W. A.; Hoffmann, C.; Grisebach, H. Phytochemistry. 1988, 27, 2501-03.
34. Doke, N.; Ohashi, Y. Physiol. Mol. Plant Pathol. 1988, 32, 163-75.
35. VanEtten, H. D. Phytochemistry 1976, 15, 655-59.

RECEIVED May 16, 1990

Chapter 24

Bioregulation of Preharvest Aflatoxin Contamination of Peanuts

Role of Stilbene Phytoalexins

Joe W. Dorner[1], Richard J. Cole[1], Boris Yagen[2], and Benedikte Christiansen[3]

[1]National Peanut Research Laboratory, Agricultural Research Service, U.S. Department of Agriculture, 1011 Forrester Drive, SE, Dawson, GA 31742
[2]School of Pharmacy, Hebrew University, Jerusalem, Israel
[3]Department of Biotechnology, Instituttet for Bioteknologi, Lyngby, Denmark

A rich and varied assortment of phytoalexins is produced by legumes. Peanuts (*Arachis hypogaea* L.) have been shown to produce various stilbene phytoalexins in response to fungal infection. Recent studies have been conducted to elucidate the role of these stilbenes in the bioregulation of preharvest aflatoxin contamination of peanuts by *Aspergillus flavus* and *A. parasiticus*. Stilbene phytoalexins appear to provide soil borne peanut seed with a degree of resistance to aflatoxin contamination by inhibiting fungal growth. Under conditions of prolonged, late-season drought stress that lead to aflatoxin contamination, phytoalexin biosynthesis breaks down, presumably due to dehydration of the peanut seed. However, seed moisture under such conditions remains adequate to support *A. flavus* growth and aflatoxin production. The chemistry and biological activity of stilbene phytoalexins as related to preharvest aflatoxin contamination of peanuts is described.

Phytoalexins, antimicrobial substances produced by plants in response to infection or stress, are important in the natural defense of plants against disease. Phytoalexins are extremely diverse chemically, and although their ubiquity throughout the plant kingdom is open to question, certain plant families have been found to contain many species capable of phytoalexin production (1). Among those is the Leguminosae, an economically important family of plants that includes the peanut (*Arachis hypogaea* L.)(2).

Peanuts are grown in many areas of the world and have a high nutritive value. However, peanuts are subject to aflatoxin contamination under certain environmental conditions, and when this happens, it nullifies their usefulness as food or feed (3).

In 1972 it was reported that peanut kernels can synthesize phytoalexins in response to fungal challenge (4), and subsequently several of these phytoalexins were chemically characterized as stilbenes (5-8). Since that time, studies have shown that these compounds possess biological activity against fungi, including *Aspergillus flavus* , one of the species that produces aflatoxin (9). Thus, it has been speculated that stilbene phytoalexins might be important in the natural defense of peanuts against aflatoxin-producing fungi.

This chapter not subject to U.S. copyright
Published 1991 American Chemical Society

The purposes of this chapter are to review the factors involved in aflatoxin contamination of peanuts, review the chemistry of stilbene phytoalexins from peanuts, discuss evidence supporting the involvement of these stilbenes in the bioregulation of aflatoxin contamination, and explore approaches to exploit or enhance such a bioregulative capacity to reduce or eliminate preharvest aflatoxin contamination of peanuts.

Aflatoxin

Aflatoxins are secondary fungal metabolites produced by *A. flavus* Link and *A. parasiticus* Speare and are recognized for their potent hepatotoxic, teratogenic, and carcinogenic effects in some animals (10).

The four naturally occurring aflatoxins are designated B_1, B_2, G_1, and G_2. Other known aflatoxins are chemical or biological products of the four naturally occurring compounds. Under certain environmental conditions the aflatoxin-producing fungi can invade various agricultural commodities, and subsequent proliferation by these fungi contaminates the commodity with aflatoxin. The commodities most affected in the United States are peanuts, corn, and cottonseed. Strict regulatory action levels for aflatoxin must be met for these commodities to be used as food or feed (10). Therefore, severe economic losses can occur when these commodities become contaminated and are diverted from edible markets.

Preharvest Aflatoxin Contamination of Peanuts

Contamination of peanuts with aflatoxin can occur during various phases of production, storage, handling, and marketing (11). Contamination that occurs after peanuts are harvested is preventable if proper steps are taken to ensure that the moisture of peanuts is maintained at a level that is unfavorable for growth of the aflatoxin-producing fungi. However, contamination that occurs in the field prior to harvest (preharvest) is much more difficult to control and usually results in the most severe aflatoxin contamination of peanuts.

Preharvest aflatoxin contamination of peanuts occurs when peanuts are subjected to severe, prolonged drought stress during the last four to six weeks of the growing season (12-17). Elevated soil temperatures that usually accompany such periods of late-season drought exacerbate the problem and produce optimum conditions for preharvest contamination (12, 18). When contamination occurs, it is not homogeneous throughout a population of peanuts. Immature peanuts are more likely to be contaminated than mature peanuts, and kernels that are damaged, particularly by insects, can contain extremely high concentrations of aflatoxin (13, 15, 18-20).

Invasion of peanuts by the aflatoxigenic fungi does not necessarily result in their being contaminated with aflatoxin. In the absence of drought stress, samples of peanuts have been shown to be colonized by *A. flavus/parasiticus* at percentages as high as 25% of the kernels without detectable levels of aflatoxin (18, 19). This indicates that in the absence of drought conditions the fungi are able to invade peanuts, but the peanuts are protected from extensive fungal proliferation by some inherent defense mechanism(s). However, when exposed to the stresses of drought and heat, a breakdown in this defense mechanism(s) allows for fungal proliferation and subsequent aflatoxin contamination.

Stilbene Phytoalexins from Peanuts

The first report of phytoalexin production by peanuts appeared in 1972 when Vidhyasekaran *et al.* found that several species of fungi, including *A. flavus*, induced production of an inhibitory principle by peanut pods (4). Although the principle was not chemically characterized, it was deemed to be a phytoalexin because it was produced by the peanut only after interaction between the fungi and the peanut. In that study, immature peanuts produced a higher quantity of the phytoalexin than mature peanuts, and the authors suggested that resistance of immature peanut pods to fungi was based on their capacity to produce phytoalexins in response to infection.

In 1975 Keen (21) reported that native microflora stimulated production of two antifungal compounds by peanut seeds that were soaked in water, sliced into sections, and incubated for 3-5 days. These compounds were judged to be phytoalexins and were subsequently identified as cis- and trans-isomers of 4-(3-methyl-but-2-enyl)-3,5,4'-trihydroxystilbene [1](5)(Figure 1). Simultaneously, Ingham (6) reported the isolation of cis- and trans-resveratrol(3,5,4'trihydroxystilbene[2]) from peanut hypocotyls. Additional stilbenes have been shown to be produced by peanut seeds in response to wounding, and these include 4-(3-methyl-but-1-enyl)-3,5,3',4'-tetrahydroxystilbene [3](7), 4-(3-methyl-but-1-enyl)-3,5,4'-trihydroxystilbene [4](7), and 3-isopentadienyl-4,3',5'-trihydroxystilbene [5](8).

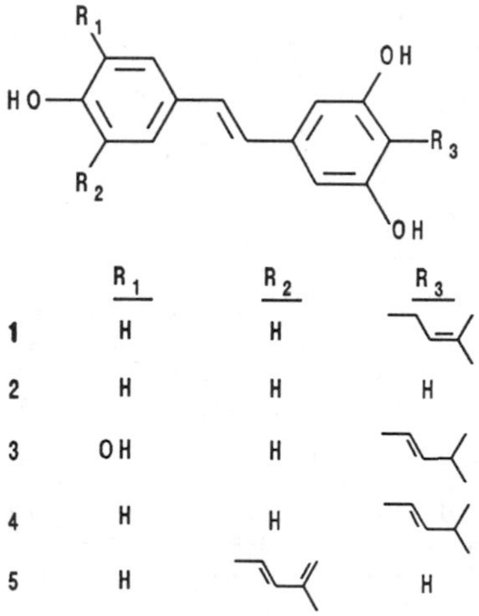

Figure 1. Chemical structures of stilbene phytoalexins from peanuts.

To stimulate peanut seeds to produce phytoalexins for laboratory studies, the dry seeds are typically allowed to imbibe water for approximately 24 hours. The seeds are then sliced into 1-3 mm sections or gently chopped to cause extensive cellular damage, and the seeds are either inoculated with a fungus and incubated in the dark for several days or incubated using the native peanut microflora to induce phytoalexin production (7, 21, 22).

The stilbenes are typically extracted from peanuts with 95% ethanol, and following partial purification the samples are subjected to either thin-layer chromatographic (TLC) or liquid chromatographic (LC) analyses. On TLC plates, the stilbenes fluoresce blue under 254 nm UV light or they can be detected in LC analysis by UV detection ranging from 290-335 nm (5-7, 21, 22).

Evidence Supporting Stilbene Involvement in Bioregulation of Preharvest Aflatoxin Contamination

Natural Occurrence of Stilbenes in Peanuts. Stilbenes are not found in sound, undamaged peanuts. However, the compounds are easily detected in peanuts that have been subjected to some type of damage in the field when they were at high water activities (unpublished data). In peanut shelling plants, damaged peanuts are removed routinely by electronic color sorting machines, and these machines efficiently eliminate discolored kernels during the processing of peanuts. Discoloration is used to identify peanuts that have been damaged (usually in the field, but also during storage) and would severely detract from the overall quality of the peanuts, particularly with regard to flavor. In many cases this discoloration is very similar to the discoloration that occurs when imbibed peanuts are stimulated to produce phytoalexins in the laboratory. When these discolored peanuts were analyzed for stilbene phytoalexins, the analyses invariably showed the presence of the compounds, suggesting that they are naturally produced in the field in response to damage (unpublished data). Since discoloration and stilbene production does not occur when peanuts are dry, it is apparent that some minimum water activity is required for the enzyme-mediated synthesis of the stilbenes.

A close examination of damaged peanuts grown under adequate moisture conditions reveals little, if any, *A. flavus* proliferation. Likewise, it is unusual to detect even low concentrations of aflatoxin in such peanuts. However, stilbene phytoalexins are easily detected in such peanuts. When similar peanuts were surface sterilized and plated out to determine counts and types of fungal colonization, the percentages of kernels colonized by *A. flavus* was as high as 25% (19). Since colonization had occurred and phytoalexins had been produced with an absence of aflatoxin contamination, phytoalexins presumably inhibited *A. flavus* growth and aflatoxin production.

Conversely, a close examination of damaged peanuts that were subjected to late-season drought conditions usually reveals several kernels with prolific *A. flavus* growth and aflatoxin concentrations that can be extremely high. Phytoalexin concentrations are typically much lower in these peanuts compared to damaged, non-stressed peanuts.

Therefore, the fact that peanuts produce stilbene phytoalexins naturally in response to damage in the field but do not become contaminated with aflatoxin (indicative of *A. flavus* growth) until subjected to prolonged drought stress points toward a presumptive role for these compounds in the natural bioregulation of aflatoxin contamination.

Biological Activity of Stilbenes Against *A. flavus* and Other Fungi. Further evidence supporting a role for stilbene phytoalexins in inhibiting fungal growth in peanuts involves the biological activity of these compounds against *A. flavus* and other fungi. Wotton and Strange tested [1], [3], and [4] for inhibition of *A. flavus* spore germination and hyphal extension (9). The ED_{50} values for inhibition of spore germination were 12.7, 12.8, and 8.9 µg/ml, respectively, in Vogel's medium. Similarly, germ tube extension was also inhibited with ED_{50} values of 6.8, 4.9, and 9.7 µg/ml, respectively, for [1], [3], and [4]. Cooksey et al. reported ED_{50} values of 14.0 and 11.3 µg/ml for inhibition of spore germination and hyphal extension of *A. flavus*, respectively, by [5] (8).

In one laboratory, a time-course study was carried out to determine the relationship of growth and aflatoxin production by *A. flavus* to phytoalexin accumulation by intact peanut kernels. Wotton and Strange (23) reported that following inoculation, *A. flavus* grew logarithmically for 2 days. However, by the third day when the phytoalexin concentration had exceeded 50 µg/g of kernels, fungal growth essentially ceased. This provided evidence for both the elicitation of phytoalexin production by *A. flavus* and the inhibition of *A. flavus* growth by the phytoalexins.

The inhibition of *A. parasiticus* growth by [4] was determined in our laboratory using a modification of the assay reported by Arnoldi and coworkers (24). The stilbene was dissolved in acetone and appropriate amounts were added to sterile potato dextrose agar (PDA) at 50°C to give concentrations of 10, 20, 50, and 100 µg/ml. The medium was dispersed in 60 mm tissue culture dishes and inoculated with 4 mm PDA plugs of actively growing *A. parasiticus* cultures. Plates were incubated in the dark at 30°C and colony diameters were measured daily. After three days the percent inhibition of *A. parasiticus* growth at 10, 20, 50, and 100 µg/ml was 15, 26, 37, and 33%, respectively (unpublished data).

Other studies were conducted in our laboratory to determine the inhibitory effects of [4] against *A. parasiticus* and several species of *Penicillium* (Christiansen, unpublished data). The germination of fungi was determined in a procedure that was similar to that reported by Wotton and Strange (9), and radial growth also was determined as previously described (24). Results presented in Table I show that both spore germination and growth of most tested species were inhibited by [4], although the degree of inhibition varied considerably among species.

Table I. Effect of [4] on spore germination and radial growth of several fungal species.

Fungus	Spore germination ED_{50} (µg/ml)	Radial growth % inhibition 25 µg/ml	50 µg/ml
Aspergillus parasiticus	94	54	54
Penicillium commune	132	36	55
P. crustosum	18000	46	61
P. verrucosum	42	33	44
P. aurantiogriseum	70	25	50
P. echinulatum var. discolor	77	13	63

Taken together, these studies show that peanut kernel stilbenes possess biological activity against species of both *Aspergillus* and *Penicillium*. This, coupled with the fact that these compounds are produced *in vivo* as a result of damage, provides evidence that stilbenes play a part in the natural defense of peanuts against fungi.

Occurrence of Aflatoxin Contamination After Cessation of Phytoalexin Production. In view of the fact that stilbene phytoalexins are naturally produced in peanut kernels in response to fungal invasion and that these stilbenes possess antifungal activity against aflatoxigenic fungi, the question of how peanuts become contaminated with aflatoxin remains. Simply stated, how does *A. flavus* overcome this apparent natural defense mechanism of peanuts?

Because of the association of preharvest aflatoxin contamination of peanuts with late-season drought stress, a study was undertaken to determine the relationship among aflatoxin contamination, drought, and phytoalexin production (22). The study involved sampling of Florunner peanuts subjected to and not subjected to late-season drought and determining the peanut kernel water activity (a_w), phytoalexin-producing capacity, and aflatoxin concentrations in all maturity stages of the sampled peanuts. It was reported that the a_w and phytoalexin-producing capacity of peanuts not exposed to drought stress remained high throughout the study period. These peanuts did not become contaminated with aflatoxin even though the final sampling took place 184 days after planting, approximately 40 days beyond the optimal harvest date.

The a_w of peanuts grown under late-season drought conditions decreased during the study (22). The moisture content was not uniform in a sample, indicating that peanuts subjected to drought stress did not dry at a uniform rate. This appeared to be associated with individual plants in that the peanuts on certain plants lost moisture more easily than those on other plants. However, the overall effect of drought stress was to reduce the moisture or a_w of peanut kernels.

As peanuts became dehydrated during the drought period, they lost the capacity to produce stilbene phytoalexins. This lost capacity was not directly due to the duration of the stress, but it was directly associated with the drop in a_w of the peanuts. Regardless of drought treatment soil temperature or peanut maturity, the phytoalexin-producing capacity of peanuts decreased as the a_w decreased, with essentially no phytoalexin production below a kernel a_w of 0.95 (Figure 2).

The onset of aflatoxin contamination did not occur until peanut kernels had lost the capacity to produce phytoalexins as an apparent result of drought-induced moisture loss. However, mature peanuts retained a higher degree of resistance to aflatoxin contamination than immature peanuts even after the loss of phytoalexin-producing capability. This indicated that immature kernels rely more heavily on a phytoalexin-based resistance than mature kernels, which apparently have some additional resistance not based on stilbene phytoalexins.

The evidence clearly supports the hypothesis that stilbene phytoalexins in peanuts are an important natural bioregulator of preharvest aflatoxin contamination. That evidence includes the facts that: (1) stilbenes are naturally produced in field-damaged peanuts; (2) stilbenes possess biological activity against *A. flavus* and *A. parasiticus*; and (3) although invasion of peanuts by *A. flavus* and *A. parasiticus* can occur under any conditions, aflatoxin contamination does not occur until peanuts lose the capacity for phytoalexin production as a result of drought-induced kernel dehydration.

Taking Advantage of the Natural Defense Mechanism

The goal of our research today is to greatly reduce or eliminate preharvest aflatoxin contamination of peanuts. As the demand for more wholesome food with less risk of exposure to toxins and carcinogens increases, the continued use of peanuts and peanut products as food becomes more dependent on effective management of the aflatoxin problem.

Many approaches are being taken to solve the aflatoxin problem in peanuts. An important question is whether the phytoalexin-based natural defense mechanism of peanuts against fungi can be exploited in some way to provide a solution; and if so, how. All data indicate that growing all peanuts with adequate late-season irrigation would essentially solve the preharvest problem, but currently this is not a feasible approach. Therefore, taking advantage of any natural defense mechanism must involve its effectiveness during periods of late-season drought stress.

Two approaches to maintaining phytoalexin-producing capacity during drought are apparent. The first would be to identify peanut genotypes that could continue producing phytoalexins as the moisture of the kernels decreased. In Florunner peanuts it appears that the approximate lower a_w limit for phytoalexin production is 0.95. Peanuts that could maintain production of phytoalexins as the a_w approached 0.90 might possess much greater protection from *A. flavus* growth and aflatoxin production. Although the lower a_w limit for aflatoxin production in peanuts is about 0.85, a practical solution

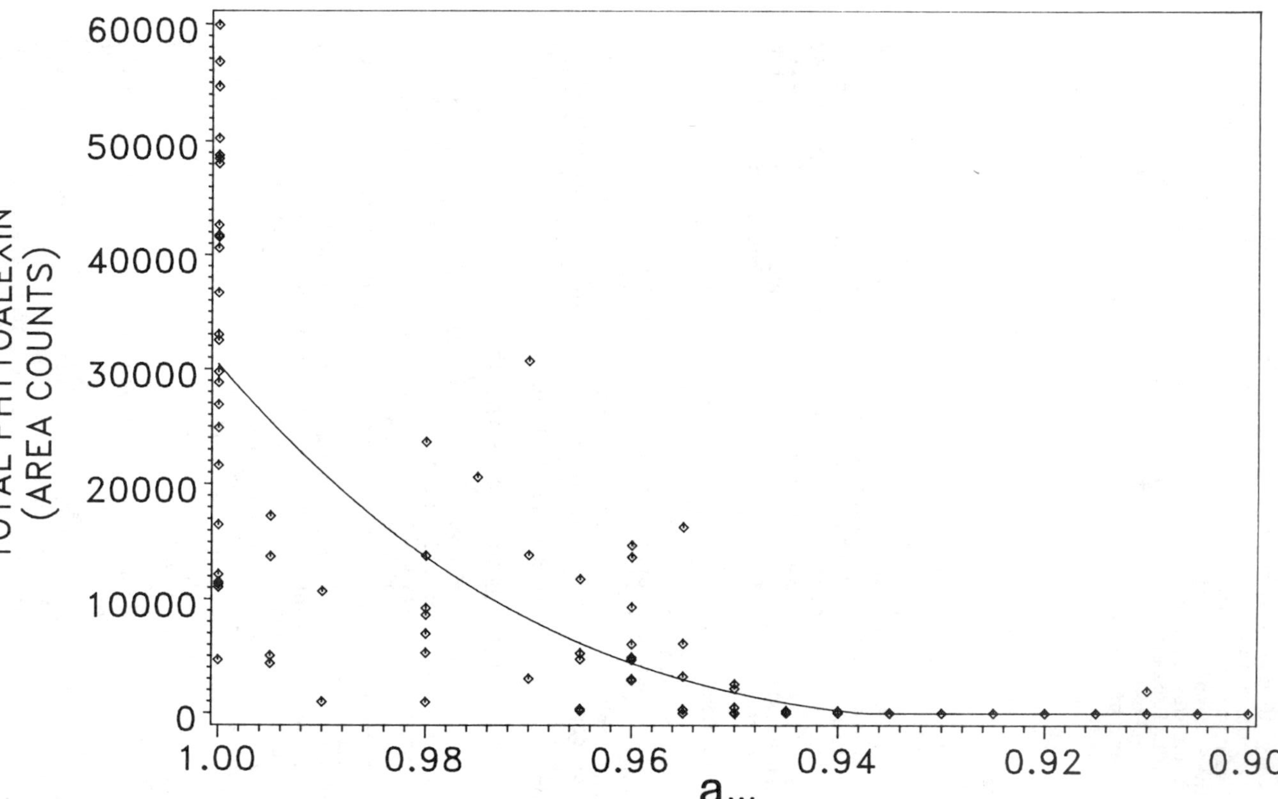

Figure 2. Relationship of phytoalexin production to peanut kernel water activity (a_w). Total phytoalexins were determined by combining areas under phytoalexin peaks from liquid chromatograms.

to the problem might not require phytoalexin production down to such a low a_w level. This is because as the a_w continues to decrease, so does the rate of fungal growth and aflatoxin production. Whether or not a genotype exists that can produce phytoalexins at lower a_w is unknown at this time. Therefore, a rigorous screening program would have to be undertaken to identify such a genetic capability in peanuts.

A second approach to maintaining phytoalexin-producing capacity during drought would be to identify drought-tolerant genotypes that can maintain a high kernel a_w for a significantly longer period during drought stress. If either of these approaches were successful, biotechnology techniques might be used to incorporate the desirable trait(s) into commercially-desirable cultivars, such as Florunner.

It is unlikely that any single approach will provide a solution to the problem of preharvest aflatoxin contamination of peanuts. However, a multifaceted approach that could include enhancement of the natural bioregulative properties of stilbene phytoalexins might ultimately yield the solution to a serious and complex problem.

Literature Cited

1. Deverall, B. J. In *Phytoalexins*; Bailey, J. A.; Mansfield J. W., Eds.; John Wiley and Sons, Inc.: New York, 1982; pp 1-20.
2. Ingham, J. L. In *Phytoalexins*; Baily, J. A.; Mansfield, J. W., Eds.; John Wiley and Sons, Inc.: New York, 1982, p 21.
3. Cole, R. J.; Sanders, T. H.; Blankenship, P. D.; Hill, R. A. In *Xenobiotics in Foods and Feeds*; Finley, J. W.; Schwass, D. E., Eds.; ACS Symposium Series No. 234; American Chemical Society: Washington, DC, 1983; pp 233-239.
4. Vidhyasekaran, P.; Lalithakumari, D.; Govindaswamy, C. V. *Indian Phytopathol*. 1972, 25, 240-45.
5. Kenn, N. T.; Ingham, J. L. *Phytochemistry* 1976, 15, 1794-95.
6. Ingham, J. L. *Phytochemistry* 1976, 15, 1971-93.
7. Aguamah, D. G.; Langcake, P.; Leworthy, D. P.; Page, J. A.; Pryce, R. J.; Strange, R. N. *Phytochemistry* 1981, 20, 1381-83.
8. Cooksey, C. J.; Gaviatt, P. J.; Richards, S. E.; Strange, R. N. *Phytochemistry* 1988, 27, 1015-16.
9. Wotton, H. R.; Strange, R. N. *J. Gen. Microbiol*. 1985, 131, 487-94.
10. Diener, U. L.; Cole, R. J.; Sanders, T. H.; Payne, G. A.; Lee, L. S.; Klich, M. A. *Ann. Rev. Phytopathol*. 1987, 25, 249-70.
11. Diener, U. L.; Pettit, R. E.; Cole, R. J. In *Peanut Science and Technology*; Pattee, H. E.; Young, C. T., Eds.; American Peanut Research and Education Society, Inc.: Yoakum, Texas, 1982; pp 486-519.
12. Blankenship, P. D.; Cole, R. J.; Sanders, T. H.; Hill, R. A. *Mycopathologia* 1984, 85, 69-74.
13. Cole, R. J.; Hill, R. A.; Blankenship, P. D.; Sanders, T. H.; Garren, K. H. *Dev. Ind. Microbiol*. 1982, 23, 229-36.
14. Dickens, J. W.; Satterwhite, J. B.; Sneed, R. E. *J. Am. Peanut Res. Educ. Soc*. 1973, 5, 48-58.
15. Hill, R. A.; Blankenship, P. D.; Cole, R. J.; Sanders, T. H. *Appl. Microbiol*. 1983, 45, 628-33.
16. Pettit, R. E.; Taber, R. A.; Schroeder, H. W.; Harrison, A. L. *Appl. Microbiol*. 1971, 22-629-34.
17. Wilson, D. M.; Stansell, J. R. *Peanut Sci*. 1983, 10, 54-56.
18. Cole, R. J.; Sanders, T. H.; Hill, R. A.; Blankenship, P. D. *Mycopathologia* 1985, 91, 41-46.
19. Sanders, T. H.; Cole, R. J.; Blankenship, P. D.; Hill, R. A. *Peanut Sci*. 1985, 12, 90-93.
20. Sanders, T. H.; Hill, R. A.; Cole, R. J.; Blankenship, P. D. *J. Am. Oil Chem. Soc*. 1981, 58, 966A-70A.
21. Keen, N. T. *Phytopathology* 1975, 65, 91-92.

22. Dorner, J. W.; Cole, R. J.; Sanders, T. H.; Blankenship, P. D. *Mycopathologia* 1989, 105, 117-28.
23. Wotton, H. R.; Strange, R. N. *Appl. Environ. Microbiol.* 1987, 53, 270-73.
24. Arnoldi, A.; Carughi, M.; Fareina, G.; Merlini, L.; Parrino, M. G. *J. Agric. Food Chem.* 1989, 37, 508-12.

RECEIVED May 16, 1990

Chapter 25

Phototoxic Metabolites of Tropical Plants

Lee A. Swain[1] and Kelsey R. Downum[2]

[1]Department of Biological Sciences, Florida International University, Miami, FL 33199
[2]Fairchild Tropical Gardens, 10901 Old Cutler Road, Miami, FL 33156

Over 400 species of tropical plants from 76 families were assayed for phototoxic activity. Furanocoumarins, an important class of phototoxins, were identified from three genera of the Moraceae (fig family). Dorstenia, an herbaceous member of this family, was particularly rich in these metabolites. The distribution of furanocoumarins in the Moraceae as well as evolutionary and ecological aspects are discussed.

Phototoxic phytochemicals, or "photosensitizers", exhibit broad spectrum biocidal activity against a range of organisms including viruses, bacteria, fungi, nematodes, insects and other plants (1-12). They are also responsible for causing serious health problems in range animals and man (13-19). After absorbtion of light (usually in the UV-A range, 320-400nm), photosensitizers undergo a transition to an excited state. Some of them (type II phototoxins) can transfer the excitation energy to molecular oxygen to form singlet oxygen, which is capable of oxidizing many biomolecules. Cell membranes and cell wall components are particularly vulnerable to these types of phototoxins (20). Others (photogenotoxins) react directly with DNA and RNA, though reactions with other cellular components are known (21-25). Although most of the information concerning the biological activity of these phototoxins has been demonstrated in vitro, it is likely that they provide a viable defensive mechanism against non-adapted organisms. Phototoxins in the wild parsnip, for example, are effective in killing or deterring generalist insects such as the armyworm (Spodoptera eridania), while they have little effect on swallowtail butterfly larvae (Papilio polyxenes) which utilize the wild parsnip as a food plant (26-31). Swallowtail caterpillars are able to detoxify the chemicals and excrete them as harmless waste products (32-37).

0097–6156/91/0449–0361$06.00/0
© 1991 American Chemical Society

Distribution of phototoxins

Various biosynthetic classes of phototoxic compounds have been isolated from over thirty plant families (5). Some plant families produce more than one type of phototoxin. Three different types of phototoxins have been isolated from the Apiaceae (celery family), for example, and eight different types have been identified so far from the Rutaceae (citrus family). Among other families that contain phototoxins are the Asteraceae (sunflower family), Euphorbiaceae (spurge family), Fabaceae (pea family) and Moraceae (fig family) (5).

It was suggested that photosensitizers would be most effective in plant species that evolved under high-light environments and that the incidence of phototoxin-containing plants might be greater in these areas (2). A recent survey for phototoxins in one high-light environment, the desert southwestern United States, suggested that light-activated phytochemicals were relatively common in families in which phototoxins had previously been described (38). Over 35% of the extracts from members of the Asteraceae exhibited phototoxic activity, with many of these belonging to a limited number of tribes of this family. Almost half of the members of the tribe Heliantheae, for example, tested positive for light-activated toxins. All of the extracts of the Pectidinae, a subtribe within the Heliantheae, were phototoxic toward the test organisms.

Photobiocides from tropical plants

We recently surveyed a cross-section of plants from many tropical regions of the world in a search for photosensitizers to further test the above hypothesis. The methods used to test for phototoxic phytochemicals are described in detail elsewhere (39). Briefly, methanolic extracts were spotted onto sterile filter-paper discs and allowed to dry. The dried discs were placed onto replicate nutrient agar plates that had been spread with E. coli B/r (a UV resistant bacterium). The plates were incubated in the dark at 37°C for 30 min. Half of the plates were irradiated for 60 min. with eight Sylvania F40BLB UVA lamps (18W m^{-2}), while the other half were kept in the dark. All plates were incubated overnight in the dark at 37°C, after which the zones of inhibition surrounding the filter paper discs were measured.

Over 400 tropical/subtropical species representing 76 families were collected at either Fairchild Tropical Gardens or Chapman Field, USDA Plant Induction Station in Miami, FL. Table I lists the number of genera and species of the families that were examined. Only five of the 76 families tested elicited phototoxic responses from E.coli. These families are listed in boldface caps in Table I. The Asteraceae has been studied extensively in regards to its phototoxic components (38), as have members of the Rutaceae (citrus family; 5,7)and the Moraceae (fig family; 40). The compound(s) responsible for phototoxicity in the Sapotaceae (sapodilla family) is not known and is the only instance of light-enhanced toxicity from this family that has been reported to date (41). Elucidation of the structure of this phytochemical is in progress and we are also examining other members of the Sapotaceae for phototoxicity.

Biocidal compounds of the Moraceae

In contrast to the Asteraceae and the Rutaceae, the Moraceae is almost

Table I. Tropical plants examined for phototoxic antimicrobial properties (#genera,#species examined).

Acanthaceae (6,9)	***HYPERICACEAE (1,2)**
Amaryllidaceae (2,2)	Iridaceae (1,1)
Angiopteridaceae (1,1)	Lamiaceae (2,2)
Annonaceae (4,5)	Lecythidaceae (1,1)
Apocynaceae (9,11)	Liliaceae (1,1)
Aquifoliaceae (1,1)	Lythraceae (1,1)
Araceae (1,1)	Malphigiaceae (3,5)
Araucariaceae (1,1)	Malvaceae (1,1)
Arecaceae (1,1)	Meliaceae (5,6)
Aristolochiaceae (1,3)	Menispermiaceae (2,2)
Asclepiadaceae (1,1)	***MORACEAE (7,86)**
***ASTERACEAE (4,5)**	Myrtaceae (1,1)
Barringtoniaceae (1,1)	Onagraceae (1,1)
Bignoniaceae (13,15)	Oxalidaceae (1,2)
Bombacaceae (3,3)	Phileiaceae (1,1)
Bromeliaceae (2,2)	Piperaceae (1,1)
Buddlejacaceae (1,1)	Pittosporaceae (1,4)
Burseraceae (1,3)	Poaceae (1,1)
Cactaceae (1,1)	Podocarpaceae (1,5)
Capparidaceae (1,1)	Polygalaceae (1,2)
Caryophyllaceae (1,1)	Polygonaceae (4,9)
Celastraceae (2,2)	Rhamnaceae (2,3)
Chrysobalanaceae (1,1)	Rosaceae (1,1)
Cistaceae (1,2)	Rubiaceae (6,6)
Clusiaceae (2,2)	***RUTACEAE (23,57)**
Combretaceae (5,11)	Sapindaceae (3,4)
Convolvulaceae (2,2)	***SAPOTACEAE (7,21)**
Costaceae (1,1)	Selaginellaceae (1,1)
Cupressaceae (2,3)	Simaroubaceae (3,3)
Cyperaceae (1,1)	Solanaceae (4,8)
Ehretiaceae (2,9)	Sterculiaceae (1,1)
Eleagnaceae (1,1)	Taxaceae (1,1)
Empetraceae (1,1)	Theophrastinaceae (1,3)
Ericaceae (1,1)	Ulmaceae (1,1)
Euphorbiaceae (4,4)	Urticaceae (2,4)
Fabaceae (20,36)	Verbenaceae (4,4)
Flacourtiaceae (7,9)	Zamiaceae (2,2)
Heliconiaceae (1,1)	Zygophyllaceae (3,4)

exclusively a tropical/subtropical family (42). Many members of this family are economically important (42-44). Besides the edible fig (mainly Ficus carica), the jakfruit (Artocarpus heterophyllus) and breadfruit (A. atilis) are important food sources throughout the tropics. In addition to their importance as food plants, many members of this family are widely used in treating diseases and other health problems (43-45). Species of Dorstenia are used for everything from mouthwash and hangover cures to emetics and diuretics. Other members provide paper (Broussonetia papyrifera) and rubber (Castilloa elastica), while still others are popular ornamentals. A summary of phototoxicity in the various genera of the Moraceae is shown in Table II (40). Of the eight genera tested, only nine

Table II. Distribution of phototoxic activity in extracts of various members of the Moraceae.

Genus	#species assayed	#species with phototoxic activity
Artocarpus	4	0
Brosimum	3	0*
Cecropia	2	0
Cudrania	1	0
Dorstenia	5	5
Fatoua	1	1
Ficus	69	2**
Morus	2	0

* 1 species listed by Murray, et al., 1982.
** 5 additional species listed by Murray, et al., 1982.

species from three genera elicited phototoxic responses from E.coli. Although 72 species of Ficus were assayed, only two, or about 3%, showed phototoxic activity. In contrast, all five species of Dorstenia were phototoxic. Only one species of Fatoua was assayed, which was the third genus that tested positive for photosensitizers.

HPLC analysis was performed on the extracts of the Moraceae, and a number of furanocoumarins were identified including psoralen (**I**) and 5-methoxypsoralen (**II**) (5-MOP). Furanocoumarins are potent photosensitizers and their presence in a small number of Ficus species has already been reported (46-47). Table III lists the distribution of furanocoumarins in Ficus. One important point that should be noted is the limited number of Ficus species from which furanocoumarins have been identified. The actual number of Ficus species tested for these compounds is unknown, but furanocoumarins have been detected in only seven species from a genus with roughly 1000 members.

Furanocoumarins were also detected in Dorstenia and Fatoua, the only

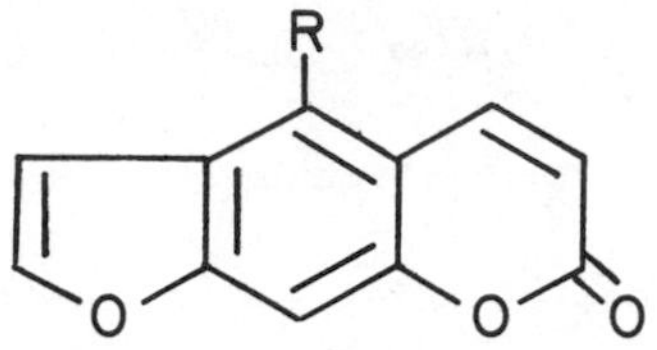

(I), R = H
(II), R = OCH_3
(III), R = $OOCCH(CH_3)\underset{O}{CHCH}CH_2CH{=}(CH_3)_2$

Table III. Distribution of furanocoumarins in Ficus determined by HPLC.

Species	psoralen	5-MOP	8-MOP
F. asprima	-	+	-
F. carica	+	+	-
F. palmata	-	+	-
F. pumila	+	+	-
F. religiosa	-	+	-
F. salicifolia	+	+	-
F. sycomorus	+	+	-

+ detected from species; - not detected from species.

two herbaceous genera in the Moraceae. In addition to psoralen and 5-MOP, a new furanocoumarin was detected. After NMR and mass spectral analysis, the compound was identified as the furanocoumarin 5-EDOP (**III**) (48). Unlike psoralen and 5-MOP which are highly phototoxic, 5-EDOP was only slightly antibiotic, and the activity was not enhanced by UVA irradiation. Table IV displays the distribution of furanocoumarins in Dorstenia and Fatoua. Psoralen and 5-MOP, the two highly phototoxic furanocoumarins, are mainly found in root and flowers, particularly in roots. 5-EDOP, the nonphototoxic furanocoumarin is the major furanocoumarin in Dorstenia and is particularly concentrated in leaf tissue. In addition, 5-EDOP was identified in all species of Dorstenia but was absent from Fatoua.

Table IV. Distribution of furanocoumarins in Dorstenia and Fatoua.

Taxon	psoralen	5-MOP	5-EDOP
Dorstenia contrajerva			
leaves	-	-	+++
flowers	+	++	++
roots	++	+++	+++
Dorstenia foetida			
leaves	-	+	++
roots	+	+	+
Dorstenia zanzibrica			
leaves	+	++	++
flowers	+	++	++
roots	++	+++	++
Dorstenia sp. (FTG 80-207)			
leaves	-	+	+++
flowers	-	++	+++
roots	+	+++	++
Dorstenia sp. (FTG 80-506)			
leaves	-	-	+++
flowers	-	+	+++
roots	+	++	++
Fatoua villosa			
leaves	-	-	-
flowers	-	+	-
roots	-	++	-

- = not detected; + = <10 ug/gdw; ++ = 10-100 ug/gdw; +++ = >100 ug/gdw.

The distribution of furanocoumarins in the Moraceae raises an important point. While all of the species of Dorstenia produce furanocoumarins, this ability is found in only a small percentage of Ficus species. The evolutionary relationships within the Moraceae are still in question (42), but it is generally agreed that Dorstenia is an evolutionary advanced genus in the family. It is possible and even probable that the ancestor to this genus possessed the ability to synthesize furanocoumarins. It may even be possible that Dorstenia evolved from a furanocoumarin-producing species of Ficus.

From an ecological perspective, the presence of furanocoumarins in Dorstenia and Fatoua is interesting as well. In contrast to the rest of the Moraceae, both Dorstenia and Fatoua are herbaceous. This might suggest that furanocoumarins serve a more important purpose in small, herbaceous plants than they do in large, woody plants. Even limited herbivory would have a severe effect on plants with small leaf areas such as Dorstenia and Fatoua due to loss of limited nutrients and photosynthetic ability. Although 5-EDOP does not demonstrate the phototoxic ability of either psoralen or 5-MOP, it may serve as feeding deterrent to potential herbivores. We are currently preparing to test this hypothesis.

A final observation brought to light by these data involves the distribution of furanocoumarins throughout the individual Dorstenia plants. The phototoxic furanocoumarins, psoralen and 5-MOP are more concentrated in the roots of Dorstenia than above ground parts while the highest concentrations of the nonphototoxic furanocoumarin, 5-EDOP, are in the leaves. A first glance, it would seem logical for phototoxic chemicals to be concentrated in an environment where light was present in order to utilize the full potential of their toxicity. Although some light is transmitted from leaves to other parts of the plant (49), the levels of activating wavelengths are probably too low to elevate furanocoumarins to their reactive state. Recently, however, it was suggested that mechanisms other than light may activate these phototoxins (50). Photochemical-type reactions can occur in the absence of light (51-52) and it has been suggested that enzymes such as peroxidase may be capable of catalyzing such reactions. Peroxidase is a common enzyme in plants and levels of peroxidase increase drastically in the roots of many plants that have been infected or wounded by invaders. This enzyme may provide the means by which furanocoumarins could be activated to their most toxic state in these tissues.

Conclusions

Plants are capable of producing a seemingly limitless number of compounds, many of which have proven to be toxic to one or more types of organisms. Phototoxic metabolites are more limited in their distribution, but have a broad spectrum of biocidal activity and appear to provide their hosts with formidable chemical defenses against potential invaders. Certain plant families such as the Rutaceae, Asteraceae, and Apiaceae are well known for their phototoxic abilities while this characteristic is less common in others.

The distribution of phototoxic furanocoumarins in the Moraceae provides us with an opportunity to study what role these chemicals may have played in the evolution of this family as well as what function they may serve today. The concentration of phototoxins in the roots of Dorstenia is by no means unique. Phototoxic thiophenes and polyacetylenes are commonly found in the roots of members of the Asteraceae, so similar roles for these compounds appears plausible. Further investigations are needed to clarify their importance to the host plants, and plants such as Dorstenia can serve as useful tools in this venture.

Acknowledgments

We thank Dr. J.M.E. Quirke, Dr. Stephen A. Winkle, Dr. Mehrzad Mehran, and Lavina Faleiro for their technical assistance. We also thank Dr. Robert Knight and the USDA Plant Induction Station at Chapman Field for many of the plant specimens. This work was supported by grants from the Whitehall Foundation and NSF (BBS-8613863) to KRD.

Literature cited

1. Towers, G.H.N. J. Chem. Ecol. 1986, 12, 813-821.
2. Downum, K.R.; Rodriguez, E. J. Chem. Ecol. 1986, 12, 823-834.
3. Towers, G.H.N. Can. J. Bot. 1984, 62, 2900-2911.
4. Arnason, T.; Swain, T.; Wat, C.-K.; Graham, E.A.; Partington, S.; Towers, G.H.N. Biochem. Syst. Ecol. 1981, 9, 63-68.
5. Downum, K.R. In Natural Resistance of Plants to Pests: Roles of Allelochemicals; Green, M.; Hedin, P., Eds.; ACS Symposium Series No. 296; American Chemical Society: Washington D.C., 1986, pp.197-205.
6. Knox, J.P.; Dodge, A.D. Phytochem. 1985, 24, 889-896.
7. Downum, K.R.; Nemec, S. In Light-Activated Pesticides; Heitz, J.R.; Downum, K.R., Eds.; ACS Symposium Series No. 339; American Chemical Society: Washington D.C., 1987, pp. 281-294.
8. Tuveson, R.W.; Berenbaum, M.R.; Heininger, E.E. J. Chem. Ecol. 1986, 12, 933-948.
9. Marston, A.; Hostettmann, K. Phytochem. 1985, 24, 639-652.
10. Hanson, C.V.; Riggs, J.L.; Lennette, E.H. J. Gen. Virol. 1978, 40, 345.
11. Specht, K.G.; Kittler, L.; Midden, W.R. Photochem. Photobiol. 1988, 47, 537.
12. Hudson, J.B. Antiviral Compounds from Plants; CRC Press: Boca Raton, 1990, pp. 67-81.
13. Oertli, E.H.; Rowe, L.D.; Lovering, S.L.; Ivie, G.W.; Bailey, E.M. Am. J. Vet. Res. 1983, 44, 1126-1129.
14. Clare, N.T. Adv. Vet. Sci. 1955, 2,182-211.
15. Martins, J.E.C.; Puzetti, G.L.; Sodre, M. Derm. Int. 1974, 13, 124.
16. Birmingham, D.J.; Key, M.; Tubich, G.E.; Perone, V.B. Arch. Dermatol. 1961, 83, 73.
17. Ivie, G.W. In Effects of Poisonous Plants on Livestock; Keeler, R.; Van Kampen, K.; James, L., Eds.; Academic Press: New York, 1977, pp.475-485.
18. Fitzpatrick, T.B.; Pathak, M.A. J. Invest. Dermatol. 1959, 32, 229.

19. Mitchell, J.; Rook, A. Botanical Dermatology; Greengrass: Vancouver, 1979, pp.471-482.
20. Girotti, A.W. Photochem. Photobiol. 1983, 38, 745-751.
21. Musajo, L.; Rodighiero, G. In Photophysiology, VII; A.C.Giese, Ed.; Academic Press: New York, 1972, pp.115-147.
22. Song, P.; Tapley, K.J. Photochem. Photobiol. 1979, 29, 1177-1197.
23. Scheel, L.D.; Perone, V.B.; Larkin, R.L.; Kupel, R.E. Biochemistry 1963, 2, 1127.
24. Veronese, F.M.; Schiavon, O.; Bevilacqua, R.; Bordin, F.; Rodighiero, G. Photochem. Photobiol. 1982, 36, 25-30.
25. Granger, M.; Helene, C. Photochem. Photobiol. 1983, 38, 563-568.
26. Berenbaum, M. Science, 1978, 201, 532-534.
27. Berenbaum, M. Ecol. Entomol. 1981, 6, 345-351.
28. Berenbaum, M. Ecology 1981, 62, 1254-1266.
29. Berenbaum, M. Evolution 1983, 37, 163.
30. Berenbaum, M. In Light-Activated Pesticides; Heitz, J.R.; Downum, K.R., Eds. ACS Symposium Series No. 339; American Chemical Society: Washington D.C., 1987, pp. 206-216.
31. Berenbaum, M.; Feeny, P. Science 1981, 212, 927-929.
32. Ivie, G.W.; Bull, D.L.; Beier, R.C.; Pryor, N.W. J. Chem. Ecol. 1986, 12, 871-884.
33. Bull, D.L.; Ivie, G.W.; Beier, R.C.; Pryor, N.W. J. Chem. Ecol. 1986, 12, 885-892.
34. Ivie, G.W.; Bull, D.L.; Beier, R.C.; Pryor, N.W. Science 1983, 221, 374-376.
35. Bull, D.L.; Ivie, G.W.; Beier, R.C.; Pryor, N.W.; Oertli, E.H. J. Chem. Ecol. 1984, 10, 893-911.
36. Ivie, G.W. In Light-Activated Pesticides; Heitz, J.R.; Downum, K.R., Eds. ACS Symposium Series No. 339; American Chemical Society: Washington D.C., 1987, pp. 217-230.
37. Larson, R.A. J. Chem. Ecol. 1986, 12, 859-870.
38. Downum, K.R.; Villegas, S.; Rodriguez, E.; Keil, D.J. J. Chem. Ecol. 1989, 15, 345-355.
39. Downum, K.R.; Hancock, R.E.W.; Towers, G.H.N. Photochem. Photophys. 1983. 6, 145-152.
40. Swain, L.A.; Downum, K.R. Biochem. Syst. Ecol. 1990. (in press).
41. Downum, K.R.; Swain, L.A. Fairchild Tropical Gard. Bull. 1988, 43, 6-11.
42. Corner, E.J.H. Gardens Bulletin (Singapore). 1962, 19, 187-252.
43. Morton, J. Fruits from Warm Climates; J.Morton: Miami, 1987, pp.47-64.
44. Usher, G. A Dictionary of Plants Used by Man; Constable & Co., Ltd.: London, 1974, pp. 62-63, 254-257.
45. Ayensu, E.S. Medicinal Plants of the West Indies; Reference Publications: Algonac, MI, 1981, pp.127-129.
46. Murray, R.D.H.; Mendez, J.; Brown, S.A. The Natural Coumarins; J.Wiley & Sons: New York, 1982.

47. Yarosh, E.A.; Nikonov, G.K. Khim. Prir. Soedin. 1973, 9, 269-270.
48. Swain, L.A.; Quirke, J.M.E.; Winkle, S.A.; Downum, K.R. Phytochem. 1990, (in review).
49. Mandoli, D.F.; Briggs, W.R. Photochem. Photobiol. 1984, 39, 709-715.
50. Gommers, F.J.; Bakker, J. In Chemistry and Biology of Naturally-occurring Acetylenes and Related Compounds; Lam J.; Breteler, H.; Arnason, T.; Hansen, L., Eds. Elsivier: Amsterdam, 1988, pp. 61-69.
51. Cilento, G. Photochem. Photobiol. Rev. 1980, 5, 199-228.
52. Cilento, G. Pure Appl. Chem. 1984, 56, 1179-1190.

RECEIVED July 31, 1990

Chapter 26

Photosensitizing Porphyrins as Herbicides

Stephen O. Duke, José M. Becerril[1], Timothy D. Sherman, and Hiroshi Matsumoto[2]

Southern Weed Science Laboratory, Agricultural Research Service, U.S. Department of Agriculture, P.O. Box 350, Stoneville, MS 38776

Several porphyrin intermediates of heme and/or chlorophyll biosynthesis are potent photosensitizers which generate high levels of singlet oxygen in the presence of molecular oxygen and light. Many compounds that affect the heme and/or chlorophyll pathways are strongly herbicidal due to accumulation of phytotoxic levels of these porphyrins in response to the chemical. For instance, several commercial and experimental herbicides inhibit protoporphyrinogen oxidase, the enzyme that converts protoporphyrinogen to protoporphyrin IX (PPIX). This leads to uncontrolled autooxidation of the substrate and results in massive accumulation of PPIX. In plants treated with these herbicides, damage is light dependent and closely correlated with the level of PPIX that accumulates. PPIX accumulation is apparently largely extraplastidic. Treatment with the porphyrin precursor δ-aminolevulinic acid (ALA), in combination with the heme and chlorophyll pathway inhibitor 2,2′-dypyridyl (DP), results in the accumulation of toxic levels of primarily Mg-PPIX monomethylester. DP deregulates porphyrin synthesis and ALA provides additional substrate. DP and other chlorophyll synthesis modulators in combination with ALA can increase the selectivity as well as enhance the efficacy of ALA as a herbicide. Exogenously applied porphyrins are far less effective as herbicides than treatment with compounds that cause plants to accumulate their own porphyrins.

Photodynamic compounds, including many natural products, have been proposed for use as herbicides by many researchers (*e.g.*, 1-3).

[1]Current address: Universidad del Pais Vasco–EHU, Facultad de Ciencias, Dept. Fisiologiá Vegetal y Ecologiá, Apartado 644, 48080 Bilbao, Spain

[2]Current address: Institute of Applied Biochemistry, University of Tsukuba, Tsukuba, Ibaraki 305, Japan

This chapter not subject to U.S. copyright
Published 1991 American Chemical Society

However, a problem with these chemicals is their indiscriminate toxicity in the presence of light and molecular oxygen. This toxicity precludes their use as pesticides. Safe alternatives are to treat target organisms with chemicals that either are selectively metabolized to photodynamic compounds or cause the target organism to produce toxic levels of natural photodynamic compounds with its own biochemical machinery. This review examines one aspect of the latter alternative - treatment of plants with compounds that cause the accumulation of herbicidal levels of photodynamic porphyrins.

Chlorophyll is a photodynamic compound when it is not a component of the photosynthetic apparatus. One way in which the plant protects itself from the photodynamic properties of chlorophyll is through linking chlorophyll to a biochemical complex that dissipates the energy of light-energized chlorophyll through splitting of water and energizing of electrons from water for energy transduction and photosynthetic reducing power. Intermediates of chlorophyll biosynthesis are photodynamic also. Since they cannot be utilized in photosynthesis and are photodynamic, there is strong selection pressure against accumulation of these compounds. Mutations that cause the accumulation of these compounds are deleterious to the plant. For instance, yellow mutants of maize have been described that accumulate high levels of protoporphyrin IX (PPIX) (4). Although these mutants have impaired chlorophyll synthesis, part of the phytotoxic effect of the mutation is due to the photodynamic effect of PPIX.

Two approaches to stimulation of porphyrin accumulation in plants have been taken. The first is to supply the plant with the porphyrin precursor δ-aminolevulinc acid (ALA) along with compounds that affect the porphyrin pathway. The second is to block porphyrin synthesis at the protoporphyrinogen oxidase step in the pathway, thereby deregulating the pathway and causing accumulation primarily of PPIX.

ALA as a Herbicide

Rebeiz et al. (5) introduced the concept of ALA in combination with various chlorophyll synthesis modulators as a herbicide. Previous literature had demonstrated that ALA treatment of plant tissues could cause the accumulation of abnormally high levels of coproporphyrin, protochlorophyllide (PChlide), PPIX, and Mg-protoporphyrin IX monomethyl ester (MgPPIXME) (6, 7). In cucumber seedlings sprayed with 10 to 20 mM ALA, and then placed in the dark for 17 h to allow chlorophyll precursors to accumulate, a two- to four-fold increase in total porphyrins (primarily PChlide) was observed (5). This led to 95 % photodynamic damage to the seedlings after they were placed in the light. The effective level of applied ALA could be reduced by spraying it in combination with 2,2′-dipyridyl (DP), a relatively inexpensive synthetic compound that stimulates porphyrin synthesis by preventing heme synthesis through chelating iron (8). The porphyrin synthesis pathway is under strong feedback control by heme (9, 10). In addition to stimulating porphyrin synthesis, DP blocks conversion of MgPPIXME to PChlide (8, 9) (Fig. 1). Thus, both the quantity and type of porphyrins that accumulate are affected.

ALA plus DP acted synergistically as herbicides, despite the fact that there was generally only an additive effect on total porphyrin accumulation (Table I). These results indicate one or more of the following: (a) one of the earlier chlorophyll intermediates (MgPPIXME or PPIX) is more herbicidal than PChlide, (b) the porphyrins act synergistically, or (3) that the synergism of ALA plus DP is not based on their effects on porphyrins.

Table I. Effects of ALA and DP, alone or in combination, on porphyrin accumulation and herbicidal damage. Cucumber seedlings (6-day-old) were assayed for porphyrins after being sprayed with the herbicides and incubated in darkness for 17 h. Herbicidal damage was assessed 10 days after porphyrins were assayed and during which they were exposed to greenhouse light conditions. Taken from ref. (5).

	Porphyrins (nmoles/100 mg protein)				
Treatment	PChlide	MgPPIXME	PPIX	Total	Damage (%)[a]
Control	17.3	0.6	0.0	17.9	0
5 mM ALA	100.7	1.6	0.0	101.3	30
15 mM DP	24.0	12.3	2.6	38.9	10
5 mM ALA + 15 mM DP	121.1	26.3	8.1	155.5	80

[a]per cent dead cotyledons

The first and second possibilities are complicated by the possibility that these porphyrins are all differentially photolabile and photodegradation products may be involved in their photodynamic action. For instance, PChlide disappears in green tissue rapidly after exposure to light (5) (Fig. 2), although the proportions converted to chlorophyll or photodegraded are not known. Some evidence indicates that most of the ALA-stimulated PChlide accumulation is photodegraded (6), although this may not always be the case. MgPPIXME levels in cucumber cotyledons decline less rapidly in light than do those of PChlide (Fig. 2). In normal etiolated plants, complete PChlide phototransformation of chlorophyllide (Chlide) in bright light is very rapid (usually less than a minute) and subsequent phytyllation of Chlide to form chlorophyll (Chl) is complete in 30 min or less (*e.g.*, 11). The prolonged decay of PChlide in tissue treated with ALA plus DP could be due to synthesis occurring more rapidly than normal rates of conversion to Chl, to slowed phototransformation and phytyllation, or to a large component of non-phototransformable (NPTF) PClide. Gassman (12) found that extended treatment of etiolated bean leaves with 10 mM ALA resulted in a greater porportion of NPTF PClide accumulation than in untreated leaves and that NPTF PClide inhibits accumulation of phototransformable (PTF) PClide. Thus, NPTF PClide may play a role as a photodynamic pigment in this system, although Rebeiz *et al.* (5) did not differentiate between PTF and NPTF PClide.

In the original system of Rebeiz *et al.* (5), a rather long post-spray dark period was required for sufficient accumulation of porphyrins for herbicidal activity occur. The herbicidal effect of ALA plus DP was age and species dependent; however, there was not always a strong correlation between the effect on porphyrin

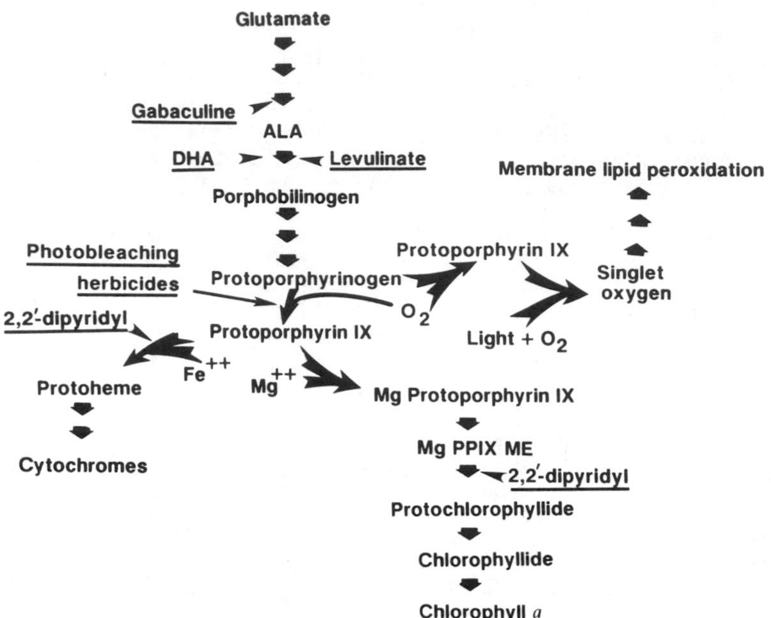

Figure 1. The porphyrin synthesis pathway and sites of inhibition of various inhibitors and modulators. Inhibitors are underlined and sites of inhibition are indicated

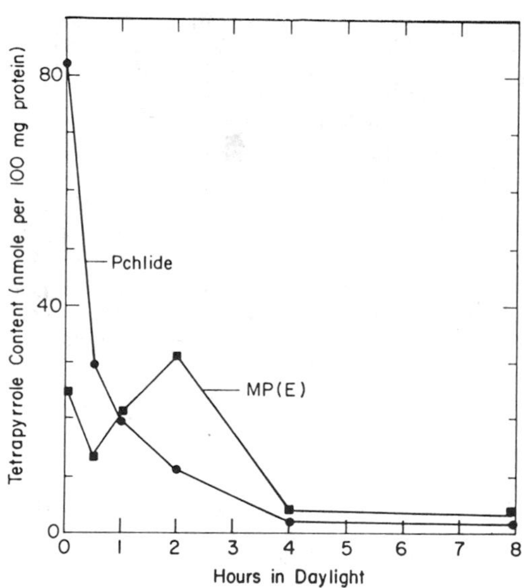

Figure 2. Time course of PChlide and Mg-PPIXME disappearance from ALA plus DP-treated cucumber seedlings in daylight after a 17-h accumulation period.

(Reprinted with permission from ref. 5. Copyright 1984 Butterworth Scientific, Ltd.)

synthesis and herbicidal damage. No herbicidal damage was observed when there was little or no effect on chlorophyll precursor levels but, in some tissues of some species, accumulation of porphyrins did not result in herbicidal damage (Table II). Generally, they found grasses, including maize, barley, oat, and wheat to be tolerant to these herbicides, while all dicot weeds tested were highly sensitive. Thus, this herbicide combination showed promise for broadleaf weed control in grain crops.

Table II. Effects of ALA (5mM) plus DP (15 mM) on porphyrin accumulation in seedling tissues of several plant species. Plant were assayed for porphyrins after being sprayed with the herbicides and incubated in darkness for 17 h. Herbicidal damage was assessed 10 days after porphyrins were assayed and subsequent exposure to greenhouse light conditions. Taken from ref. (5).

Species (Seedling age)	Porphyrins (nmoles/100 mg protein) PChlide		MgPPIXME		PPIX		Herbicidal Damage (%)
	Con	Treated	Con	Treated	Con	Treated	
Mustard leaves (12)	30	201	12	36	29	24	90
Cotton cotyledon (14)	18	37	4	9	0	0	63
Cotton stem (14)	4	4	1	1	0	0	0
Kidney bean leaf (9)	117	439	3	430	5	22	100
Kidney bean stem (9)	37	82	4	76	3	14	0
Giant foxtail (6)	8	79	0.4	12	0	14	S.N.[a]
Maize (9)	79	85	5	15	12	0	S.N.

[a]small necrotic areas

Rebeiz *et al*. (13) classified species into four different greening groups, based on their Chl synthesis heterogeneity. The four groups were categorized according to the predominance of divinyl or monovinyl porphyrin synthesis and under what conditions (dark or light) synthesis of each porphyrin type occurred. The four groups are described with representative species in Table III.

Table III. Greening groups and representative plant species.

Greening Group	Representative Species
1. Dark divinyl/light divinyl (DDV/LDV)	Cucumber, mustard, common purslane
2. Dark monovinyl/light divinyl (DMV/LDV)	Maize, wheat, barley, common bean, soybean, pigweed
3. Dark divinyl/light monovinyl (DDV/LMV)	Ginkgo
4. Dark monovinyl/light monovinyl (DMV/LMV)	Apple, johnsongrass

They hypothesized that the sensitivity of a species to ALA plus chlorophyll modulators is due to both extent of porphyrin synthesis and the chemical nature of the accumulated porphyrins. Thus, the greening group to which a species belongs could strongly influence its susceptibility. It followed that since some species accumulate relatively large amounts of porphyrins in response to ALA, but display little herbicidal damage, some porphyrins might be relatively poor photosensitizers. This hypothesis was tested by using ALA in combination with chlorophyll synthesis modulators other than DP (13).

Although the results of these extensive experiments are not easily interpreted, the hypotheses were made (a) that PChlide is the most important and ubiquitous photodynamic species caused to accumulate by ALA-based treatments, (b) that MV PChlide is a more effective photodynamic pigment than DV PChlide in DDV/LDV and DMV/LDV species, and (c) that both DDV/LDV and DMV/LDV species are highly susceptible to a mixture of Mg-PPIX and Mg-PPIXME (14). The results are difficult to interpret because equimolar levels of different porphyrins were not produced and the combinations of porphyrins produced by different modulators varied with species. Potential differences in tolerance to toxic oxygen species between species were not considered. Others have attempted to explain differential sensitivity to porphyrin-generating herbicides between species (15) and between herbicide-sensitive biotypes within species (16) by differences in ability to detoxify toxic oxygen species. As with other herbicides, penetration of the leaf cuticle by ALA and/or DP can also play a role in differences in efficacy of this herbicide combination (17).

The Chl synthesis modulators that Rebeiz *et al.* (13, 14) used in conjunction with ALA could be divided into three categories: A) enhancers of ALA conversion to porphyrins (2-pyridine aldoxime, 2-pyridine aldehyde, picolinic acid, 2,2′dipyridyl disulfide, 2,2′-dipyridyl amine, 4,4′dipyridyl, and phenanthridine), B) inducers of ALA biosynthesis and porphyrin accumulation (2,2′-dipyridyl and 1,10-phenanthroline), and C) inhibitors of MV PChlide synthesis (2,3-dipyridyl, 2,4-dipyridyl, 1,7-phenanthroline, and 4,7-phenanthroline). Compounds in group A did not cause significant porphyrin accumulation alone; however, they enhanced dark conversion of exogenous ALA to porphyrins. This group was further subdivided into compounds that enhanced conversion of ALA to MV PChlide (2-pyridine aldoxime, 2-pyridine aldehyde, picolinic acid, and 2,2′-dipyridyl disulfide) and those that stimulated conversion to DV PChlide (4,4′dipyridyl, 2,2′dipyridyl amine, and phenanthridine). To qualify as an ALA biosynthesis and porphyrin accumulation inducer (category B), the compound had to cause these effects in the absence of ALA. Compounds in category C had to inhibit accumulation of MV PChlide with or without ALA. In most cases, in conjunction with ALA, the compounds stimulated DV PChlide accumulation compared to the ALA-treated control.

With knowledge of greening group characteristics and modulator type, one can theoretically manipulate the selectivity of ALA-modulator combinations. This approach was used to design a combination for control of creeping charlie (DDV/LDV) in Kentucky bluegrass (DMV/LDV)(13). Since the greening type of the two species diverged at night, a "dark spray" applied near dusk was theorized to be most selective. However, a final choice of the ALA plus DP combination was made since this combination led to the generally lethal accumulation of DV MgPPIXME and DV MgPPIX in most species.

However, Kentucky bluegrass was generally much less sensitive, despite these accumulations. The mechanism of tolerance was not explained.

Recently, Averin *et al.* (18) found that more ALA supplied to barley accumulated as porphobilinogen, rather than being converted to porphyrins, than in bean. Also, the porphyrins in barley were degraded in light much more rapidly than in bean. Thus, some of the selectivity of ALA plus DP may be influenced by many factors other than greening type. The ultrastructural development of damage from treatment with ALA plus DP in bean cotyledons occurs rapidly (19). Within 1 h of exposure to light chloroplasts swelled and became spherical. Subsequently, grana and stromal thylakoids swelled and destruction of intrachloroplast membranes was observed by 2 h. Within 6 to 24 h, chloroplasts and mitochondria were broken.

To date, only six publications (5, 13, 14, 17-19) exist on ALA as a herbicide. Another paper has been published on ALA as an insecticide (20). Although most of these are highly substantial papers, several questions remain regarding results of these studies. The actual relative phytotoxicity of various porphyrins is not clear. The intracellular site(s) of porphyrin accumulation are also not known. Furthermore, the complex interactions between greening type, tolerance to toxic oxygen species, and capacity to synthesize porphyrins is poorly understood.

Although ALA in combination with various Chl synthesis modulators has been patented for herbicide use, none of the combinations is presently commercially available. However, a large number of synthetic herbicides that act by causing the accumulation of photodynamic porphyrins are sold throughout the world.

Synthetic Herbicide-Induced Accumulation of Porphyrins

Historical background. Diphenyl ethers of the general structure shown below were introduced as commercial herbicides in the 1960's (21) and since that time many members of this herbicide class have been commercialized (22). All of the commercialized versions have been *para*-nitro substituted.

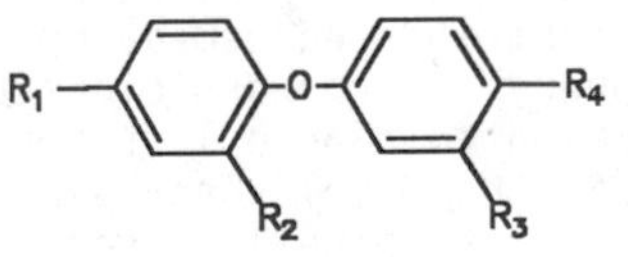

R_1 = CF_3, Cl

R_2 = Cl, NO_2

R_3 = OCH_3, $COOCH_3$, OC_2H_5, H

R_4 = NO_2, Cl, I, NO

These herbicides cause rapid bleaching and dessication of green tissues, similar to the effects of paraquat. Like paraquat, light is required for activity (22), however, unlike paraquat, photosynthesis is not a requirement for activity (23-27), except when photosynthesis is indirectly required for substrate (28) or generation of oxygen for lipid peroxidation (27). The development of injury to plant tissues affected by these compounds is much like that caused by photodynamic pigments. The first measureable effect is cellular leakage, followed sequentially by inhibited photosynthesis, ethylene evolution, ethane and malondialdehyde evolution, and finally bleaching of chloroplast pigments - all characteristic of photodynamic membrane lipid peroxidation (29). Furthermore, after a sufficiently long incubation in darkness, the herbicidal activity is almost entirely independent of temperature, like a photodynamic dye (30). It was obvious from several studies that the herbicide itself is not the photodynamic dye.

The most compelling evidence for this is that the action spectrum indicated that the photoreceptor for photodynamic damage is a visible pigment (31-33). Diphenyl ether herbicides absorb in the ultraviolet rather than the visible spectrum. Furthermore, there was no strong evidence that the diphenyl ether herbicide acted as a lipid-peroxidizing radical as the result of energy transfer from a photoreceptor, even though some nitrodiphenyl ether herbicides can be photoreduced to nitro radical anions by β-carotene (34). This, coupled with apparent evidence that carotenoids are involved in the mode of action of these herbicides (*e.g.*, 23, 24, 26, 35-37), led some to hypothesize that a carotenoid-diphenyl ether exciplex might be involved in the mechanism of action of these herbicides (38). In fact, oxyfluorfen-treated thylakoid membranes will generate singlet oxygen when exposed to light (39). Although some of these compounds can form radicals, there are diphenyl ether herbicides that do not form radicals that are quite effective as lipid-peroxidizing herbicides (40-42).

Despite investigations by many laboratories, the nature of the photoreceptor for the photodynamic damage remained an enigma for more than two decades. Studies demonstrating that there was a metabolic requirement before the herbicide could cause effects like a photodynamic dye (24, 28, 30, 43) should have provided a clue to the actual mechanism - the induction of the accumulation of a natural photodynamic compound.

Site of action. Matringe and Scalla (44, 45), followed closely by others (46, 47) reported that diphenyl ether herbicide-treated tissues accumulated abnormally high levels of PPIX. Furthermore, specific inhibitors of porphyrin synthesis could completely or almost completely prevent herbicidal damage from diphenyl ether herbicides (44-48) (Figs. 1 and 3) and the absorption spectrum of PPIX roughly fit the action spectra for the light-induced damage by these herbicides (31-33, 44). Non-diphenyl ether herbicides that had been observed to act in a similar fashion to diphenyl ethers (oxadiazon, the pyridine derivative LS 82-556, the novel phenylpyrazole TNPP-ethyl, and the cyclic imide chlorophthalim - see below) also caused treated plants to accumulate high levels of PPIX (44-46, 48-52). The cyclic imides, diphenyl ethers, and oxadiazoles had previously been demonstrated to inhibit chlorophyll synthesis (53-56), however, the connection to their lipid- peroxidizing action was not clear. To date, common structure/activity relationships between these diverse herbicide groups have not been determined.

Protoporphyrin IX-magnesium chelatase synthesizes Mg-PPIX from PPIX. Thus, it seemed likely that inhibition of this enzyme would lead to accumulation of PPIX (47, 49). However, Matringe *et al.* (57, 58) found that the accumulation of PPIX was due, in fact, to strong inhibition of protoporphyrinogen oxidase (Protox), the enzyme that converts protoporphyrinogen to PPIX (Fig. 1). These results were confirmed by Witkowski and Halling (59). Apparently, blockage at this site leads to autooxidation of the substrate to form PPIX, as has been observed when this enzyme is inactive due to genetic lesions in yeast and humans (60-62).

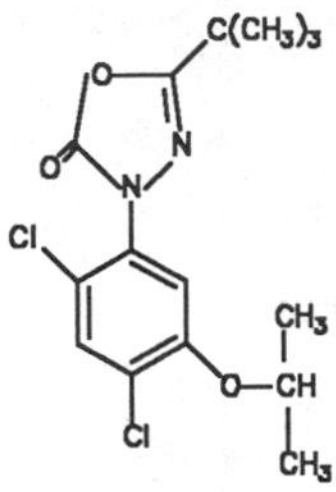

OXADIAZON 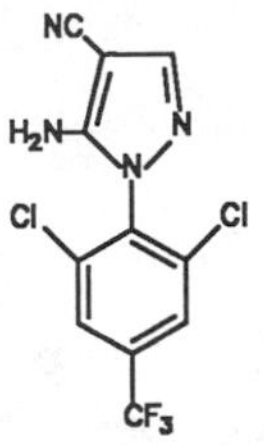 M & B 39279 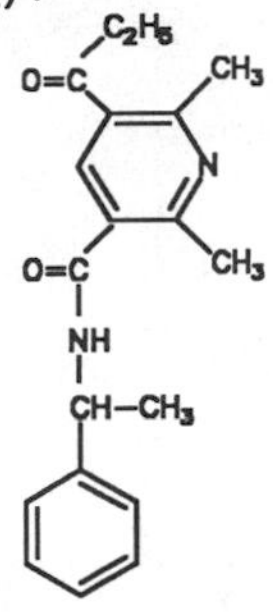 LS 82–556 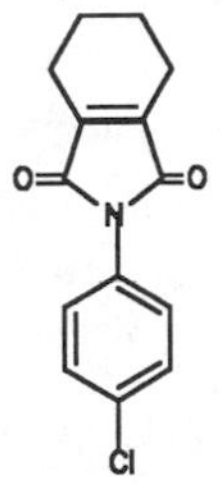 CHLOROPHTHALIM 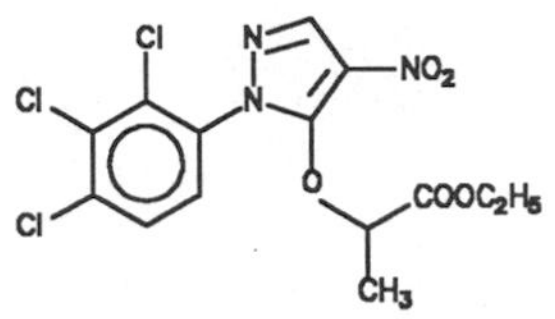 TNPP–ETHYL

Acifluorfen-methyl and LS 820340, *p*-nitro and *p*-chloro diphenyl ethers, respectively, had I_{50}'s of less than 1 μM for the synthesis of Mg-PPIX from ALA in a maize etioplast preparation (57). A herbicidally inactive analog of acifluorfen-methyl (RH 5348) had an I_{50} almost three orders of magnitude higher than that of acifluorfen-methyl. None of the compounds inhibited Mg-PPIX formation from PPIX. The I_{50}'s for maize etioplast Protox were about 10 nM for the herbicidal diphenyl ether. Similar results were obtained with potato, yeast, and mouse liver mitochondria. No significant inhibition of the PPIX ferrochelatase was measured. Similar results were observed with oxadiazon, LS 82-556, and M&B 39279 (58). M&B 39279 (see above) is a phenyl pyrazole that appears to have a mechanism of action similar to that of diphenyl ethers (63). The results of Witkowski and Halling (59) with acifluorfen-methyl on cucumber Protox were similar, although they found an I_{50} of about 30 nM. To date, no studies have been published on the type(s) of inhibition of Protox caused by these herbicides or on whether the binding sites for the different chemical types overlap.

Mode of action. PPIX is a photolabile compound, so the question of whether sufficient levels of it can exist *in vivo* for it to exert its effect is important. We found the half-life of PPIX in acifluorfen-treated cucumber tissue during exposure to bright light to be about 2.5 h (41) - sufficient time for it to be an effective herbicide. Furthermore, PPIX accumulated rapidly in bright light and did not begin to decrease until cellular damage was nearly complete.

Others have found little correlation between the herbicidal effects and the amount of PPIX accumulated by Protox-inhibiting herbicides (50, 64). In cucumber (Fig. 4), pigweed, and velvetleaf, we found a strong correlation between the amount of PPIX accumulated in response to acifluorfen and the amount of ensuing herbicidal damage (41). Also, there was an excellent correlation between the PPIX and the resulting herbicidal damage caused by a variety of diphenyl ether and oxadiazole herbicides (41).

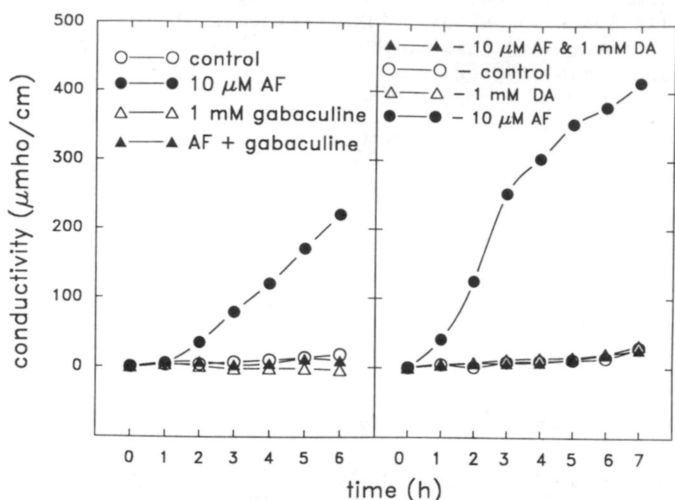

Figure 3. Effects of gabaculine and dioxoheptanoic acid (DA) on efficacy of acifluorfen (AF) on cellular leakage as measured by electrolyte increase in the bathing media of cucumber cotyledon discs incubated in the various treatment solutions for 20 h in darkness before exposure to light (time 0).

(Reprinted with permission from ref. 46. Copyright 1988 Academic Press.)

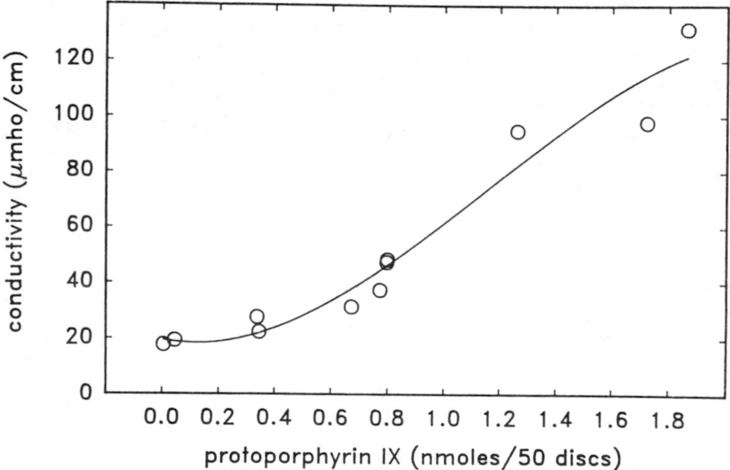

Figure 4. Relationship between cellular damage in green cucumber cotyledon discs and PPIX accumulation caused by exposure to various concentrations of acifluorfen. Cellular damage was assayed as in Fig. 3, 1 h after exposure to light. PPIX was assayed just before exposure to light and after a 20 h incubation in darkness.

(Reprinted with permission from ref. 41. Copyright 1989 Plant Physiology.)

Some laboratories (64-66) have found PChlide to be the primary porphyrin to accumulate in diphenyl ether-treated tissues. In our green cucumber cotyledon disc system, we have found only PPIX levels to be increased by these herbicides (67), however, in intact cucumber seedlings (Fig. 5) and tentoxin-affected cucumber cotyledon discs (68) we found diphenyl ether-enhanced PChlide levels. PPIX levels accumulate to many (as much as several hundred) times the control levels in herbicide-treated tissues, whereas the maximum PChlide accumulation is only four-fold that of the control. Since the metabolic block is at Protox, there is no inhibition of conversion of PChlide to Chl after exposure to light. Therefore, it seems unlikely that PChlide plays a significant part in the mechanism of action of these herbicides. Indeed, of the Chl precursors assayed, only PPIX significantly correlated with herbicidal damage caused by a several different acifluorfen/DP/ALA treatment combinations (67). No significant correlations were found with accumulated PChlide, Mg-PPIX, or Mg-PPIXME. Furthermore, neither of the photodynamic precursors of PPIX, coproporphyrinogen nor uroporpophyrinogen (extracted as coproporphyrin III and uroporphyrin III) accumulate in diphenyl ether herbicide-treated plant tissues (Table IV). All of our data are consistent with the view that PPIX is the primary photodynamic pigment involved in the mechanism of action of these herbicides.

Table IV. Effect of 10 µM acifluorfen on accumulation of PPIX its immediate precursors in green cucumber cotyledon discs during 20 h of dark incubation. Previously unpublished data of Matsumoto and Duke.

Porphyrin	Control	Treated
	-(nmoles/g fresh weight)-	
Uroporphyrin III	0.016 ± 0.005	0.025 ± 0.001
Coproporphyrin III	0.004 ± 0.001	0.014 ± 0.004
Protoporphyrin IX	0.008 ± 0.001	0.207 ± 0.015

Our laboratory (46) and that of Yanase and Andoh (52) found the herbicidal effect of ALA to be short-lived in the light compared to that of Protox-inhibiting herbicides. This is probably due to the fact that PChlide levels are rapidly reduced by conversion to Chl or by photodestruction in the light in ALA-treated plant tissues. The half-life of PChlide in cucumber in light was a half hour or less (Fig. 2), whereas, the half-life of PPIX in cucumber cotyledons discs was more than 2 h (41).

The total amount of porphyrins that accumulate in diphenyl ether-treated tissues is considerably higher than that which accumulates in untreated tissues (67). Therefore, these herbicides appear to increase carbon flow into this pathway. Heme is known to feedback inhibit this pathway and PPIX is required for heme synthesis (9). Feeding heme to diphenyl ether-treated plant cells reduces the amount of porphyrin-caused lipid peroxidation of cell homogenates as measured by oxygen consumption (10).

Why PPIX formed by nonenzymatic oxidation of protoporphyrinogen does not immediately reenter the chlorophyll pathway is probably due to a requirement for accumulation of a threshold level or saturation of a non-available pool before reentry via a non-pathway route can occur. If substrate channeling, as in heme synthesis in mouse mitochondria (69), occurs with conversion of protoporphyrinogen to PPIX, PPIX formed by autooxidation may be outside the normal metabolic channel. Thus, only after a sufficient concentration of PPIX builds up outside the normal metabolic channel will it reenter the pathway. The simultaneous kinetics of PPIX and PClide accumulation in acifluorfen-treated, yellow (tentoxin-treated) cucumber tissues support this hypothesis (68). High levels of PPIX accumulate before the herbicide enhances PClide accumulation.

Protox is thought to be a membrane-bound enzyme; probably in or on the plastid envelope (Fig. 6). Fluorescence microscopy of achlorophyllous tissues treated with acifluorfen in darkness, results in strong porphyrin fluorescence in both plastids and cytoplasm, whereas, fluorescence was localized almost exclusively in plastids of untreated cells (68). PPIX concentrations were almost 200-fold greater in treated than untreated tissues. These data suggest that PPIX leaks from plastids or plastid envelopes of treated tissues and that this leakage is independent of membrane damage due to lipid peroxidation. Thus, as in our previous model (38), the cellular site of action of these herbicides may be the plastid envelope.

The action spectra for herbicidal damage caused by Protox inhibitors has a strong component in the red (31, 32, 70), whereas, PPIX absorbs relatively weakly in this region of the spectrum. Sato (70) has speculated that, in green cucumber tissue, inactivation of photosynthesis as a secondary effect of peroxidative damage results in chlorophylls being involved in the photodynamic process. Sato (70) found a porphyrin-protein complex in chloroplasts of cucumber tissues treated with S-23142, a chlorophthalim analog. The complex was not found in untreated plants. The protein was a 63-66 kD membrane protein and PPIX was identified as one of two porphyrins that complex it. Apparently this protein is not Protox, since it has been determined to have a molecular mass of 36 kD (71). The role of this porphyrin-protein complex in the mode of action of Protox-inhibiting herbicides has yet to be determined.

Applied alone, PPIX is relatively ineffective as a herbicide (46). Even if it were herbicidally effective, its cost and toxicological properties would probably prohibits its use. The low activity of PPIX supplied exogenously could be due to poor cellular absorption of this relatively complex molecule.

The selectivity of Protox-inhibiting herbicides is probably due to many factors, including: differential degradation of the herbicides (72), differential sensitivity to toxic oxygen species (15, 16, 73), and, perhaps, differential susceptibility at the site of action. No data are available to support the latter possibility. However, mutants of the unicellular green alga *Chlamydomonas reinhardtii*, produced by selection with a Protox inhibitor after mutagenesis, were cross resistant to a variety of Protox-inhibiting herbicides, but not to PSII inhibitors or to paraquat, indicating resistance a the molecular site of action (70).

Summary and Conclusions

Porphyrins cannot be used directly as photosensitizing herbicides because of the cost and possible toxicological dangers. However, a

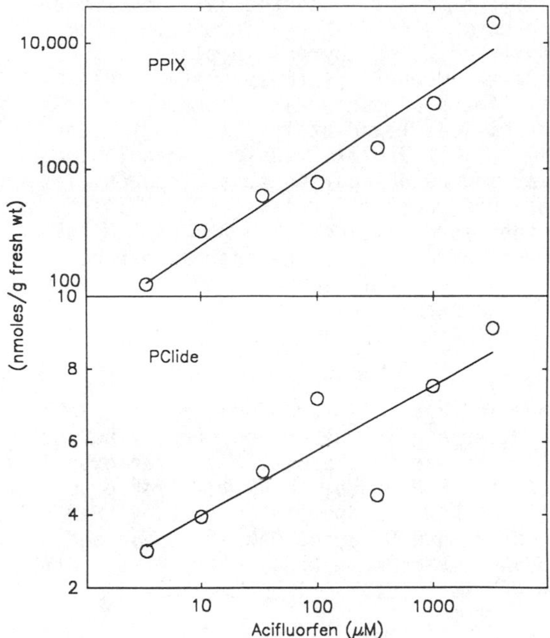

Figure 5. Effect of various concentrations of acifluorfen on PPIX and PChlide accumulation in cotyledons of intact, green cucumber seedlings. The plants were sprayed to runoff with the herbicide in 0.5 % Tween 80 (v/v) and incubated in darkness for 12 h before assays were conducted. Previously unpublished data of Becerril and Duke.

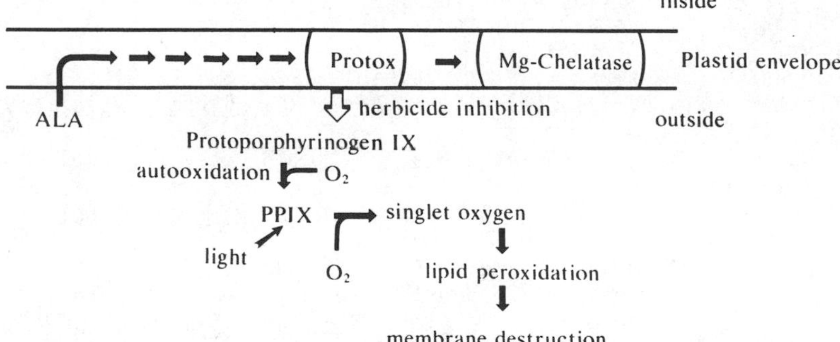

Figure 6. Hypothetical model of the phytotoxic mechanism of action for herbicides that inhibit Protox. Protoporphyrinogen which accumulated as a result of Protox inhibition leaves the membrane-bound, channeled porphyrin pathway and is autooxidized to PPIX. Redrawn from (59).

variety of both naturally-occurring and synthetic compounds can effectively stimulate plants to synthesize lethal amounts of photosensitizing porphyrins. ALA, a porphyrin precursor, is readily absorbed and converted to porphyrins in plant tissues. Compounds that modulate the chlorophyll synthesis pathway can be used to synergize ALA's effects on porphyrin accumulation. Several classes of commercial herbicides (oxadiazoles, *N*-phenylimides, and diphenyl ethers) inhibit Protox, causing deregulation of the porphyrin pathway and autooxidation of protoporphyrinogen to form photosensitizing, destructive levels of PPIX. These compounds are much more efficient as herbicides than ALA plus modulators. Future research on structure-activity relationships between these herbicides and inhibition of Protox could improve the activity and/or selectivity of these herbicides.

Acknowledgments

We thank Rohm and Haas Co. for providing technical grade chemicals used in our studies described here. The technical assistance of Al Lane was invaluable in the our experiments reported here. J.L. Wickliff, S.H. Duke, and G.W. Elzen provided helpful criticisms. This work was supported in part by the Fulbright/ Ministry of Education and Science, Madrid, Spain and the Japanese Ministry of Education, Science, and Culture. Ryo Sato generously provided a copy of his Ph.D. Thesis.

Literature Cited

1. Towers, G. H. N.; Arnason, J. T. Weed Technol. 1988, 2, 545-9.
2. Knox, J. P.; Dodge, A. D. Plant Cell Environ. 1985, 8, 19-25.
3. Knox, J. P.; Dodge, A. D. Plant Sci. Lett. 1984, 37, 3-7.
4. Mascia, P. N.; Robertson, D. S. Planta 1978, 143, 207-11.
5. Rebeiz, C. A.; Montazer-Zouhoor; Hopen, H. J.; Wu, S.-M. Enzyme Microb. Technol. 1984, 6, 390-401
6. Sisler, E. C.; Klein, W. H. Physiol. Plant. 1963, 16, 315-22.
7. Rebeiz, C. A.; Haidar, A.; Yaghi, M.; Castelfranco, P. A. Plant Physiol. 1970, 46, 543-9.
8. Duggan, J.; Gassman, M. Plant Physiol. 1974, 53, 206-15.
9. Castelfranco, P. A.; Beale, S. I. Annu. Rev. Plant Physiol. 1983, 34, 241-278.
10. Masuda, T.; Kouji, H.; Matsunaka, Pestic. Biochem. Physiol. 1990, 36, In Press.
11. Wickliff, J. L.; Duke, S. O.; Vaughn, K. C. Physiol. Plant. 1982, 56, 399-406.
12. Gassman, M. A. Plant Physiol. 1973, 52, 590-594.
13. Rebeiz, C. A.; Montazer-Zouhoor; Mayasich, J. M.; Tripathy, B. C.; Wu, S.-M.; Rebeiz, C. C. Amer. Chem. Soc. Symp. Ser. 1987, 339, 295-328.
14. Rebeiz, C. A.; Montazer-Zouhoor; Mayasich, J. M.; Tripathy, B. C.; Wu, S.-M.; Rebeiz, C. C. CRC Crit. Rev. Plant Sci. 1988, 6, 385-436.
15. Finckh, B. F.; Kunert, K. J. J. Agric. Food Sci. 1985, 33, 574-7
16. Shaaltiel, Y.; Glazer, A.; Bocion, P. F.; Gressel, J. Pestic. Biochem. Physiol. 1988, 31, 13-23.

17. Yaronskaya, E. B.; Shalygo, N. V.; Averina, N. G. Vestsi Akad. Navuk BSSR, Ser. Biayal. Navuk 1989, (4), 38-40.
18. Averina, N. G.; Shalygo, N. V.; Yaronskaya, E. B.; Rassadina, V. V. Vestsi Akad. Navuk BSSR, Ser. Biyal. Navuk 1989, 100-2.
19. Averina, N. G.; Radyuk, M. S. Dokl. Akad. Nauk BSSR 1989, 33, 471-4.
20. Rebeiz, C. A.; Juvik, J. A.; Rebeiz, C. C. Pestic. Biochem. Physiol. 1988, 30, 11-27.
21. Matsunaka, S. in Herbicides: Chemistry, Degradation, and Mode of Action, Vol. 2.; Kearney, P. C.; Kaufman, D. D., Eds.; Marcel Dekker: New York, 1976; pp. 709-39.
22. Kunert, K. J.; Sandmann, G.; Böger, P. Rev. Weed Sci. 1987, 3, 35-55.
23. Matsunaka, S. J. Agric. Food Chem. 1969, 17, 171-5.
24. Duke, S. O.; Vaughn, K. C.; Meeusen, R. L. Pestic. Biochem. Physiol. 1984, 21, 368-76.
25. Ensminger, M. P.; Hess, F. D. Plant Physiol. 1985, 78, 46-50.
26. Duke, S. O.; Kenyon, W. H. Plant Physiol. 1986, 81, 882-8.
27. Bowyer, J. R.; Hallahan, B. J.; Camilleri, P.; Howard, J. Plant Physiol. 1989, 89, 674-80.
28. Nurit, F.; Ravanel, P.; Tissut, M. Pestic. Biochem. Physiol. 1988, 31, 67-73.
29. Kenyon, W. H.; Duke, S. O.; Vaughn, K. C. Pestic. Biochem. Physiol. 1985, 24, 240-50.
30. Kenyon, W. H.; Duke, S. O.; Paul, R. N. Pestic. Biochem. Physiol. 1988, 30, 57-66.
31. Ensminger, M. P.; Hess, F. D. Plant Physiol. 1985, 77, 503-5.
32. Sato, R.; Nagano, E.; Oshio, H.; Kamoshita, K.; Furuya, M. Plant Physiol. 1987, 85, 1146-50.
33. Gaba, V.; Cohen, N.; Shaaltiel, Y.; Ben-Amotz, A; Gressel, J. Pestic. Biochem. Physiol. 1988, 31, 1-12.
34. Rao, D. N. R.; Mason, R. P. Photochem. Photobiol. 1988, 47, 791-5.
35. Kenyon, W. H.; Duke, S. O. Plant Physiol. 1985, 79, 862-6.
36. Fadayomi, O.; Warren, G. F. Weed Sci. 1976, 24, 598-600.
37. Orr, G. L.; Hess, F. D. Plant Physiol. 1982, 69, 502-7.
38. Duke, S. O.; Kenyon, W. H. Z. Naturforsch. 1987, 42c, 813-8.
39. Haworth, P.; Hess, F. D. Plant Physiol. 1988, 86, 672-6.
40. Orr, G. L.; Elliott, C. M.; Hogan, M. E. Plant Physiol. 1983, 73, 939-44.
41. Becerril, J. M.; Duke, S. O. Plant Physiol. 1989, 90, 1175-81.
42. Ensminger, M. P.; Hess, F. D.; Bahr, J. T. Pestic. Biochem. Physiol. 1985, 23, 163-70.
43. Matringe, M.; Scalla, R. Pestic. Biochem. Physiol. 1987, 27, 267-74.
44. Matringe, M.; Scalla, R. Proc. Brit. Crop Protect. Conf. 1987, 9B, 981-8.
45. Matringe, M.; Scalla, R. Plant Physiol. 1988, 86, 619-22.
46. Lydon, J.; Duke, S. O. Pestic. Biochem. Physiol. 1988, 31, 74-83.
47. Witkowski, D. A.; Halling, B. P. Plant Physiol. 1988, 87, 632-7.
48. Matringe, M.; Scalla, R. Pestic. Biochem. Physiol. 1988, 32, 164-72.
49. Duke, S. O.; Lydon, J.; Paul, R. N. Weed Sci. 1989, 37, 152-60.
50. Sandman, G.; Böger, P. Z. Naturforsch. 1988, 43c, 699-704.

51. Nicolaus, B.; Sandmann, G.; Watanabe, H.; Wakabayashi, K.; Böger, P. Pestic. Biochem. Physiol. 1989, 35, 192-210.
52. Yanase, D.; Andoh, A. Pestic. Biochem. Physiol. 1989, 35, 70-80.
53. Sandmann, G.; Reck, H.; Böger, P J. Agric. Food Chem. 1984, 32, 868-72.
54. Wakabayashi, K.; Matsuya, K.; Teraoka, T.; Sandmann, G.; Böger, P. J. Pestic. Sci. 1986, 11, 635-40.
55. Teraoka, T.; Sandmann, G.; Böger, P.; Wakabayashi, K. J. Pestic. Sci. 1987, 12, 499-504.
56. Halling, B.P.; Peters, G. R. Plant Physiol. 1987, 84, 1114-20.
57. Matringe, M.; Camadro, J.-M.; Labbe, P.; Scalla, P. Biochem. J. 1989, 260, 231-5.
58. Matringe, M.; Camadro, J.-M.; Labbe, P.; Scalla, P. FEBS Lett. 1989, 245, 35-8.
59. Witkowski, D. A.; Halling, B. P. Plant Physiol., 1989, 90, 1239-42.
60. Camadro, J. M.; Urban-Grimal, D.; Labbe, P. Biochem. Biophys. Res. Commun. 1982, 106, 724-30.
61. Brenner, D. A.; Bloomer, J. R. New Engl. J. Med. 1980, 302, 765-8.
62. Deybach, J. C.; de Verneuil, H.; Nordmann, Y. Hum. Genet. 1981, 58, 425-8.
63. Derrick, P. M.; Cobb, A. H.; Pallett, K. E. Pestic. Biochem. Physiol. 1988, 32, 153-63.
64. Mayasich, J. M.; U. B. Nandihalli; R. A. Leibl; C. A. Rebeiz WSSA Abstr. 1989, 29, 90.
65. Kouji, H.; Masuda, T.; Matsunaka, S. J. Pestic. Sci. 1988, 3, 495-9.
66. Kouji, H.; Masuda, T.; Matsunaka, S. Pestic. Biochem. Physiol. 1989, 33, 230-8.
67. Beccerril, J. M.; Duke, S. O. Pestic. Biochem. Physiol. 1989, 35, 119-26.
68. Lehnen, L. P.; Sherman, T. D.; Becerril, J. M.; Duke, S. O. 1990, Pestic. Biochem. Physiol. - In Press.
69. Ferreira, G. C.; Andrew, T. L.; Karr, S. W.; Dailey, H. A. J. Biol. Chem. 1988, 263, 3835-9.
70. Sato, R., Ph.D. Thesis, University of Tokyo, Tokyo, Japan, 1989.
71. Jacobs, J. M.; Jacobs, N. J. Biochem. J. 1987, 244, 219-224.
72. Frear, D. S.; Swanson, H. R.; Mansager, E. R. Pestic. Biochem. Physiol. 1983, 20, 299-310.
73. Sandmann, G.; Böger, P. Amer. Chem. Soc. Symp. Ser. 1990, 421, 407-18.

RECEIVED May 16, 1990

ALLELOCHEMICALS AS PLANT DISEASE CONTROL AGENTS

Chapter 27

Black Shank Disease Fungus

Inhibition of Growth by Tobacco Root Constituents and Related Compounds

Maurice E. Snook[1], Orestes T. Chortyk[1], and Alex S. Csinos[2]

[1]Russell Research Center, Agricultural Research Service, U.S. Department of Agriculture, P.O. Box 5677, Athens, GA 30613
[2]Department of Plant Pathology, University of Georgia, Coastal Plain Experiment Station, Tifton, GA 31793

Tobacco root phenolics were investigated for their possible role in resistance of tobacco to Phytophthora parasitica var. nicotianae (black shank). Chlorogenic acid (CA), scopolin, and scopoletin were found to be significantly increased in apparently healthy root tissue, adjacent to infected tissue. In contrast, CA and scopolin concentrations remained relatively constant throughout the length of control roots, while scopoletin increased slightly from the proximal to the distal end. Roots of resistant and susceptible varieties showed similar trends in concentrations of phenolics. Chlorogenic acid, scopolin, scopoletin, and structurally related compounds were evaluated for inhibition of growth of black shank fungus in a laboratory bioassay. At a 4000 ppm dosage level, CA produced 25% inhibition of fungal growth, while scopoletin gave 39% inhibition at 1000 ppm dosage. The activity of CA was shown to reside in the caffeic acid portion of the molecule. Free phenolic acids were also investigated and were found to be very active against black shank growth, with mono-hydroxycinnamic acids being more active than the 3,4-di-hydroxy acid and the o-hydroxy acid being the most active. Dihydro-cinnamic acids were slightly more active than the corresponding cinnamic acids. While scopoletin and esculetin were active in the laboratory bioassay, their glucose-derivatives were found to be completely inactive.

Phytophthora parasitica var. nicotianae (black shank) is a fungus that only attacks tobacco and also causes extensive physical damage, resulting in large economic losses for farmers. It affects both flue-cured and burley types of tobaccos. The fungus attacks the roots of the plant and eventually blocks water and nutrient transport, resulting in the death of the plant. Several varieties of tobacco are known to be highly resistant to the black shank fungus. Several reports have shown that phenolic compounds of tobacco increase in tissues that have been exposed to pathogens, such as tobacco mosaic virus (1), blue mold (Peronospera tabacina) (2), and black root rot (Thielaviopsis

This chapter not subject to U.S. copyright
Published 1991 American Chemical Society

basicola) (3). Recently, Gasser, et al. (3) reported that chlorogenic acid (3-O-caffeoylquinic acid, CA) and scopolin (6-methoxy-7- glucosylcoumarin) increased in callus tissue cultures of black root rot-resistant tobacco varieties. They also reported that CA and scopoletin (6-methoxy-7-hydroxycoumarin), phenolics present in tobacco root tissue, were toxic to the black root rot fungus in laboratory bioassays. We, therefore, have investigated the levels of the major tobacco phenolics in roots of susceptible and resistant varieties grown in black shank infected and disease free fields. Also, a number of tobacco root phenolics and related compounds were tested for inhibition of growth of the black shank fungus in a laboratory bioassay.

MATERIALS AND METHODS

Tobacco was grown in 1989 at the University of Georgia Coastal Plain Experiment Station, Tifton, GA, under standard cultural practices. Varieties (NC 2326, NC 82, McNair 944, Coker 371, and VA 509) were grown in both a black shank infected field and in a disease free field. Susceptible varieties were sampled in July, while resistant varieties were sampled in August. The soil was gently washed from the roots. Single roots, attached to the main tap root and possessing visible signs of disease incidence (as indicated by brown discoloration), were detached. Root hairs were removed and the entire root length was cut into 1-cm segments and each segment was immediately frozen in dry-ice. Roots ranged from 4-7 mm in diameter at their point of attachment to 2-3 mm at their tip end (distal end). The segments were freeze-dried and chopped with a sharpened spatula. Each segment was extracted with 2.5 mL MeOH (containing 0.14 mg 5,7-dimethoxycoumarin as ISTD) by ultrasonication for 10 min. Then, 2.5 mL water were added and the solution ultrasonicated for an additional 10 min. The samples were filtered and analyzed by high performance liquid chromatography, as described before (4). Briefly, a Waters uBondapak C18 column was used with a concave gradient solvent program from 13% MeOH/H_2O (containing 0.08 M KH_2PO_4 buffer adjusted to pH 4.45) to 50% MeOH/H_2O. A Hewlett-Packard 1040 Diode Array Detector, set at 340 nm, was employed. A typical chromatogram of a tobacco root extract is shown in Figure 1.

The laboratory bioassay of Phytophthora parasitica var. nicotianae was similar to that already described (5). Stock P. parasitica var. nicotianae was grown on V-8 juice agar. Additional V-8 juice agar was prepared, autoclaved, cooled to 45°C, and the test compounds were added in quantities to prepare 250 mL solutions of each concentration of test compound. The agar and chemicals were mixed well, dispensed into 60X15 mm plastic petri plates, and allowed to solidify. Each concentration of test material was replicated ten times. Plugs of the fungus culture, cut out with a 5mm #2 cork borer, were placed with fungal culture side down on the edge of the test plates. Plates were placed in Ziploc bags and incubated at 27°C for 14 days or until fungal growth had reached the opposite edge of the agar control plates. Radial mycelial growth was measured and the percent inhibition of growth determined.

Scopolin was isolated from freeze-dried, flue-cured tobacco roots. The ground roots were extracted with MeOH and the extract was separated by silicic acid column chromatography, eluted first with ethyl acetate and then with acetone. The acetone eluent contained the scopolin, which was purified by preparative reverse-phase C18 chromatography on a Waters PrepPak 500 C18 column, using a 35% MeOH/H_2O solvent.

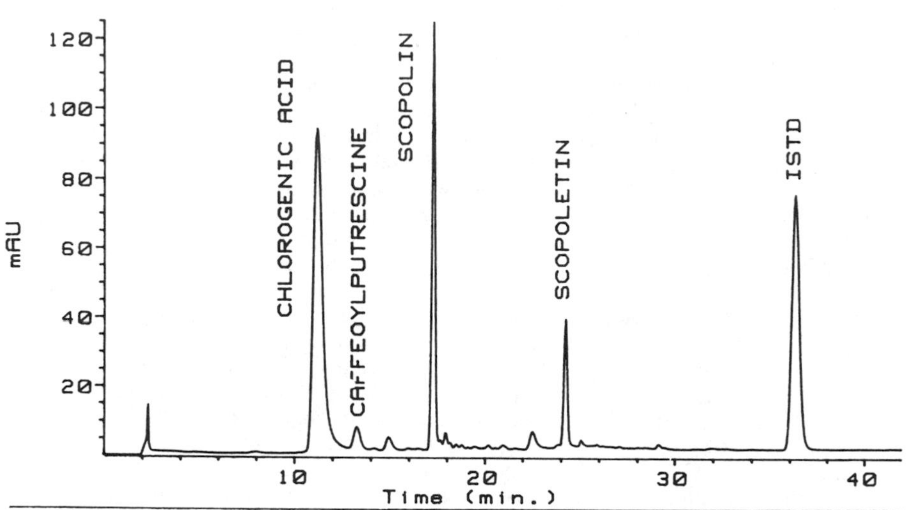

Figure 1. High Performance Liquid Chromatogram of Tobacco Root Phenolics.

RESULTS AND DISCUSSION

Tobacco Root Chemistry Versus Fungal Attack

The major phenolic compound in tobacco roots (Figure 1) is chlorogenic acid (CA), correctly termed 3-O-caffeoylquinic acid (3-O-CQA). Only minor to trace amounts of the 4- and 5-O-CQA isomers are found in the roots, but occur in larger quantities in the leaves. Two coumarin derivatives are found in major amounts in root tissue: scopoletin (6-methoxy-7-hydroxycoumarin) and its glucoside, scopolin. Both CA and scopolin can reach levels of 1% of the dry weight of the root. Very little scopolin or scopoletin is found in the leaves.

Roots from resistant and susceptible plants, grown in disease free and infected fields were analyzed for their phenolic contents. Whole roots were removed from the tap root, cut into 1-cm segments and analyzed for CA, scopolin, and scopoletin (Figures 2 and 3). It was determined that CA and scopolin remained relatively constant throughout the entire length of a root, from point of attachment at the tap root to the distal end (Figures 2a and 2b). Scopoletin however, tended to increase towards the distal end of the root. Both resistant and susceptible tobacco roots gave these same trends.

The fungus attacks the roots of the plant by first entering the root hairs and then advancing towards the tap root. The disease will eventually engulf the tap root and lower stalk, becoming visible as a brown to black coloration on the stalk surface; hence, the name "black shank". Frequently, one observes apparently healthy roots attached to an infected tap root with infection advancing from the tap root outward to the root tip. Roots, with varying degrees of fungal attack (as indicated in Figures 2c and 3a-3d), were analyzed for their phenolics. Roots, under pressure of fungal attack, showed different patterns of phenolics in their tissues, than roots grown in disease-free fields. Chlorogenic acid, scopolin, and scopoletin all increased significantly in tissue adjacent to visibly-infected tissue. This trend was consistent whether the infection was progressing from the tip end (Figure 2c), from the tap root outward (Figures 3a,3b,3c), or had entered a root hair at the midsection of the root (Figure 3d). Both susceptible (Figures 2c,3a) and resistant (Figures 3b-3d) varieties showed similar increases in these phenolics in root tissues adjacent to infection. Also, there was no difference in flue-cured (Figures 3b,3c) versus burley (Figure 3d) resistant varieties. Gasser et al. (3) found that callus cultures of resistant tobaccos gave an increase in phenolics, while susceptible tobaccos gave a decrease, when challenged by black root rot. In contrast, both resistant and susceptible whole root tissue showed increases in phenolics, when challenged by black shank. Perhaps, callus cultures infected with black shank would show the same effect observed by Gasser for black root rot.

Laboratory Bioassays of Tobacco Root Compounds

The reported activity of CA and scopoletin in laboratory bioassays of black root rot (3) prompted us to examine these and related compounds for activity towards black shank. CA and scopoletin were tested for inhibition of growth of black shank in our laboratory bioassay (Table I). Chlorogenic acid produced a 25% inhibition of growth at the highest level tested (4000 ppm), while scopoletin gave 39% inhibition of growth at a 1000 ppm dosage rate. The constituent parts of chlorogenic acid (caffeic and quinic acids) were also tested (Table I) and showed that the activity of chlorogenic acid lies in the caffeoyl moiety of the molecule.

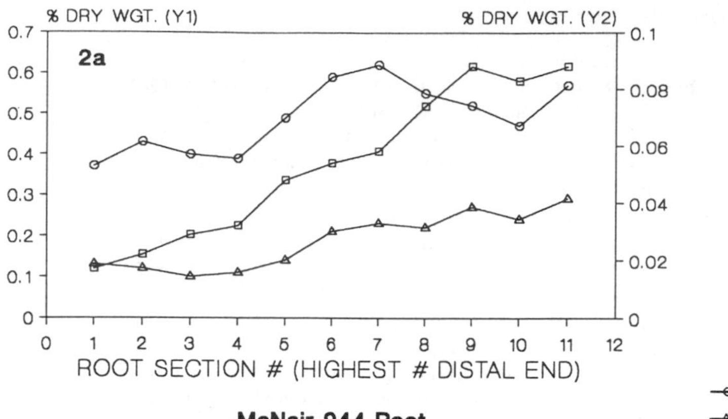

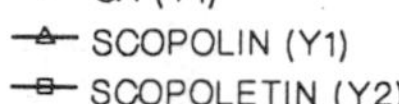

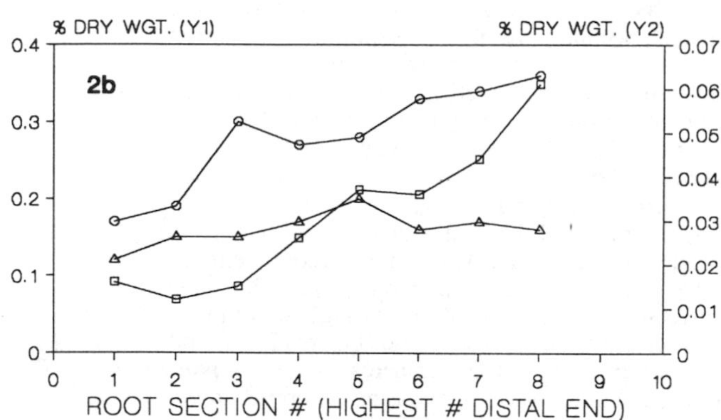

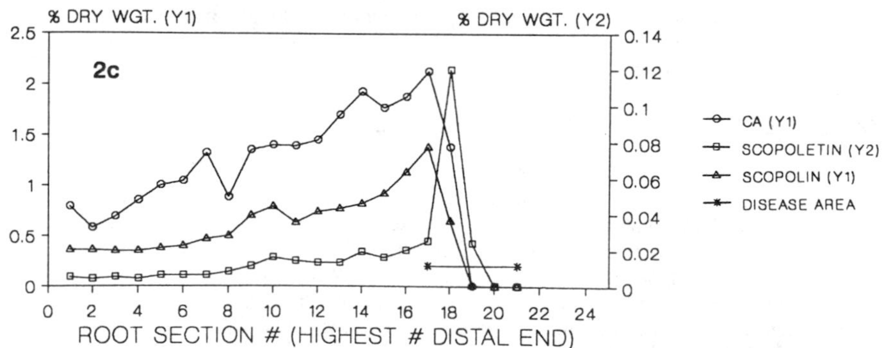

Figure 2. Root Levels of Phenolics in Healthy Resistant Varieties (2a-2b) and a Black Shank Infected Susceptible Tobacco Variety (2c).

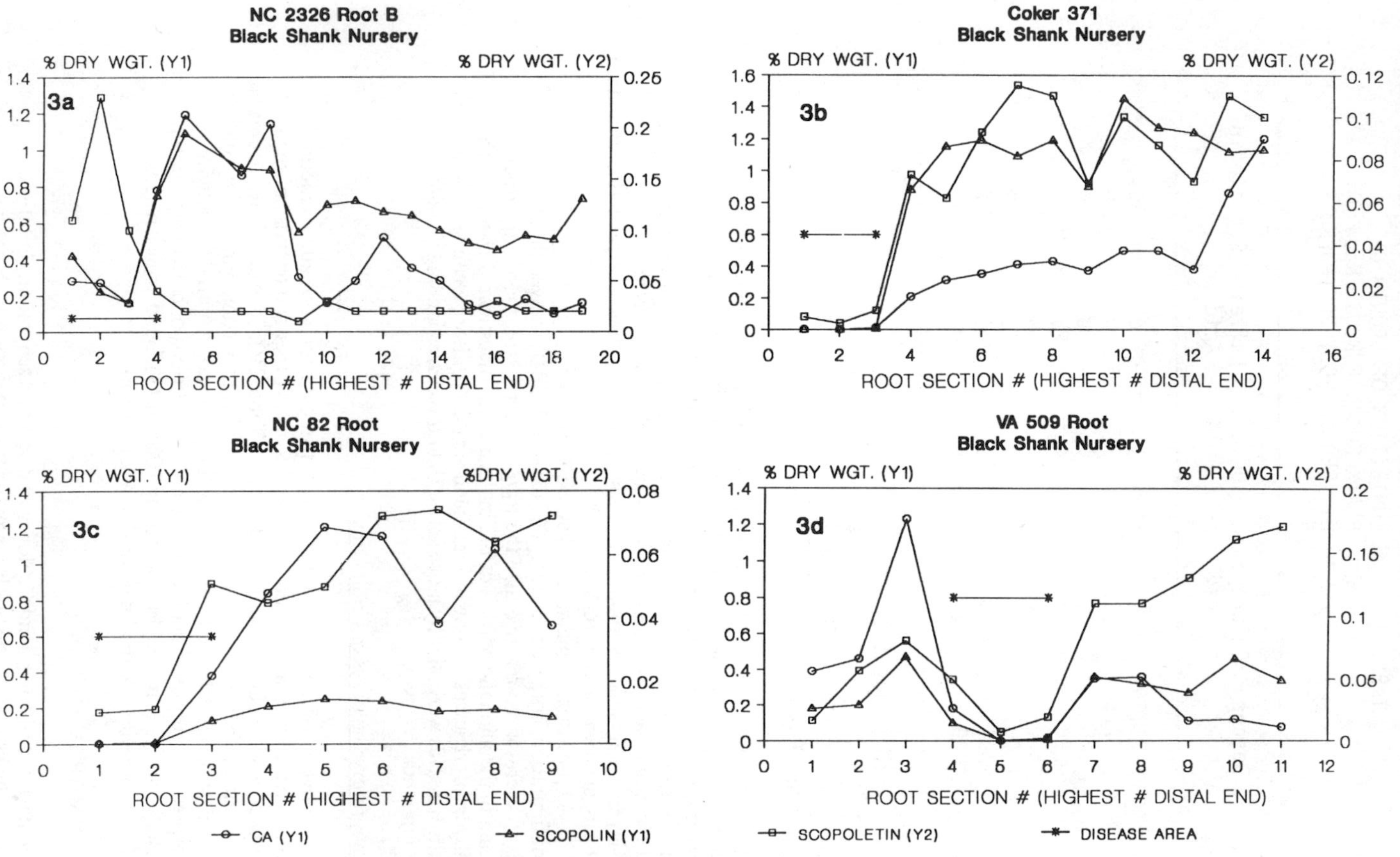

Figure 3. Levels of Phenolics in Black Shank Infected Tobacco Roots of Different Varieties.

TABLE I % INHIBITION OF GROWTH OF BLACK SHANK FUNGUS (Race 0) BY TOBACCO ROOT POLYPHENOLS

Dosage (ppm)	Chlorogenic Acid	Quinic Acid	Caffeic Acid	Scopoletin
31.25	1	1	0	0
62.5	1	2	0	0
125	2	0	0	0
250	2	4	25	17
500	6	11	59	38
1000	11	4	71	39
2000	14	11	74	NT[a]
4000	25	NT	NT	NT

[a]NT - not tested

Laboratory Bioassays of Phenolic Acids

The high activity observed for caffeic acid was further investigated by studying the activity of a number of related compounds, in order to determine structure-activity relationships. Table II compares the activities of mono-hydroxycinnamic acids to caffeic acid (3,4-dihydroxy-cinnamic acid). All of the mono-hydroxy acids were more active than caffeic acid. o-Hydroxycinnamic (o-coumaric) acid was the most active and gave 91% inhibition of fungal growth at only 62.5 ppm.

As p-hydroxy-cinnamic acid can exist in either the cis or trans configuration, both were tested. The data (Table III) showed that the cis-isomer was more active than the trans-isomer. Saturation of the double bond in the aliphatic portion of the molecule was investigated by testing the dihydro-derivatives of p-hydroxycinnamic and 3,4-dihydroxycinnamic acids. The dihydro-acids were slightly more active than the corresponding cinnamic compounds.

Laboratory Bioassays of Coumarins

The activities of scopoletin (6-methoxy-7-hydroxycoumarin) and esculetin (6,7-dihydroxycoumarin) were compared to each other and to those of their respective glucosides, scopolin and esculin (Table IV). The dihydroxy-coumarin was found to be just as active as the methoxy-hydroxy-compound. Thus, the presence of a free hydroxyl group at position-6 is not required for activity. Surprisingly, substitution of a glucose on the hydroxyls of both compounds resulted in the complete loss of activity. Gasser (3) found similar results with the activity of scopolin towards black root rot.

SUMMARY

We have found compounds (chlorogenic acid and scopoletin) in tobacco roots that inhibit the growth of the black shank fungus in a laboratory bioassay.

TABLE II. % INHIBITION OF GROWTH OF BLACK SHANK FUNGUS BY HYDROXY-CINNAMIC ACIDS

Dosage (ppm)	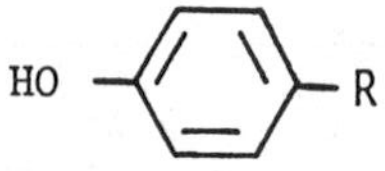 p-Coumaric	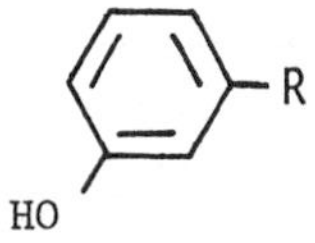 m-Coumaric	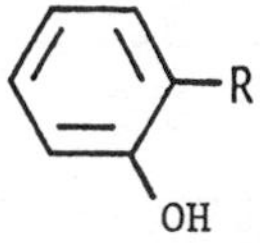 o-Coumaric	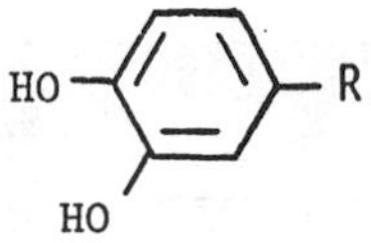 Caffeic
31.25	0	17	65	0
62.5	14	37	91	0
125	39	57	96	2
250	87	75	100	22
500	100	100	100	75
1000	100	100	100	78

R= -CH=CH-COOH

TABLE III. % INHIBITION OF GROWTH OF BLACK SHANK FUNGUS BY HYDROXY-CINNAMIC ACIDS

Dosage (ppm)	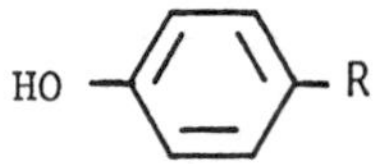 HO–⌬–R trans	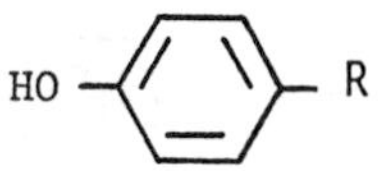 HO–⌬–R cis	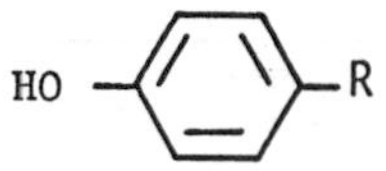 HO–⌬–R Dihydro	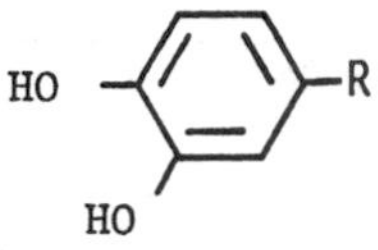 HO–⌬(HO)–R trans	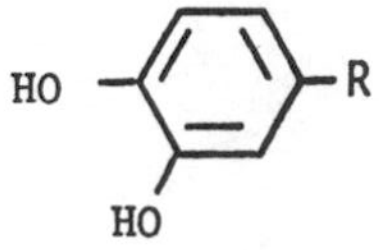 HO–⌬(HO)–R Dihydro
31.25	0	0	0	0	3
62.5	0	0	17	0	12
125	0	33	24	2	16
250	21	66	48	22	42
500	39	84	71	75	60
1000	60	100	99	78	70

R= -CH=CH-COOH

TABLE IV. % INHIBITION OF GROWTH OF BLACK SHANK FUNGUS BY COUMARINS

Dosage (ppm)	Esculetin	Esculin	Scopoletin	Scopolin
31.25	0	0	0	0
62.5	0	0	0	0
125	7	0	0	0
250	23	0	17	0
500	40	0	38	0
1000	32	16	39	0
2000	NT	22	NT	NT

NT - not tested.

However, levels of these compounds in roots of disease-free, resistant, and susceptible varieties were comparable. Although levels of these compounds have been shown to increase in roots in response to infection, the levels of increase were similar in resistant and susceptible roots. Free phenolic acids were shown to be very active in inhibiting the growth of black shank. Free phenolic acids do not occur in tobacco, but could play a role in resistance. Perhaps, they are produced at the narrow point of attack of the fungus. Because of their high activity, only small amounts would be needed to block the advance of the fungus through the root. Further research is needed to determine if this is a viable mode of resistance and to search for other chemicals that may be responsible for the observed black shank resistance of certain tobacco varieties.

LITERATURE CITED

1. Fritig, B.; Hirth, L. Acta Phytopathol. Acad. Sci. Hung. 1971, 6, 21-29.
2. Cohen, Y.; Kuc, J. J. Phytopathology 1981, 71, 209.
3. Gasser, R.; Kern, H.; Defago, G. J. Phytopathology 1988, 123, 115-123.
4. Snook, M. E.; Mason, P. F.; Sisson, V. A. Tob. Sci. 1986, 30, 43-49.
5. Csinos, A. S.; Fortnum, B. A.; Gayed, S. K.; Reilly, J. J.; Shew, H. D. In Methods for Evaluating Pesticides for Control of Plant Pathogens; Hicky, K. D., Ed.; APS Press: St. Paul, MN; pp 231-236.

RECEIVED July 18, 1990

Chapter 28

A Search for Agrochemicals from Peruvian Plants

D. Howard Miles[1], Jorge Meideros[1], Liza Chen[1], Vallapa Chittawong[1], Colin Swithenbank[1], Zev Lidert[2], Allen Matthew Payne[1], and Paul A. Hedin[3]

[1]Department of Chemistry, University of Central Florida, Orlando, FL 32826
[2]Rohm and Haas Company, Spring House, PA 19477
[3]Crop Science Laboratory, U.S. Department of Agriculture, Mississippi State, MS 39762

The objective of this study was the isolation and identification of the antifeedant, antifungal, and antibacterial active constituents from Peruvian plants. 5,7-Dihydroxy-6,8-dimethyl-4′-methoxyflavanone (matteucinol) was isolated from the root of *Miconia cannabina.* This is the first report of the spectral data of matteucinol. Matteucinol showed significant antifungal activity (*Pythium ultimum* 135%). Also reported is the isolation of sitosterol and medicarpin from the toluene extract of the wood of *Dalbergia monetaria.* Medicarpin possessed high antifungal (*Rhizoctonia solani*, 145% and *Helminthosporium teres*, 100%) and antibacterial activity (*Xanthomonas campestris*, 100%) along with high antifeedant activity against the cotton boll weevil (*Anthonomus grandis*, 98%).

As long as mankind continues to cultivate and grow plants for food, feed, and fiber, crops will be threatened by various enemies and agents such as insects, fungi, and bacteria. Prevention and control of insect pests and plant diseases will remain an important objective of agrochemical research. In this study, the isolation and structure determination of several compounds showing insect antifeedant activity and antifungal activity from Peru plants are presented.

The plant *Miconia cannabina* Markgr. (Melastomaceae) is common to Iquitos, Peru. No previous chemical study has been undertaken on this plant; however, examples of studies on related plants are known. A study of the protein and water contents of several parts of the plant *M. theaezans* was undertaken by Gaulin and Craker (1). In another study Marini-Bettolo (2) isolated primin and miconidin from the ethyl acetate extract of a nonidentified species of *Miconia.* Primin has biological activity against various Gram positive and negative microorganisms and fungi as well as antitumor activity against sarcoma of mice (Swiss). Miconidin is less active against common microorganisms, but inhibits *Mucobacterium* sp and *Gibberella fulikuroi* (2). These two compounds also showed antifeedant activity (3) against six insect species - desert locust *Schistocerca gregaria*, the migratory locust *Locusta migratoria*, the armyworms *S. littoralis* and *S. extemota*, the budworm *Heliothis amigera* and caterpillars of the cabbage white butterfly *Pieris brassicae.*

Dalbergia, a wood climber, belongs to the family Leguminosae and subfamily Papilionaceae. Nearly 120 *Dalbergia* species are known to occur in nature (4). The chemical components of this genus of plants provides a great number of structures belonging to widely different chemical categories such as, neoflavonoids, isoflavonoids, flavonoids, furans, isoflavans, benzophenones, styrenes, sterols, and terpenoids.

0097–6156/91/0449–0399$06.00/0
© 1991 American Chemical Society

Materials and Methods

Sample collection. — Roots of *M. canabina* (voucher number 7261) and wood of *D. monetaria* (voucher number 7000) were collected in Iquitos, Peru in 1982, and identified

by Dr. Sidney McDaniel, Department of Biological Sciences, Mississippi State University. The plant material was air dried and ground in a Wiley mill before extraction.

Extraction. — *M. Canabina* (6.8 kg dry weight) was extracted in a soxhlet apparatus with methylene chloride (16 h.) and then ethanol (16 h.). The solvent was evaporated under reduced pressure to yield 6.7 g of methylene chloride extract (fraction A, Figure 1) and 22.6 g of ethanol extract (fraction B, Figure 1).

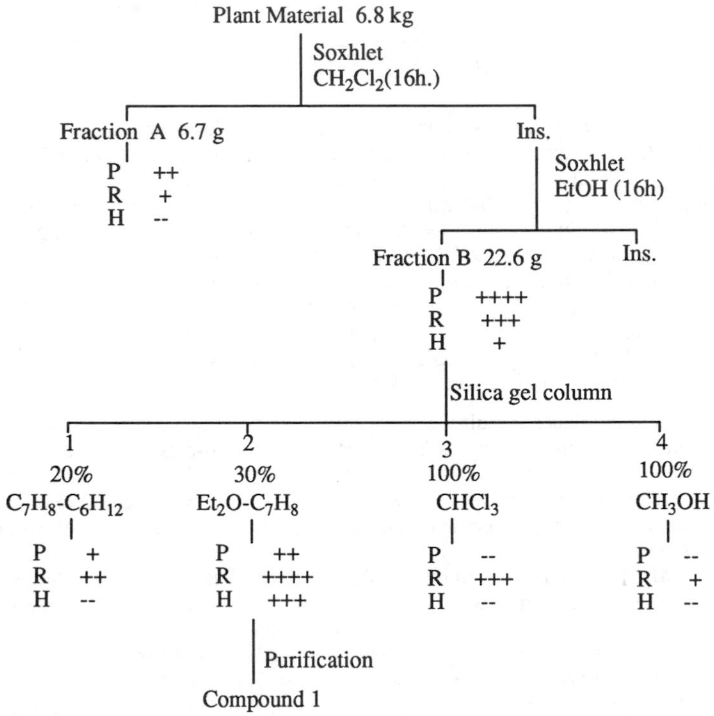

P - *Pythium ultimum*; R - *Rhizoctonia solani*; H - *Helminthosporium teres*
diameter of the clear zone of inhibition around the disk:
+ - 1-2 mm; ++ - 2-5 mm; +++ - 5-10; ++++ - > 10 mm

Figure 1. Fractionation Scheme for *Miconia cannabina*

D. monetaria (25.6 kg dry weight) was extracted in sequence with methylene chloride (16 h.) and ethanol (16 h.) in a Soxhlet apparatus. The solvent was removed in vacuo to yield 211.1 g of methylene chloride extract (fraction A, Figure 2) and 174.4 g of extract ethanol (fraction B, Figure 2). Fraction A was sequentially extracted with hot hexane (fraction A1, Figure 2) and toluene (fraction A2, Figure 2). Fraction B was extracted with ether. The ether-insoluble fraction (72.2 g) (fraction C, Figure 2) was sequentially extracted with hexane : ether (2:1), and chloroform. The ether-soluble fraction (60.9 g) was partitioned between chloroform and water (1:1).

Fraction A2 (73.8 g) was extracted with chloroform. The chloroform solution was extracted with a 5% hydrochloric acid solution. The aqueous layer was neutralized with a 5% sodium hydroxide solution and shaken several times with chloroform. The chloroform was removed in vacuo to yield 4 g of fraction A2-b (Figure 3).

The original chloroform layer was partitioned between chloroform and 5% aqueous sodium hydroxide solution (1:1). After separation of the layers, the aqueous layer was neutralized with 5% hydrochloric acid and extracted repeatedly with chloroform. The combined chloroform extracts were then dried over anhydrous sodium sulfate, filtered, and evaporated in vacuo to yield 12.0 g of residue (fraction A2-a).

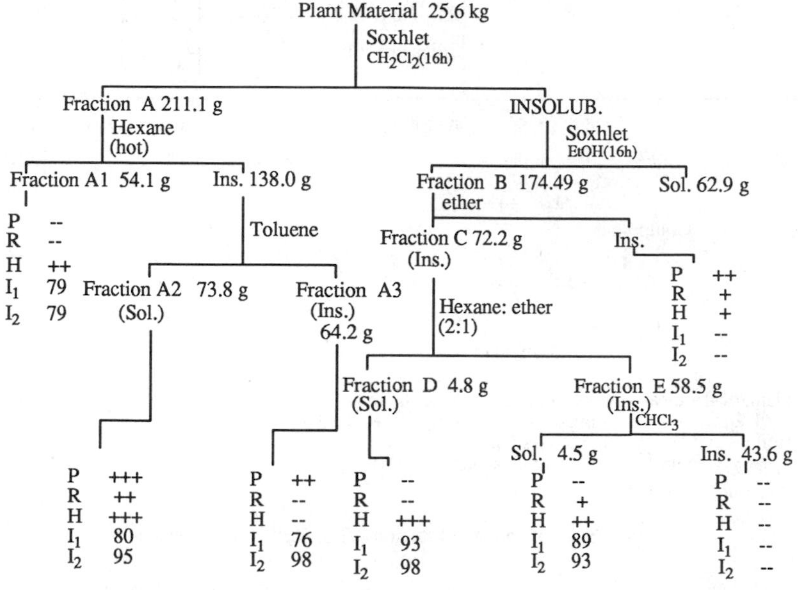

diameter of the clear zone of inhibition around the disk:
+ - 1-2 mm; ++ - 2-5 mm; +++ - 5-10 mm; ++++ - >10 mm
Boll weevil antifeedant activity (I):
$I_n\% = [(\#\text{ holes blank} - \#\text{ holes sample}) / (\#\text{ holes blank})] \times 100$
n = # mg / cm^2 paper

Figure 2. Fractionation Scheme for *Dalbergia monetaria*

Fraction A2-a was chromatographed on silica gel (60-200 mesh, 800 g) with eluents of increasing polarity from hexane through toluene to chloroform, and then to methanol. Fractions of 250 ml were collected, the solvent was removed in vacuo, and those fractions exhibiting similar TLC profiles were combined.

Bioassay

Antifungal and Antibacterial Bioassay. — Fractions were screened for activity against the fungi *Helminthosporium teres*, *Pythium ultimum*, and *Rhizoctonia solani* and the bacteria *Xanthonomas campestris* by the paper disc method (5). Each fraction to be tested was

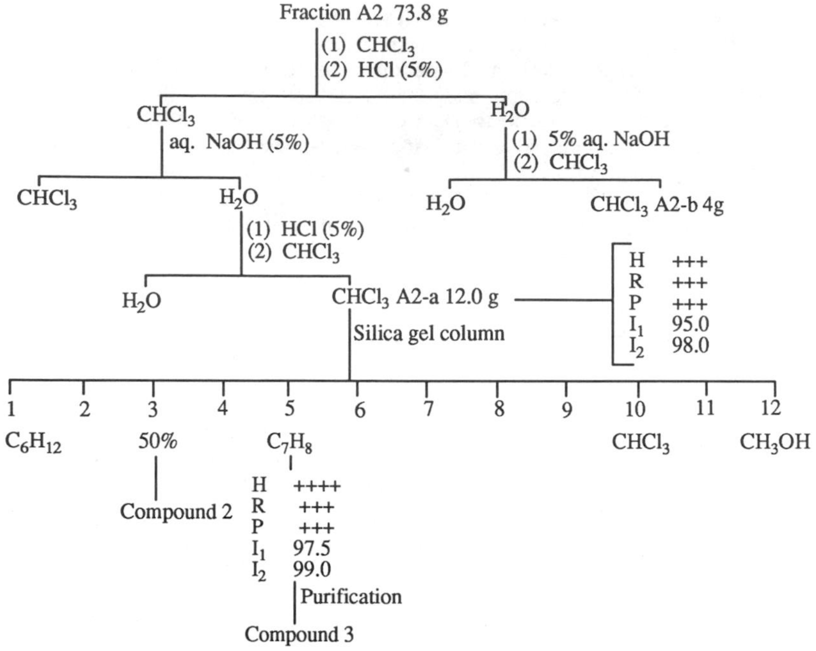

diam. of the clear zone of inhibition around the disk:
+ - 1-2 mm; ++ - 2-5 mm; +++ - 5-10 mm; ++++ - >10 mm
Boll weevil antifeedant activity (I):
I_n % = [1-(# holes sample / # holes blank)] x 100
n = # mg / cm^2 paper

Figure 3. Purification of Fraction A2 of *Dalbergia monetaria*

dissolved in an appropriate solvent. In the preliminary tests of semi-purified plant extracts, a 100,000 ppm solution was prepared, while in the final test of pure compounds a 10,000 ppm solution was used. Filter paper discs (6 mm dia.) were soaked in the test solution and allowed to dry overnight.

Petri plates containing solidified agar (DIFCO Potato-Dextrose for *H. teres* and *R. solani*, DIFCO cornmeal for *P. ultimum*, and DIFCO nutrient agar for *X. campestris*) were inoculated with fungi. The inoculation with *H. teres* and *X. campestris* was performed by mixing 100 ml of the appropriate agar with the filtrate of an aqueous suspension of their spores and then placing 5 ml of the mixture on top of the solidified agar layer. For *R. solani* and *P. ultimun*, pieces of the mycelium were cut from the culture plate and placed in the center of the bioassay plate. Three of the prepared paper disks were then placed on each Petri plate. A paper disk soaked only in the solvent used and dried by the usual method was also placed on each plate as a blank. As a control, a 10,000 ppm solution of Dithane M-45 (from Rhom & Hass Co.), a commercially available fungicide, was used. Inhibition was determined by measuring the diameter of the clear zone (if present) around the disc.

Boll weevil antifeedant bioassay. — The bioassay was performed according to the procedure of Hedin et. al. (6, 7). Agar plugs were formed by boiling 3 grams of agar and

3 grams of freeze-dried cotton bolls in 100 mls of distilled water to effect a viscous sol. The sol. was poured into 18 mm diameter tubing, and gelation occurred after cooling. These gelatinous rods were cut into individual 3.6 cm plugs.

The extracts of the plant samples were applied to preweighed 4.5 x 4.5 cm squares of Whatman #1 chromatography paper by dipping the paper into a solution of the test sample. After air-drying, the paper was weighed. A blank was prepared by dipping one of the papers into the solvent used on the test sample. The papers were wrapped around the agar-cotton plugs and fastened with four staples. The ends of the plugs were sealed with corks. The plugs were then placed staple-side down in the petri dishes. Twenty newly emerged boll weevils were placed in 14 x 2 cm petri dishes containing the test and control plugs. The bioassay was carried out in the dark at 25°C for 4 hours. When the bioassay was finished the paper was removed from the plugs and the number of punctures counted.

Antifeedant activity (I_n) was expressed as: (I_n) % = [1-(# holes sample/# holes blank] x 100 where n indicates the amount used (in grams) of the compound per surface units (cm^2) of the filter paper.

Results and Discussion

The ethanolic extract of *M. cannabina* revealed the highest antifungal activity when tested against *P. ultimum* (P), *R. solani* (R), and *H. teres* (H), using the preliminary "paper disc" method (5). Fractionation of the ethanolic extract was performed by column chromatography (Figure 1) with silica gel as the support. The resulting chromatographic fractions were monitored by the "paper disc" bioassay. The material which was eluted with 30% (v/v) ether in toluene was further purified by repetitive crystallization from methanol to yield the compound 1.

The high resolution mass spectrum suggested a molecular formula of $C_{18}H_{18}O_5$ for compound 1. This compound developed the characteristic color reactions of flavonoids (8,9). The UV spectrum of compound 1 in methanol, consisted of two major absorption maxima, one of which occurred at 294 nm. The other band at 328 nm was less intense and was suggestive of the presence of a flavonoid lacking conjugation between the A- and B-rings (9,10,11). A doublet of doublets at δ 2.83 and 3.08 in the ^{1}H NMR spectrum of compound 1 suggested the presence of a flavanone.

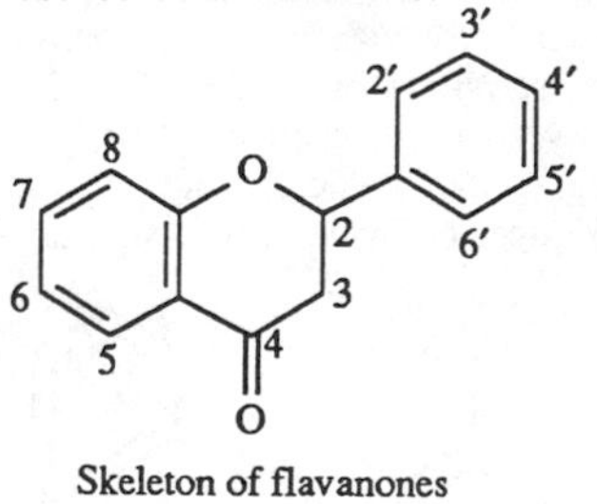

Skeleton of flavanones

Examination of the UV spectra of compound 1 in methanol with the shift reagents NaOMe, $AlCl_3$, $AlCl_3$ + HCl, NaOAc, and NaOAc + H_3BO_3 indicated that compound 1 had the basic skeleton of 5,7-dihydroxyflavanone. The absence of signals between δ 5.7 - 6.9 in the ^{1}H NMR spectrum indicated that the C-6 and C-8 positions in the A-ring were substituted. The presence of protons, at C-3′, C-5′, C-2′, and C-6′ was indicated by two pairs of *ortho* coupled doublets at δ 6.98 and 7.42 (J = 10 Hz). A mass spectral fragment at m/z 180, suggested that the two methyl groups observed in the ^{1}H NMR spectrum (δ 2.06 and 2.08 were at C-6 and C-8; therefore, the methoxy group at δ 3.84 ppm in the ^{1}H NMR spectrum could be located at C-4′. These spectral properties are consistent with the assignment of the structure of compound 1 as 5,7-dihydroxy-6,8-dimethyl-4′-methoxyflavanone.

A compound of this structure was first isolated from *Matteucia orientalis* by Munesada (12) and was named (-)-matteucinol. Later the isolation of matteucinol was reported from two species of *Rhododendron*, by Arthur and Hui (13) and by Tanabe, Kondo and Takahashi (14); however, the structural assignments were not unequivocal since no spectral data was reported. Thus, this paper provides the first report of the spectral data of (-)-matteucinol and confirms the structure assignment as 5,7-dihydroxy-α,δ-dimethyl-4′-methoxyflavanone.

Matteucinol demonstrated excellent antifungal and antibacterial activity (Table I). The highest antifungal activity was observed against *R. solani* for which a threshold concentration of 15.1 μg/cm^2 was determined by regression analysis. A threshold concentration of 7 x 10^{-2} μg/cm^2 was determined for the activity against the bacterium *X. campestris*.

Table I. Antifungal and antibacterial activity* of Matteucinol and (-)-Medicarpin

	MICROORGANISM			
	P. ultimum	*R. solani*	*H. teres*	*X. campestris*
Matteucinol	25	135	30	69
(-)-Medicarpin	75	145	100	80

*relative activity of compound (10,000 ppm) to the standard (10,000 ppm):

$$\text{Relative activity (\%)} = \frac{\text{diam. of clear zone for test compound}}{\text{diam. of clear zone for standard}} \times 100$$

In a continuing search for compounds with antifungal and insect antifeedant activity, the active constituents from the wood of *D. monetaria* were also examined. The toluene soluble material (fraction A-2) from the methylene chloride extract of the wood of this plant showed both potent insect antifeedant activity against the boll weevil (*A. grandis* Boheman) and high antifungal activity (Figure 2). Partitioning of fraction A-2 between chloroform and 5% aqueous hydrochloric acid resulted in the concentration of the basic components in the aqueous layer. The basic fraction (fraction A-2-b), obtained as shown in Figure 3 presented no antifungal or insect antifeedant activity.

OMe
HO
O
OH O

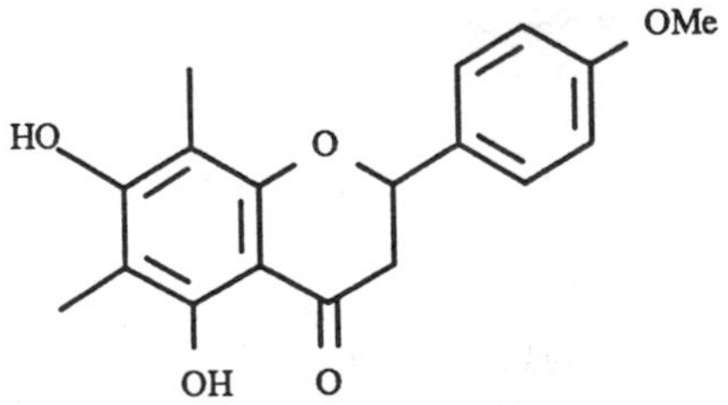

Matteucinol

Partitioning the chloroform layer between chloroform and 5% aqueous sodium hydroxide (Figure 3) resulted in concentration of the acid components in the aqueous layer. The acid fraction (fraction A-2-a) showed high antifungal activity (*P. ultimum* - +++;*R. solani* - +++; *H. teres* - +++) and antifeedant activity against boll weevils (feeding inhibition activity of 95.0 and 98.0% at the dose level of 1 and 2 mg/cm^2 of the filter paper). Fraction A-2-a was fractionated on a silica gel column with eluants of increasing polarity from hexane through toluene to chloroform and then to methanol as shown in Figure 3. The material was eluted with toluene-hexane (1:1) and was further purified by recrystallization from benzene to produce compound 2. A more polar material was eluted with toluene and purified by flash chromatography using hexane-acetone (2:1) as the eluant. Fractions 13-18 were combined and further purified by preparative TLC using 10% methanol in chloroform (v/v) to obtain compound 3.

Compound 2, m.p. 134 - 135°C, showed a M^+ at m/e 414. The ^1H NMR, UV, and IR spectra suggested the presence of a steroid. The color reactions developed with the Liebermann-Burchard and Carr-Price reagents were also in accord with a steroid structure. Compound 2 was identified as sitosterol by comparing its m.p. and ^1H NMR, IR, UV, and MS spectra data with those of an authentic sample (Aldrich Chemical Company, Inc., Milwaukee, WI).

Compound 3 had a molecular formula of $C_{16}H_{14}O_4$, as determined by high resolution mass spectrum. The ^1H NMR spectrum was similar to those observed for pterocarpans (13). Compound 3 was identified as (-)-Medicarpin by comparing its m.p. and ^1H NMR, IR, UV, and MS with those of an authentic sample (Dr. H. D. Van Etton, Cornell University).

(-)-Medicarpin showed excellent antifungal and antibacterial activity (Table I). The highest antifungal activity (14.5%) was observed against *R. solani*.

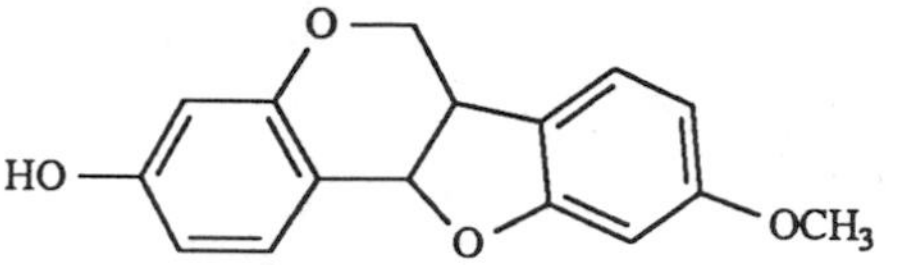

(-)-3-Hydroxy-9-methoxypterocarpan
[(-)-medicarpin]

Conclusion

This is the first report of the isolation of matteucinol from *M. cannabina* and (-)-medicarpin from *D. monetaria*. Matteucinol demonstrated excellent antifungal activity (*R. solani* 135%). (-)-Medicarpin showed activity against the fungi *P. ultimum* (75%), *R. solani* (145%), *H. teres* (100%), the bacteria *X. campestris* (80%), and the boll weevil (100% antifeedant at a dose level of 2 mg/ml). Thus matteucinol and (-)-medicarpin have potential as new agrochemicals.

Acknowledgments

This study was supported by the Rohm & Haas Co. We are indebted to Drs. Catherine Costello and Thomas Dorsey of the MIT mass spectrometry facility for their invaluable assistance in obtaining the mass spectrum of compounds. We thank Dr. Jim Dechter and Vicki Farr of the University of Alabama NMR facility for their help in obtaining ^1H NMR data.

Literature Cited

1. Gaulin, S.J.C.; Carker, L.E. *J. Agric. Food Chem.* **1974**, *27(4)*, 791.
2. Marini Bettolo, G.B.; Delle Monache, F.; Goncalves Da Lima, O.; de Barros Coelho, S. *Gazz. Chim. Ital.* **1971**, *101*, 41.
3. Benays, E.; Lupi, A.; Marini Bettolo, R.; Mastrofrancesco, C.; Tagliatesta, *P. Experientia*, **1984**, *41*, 1010.
4. Chawla, H.M.; Chibber, S.S. *J. Scientific and Industrial Research* **1981**, *40*, 313-325.

5. The Official Methods of Analysis of the Association of Official Analytical Chemist, Official First Action, Eleventh Ed., 1970, p. 843.
6. Hedin, P.A.; Thompson, A.C.; Minyard, J.P. *J. Econ. Entomol.* **1966**, *59*, 181.
7. Struck, R.F.; Frye, J.; Shealy, Y.F.; Hedin, P.A.; Thompson, A.C.; Minyard, J.P. J. *Econ. Entomol.* **1961**, *61*, 270.
8. Zweig, G.; Sherma, J., (Ed.s), Handbook of Chromatography, Vol. 2, The Chemical Rubber Co., Clevelent, 1972.
9. Mabry, T.J.; Markham, K.R.; Thomas, M.B. In: The Systematic Identification of Flavonoids, Springer-Verlag, New York, 1970.
10. Jurd, L., In: "The Chemistry of Flavonoid Compounds", Geissman, T.A., (Ed.), MacMillan, New York, 1962.
11. Harbone, J.B.; Mabry, T.J. In: "The Flavonoid Compounds", Academic: New York, **1975**, Vol.1.
12. Munesda, *J. Pharm. Soc.* Japan, **1924**, *505*, 185.
13. Arthur, H.R.; Hui, W.K.; *J. Chem. Soc.*, **1954**, 2782.
14. Yoshihisa, T.; Kondo, S.; Takahashi, K.; *Kanazawa Daigaku Yakugakubu Kenkyu Nempo*, **1962**, *12*, 7-14.

RECEIVED May 16, 1990

Chapter 29

Suppression of Fusarium Wilt of Adzukibean by Rhizosphere Microorganisms

Shinsaku Hasegawa[1], Norio Kondo[2], and Fujio Kodama[2]

[1]Hokkaido Institute of Public Health, North 19, West 12, North-ward, Sapporo 060, Japan
[2]The Hokkaido Central Agricultural Experiment Station, Naganuma 069–13, Japan

During a study on the biological contorol for Fusarium wilt of adzukibean caused by *Fusarium oxysporum* f. sp. *adzukicola*, we isolated antagonistic microorganisms from the rhizoshere soil of adzukibean root by the improved triple layer method. Hemipyocianine, chrororaphin and phenazine-1-carboxylic acid were isolated from *Pseudomonas aeruginosa* S-7 and *P. fluorescens* S-2; pyrrolnitrin from *P. cepacia* B-17; antibiotic Y-1 (monazomycin like) from *Streptomyces flaveus* Y-1. These microbial cultures and antifungal agents strongly inhibited growth of *F. oxysporum* f. sp. *adzukicola* FA-3. In greenhouse, the treatment of adzukibean seed with *P. cepacia* B-17 in *F. oxysporum* FA-3-infested soil decreased the diseases incidence from 76.3 to 8.8%; *St. flaveus* Y-1, 9.0%; *P. aeruginosa* S-7, 11.6%; *P. fluorescens* S-2, 13.3%. At a crop filed, the incidence decreased from 88% to 58%, 38%, 63% and 59% respectively. The isolates may be useful as antagonists to Fusarium wilt of adzukibean.

Fusarium wilt of the adzukibean caused by *Fusarium oxysporum* f. sp. *adzukicola* is a serious and widespred disease in Hokkaido, northern part of Japan. This pathogen is a root-infecting fungus that spreads relatively slowly along the vascular bundle and causes necrosis, yellowing and wilt(1-3). In our work on Fusarium wilt of adzukibean, we have isolated microorganisms antagonistic to *F. oxysporum* f. sp. *adzukicola* FA-3 from the adzukibean rhizosphere, and report here: the isolation and identification of antagonists and their producing antifungal agents; and the efficacy of isoates or these antifungal agents as seed treatment to Fusarium wilt by *F. oxysporum*. f. sp. *adzukicola*(1,4-6).

0097–6156/91/0449–0407$06.00/0
© 1991 American Chemical Society

Isolation and Identification of Antagonistic Microorganisms and these Producing Antifungal Agents

We isolated antagonists to *Fusarium oxysporum* FA-3 from the rizosphere of adzukibean and other materials by the improved triple layer

Table I. Isolated Antagonists and Their Producing Antifungal Agents

Antagonist	Antifungal agents / Source( )*1	Hemi-pyocianine	Chlororaphin	Phenazine-1-carboxylic acid	Pyrrolnitrin	Pyoluteorin
Pseudomonas						
aeruginosa S-1	Adzukibean(Rl)	(+)*2	+	+	-	-
S-7	Adzukibean(Rl)	+	+	+	-	-
SH-6	Soil	-	+	+	-	-
fluorescens S-2	Adzukibean(Rl)	-	+	(+)	-	-
Y-15	Melone(Rl)	-	-	-	-	+
NS-7371	lily(Rl)	-	-	+	+	-
cepacia B-17	Adzukibean(Rh)	-	-	-	+	-
1218	Beet(Rh)	-	-	-	+	-
Streptomyces						
sp. No. 2	Adzukibean(Rl)	Water soluble				
sp. B-6	Adzukibean(Rl)	Water soluble				
flaveus Y-1	River water	Antibiotic Y-1 (monazomycin like)				

*1 Sample source : (Rl), Rhizoplane; (Rh), Rhizosphere
*2 Antifungal agent: producing, +; trace, (+); no producing, -

Hemipyocianine

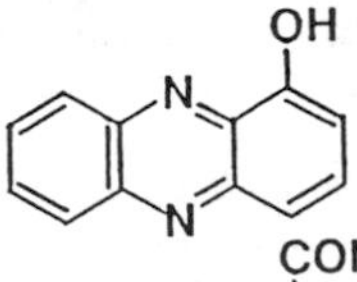

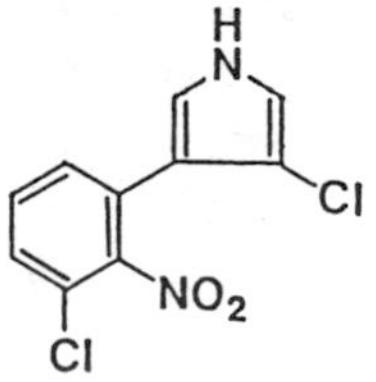

Chlororaphin

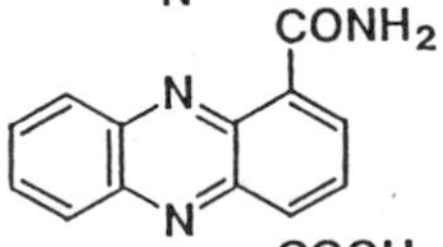

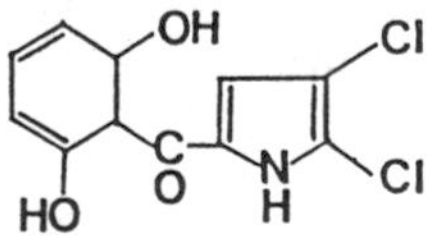

Phenazine-1-carboxylic acid

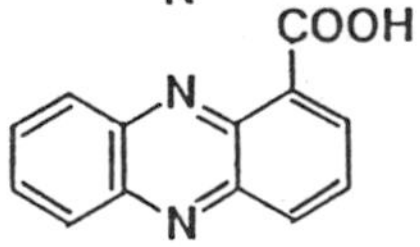

Pyrrolnitrin

Pyoluteorin

Antibiotic Y-1 (Monazomycin like)

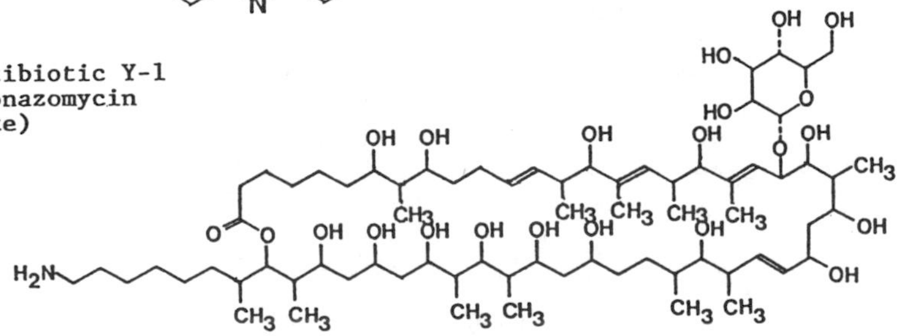

method(4). One ml of diluted(10^2-10^5w/v) rhizosphere soil of adzukibean were spread in plates with 1.5% agar and added the spore suspension of *F. oxysporum* FA-3 on the plates. These plates were incubated at 28 C for 2 to 4 days, and then the colonies with clear zone were picked up. These isolates were identified as *Pseudomonas aeruginosa*, *P. fluorescens*, *P. cepacia*, *Streptomyces flaveus* and *Streptomyces* spp.(1,2,4-8, Table I). These produced hemipyocianine, chlororaphin, phenazine-1-carboxylic acid(1,6), pyrrolnitrin(1,4), pyoluteorin(5,6), Antibiotic Y-1(monazomycin like, 7) and water soluble unknowns(Table I).

in vitro Antifungal Activities of Microbial Cultures and Antifungal Agents against *Fusarium oxysporum* f. sp. *adzukicola* FA-3

Bioassay by reversed layer method(4, Fig. I) showed that 3 strains of *Pseudomonas* and *St. flaveus* Y-1(8) were highly inhibitory to *F. oxysporum* FA-3(Table II). On agar spot inoculation plates(4, Fig. I), *P. aeruginosa* S-7, *St. flaveus* Y-1 and *Streptomyces* sp. No.2 showed highly inhibition and continuance of antifungal activities. The minimum inhibitory concentration(MIC) of pyrrolnitrin against *F. oxysporum* FA-3 was 3.13 μg/ml; Antibiotic Y-1, 6.25 μg/ml; hemipyocianine, 50 μg/ml; chlororaphin and phenazine-1-carboxylic acid, 100 μg/ml; and water soluble unknowns, 25-50 μg/ml (Table II).

Antifungal Spectrum of Microbial Cultures and Antifungal Agents

The microbial cultures of isolates and their antifungal agents strongly inhibited growth of *F. oxysporum* and also many other plant pathogenic fungi: *F. solani*, *F. moniliforme*, *F. roseum*, *Biporaris sorokiniana*, *Alternaria alternata*, *Cladosporium cucumerinum*, *Pyricularia oryzae*, *Pythium graminicolum*, *Verticillium dahliae*, *Cylindrocarpon* sp., *Rhizoctonia solani*(Table III, IV).

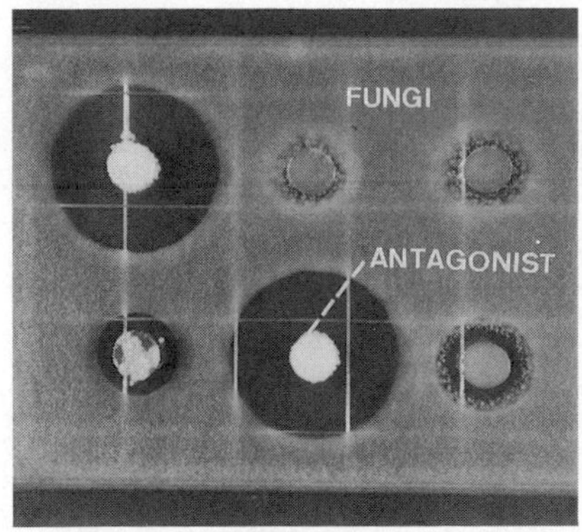

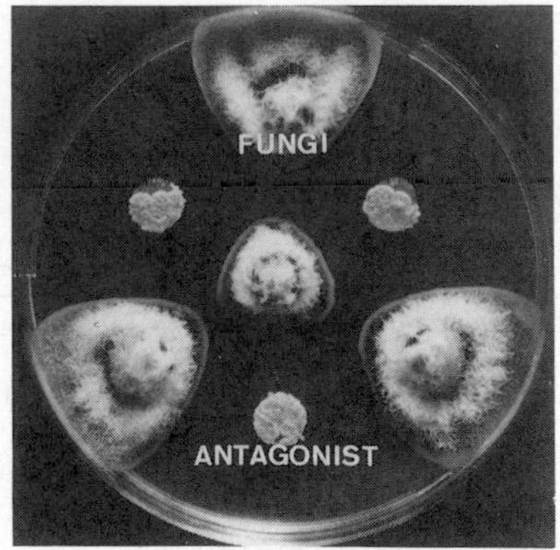

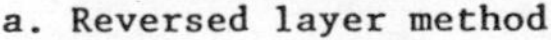
a. Reversed layer method b. Agar spot inoculation method

Fig. I Antifungal activities of antagonists against Fusarium oxysporum FA-3

Antibiotic Y-1 was an effective inhibitor of all tested pathogens in culture. But pyrrolnitrin was much less effective against *Fusarium* species than any other genera, and was totally ineffectives against *F. oxysporum* f.sp. *cepae*. Also, phenazines, such as hemipyocianine did not inhibit strongly the growth of *Fusarium* species(Table IV). But the strains which produced phenazines were active against *Fusarium* species and continued these strength on the agar spot inoculation method(Table III). The specificity of the strains for susceptibility against antifungal agents were recognized.

Table II The Inhibition effect of isolated antagonists and their antifungal agents against Fusasrium oxysporum FA-3

Antagonist	I. Antagonist agar disc		II. Antifungal agent	
	Reversed layer method*1, 2days (inhibition zone, mm)	Agar spot inoculation method*2 (inhibition zone, mm)	Isolated antifungal agent	Minimum inhibitory concentration (μg/ml, 7days)
Pseudomonas aeruginosa S-7 (Brucella agar, 30 °C, 4days)	30.0	9.5 +++(4days) ↓ 9.5 +++(7days)	Hemipyocyanine Chlororaphin Phenazine-1-carboxylic acid	100 200 200
Pseudomonas cepacia B-17 (Brucella agar, 30 °C, 4days)	33.8	7.0 ++ (4days) ↓ 3.5 + (7days)	pyrrolnitrin	3.13
Pseudomonas fluorescens NS-7371 (Brucella agar, 30 °C, 4days)	31.5	7.2 ++ (4days) ↓ 4.8 + (7days)	Pyrrolnitrin Phenazine-1-carboxylic acid	3.13 200
Streptomyces sp. No. 2 (YM agar*3, 30 °C, 7days)	24.0	11.2 +++(4days) ↓ 9.5 +++(7days)	Crude--- water soluble	25
Streptomyces sp. B-6 (PDA*4, 30 °C, 4days)	25.0	4.7 ++ (4days) ↓ 0.0 - (7days)	Crude--- water soluble	50
Streptomyces flaveus Y-1 (GS agar*5, 30 °C, 7days)	31.2	10.7 +++(4days) ↓ 10.1 +++(7days)	Antibiotic Y-1 (Monazomycin like)	3.13

*1 Reversed layer method: Fusarium oxysprum FA-3 spores were used for bottom layer and 4 or 7 days culture agar discs of antagonists were used.
*2 Agar spot inoculation method: 7 days culture agar discs of Fusarium oxysporum FA-3 and 4 or 7 days culture agar discs of antagonists were used. Inhibition zone between the pathogen and the antagonist: +++≥ 9.0mm, 9.0>++≥5.0, 5.0>+≥2.0, 2.0>(+)>0.0, -=0.0
*3 YM agar: Yeast-malt agar
*4 PDA: Potato dextrose agar
*5 GS agar: Glycerol-soybean meal agar

Table III Growth inhibition of the plant pathogens by the microbial cultures of antagonists*1

Phytopathgenic fungi*2 / Culture days	Antagonists											
	Pseudomonas aeruginosa S-7		Pseudomonas cepacia B-17		Pseudomonas fluorescens NS-7371		Streptomyces sp. No. 2		Streptomyces sp. B-6		Streptomyces flaveus Y-1	
	4	7	4	7	4	7	4	7	4	7	4	7
Fusarium oxysporum FA-3(adzukibean)*3	+++	+++*5	++	+	++	++	+++	+++	++	-	+++	+++
Fusarium oxysporum FGH(adzukibean)	++	++	++	+	++	+	+++	+++	+	-	+++	++
Fusarium oxysporum F-1(adzukibean)	++	++	++	-	++	++	+++	+++	++	-	+++	+++
Fusarium oxysporum f.sp.*4 spinacea H-18	++	++	++	++	++	++	+++	++	+	-	+++	+++
Fusarium oxysporum f.sp.spinacea H-29	++	-	++	+	++	+	+++	++	-	-	+++	+++
Fusarium oxysporum H-32(green pepper)	++	++	++	+	++	+	+++	+++	-	-	+++	+++
Fusarium oxysporum f.sp.fragariae H-34	+++	++	++	+	++	+	+++	++	++	-	+++	+++
Fusarium oxysporum f.sp.melonis H-34	+++	++	+	+	++	++	++	++	-	-	+++	+++
Fusarium oxysporum f.sp.lycopersici KF-244	++	-	++	-	++	-	+	+	-	-	+++	++
Fusarium oxysporum f.sp.cepae KF-228	++	++	-	-	++	++	+	+	+	-	+++	+++
Fusarium oxysporum KF 854(lily)	++	++	++	-	++	-	+	+	+	-	+++	+++
Fusarium solani H-41(rice)	+++	+++	++	-	++	-	++	++	-	-	+++	+++
Fusarium moniliforme H-24(rice)	+++	+++	++	-	++	-	+++	+++	++	-	+++	+++
Fusarium moniliforme H-36(rice)	+++	+++	++	+	++	+	+++	+++	+	-	+++	+++
Fusarium roseum H-35(wheat)	++	-	+++	-	+++	-	++	++	-	-	+++	+++
Biporaris sorokiniana H-21(rice)	+++	+++	+++	+++	+++	+++	+++	+++	++	++	+++	+++
Alternaria alternata H-25(rice)	+++	+++	+++	+++	+++	+++	+++	+++	+++	+++	+++	+++
Cladosporium cucumerinum H-23(melon)	+++	+++	+++	+++	+++	+++	+++	+++	+++	+++	+++	+++
Pyricularia oryzae H-22(rice)	+++	+++	+++	++	+++	+++	+++	++	ND*6	++	+++	+++
Pythium graminicolum H-30(rice)	+++	+++	+++	+++	+++	+++	+++	+++	+++	++	NT*7	
Verticillum dahliae H-28(strawberry)	+++	+++	+++	+++	+++	+++	+++	+++	NT		+++	+++
Cylindrocarpon sp. KF 846(lily)	+++	+++	+++	+++	+++	+++	+++	+++	+++	++	+++	+++
Rhizoctonia solani H-31(potato)	-	-	+++	++	+++	++	+++	+++	++	++	+++	+++
Rhizoctonia solani KF 852(lily)	++	++	++	+	-	-	+++	+++	++	++	+++	+++

*1 Agar spot inocuration method: 7 days culture agar discs of phytopathogenic fungi and 4 days culture agar discs of antagonists, 7 days of actinomycetes were used.
*2 Microbial culture(culture agar disc): Potato dextrose agar(Difco) for phytopathogenic fungi, Brucella agar(Difco) for bacteria, Yeadt extract-malt extract agar(ISP medium No. 2) for actinomycetes, SG(Soybean-glycerol) agar for <u>Streptomyces</u> <u>flaveus</u> Y-1.
*3 Plant Source
*4 f. sp.: forma specialis
*5 Inhibition zone between the pathogen and the antagonist: +++≧9.0mm, 9.0>++≧5.0, 5.0>+≧2.0, 2.0>(+)>0.0, -=0.0
*6 ND: not detected.
*7 NT: not tested.

Table IV Growth inhibition of the plant pathogens by the microbial cultures of antifungal agents

Antifungal agents / Phytopathgenic fungi	MIC(μg/ml)*1				
	Pyrrolnitrin	Pyoluteonin	Hemipyocianine	Phenazine-1-carboxylic acid	Antibiotic Y-1
Fusarium oxysporum FA-3(adzukibean)*2	3.13	25	50	100	6.25
Fusarium oxysporum FGH(adzukibean)	3.13	25	100	100	25
Fusarium oxysporum F-1(adzukibean)	3.13	25	100	50	12.5
Fusarium oxysporum f.sp.*3spinacea H-18	3.13	25	100	100	6.25
Fusarium oxysporum f.sp.spinacea H-29	3.13	25	100	100	6.25
Fusarium oxysporum H-32(green pepper)	3.13	12.5	200	50	12.5
Fusarium oxysporum f.sp.fragariae H-34	6.25	12.5	200	50	12.5
Fusarium oxysporum f.sp.melonis H-34	6.25	12.5	200	100	6.25
Fusarium oxysporum f.sp.lycopersici KF-244	6.25	25	200	100	25
Fusarium oxysporum f.sp.cepae KF-228	200	>200	>200	100	25
Fusarium oxysporum KF 854(lily)	12.5	25	>200	>200	12.5
Fusarium solani H-41(rice)	25	50	>200	>200	25
Fusarium moniliforme H-24(rice)	25	25	200	100	25
Fusarium moniliforme H-36(rice)	3.13	25	50	50	50
Fusarium roseum H-35(wheat)	3.13	12.5	>200	100	12.5
Biporaris sorokiniana H-21(rice)	0.78	6.5	100	100	0.39
Alternaria alternata H-25(rice)	0.78	3.13	>200	25	0.39
Cladosporium cucumerinum H-23(melon)	0.78	3.13	100	25	0.39
Pyricularia oryzae H-22(rice)	0.78	25	100	50	1.57
Pythium graminicolum H-30(rice)	3.13	12.5	>200	50	NT*4
Verticillium dahliae H-28(strawberry)	0.78	12.5	>200	25	3.13
Cylindrocarpon sp. KF 846(lily)	0.78	12.5	>200	25	1.57
Rhizoctonia solani H-31(potato)	0.39	400	>200	100	0.39
Rhizoctonia solani KF 852(lily)	1.56	400	>200	100	0.39

*1 MIC: Minimum inhibitory concentration(μg/ml), 7 or 14 days(27 °C) culture on Potato dextrose agar(Difco)
*2 Plant Source
*3 f. sp.: forma specialis
*4 NT: not tested.

Efficacy of Seed Bacterization with Microbial Cultures to Control of Fusarium Wilt of Adzukibean

In greenhouse, treatment of the adzukibean seed with washed bacterial cells of *P. cepacia* B-17 in *F. oxysporum* FA-3-infested soil(Fig. II) decreased the disease incidence from 76.3 to 8.8% (Fig. III); *St. flaveus* Y-1, 9.0%; *P. aeruginosa* S-7, 11.6%;

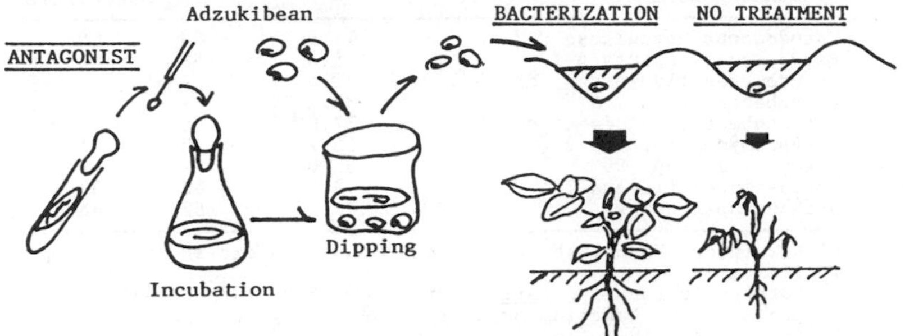

Fig. II Biological control of Fusarium wilt

F.oxy. FA-3	-	-	+	+
P.cep. B-17	-	+	+	-

Fig. III Effect of seed bacterization with Pseudomonas cepacia B-17 on the control of Fusarium wilt of adzukibean caused by Fusarium oxysporum FA-3*1

*1 Adzukibean: Hayate-syouzu,
Treatment: Pathogenic fungi, Fusarium oxysporum FA-3(10^5spores/g soil); Antagonist, Pseudomonas cepacia B-17(10^8cells/ml solution).

Table V Effects of seed bacterization with antagonists on the control of adzukibean wilt caused by <u>Fusarium oxysporum</u> in green house and crop field*1

	Disease incidence(%)		
	Green house (36days)	Crop field (45days)	
Antagonist*2	-	-	Me-cellulose
Pseudomonas aeruginosa S-7	11.6	63	60
Pseudomonas cepacia B-17	8.8	58	46
Pseudomonas fluorescens NS-7371	20.5	-*3	-
Flavobacterium sp. AB7	30.2	-	-
Acinetobacter calcoaceticus B28	77.0	85	-
Streptomyces sp. No. 2	15.2	92	41
Streptomyces sp. B6	65.0	82	-
Streptomyces flaveus Y-1	9.0	50	38
Without antagonist	76.3	88	88

*1 Adzuibean: Hayate-shozu in green house; Takara-shozu in crop field,
Pathogenic fungi: <u>Fusarium oxysporum</u> FA-3, 10^5spores/g soil in green house; <u>Fusarium oxysporum</u> 10^3cells/g soil in crop field
*2 Antagonist: 10^8cells/ml solution
*3 Not tested.

Streptomyces sp. No.2, 15.2%; and *P. fluorescens* NS-7371, 20.5% (Table V).

In crop filed, treatment with *St. flaveus* Y-1 cultures decreased the disease incidence from 88 to 38%; *Streptomyces* sp. No.2, 41%; and *P. cepacia* B-17, 46%(Table V). Methyl cellulose was used for injection of microbial cells at the time of seeding in field.

Growth of Microorganisms and Production of Antifungal Agents in Soil on the Control of Fusarium Wilt of Adzukibean

In the case of low disease incidence, such as *P. cepacia* B-17 and *St. flaveus* Y-1, the increase of the number of antagonistic microorganisms and the decrease of the number of pathogen, *F. oxysporum* FA-3, in the rhizosphere soil of adzukibean were recognized(Fig.IV, Table VI). The tightly relationship between the increase of antifungal agents and the decrease of disease incidence was not recognized. Washed microbial cells were as efficacious as whole cultures or culture filtrates, even though no or a few amount of antifungal agents were present when the seeds were treated. None of the treatments was phytotoxic to adzukibean.

The protective effect exhibited by treatment of adzukibean seed with cultures of the antagonistic microorganisms may be due to release of the antifungal agents by the gradual growth of the microbial cells. Even though the concentration may be low, the lysing cells may effect prolonged release and availability of the antifungal agents during the critical period of seeding growth.

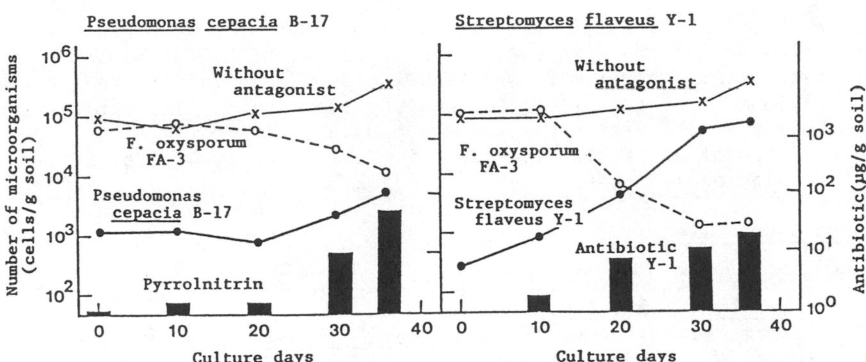

Fig. IV Time course comparison of growth of microorganisms and production of antifungal agents on the control of Fusarium wilt of adzukibean in green house

Table VI Growth of microorganisms and production of antifungal agents in rizoshere soil of adzukibean in green house

Antagonist	Number of cells/g soil: Antagonist*1 (days) 0 10 20 30 36	Number of cells/g soil: F. oxysporum*2 0 10 20 30 36	Production of antifungal agent (µg/g soil) 0 10 20 30 36	Disease incidence (%)
Pseudomonas aeruginosa S-7	10^5 10^4 10^3		10^3 10^2 10^1 10^0	11.6
Pseudomonas cepacia B-17				8.8
Pseudomonas fluorescens NS-7371				20.5
Acinetobacter calcoaceticus B-28				77.0
Streptomyces sp. No. 2				15.2
Streptomyces sp. B-6				65.0
Streptomyces flaveus Y-1				9.0
Without antagonist				76.3

*1 10^8 cells/ml solutions of antagonists were used for seed bacterization.
*2 10^5 cells/g soil of Fusarium oxysporum FA-3 were used.

Conclusion

These results suggested that the isolates such as *P. cepacia* B-17, *St. flaveus* Y-1 and *Streptomyces* sp. No.2 may be useful as antagonists to *F. oxysporum* f. sp. *adzukicola* and may facilitate establishment of cultivation of healthy adzukibean. The antagonisms exhibited by the microorganisms are possibly the result of production of antifungal agents, which are themselves effective in protecting agaist Fusarium wilt of adzukibean.

Literature Cited

1. Hasegawa, S.; Kodama, F.; Kaneshima, H.; Akai, J. J. Pharmacobio-Dyn. 1987, 10S, 57.
2. Kodama, F.; Hasegawa, S. Kongetu no Nouyaku 1985, 30, 22-27
3. Kondo, N.; Kodama, F. Shokubutu Boeki 1989, 43, 28-33.
4. Hasegawa, S.; Kamneshima, H.; Kodama, F.; Akai, J. Report of the Hokkaido Institute of Public Health 1986, 36, 16-23.
5. Hasegawa, S.; Kondo, N.; Kodama, F. Proceedings of 7th Symposium on the Development and Application of Naturally Occurring Drug Materials, 1989, 7, 67-70.
6. Hasegawa, S.; Kondo, N.; Kodama, F. J. Pharmacobio-Dyn. 1990, in press.
7. Hasegawa, S.; Kamneshima, H. Report of the Hokkaido Institute of Public Health 1987, 37, 6-17.
8. Hasegawa, S.; Kodama, F.; Nakajima, M.; Akai, J.; Murooka, H. Bull. Agr. & Vet. Med., Nihon Univ. 1987, 44, 80-86.

RECEIVED May 16, 1990

Chapter 30

Soil-Borne Diseases in Japan

Practical Methods for Biological Control

Shinsaku Hasegawa[1], Fujio Kodama[2], and Norio Kondo[2]

[1]Hokkaido Institute of Public Health, North 19, West 12, North-ward, Sapporo 060, Japan
[2]The Hokkaido Central Agricultural Experiment Station, Naganuma 069–13, Japan

The interest and research in biological control have increased in this decade. The following are the promising procedures which will be adopted in the near future. **1.** Sweet potato sprouts which were previously inoculated with non-pathogenic *Fusarium oxysporum* isolate showed high resistance against Fusarium wilt caused by *F. oxysporum* f.sp. *betatas*. **2.** *Pseudomonas gladioli* suppressed some Fusarium diseases. Seedling of associate crops(*Allium* spp.) which had been dipped in suspension of this strain, and mix-cropped with commercial crops in *Fusarium* infected field. Fusarium diseases of tomato and bottle gourd(*Lagenaria cicereria*) were suppressed. **3.** Edible lily root rot caused by *Cylindrocarpon destructans* was controlled when the mother bulbs were coated with *P. fluorescens* S-2 or *P. aeruginosa* S-7. These bulbs were much bigger than the non-treated ones.

Biological control of soil borne diseases is keenly being promoted by the government and agricultural industries in Japan. This is in part a response to public concern about hazards associated with chemical pesticides. Interest and research on the issue have increased this decade in Japan. Biological control has been studied for over 65 years, but few successes have been made in the commercial field. The following are the promising articles which will be adopted in the field in near future.

The authors discuss three examples of biological control of soil borne diseases. The first is the control of Fusarium wilt of sweet potato by cross-protection that involves a prior inoculation of nonpathogenic *Fusarium oxysporum*(Ogawa *et al.*, 1984)(1). The second is the control of Fusarium wilt of bottle gourd by mixed-cropping with associate crops(Kijima *et al.*, 1986)(2). The third is the seed bulb bacterization for the control of root rot of edible lily(Hasegawa *et al.*, 1990)(3).

0097–6156/91/0449–0417$06.00/0
© 1991 American Chemical Society

Biological Control of Fusarium Wilt of Sweet Potato with Cross-protection by Prior Inoculation with Nonpathogenic *Fusarium oxysporum*

The causal fungus of Fusarium wilt of sweet potato is *Fusarium oxysporum* f. sp. *betatas*. Some isolates of *F. oxysporum* which are obtained from healthy sweet potato plants showed remarkable cross-protection against the disease, when they were previously inoculated in the sweet potato sprouts before being planted in the infested soil. These isolates of *F. oxysporum* were not pathogenic to sweet potato, and also not to other plants such as tomato, cucumber etc.

In naturally infested experimental or commercial fields, cross-protection by pre-inoculation with nonpathogenic isolates of *F. oxysporum* has always brought remarkable decreases in wilt incidence and increases in the yield of sweet potato. The effects were equivalent to those obtained with chemical treatment in which cut-ends of the sprouts were dipped into a benomyl suspension(500 times of 50% w.p.) for 30 min., as shown Fig. I.

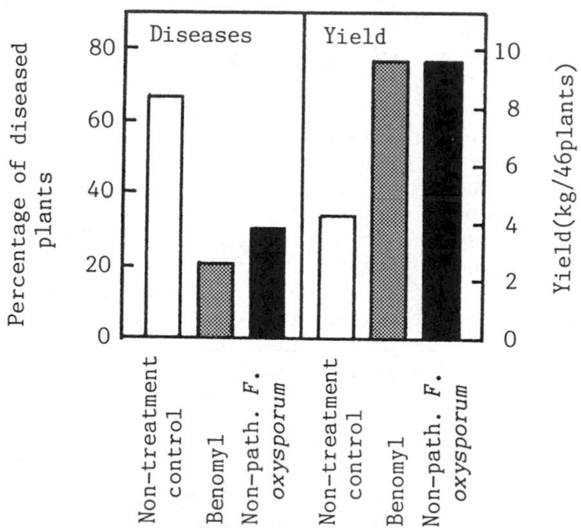

Fig. I Effect of pre-inoculation with nonpathogenic *F. oxysporum* on disease incidence of Fusarium wilt and yield of sweet potato in comparison with dipping the sprouts in benomyl suspension (500 times of 50% w.p.) in a naturally infested field.

No antagonisms were observed between the cross-protective isolates and the pathogen in controlled plate culture. When the cut-ends of sprout were dipped in liquid paraffin, no protection was observed against the disease. Even if the previously inoculated sprouts were bent and the bent portion was planted, cross-protection was not satisfactorily obtained. The cross-protective isolates were not pathogenic to sweet potato; however, colonization of the cut-ends was followed by a remarkable development of local lesions there. From these results, Ogawa *et al.*(1) suggest that the cross-protection is due to the resistance which is induced by the non-pathogenic *F. oxysporum* colonizing and bringing about local, but severe infection at the cut-end of the sprouts. They also suggested that the plants react to infection by producing resistance product(s) which translocate systemically from the basal to the upper stem of the plant.

Biological Control of Fusarium Wilt on Bottle Gurd by Mixed Cropped Plant

Bottle gourd is a cucurbitaceaeous plant. Fusarium wilt of bottle gourd is serious commercial problem in Japan. In Tochigi prefecture, the central section of Japan, the farmers have mixed-cropped welsh onion(*Allium fistulosum*), customarily, as an associate crop with bottle gourd. These commercial fields showed little occurrence of the disease in spite of continuous cropping and disease suppression. Kijima *et al.*(2) studied the possible role of microorganisms associated with welsh onion. *Pseudomonas gladioli* is frequently associated with roots of welsh onion. For practical use, beneficial isolates of the bacterium should strongly antagonize *Fusarium oxysporum* f. sp. *lagenariae*, but they should not be pathogenic to welsh onion. After screening more than 300 isolates, *Pseudomonas gladioli* M-2196 was selected for practical use. Table I shows that antifungal activity to pathogenic fungus and affinity to welsh onion of *P. gladioli* M-2196 occurred. This strain had strong antifungal

Table I Antifungal activity to *Fusarium oxysporum* f. sp. *lagenariae* and affinity to welsh onion of *Pseudomonas gladioli*

Bacteria number of isolate*1	Antifungal activity*2	Affinity to welsh onion*3
M-2196	++	+
M-2197	+	+
W-2443	++	++
W-2444	++	++
W-2445	++	++
Ca-0550	–	+
V-0560	+	+
Cy-0617	–	++
S-2258	–	+

*1 Source of isolate, M, *Miltonia* sp.; W, welsh onion (*Allium fistulosum*); Ca, *Cattleya* sp.; V, *Vuylstekeara*; Cy, *Cymbidium* sp.; S, *Dendrobium* sp..

2 Activities were investigated on bouillon-peptone agar plate.

3 +, Good affinity(Did not damaged welsh onion, but multiplied well on or in it); ++, +, Damaged welsh onion.; –, Did not multiply on or in welsh onion.

activity, colonized welsh onion roots, but was not pathogenic to welsh onion.

Remarkable biological control by *P. gladioli* M-2196 against Fusarium wilt of bottle gourd was observed in the field(Table II). Roots of welsh onion or chinese chive(*Allium tuberosum*) were dipped in the suspension of strain M-2196 for 5 min. before *Allium* plants were transplanted. The disease was completely controlled by mixed-cropping with chinese chive or welsh onion. By the use of a scanning electron micrograph, it was observed that this strain was present near the root tip, from 10 to 90 days after dipping. It was reisolated from the roots of chinese chive more frequently than other bacteria. This strain produced an antifungal substance, pyrrolnitrin 1(Fig. II).

Table II Control of Fusarium wilt of bottle gourd (*Lageneria siceraria*)*1 by mix-cropping with welsh onion or chinese chive inoculated with *Pseudomonas gladioli* M-2196*2 (field test)

Treatment	Percentage of diseased plants
Mix-cropped bottle gourd with	
uninoculated chinese chive*3	26.7
uninoculated welsh onion*3	40.0
M-2196 inoculated*4 chinese chive*3	0.0
M-2169 inoculated*4 welsh onion*3	0.0
Mono-cropped	
M-2196 inoculated*4 bottle gourd	100.0
uninoculated bottle gourd	100.0
Mono-cropped bottle gourd in sterilized soil	0.0

*1 Cultivar, Shimotsukeshiro, grown in sterilized nurseries.
2 Incubated in bouillon-peptone broth for 5 days at 25 C.
3 Root systems were washed with 0.1% benzalkonium chloride solution before inoculation.
4 Root systems were dipped in *Pseudomonas gladioli* M-2196 culture solution for 5 min.

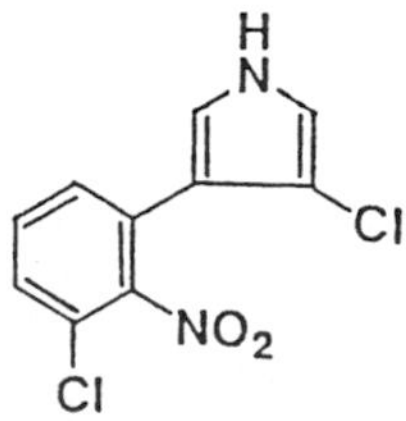

H N Cl NO_2 Cl

Pyrrolnitrin 1

Fig. II Structure of pyrrolnitrin

Based on their research(2), the mechanism of biological control by mixed-cropping technique was proposed(Fig. III). The antagonist, strain M-2196, grows on the roots of welsh onion or chinese chive and produces an antifungal agent which is spread into the soil. Pathogens which cause soil borne diseases are antagonized by the antifungal agents. Therefore, pathogens cannot infect plants, such as bottle gourd and tomato grown in association with treated plants.

The findings above are based on the work of Kijima *et al*.(2). Hasegawa *et al*.(4) recently found that *Pseudomonas cepacia* and some other *Pseudomonas* species which produce antifungal agents such as pyrrolnitrin were specially aggregated on, or in the roots of *Allium* species. This suggests that biological control of Fusarium diseases by *Allium* mixed-cropping without inoculating M-2196 is possible.

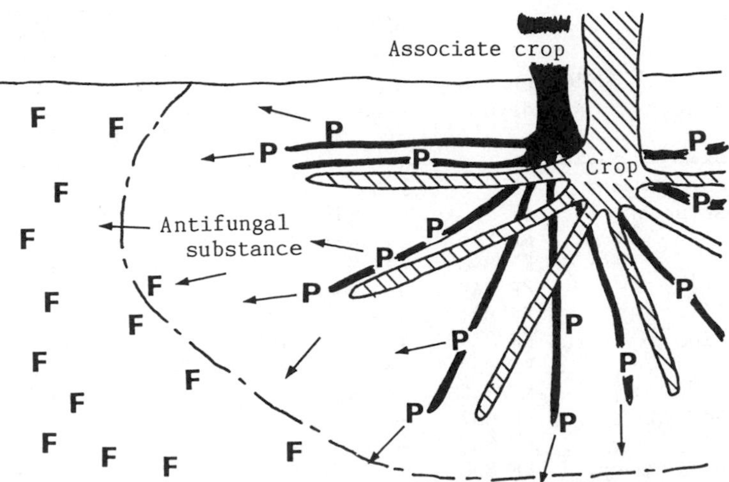

Fig. III Proposed mechanism of the biological control
F, *Fusarium* species(Pathogens);
P, *Pseudomonas gladioli* M-2196(Antagonist)

Biological Control of Root Rot of Edible Lily(*Lillium lancifolium*) by Seed Bulb Bacterization

Edible lily(*Lillium lancifolium*) is one of the important commercial crops in Japan. *Cylindrocarpon destructans* is a causal fungus of root rot of the lilies of which the symptoms on the ground are not obvious. The fungus attacks the juvenile tissue, i.e. root hairs and tips, of the plant. So the disease is considered chronic. Yields are reduced by this disease because the weight of each bulb is much smaller than that of healthy ones. The seed bulbs that are nursed for two years are transplanted in October and harvested in September next year.

Biological control of edible lily was carried out by Hasegawa *et al*.(3). The antagonist shows growth inhibition against the fungi isolated from rhizoplane of *Lillium lancifolium* by the agar spot inoculation method(Table III)(5). *Cylindrocarpon destructans* was inhibited drastically by these three *Pseudomonas* species(4-6). *Pseudomonas fluorescens* S-2 produces phenazine-1-carboxylic acid 2, chlororaphin 3, *P. aeruginosa* S-7 produces 2, 3, hemipyocianine(Fig. IV, 5,6) and P. cepacia B-17 produces pyrrolnitrin 1(Fig II). It is thought that these antifungal activities were caused by the antifungal agents.

Table III Antifungal Activity of *Pseudomonas* species to the fungi isolated from lily root*1

Fungi*2	Incubation days	Antagonist*2								
		Pseudomonas aeruginosa S-7			*Pseudomonas fluorescens* S-2			*Pseudomonas cepacia* B-17		
		4	7	14	4	7	14	4	7	14
Cylindrocarpon destructans	KF 845	+++	+++	+++*3	+++	+++	+++	+++	+++	+++
	KF 846	+++	+++	+++	+++	+++	+++	+++	+++	+++
	KF 848	+++	+++	+++	+++	+++	+++	+++	+++	+++
	KF 849	+++	+++	+++	+++	+++	+++	+++	+++	+++
	KF 851	+++	+++	+++	+++	+++	+++	+++	+++	+++
Fusarium oxysporum	KF 854	++	++	++	++	+	-	++	-	-
	KF 855	+++	++	++	++	-	-	++	-	-
	KF 856	+++	++	+	++	-	-	++	-	-
Trichoderma sp. 1	KF 857	-	-	-	-	-	-	-	-	-
	KF 862	-	-	-	-	-	-	-	-	-
Trichoderma sp. 2	KF 858	+	+	+	-	-	-	-	-	-
	KF 859	+	+	+	-	-	-	-	-	-
Rhizoctonia solani	KF 852	++	++	++	+	-	-	++	+	+
	KF 853	++	++	++	+	-	-	++	+	+

*1 Agar spot inoculation method(5).
2 7 days culture agar discs(potato dextrose agar) of fungi and 4 days culture agar discs(brucella agar) of bacteria were used.
3 Inhibition zone between the fungi and the bacteria, +++ $\geqq$ 12.0 mm, 12.0 > ++ $\geqq$ 6.0, 6.0 > + $\geqq$ 3.0, 3.0 > trace > 0.0, - = 0.0.

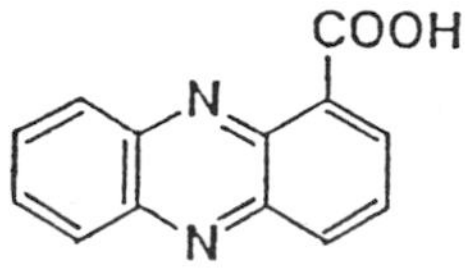

Phenazine-1-carboxylic acid 2 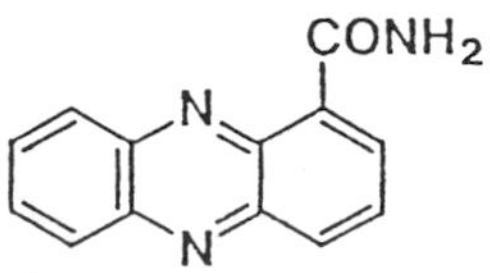 Chlororaphin 3 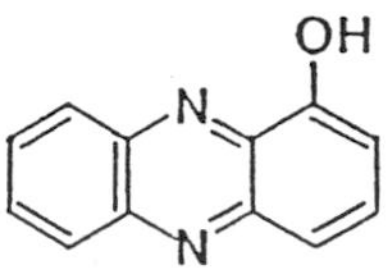 Hemipyocianine 4

Fig. IV Structure of phenazines.

After dipping into the suspension of the biocontrol agent, i.e. strain S-2, the seed bulbs were transplanted into the field that was infested with *Cylindrocarpon destructans*. Table IV shows the effect of bacterization on the yield of edible lily. The growth on the soil of the treated plants was obviously higher than the non-treated check. Yields of strain S-2 and S-7 bulbs treated by dipping were increased 167 and 145%, respectively. Strain S-2 was the most effective with an average bulb weight of 114.6 gm.

Table IV Effect of bacterization on yield of lily

Treatment	Yield ($kg/4.8m^2$)	Weight of bulb(g)	Length of stem(cm)
Pseudomonas aeruginosa S-7	9.6 b*1	100.9 b	72.2
Pseudomonas fluorescens S-2	11.0 a	114.6 a	68.0
Pseudomonas cepacia B-17	7.9 c	85.0 c	64.5
Streptomyces sp. B-2	7.1 cd	76.7 d	64.2
Control	6.6 d	74.4 d	64.9

*1 Values within a column followed by the same letter are not significantly different according to Duncan's multiple range test(P=0.05).

In order to examine whether the antagonist played a role on the roots of the lily, we isolated microorganisms from lily roots treated with antagonists at harvest time(Table V). The typical microflora was recognized. In the plot of strain S-2, no *C. destructance* and no *Fusarium oxysporum* were isolated, and the plot of strain S-7 showed the same indication as strain S-2. Also, there is a interesting phenomenon that the number of *Trichderma* sp. increased in the plot of strains S-2 and S-7. The antagonists were re-isolated from the treated plants, but not from the non-treated one. The antagonisms exhibited by the treated microorganisms are possibly the result of the production of antifungal agents, which are themselves an effective protectant against *C. destructance*. These results indicate that strain S-2 and strain S-7 may facilitate establishment of stands of healthy edible lily.

We assume the mechanism of bacterization by antagonists showed in Fig. V, and expect that two types of effects may occur. The first is antifungal agents being produced by antagonists at the rhizosphere of plants. We have already isolated and identified antifungal agents such as phenazines(Fig. IV), and confirmed that these agents showed strong *in vitro* antagonistic abilities against *Cylindrocarpon destructans*(Table III). The other is the production of growth promoting factors. We have not identified the growth promoting substance from strains S-7 and S-2. But, *P. cepacia* B-17 produced auxin, a typical plant growth hormone. These actions are effective in protecting against root rot and growing bulbs of edible lily.

Table V Isolation of microorganisms from lily roots treated with antagonistic bacteria

Isolated microorganism	Treatment		Control
	Pseudomonas aeruginosa S-7	*Pseudomonas fluorescens* S-2	
Fungi			
Cylindrocarpon destructans	2/25*1	0/25	15/25
Fusarium oxysporum	1/25	0/25	5/25
Trichoderma spp.	10/25	16/25	1/25
Rhizoctonia solani	0/25	0/25	2/25
Pseudomonas spp.			
Total *Pseudomonas* spp.	$3.4x10^5$ *2	$9.3x10^4$	$1.5x10^4$
Pseudomonas aeruginosa S-7	$5.0x10^2$	0	0
Pseudomonas fluorescens S-2	0	$5.0x10^2$	0

*1 Number of rootlets isolated/Number of rootlets tested.
*2 Number of Pseudomonas spp./g of rhizosphere soil.

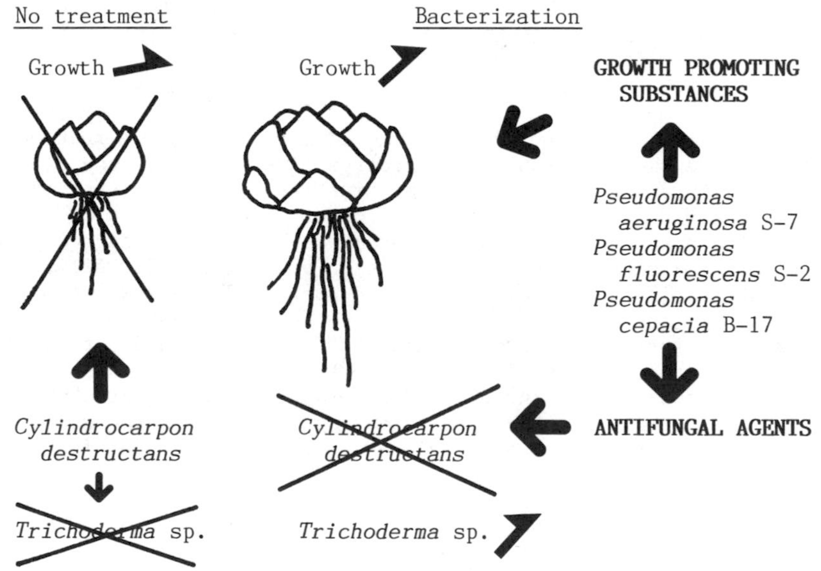

Fig. V Proposed mechanism of biological control of root rot of edible lily by *Pseudomonas* species

Conclusion

In Japan, it is becoming more and more difficult for growers to achieve satisfactory crop yields using traditional biological agricultural treatments such as crop rotation, tillage and organic amendments. On the other hand, chemical control of pests is often too expensive. That is the reason why new biological control procedures are being keenly demanded. The three types of biological control mentioned above will surely be widely applied in the near future. It is true that obviously, much more work remains to be done to clarify the mechanisms of disease suppression, but successful answers can be obtained from the fields. The authors will continue to try several approaches, especially the effect of antagonists on edible lily root rot biological control.

Literature Cited

1. Ogawa, K.; Komada, H. Ann. Phytopath. Soc. Japan 1984, 50, 1-9.
2. Kijima, T.; Tezuka, T; Arie, T; Namba, S; Yamashita, S; Doi, Y Ann. Phytopath. Soc. Japan 1986, 52, 542.
3. Hasegawa, S.; Kondo, N.; Kodama, F. J. Pharmacobio-Dyn. 1990, 13, in press.
4. Hasegawa, S.; Kondo, N.; Kodama, F. Ann. Phytopath. Soc. Japan 1990, in press.
5. Hasegawa, S.; Kaneshima, H.; Kodama, F.; Akai, J. Report of the Hokkaido Institute of Public Health 1986, 36, 16-23.
6. Hasegawa, S.; Kondo, N.; Kodama, F. Naturally Occurring Pest Bioregulators; American Chemical Society: Washington, DC,

RECEIVED July 31, 1990

Chapter 31

Preparative Separation of Complex Alkaloid Mixture by High-Speed Countercurrent Chromatography

Richard J. Petroski and Richard G. Powell

Northern Regional Research Center, Agricultural Research Service, U.S. Department of Agriculture, 1815 North University Street, Peoria, IL 61604

Tall fescue (Festuca arundinacea), infected with the fungal endophyte Acremonium coenophialum, contains both ergot-type and saturated-pyrrolizidine (loline) alkaloids. These alkaloids are believed to be at least partially responsible for the insect pest resistance of tall fescue and for periodic or seasonal toxicity of endophyte-infected tall fescue to grazing cattle. Attempts to separate the relatively minor amounts of ergot-type alkaloids (1-10 ppm) from the much more abundant saturated-pyrrolizidine alkaloids (1000-6000 ppm) using ordinary chromatography systems, have been generally unsatisfactory. High-speed countercurrent chromatography was examined as an alternative separation method. A crude fescue alkaloid mixture (1 g) was separated, in less than 8 hours, using a two-phase solvent system composed of chloroform-methanol-water (5:4:3, v/v/v). Alkaloids such as N-methylloline, N-acetylloline, and N-formylloline were well separated from each other and recovery of these alkaloids was quantitative. Relatively pure ergonovine was isolated from the mixture in one pass. Other minor alkaloids, including lysergic acid amide, were either separated from each other or were highly enriched in certain fractions.

This chapter not subject to U.S. copyright
Published 1991 American Chemical Society

Tall fescue, Festuca arundinacea Schreb, is a cool season pasture grass that is used extensively in the southeastern United States. Most existing pastures are infected with an endophytic fungus, Acremonium coenophialum Morgan-Jones and Gams (1), that has been associated with insect pest resistance of tall fescue (2-5). Endophyte-infected tall fescue contains both ergot-type (Figure 1) and saturated-pyrrolizidine (loline-type) alkaloids (Figure 2); neither type of alkaloid is found in tall fescue when the endophyte is absent (6-8). These two classes of alkaloids have also been associated with production losses in cattle (9), and economic losses to cattle producers have been estimated at $50 to $200 million annually (10). Ergot alkaloids present the greater toxic danger (8,11). The objective of this study was to isolate and identify the simpler amides of lysergic acid present in endophyte-infected tall fescue, including the two unknown compounds reported previously by Yates et al. (8).

Attempts to separate ergot-type alkaloids from the much more abundant loline-type alkaloids using ordinary chromatography systems, have been generally unsuccessful. The number of alkaloids present, the widely different concentrations and structural types of alkaloids present, and tailing of alkaloid peaks into one another, have been major problems preventing isolation and characterization of the minor alkaloids in tall fescue. High-speed countercurrent chromatography enables rapid and efficient separation of complex mixtures without adsorptive loss or degradation caused by solid supports such as alumina, silica, or cellulose (12-15). We report the gram-scale separation of a complex fescue alkaloid mixture using a commercial preparative cross-axis high-speed countercurrent instrument. The procedure has allowed us to isolate and to identify several alkaloids previously not known to occur in tall fescue.

Materials and Methods

Apparatus. A Pharma-Tech (Baltimore, MD, USA) Model CCC-600CX cross-axis preparative countercurrent chromatograph was used in our experiments. The complete instrument includes a dual coil column (2.6 mm I.D. polytetraflouroethylene tubing) planet centrifuge, LDC Milton Roy liquid pump, digital revolution speed monitor, Rheodyne HPLC injector, and a pressure monitor. The axis of the column holder is positioned perpendicular to, and at a fixed distance away from the centrifuge axis to promote more vigorous mixing of stationary and mobile phases. Replacement of the counterweight by a second identical column eliminates the need for tedious balancing of the centrifuge system and doubles the total column capacity. The total column capacity is 1260 mL and the optimum revolutional speed of this apparatus is 600 rpm.

Preparation of solvent system. A two-phase solvent system composed of $CHCl_3/MeOH/H_2O$ (5:4:3, v/v/v) was used. The solvent mixture was thoroughly equilibrated in a separatory funnel at 25°, and the two phases were separated before use. The upper phase was the stationary phase and the bottom phase was the mobile phase.

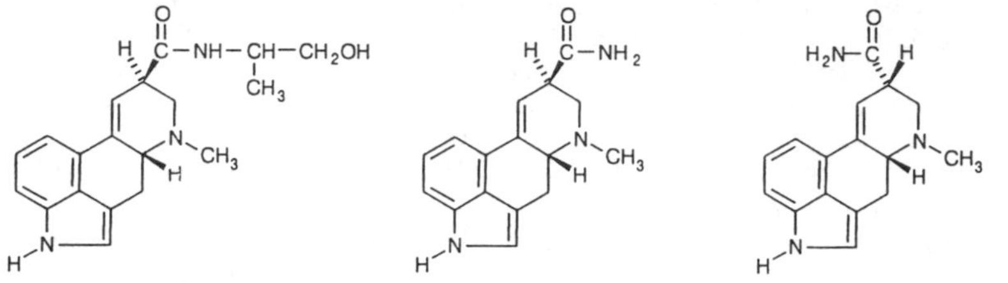

Ergonovine Lysergic acid amide Isolysergic acid amide

Figure 1. Ergot-Type Alkaloids Isolated From Endophyte-Infected Tall Fescue.

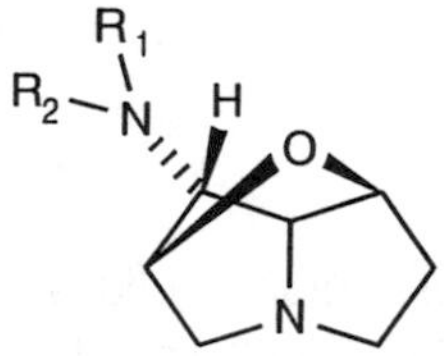

	R_1	R_2
Norloline	H	H
Loline	H	CH_3
N-Methyl Loline	CH_3	CH_3
N-Formyl Norloline	H	HCO
N-Acetyl Norloline	H	CH_3CO
N-Formyl Loline	CH_3	HCO
N-Acetyl Loline	CH_3	CH_3CO

Figure 2. Saturated-Pyrrolizidine (Loline-Type) Alkaloids Reported in Endophyte-Infected Tall Fescue.

Preparation of sample. Although the endophyte infects the whole fescue plant, seed is a rich source of both loline-type and ergot-type alkaloids (7,8), and presents fewer isolation problems. Endophyte-infected tall fescue seed (Lambert Seed Co., Camden, AL) was defatted with hexane (750 ml per 500 g seed), air-dried, ground in a milling machine to 2 mm mesh, and extracted with methanol (3x, 1500 mL). The methanol extract was concentrated *in vacuo* approx. 50-fold, to 90 mL, and diluted with 1% aqueous citric acid (500 mL).

The acidic solution was extracted with $CHCl_3$ (3x, 500 mL) and the $CHCl_3$ extracts were combined to give a fraction that was saved for future study. This fraction contained a mixture of ergopeptine alkaloids, including ergovaline. The aqueous phase was collected and the pH was adjusted to 10 with NaOH. The resulting alkaline solution was carefully extracted with $CHCl_3$ (5x, 500 ml) using thorough but gentle mixing to avoid troublesome emulsions. The $CHCl_3$ extracts were combined, concentrated *in vacuo* to 250 ml, and extracted with 0.2N aqueous HCl (3x, 80 ml). The $CHCl_3$ phase was discarded because it contained only small amounts of residual, relatively non-basic alkaloidal material. The aqueous HCl phases, containing alkaloids as their HCl salts, were combined and adjusted to pH 10 with NaOH. The resulting alkaline solution was extracted with $CHCl_3$ (5x, 250 ml) to afford a fraction that contained both loline-type and ergot-type alkaloids as free bases. The alkaloid fraction was concentrated *in vacuo* to 125 ml, and dried over anhydrous Na_2SO_4. Chloroform was removed *in vacuo* yielding an oily residue (1 g). The alkaloidal residue was then dissolved in a mixture of mobile phase (3 mL) and stationary phase (1 mL).

Countercurrent Chromatography Procedure. The entire column (pair of coiled multilayer columns connected in series) was filled with the stationary phase. The apparatus was then rotated counterclockwise at 600 rpm in planetary motion while the mobile phase was pumped into the inlet of the column at a flow-rate of 2.2 mL/min (head to tail elution mode). Maximum pressure at the outlet of the pump measured 80 psi. After a 1-hour equilibration period, the sample was loaded into the Rheodyne injector loop and injected. Effluent from the outlet of the column was continuously monitored with a Shimadzu UVD-114 detector at 312 nm and fractions collected with a Gilson FC-100 fraction collector to obtain approximately 8.8 mL of eluant in each tube (during a 4-min interval). Retention of the stationary phase was estimated to be 930 mL (74%) by measuring the volume of stationary phase eluted from the column before the effluent changed to mobile phase (330 mL) and subtracting this volume from the total column capacity of 1260 mL.

Analysis of fractions for loline alkaloids. Loline alkaloids present in the various fractions were determined by the quantitative capillary gas chromatographic method of Yates *et al.* (7). Prior to GC analysis, 50-μL or 100-μL aliquots of each fraction to be analyzed were diluted to 0.98 mL with MeOH, and phenyl morpholine was added as an internal standard (200 μg in 20 μL MeOH); 1-μL injections were used.

Analysis of fractions for ergot alkaloids. Ergot-type alkaloids were determined by HPLC with fluorescence detection (8). The HPLC system consisted of a Spectra-Physics SP8800 ternary solvent delivery system, Rheodyne injector, Varian Fluorichrom detector, and a Spectra-Physics SP4290 integrator. HPLC analysis was performed with a du Pont Zorbax ODS C-18 column (4.6 mm ID x 250 mm, 5 μm ODS) fitted with a Supelco 5-8954 prepacked disposable 2-cm guard column (LC-18 cartridge). The chromatography solvent system was composed of an 0.1 N ammonium acetate buffer (pH 7.6) and acetonitrile, at volume ratios of 65:35 or 80:20. All runs were isocratic with a flow-rate of 0.8 mL/min. For fluorescence detection of ergot alkaloids, the excitation wavelength was 310 nm; and the emission wavelength band passed, by Varian 3-71 and 4-76 emission filters, was between 375 and 460 nm. Alkaloids were determined by measurement of fluorescence peak height and comparison with ergotamine tartrate standard curves, and ergot alkaloid amounts were expressed as μg of ergotamine tartrate. Ergotamine tartrate and ergonovine were purchased from Sigma Chemical Co., St. Louis, MO.

Isolation of ergot alkaloids. Preparative TLC of countercurrent fractions was carried out on silica gel 60 F-254 plates (E. Merck) developed with CH_2Cl_2/MeOH (4:1). R_f values were 0.42, 0.50, and 0.68 for lysergic acid amide, ergonovine, and isolysergic acid amide, respectively. Fluorescent bands on TLC plates were separately scraped from the plates, transferred to disposable Pasteur pipettes, and ergot alkaloids were then eluted with the same solvent system. Mass spectra were obtained with a Finnigan MAT 4535/TSQ instrument equipped with a DEP probe.

Results and Discussion

Although sample is normally injected into a countercurrent chromatograph when the mobile phase is initially pumped into the column (12-15), it was found that better retention of stationary phase, a more stable absorbencies baseline, and better resolution of peaks resulted when samples were injected after a 1-hour equilibration period. Unstable baselines were observed during the time in which effluent changed from stationary phase to mobile phase; however, baselines stabilized by the time the first sample component appeared in the effluent. An alkaloid extract from endophyte-infected tall fescue seed prepared by a series of acid/base extractions was injected into the countercurrent chromatograph, rather than a crude ethanolic seed extract, because injection of the latter resulted in emulsions. When emulsions were formed, the stationary phase was not retained and separations were unsuccessful. Attempts to process more than 1 g of alkaloid concentrate per injection resulted in reduced resolution of sample components. The optimum flow-rate was 2.2 mL/min.

A typical countercurrent chromatogram of an alkaloid extract from endophyte-infected tall fescue seed is shown in Figure 3. Two curves are superimposed; the solid line represents mg of loline-type alkaloids determined by GC analysis, and the dotted line represents absorbencies at 312 nm, which is characteristic of

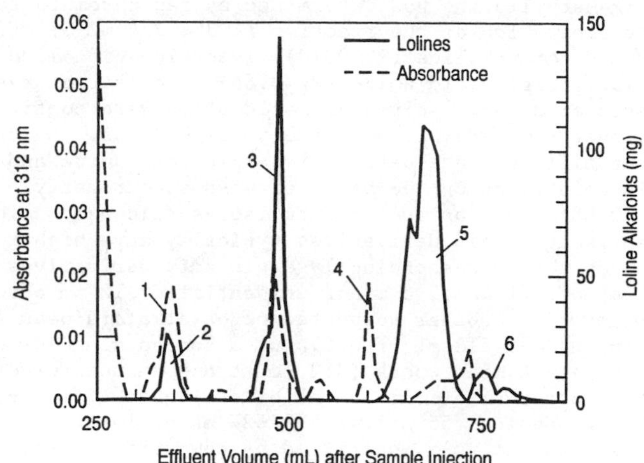

Figure 3. Countercurrent Chromatogram of an Alkaloid Concentrate from Endophyte-Infected Tall Fescue, with $CHCl_3/MeOH/H_2O$ (5:4:3) Lower Phase Mobile. Identified peaks: 1, lysergic acid amide (670 μg) plus isolysergic acid amide (520 μg); 2, N-methylloline (61 mg); 3, N-acetylloline (305 mg); 4, ergonovine (55 μg); 5, N-formylloline (601 mg); 6, loline (31 mg) plus N-acetylnorloline (28 mg). Loline-type alkaloids determined by GC. Ergot-type alkaloids determined by HPLC with fluorescence detection and expressed as μg ergotamine tartrate.

the cross-conjugated styrene-indole system of lysergic acid derivatives (16). Although 8.8-mL fractions were collected, numbers (mL effluent after sample injection) have been rounded off to integers in order to simplify the discussion. The first sample component appeared, at an effluent volume of 250 mL, as a non-alkaloidal yellow material. A broad peak at 304-374 mL of effluent contained two highly-fluorescent ergot alkaloids (670 μg of lysergic acid amide and 520 μg of isolysergic acid amide); N-methylloline (61 mg) was present at 312-365 mL of effluent. N-Methylloline was determined by gas chromatography. Preparative separation of the fraction at 304-374 mL of effluent by TLC, yielded N-methylloline (R_f 0.10), lysergic acid amide (R_f 0.42) and isolysergic acid amide (R_f 0.68). The structures of lysergic acid amide and isolysergic acid amide were confirmed by mass spectrometry via comparison with published mass spectra (17). We could not differentiate between lysergic acid amide and isolysergic acid amide on the basis of mass spectrometry. The higher R_f (0.68) band was assigned to isolysergic acid amide because isolysergic acid derivatives typically have higher R_f values than do the corresponding lysergic acid derivatives (18). At 374-392 mL of effluent, a minor unidentified 312 nm absorbance peak was observed. Another suspected ergot-alkaloid peak (25 μg) occurring at 453-513 mL of effluent also contained N-acetyl loline (305 mg). A minor peak (312 nm, trace amounts of ergot alkaloid) was observed at 513-557 mL of effluent. This peak was followed by ergonovine (55 μg) at 583-627 mL effluent. Ergonovine was tentatively identified by comparison with, and co-chromatography with an ergonovine standard using two different HPLC chromatographic solvent systems. Ergonovine was then isolated by TLC, and the structure confirmed by mass spectrometry via comparison with authentic ergonovine. Subsequent fractions with absorbencies at 312 nm contained only trace amounts of ergot alkaloids, expressed as μg ergotamine tartrate. N-formylloline (601 mg) eluted at 627 to 733 mL of effluent. Some suspected ergot-alkaloid overlap occurred from 654-750 mL of effluent. Loline (31 mg) and N-acetylnorloline (28 mg) were found together at 733-812 mL of effluent. No absorbencies at 312 nm or loline alkaloids were found after 812 mL of effluent.

Partition efficiencies of the major loline-type alkaloids were computed according to the conventional gas chromatographic formula (13):

$$N = (4R/w)^2$$

where N denotes the partition efficiency expressed in terms of theoretical plate number, R, the retention volume (mL of effluent) after injection of the peak maximum, and w, the peak width expressed in the same units as R. The present separation yielded partition efficiencies of 680, 1050, and 560 theoretical plates for N-methylloline, N-acetylloline, and N-formylloline, respectively.

In summary, N-methylloline, N-acetylloline, and N-formylloline were well separated from each other using gram-scale high-speed countercurrent chromatography, and recovery of loline alkaloids was quantitative. Ergot alkaloids were well separated from each other and were enriched in certain fractions. The preparative TLC system was not sufficient, in itself, to isolate ergot alkaloids from the alkaloid concentrate used in this study but was successful on the greatly simplified mixtures contained in countercurrent chromatographic fractions. This is the first reported occurrence of lysergic acid amide and isolysergic acid amide in endophyte-infected tall fescue, and HPLC analysis showed that these are the two unknown compounds reported previously (8).

The combination of preparative high-speed countercurrent chromatography with other separation methods, such as HPLC, and TLC, will enable chemists to isolate minor components of complex alkaloid mixtures more efficiently. This technique is not limited to alkaloid separations and, in theory, other complex mixtures of compounds having only minor differences in their partition coefficients should be efficiently separated by high-speed countercurrent chromatography.

Acknowledgment
We thank R. D. Plattner for mass spectra and T. Wilson for technical assistance.

Mention of companies or products by name does not imply their endorsement by the U.S. Department of Agriculture over others not cited.

Literature Cited

1. Morgan-Jones, G.; Gams, W. Mycotaxon. 1982, 15, 311.
2. Funk, C. R.; Halisky, P. M.; Johnson, C.; Siegel, M. R.;, Stewart, A. V.; Ahmad, S.; Hurley, R. H.; Harvey, I. C. Biotechnology 1983, 1(2), 189.
3. Hardy, T. D.; Clay, K.; Hammond, A. M. Jr. Environ. Entomol. 1986, 15, 1083.
4. Johnson, M. C.; Dahlman, D. L.; Siegal, M. R.; Bush, L. P.; Latch, G. C.; Potter, D. A.; Varney, D. R. Appl. Environ. Microbiol. 1985, 49, 568.
5. Latch, G. C.; Christensen, M. J.; Gaynor, D. L. N. Z. J. Agr. Res. 1985, 28, 129.
6. Petroski, R. J.; Yates, S. G.; Weisleder, D.; Powell, R.G. J. Nat. Prod. 1989, 52(4), 810.
7 Yates, S. G.; Petroski, R. J.; Powell, R. G. J. Agr. Food Chem. 1990, 38, 182.
8. Yates, S. G.; Powell, R. G.; J. Agr. Food Chem. 1988, 36, 337.
9. Sanchez, D. Agr. Res. 1987, 35(8), 12.
10. Siegel, M. R.; Johnson, M. C.; Varney, D. R.; Nesmith, W. C.; Buckner, R. C.; Bush, L. P.; Burrus, P. B.; Jones, T. A.; Boling, J. A. Phytopathology 1984, 74, 932.
11. Berde, B., Schild, H. O., Eds. "Ergot Alkaloids and Related Compounds"; Springer-Verlag, New York, 1978.
12. Ito, Y; Oka, H; Slemp, J. L. J. Chromatogr. 1989, 463, 305.
13. Bhatnager, M.; Oka, H.; Ito, Y. J. Chromatogr. 1989, 463, 317.
14. Zhang, T.-Z.; Pannell, L. K.; Pu, Q.-L.; Cai, D.-G.; Ito, Y. J. Chromatogr. 1988, 455.
15. Zhang, T.-Y.; Pannell, L. K.; Cai, D.-G.; Ito, Y. J. Liq. Chromatogr. 1988, 11(8), 1661.

16. Scott, A. I. "Interpretation of the Ultraviolet Spectra of Natural Products"; Pergamon Press: Frankfurt, Germany, 1964; pp. 172-178.
17. Inoue, T.; Nakahara, Y,; Niwaguchi, T. Chem. Pharm. Bull. 1972, 20 (2), 409.
18. Stahl, E. Ed. "Thin-Layer Chromatography, A Laboratory Handbook"; Springer-Verlag, New York, 1969; pp. 452-453.

RECEIVED May 16, 1990

INDEXES

Author Index

Affiliation Index

Subject Index

A

B

C

D

E

I

J

K

L

M

N

Q

T

V

Production: Peggy D. Smith
Indexing: Colleen P. Stamm
Acquisition: Barbara C. Tansill

Books printed and bound by Maple Press, York, PA

Paper meets minimum requirements of American National Standard for Information Sciences—Permanence of Paper for Printed Library Materials, ANSI Z39.48–1984 ∞

Other ACS Books

Chemical Structure Software for Personal Computers
Edited by Daniel E. Meyer, Wendy A. Warr, and Richard A. Love
ACS Professional Reference Book; 107 pp;
clothbound, ISBN 0–8412–1538–3; paperback, ISBN 0–8412–1539–1

Personal Computers for Scientists: A Byte at a Time
By Glenn I. Ouchi
276 pp; clothbound, ISBN 0–8412–1000–4; paperback, ISBN 0–8412–1001–2

Biotechnology and Materials Science: Chemistry for the Future
Edited by Mary L. Good
160 pp; clothbound, ISBN 0–8412–1472–7; paperback, ISBN 0–8412–1473–5

Polymeric Materials: Chemistry for the Future
By Joseph Alper and Gordon L. Nelson
110 pp; clothbound, ISBN 0–8412–1622–3; paperback, ISBN 0–8412–1613–4

The Language of Biotechnology: A Dictionary of Terms
By John M. Walker and Michael Cox
ACS Professional Reference Book; 256 pp;
clothbound, ISBN 0–8412–1489–1; paperback, ISBN 0–8412–1490–5

***Cancer: The Outlaw Cell*, Second Edition**
Edited by Richard E. LaFond
274 pp; clothbound, ISBN 0–8412–1419–0; paperback, ISBN 0–8412–1420–4

Practical Statistics for the Physical Sciences
By Larry L. Havlicek
ACS Professional Reference Book; 198 pp; clothbound; ISBN 0–8412–1453–0

The Basics of Technical Communicating
By B. Edward Cain
ACS Professional Reference Book; 198 pp;
clothbound, ISBN 0–8412–1451–4; paperback, ISBN 0–8412–1452–2

The ACS Style Guide: A Manual for Authors and Editors
Edited by Janet S. Dodd
264 pp; clothbound, ISBN 0–8412–0917–0; paperback, ISBN 0–8412–0943–X

Chemistry and Crime: From Sherlock Holmes to Today's Courtroom
Edited by Samuel M. Gerber
135 pp; clothbound, ISBN 0–8412–0784–4; paperback, ISBN 0–8412–0785–2

For further information and a free catalog of ACS books, contact:
American Chemical Society
Distribution Office, Department 225
1155 16th Street, NW, Washington, DC 20036
Telephone 800–227–5558

DATE DUE

AUG 0 8 1995

MAY 0 8 1996

Demco, Inc. 38-293